T0211748

AIRO Springer Series

Volume 5

The AIRO Springer Series focuses on the relevance of operations research (OR) in the scientific world and in real life applications.

The series publishes peer-reviewed only works, such as contributed volumes, lectures notes, and monographs in English language resulting from workshops, conferences, courses, schools, seminars, and research activities carried out by AIRO, Associazione Italiana di Ricerca Operativa - Optimization and Decision Sciences: http://www.airo.org/index.php/it/.

The books in the series will discuss recent results and analyze new trends focusing on the following areas: Optimization and Operation Research, including Continuous, Discrete and Network Optimization, and related industrial and territorial applications. Interdisciplinary contributions, showing a fruitful collaboration of scientists with researchers from other fields to address complex applications, are welcome.

The series is aimed at providing useful reference material to students, academic and industrial researchers at an international level.

Should an author wish to submit a manuscript, please note that this can be done by directly contacting the series Editorial Board, which is in charge of the peer-review process.

More information about this series at http://www.springer.com/series/15947

Claudio Gentile • Giuseppe Stecca • Paolo Ventura
Editors

Graphs and Combinatorial Optimization: from Theory to Applications

CTW2020 Proceedings

Editors
Claudio Gentile
Consiglio Nazionale delle Ricerce
Istituto di Analisi dei Sistemi ed
Informatica "Antonio Ruberti"
Roma, Italy

Giuseppe Stecca
Consiglio Nazionale delle Ricerche
Istituto di Analisi dei Sistemi ed
Informatica "Antonio Ruberti"
Roma, Italy

Paolo Ventura
Consiglio Nazionale delle Ricerche
Istituto di Analisi dei Sistemi ed
Informatica "Antonio Ruberti"
Roma, Italy

ISSN 2523-7047   ISSN 2523-7055   (electronic)
AIRO Springer Series
ISBN 978-3-030-63074-4   ISBN 978-3-030-63072-0   (eBook)
https://doi.org/10.1007/978-3-030-63072-0

This Springer imprint is published by the registered company Springer Nature Switzerland AG.
The registered company address is: Gewerbestrasse 11, 6330 Cham, Switzerland

Preface

The Cologne-Twente Workshop (CTW) on Graphs and Combinatorial Optimization is a workshop series initiated by Ulrich Faigle, around the time he moved from the Twente University to the University of Cologne. After many CTW editions in Twente and Cologne, it was decided that CTWs were mature enough to move about: in 2004, the CTW was organized in Villa Vigoni (Menaggio, Como, Italy) by Francesco Maffioli (Politecnico di Milano) and Leo Liberti (CNRS-LIX). Since then, the CTW visited again Italy for three times, beyond France, Germany, the Netherlands, and Turkey. This edition is the first time that the CTW is organized with the contribution of CNR-IASI members (Claudio Gentile, Giuseppe Stecca, Paolo Ventura, Giovanni Rinaldi, and Fabio Furini) in addition to the members of University of Rome "Tor Vergata" (Andrea Pacifici), Roma Tre University (Gaia Nicosia), and CNRS & LIX Polytechnique Palaiseau (Leo Liberti).

Having been initially set up by discrete applied mathematicians, the CTW still follows the mathematical tradition. In this CTW edition (hereafter, CTW2020), for the first time we adopted two submission tracks: standard papers of at most 12 pages and traditional CTW extended abstracts of at most 4 pages.

This volume collects the standard papers that were submitted to the CTW2020. The papers underwent a standard peer review process performed by a Program Committee consisting of 30 members:[1] 17 CTW steering committee members and

---

[1] Ali Fuat Alkaya (Marmara U., Turkey), Christoph Buchheim (TU Dortmund, Germany), Francesco Carrabs (U. Salerno, Italy), Alberto Ceselli (U. Milano, Italy), Roberto Cordone (U. Milano, Italy), Ekrem Duman (Ozyegin U., Turkey), Yuri Faenza (Columbia U., USA), Bernard Gendron (IRO U. Montreal & CIRRELT, Canada), Claudio Gentile (CNR-IASI, Italy), Johann Hurink (U. Twente, The Netherlands), Ola Jabali (Politecnico di Milano, Italy), Leo Liberti (CNRS & LIX Polytechnique Palaiseau, France), Frauke Liers (FAU Erlangen-Nuremberg, Germany), Bodo Manthey (U. Twente, The Netherlands), Gaia Nicosia (U. Roma Tre, Italy), Tony Nixon (U. Lancaster, UK), Andrea Pacifici (U. Roma Tor Vergata, Italy), Ulrich Pferschy (U. Graz, Austria), Stefan Pickl (U. Bundeswehr München, Germany), Michael Poss (LIRMM U. Montpellier & CNRS, France), Bert Randerath (U. Koeln, Germany), Giovanni Righini (U. Milano, Italy), Heiko Roeglin (U. Bonn, Germany), Oliver Schaudt (RTWH Aachen U., Germany), Rainer Schrader (U. Koeln, Germany), Giuseppe Stecca (CNR-IASI, Italy), Frank Vallentin (U. Koeln, Germany),

13 guest members. PC members came from Italy, Germany, France, the USA, Canada, the Netherlands, the UK, Austria, and Turkey. We received 46 submissions of which we accepted 31 for publication in this volume with a rate of success of 67%.

The chapters of this volume present works on graph theory, discrete mathematics, combinatorial optimization, and operations research methods, with particular emphasis on coloring, graph decomposition, connectivity, distance geometry, mixed-integer programming, machine learning, heuristics, meta-heuristics, mathheuristics, and exact methods. Applications are related to logistics, production planning, energy, telecommunications, healthcare, and circular economy.

The scientific program of the CTW2020 includes the 31 standard papers in this volume, 33 extended abstracts, and two plenary invited talks. As usual for the CTW, extended abstracts were subject to a high acceptance level, allowing also papers presenting preliminary results with a particular accent to works presented by MScs, PhDs, or Postdocs. The traditional CTW extended abstracts will be published on the conference's website http://ctw2020.iasi.cnr.it, where also additional material collected during the conference will be posted.

We thank all the PC members and the subreviewers for the complex work performed to select the papers and to improve their quality considering also a possible second round of revision.

Following the CTW tradition, a special issue of Discrete Applied Mathematics (DAM) dedicated to this workshop and its main topics of interest will be edited.

Not every CTW edition features invited plenary speakers, but this one does. Two very well-known researchers accepted our invitation: Prof. Dan Bienstock (Columbia University) and Prof. Marco Sciandrone (University of Florence). Prof. Dan Bienstock works in many topics of Combinatorial Optimization, Integer and Mixed-Integer Programming, and Network Design. He is the author of many journal and conference papers and of two textbooks: "Electrical Transmission System Cascades and Vulnerability: An Operations Research Viewpoint," ISBN 978-1-611974-15-7, SIAM-MOS Series on Optimization (2015), and "Potential Function Methods for Approximately Solving Linear Programming Problems: Theory and Practice," ISBN 1-4020-7173-6, Kluwer Academic Publishers, Boston (2002). Prof. Marco Sciandrone works in Nonlinear Programming with a particular expertise in Machine Learning, Neural Networks, Multiobjective Optimization, and Nonlinear Approximation of Discrete Variables.

Finally, we thank AIRO for hosting this volume in its AIRO-Springer series. We thank both AIRO and CNR-IASI for their support to the realization of the conference.

Paolo Ventura (CNR-IASI, Italy), Maria Teresa Vespucci (U. Bergamo, Italy), and Angelika Wiegele (Alpen-Adria U. Klagenfurt, Austria).

This conference was originally supposed to take place in the wonderful Ischia island on 15–17 June, 2020. Due to the Covid-19 pandemic, we were first obliged to reschedule the conference in September 14–16, 2020, and then to move it online as the majority of conferences in 2020. Nevertheless, we very much hope you will all enjoy the CTW2020.

Rome, Italy   Claudio Gentile
Rome, Italy   Giuseppe Stecca
Rome, Italy   Paolo Ventura
September 2020

Contents

About the Editors

Claudio Gentile is a Research Director at the Institute of Systems Analysis and Computer Science "Antonio Ruberti" of the Italian National Research Council (CNR-IASI). From 2006 to 2015, he directed the CNR-IASI research unit "Control and Optimization of Complex Systems," and since 2016, he directs the CNR-IASI research unit "OPTIMA: Optimization and Discrete Mathematics." His main research interests are in Combinatorial Optimization, Polyhedral Theory for Linear and Nonlinear Mixed-Integer Programming Problems, Interior Point Methods with applications in Power Energy Production and Distribution, Logistics, Network Design, Staff Management, and Ship Scheduling. He is the author of many scientific publications among journal papers, book chapters, and articles in conference proceedings.

Giuseppe Stecca is a Research Scientist at the Institute of Systems Analysis and Computer Science "Antonio Ruberti" of the Italian National Research Council (CNR-IASI). He holds the Chair of Supply Chain Management at the University of Rome "Tor Vergata," where he also teaches Operations Research. He is a member of the board of the Italian Association for Operations Research (AIRO). His main research interests are related to the optimization of sustainable production and logistic systems. He works actively in research projects and also as an evaluator for the Italian Ministry of Economic Development in the area of logistics and industry 4.0.

Paolo Ventura is a Research Scientist at the Institute of Systems Analysis and Computer Science "Antonio Ruberti" of the Italian National Research Council (CNR-IASI). His main research interests are Integer Programming and Combinatorial Optimization with applications in logistics and transportation. He is the author of many articles in the most relevant international journals of the area. Since 2004, he teaches Operations Research at the University of Rome "Tor Vergata." He is a member of the organizing committee of the yearly "Cargese Workshop on Combinatorial Optimization" and, in the odd years, of the "Aussois Combinatorial Optimization Workshop."

The Chromatic Polynomial of a Digraph

Winfried Hochstättler and Johanna Wiehe

Abstract An acyclic coloring of a digraph as defined by V. Neumann-Lara is a vertex-coloring such that no monochromatic directed cycles occur. Counting the number of such colorings with k colors can be done by counting so-called Neumann-Lara-coflows (NL-coflows), which build a polynomial in k. We will present a representation of this polynomial using totally cyclic subdigraphs, which form a graded poset Q. Furthermore we will decompose our NL-coflow polynomial, which becomes the chromatic polynomial of a digraph by multiplication with the number of colors to the number of components, using the geometric structure of the face lattices of a class of polyhedra that corresponds to Q. This decomposition leads to a representation using certain subsets of edges of the underlying undirected graph and will confirm the equality of our chromatic polynomial of a digraph and the chromatic polynomial of the underlying undirected graph in the case of symmetric digraphs.

Keywords Dichromatic number · Chromatic polynomial · Flow polynomial · Totally cyclic subdigraphs · Face lattice

1 Introduction

The notion of classic graph coloring deals with finding the smallest integer k such that the vertices of an undirected graph can be colored with k colors, where no two adjacent vertices share the same color. The chromatic polynomial counts those proper colorings a graph admits, subject to the number of colors. William T. Tutte developed a dual concept [17], namely his nowhere-zero flows (NZ-flows), which build a polynomial, the flow polynomial, too.

W. Hochstättler · J. Wiehe (✉)
FernUniversität in Hagen, Hagen, Germany
e-mail: winfried.hochstaettler@fernuni-hagen.de; johanna.wiehe@fernuni-hagen.de

© The Author(s), under exclusive license to Springer Nature Switzerland AG 2021
C. Gentile et al. (eds.), *Graphs and Combinatorial Optimization: from Theory to Applications*, AIRO Springer Series 5,
https://doi.org/10.1007/978-3-030-63072-0_1

1

We turn our attention to directed graphs, or digraphs for short. In 1982 Víctor Neumann-Lara [12] introduced the dichromatic number of a digraph D as the smallest integer k such that the vertices of D can be colored with k colors and each color class induces an acyclic digraph. This seems to be a reasonable generalization of the chromatic number since both numbers coincide in the symmetric case, where we have all arcs in both directions.

Moreover Neumann-Lara conjectured in 1985, that every orientation of a simple planar graph can be acyclically colored with two colors [13]. Regarding the dichromatic number this is not the only conjecture remaining widely open. Up to some relaxations, for instance Mohar and Li [10] affirmed the two-color-conjecture for planar digraphs of digirth four, it is known [4], that deciding whether an arbitrary digraph has dichromatic number at most two is NP-complete.

Although some progress has been made according thresholds (see e.g. [8]), even the complete case seems to be quite hard. To our knowledge it is not known how many vertices suffice to build a tournament which has dichromatic number five [14].

Nevertheless, Ellis and Soukup determined [6] thresholds for the minimum number of cycles, where reversing their orientation yields a digraph resp. tournament that has dichromatic number at most two.

Comparing the chromatic and the dichromatic number Erdős and Neumann-Lara conjectured [7] in 1979 that if the dichromatic number of a class of graphs is bounded, so is their chromatic number. While Mohar and Wu [11] considered the fractional chromatic number of linear programming proving a fractional version, this is another conjecture remaining unsolved.

With our work we hope to contribute to a better understanding of the dichromatic number. W. Hochstättler [9] developed a flow theory for the dichromatic number transferring Tutte's theory of NZ-flows from classic graph colorings. Together with B. Altenbokum [2] we pursued this analogy by introducing algebraic Neumann-Lara-flows (NL-flows) as well as a polynomial counting these flows. The formula we derived contains the Möbius function of a certain poset. Here, we will derive the values of the Möbius function by showing that the poset correlates to the face lattice of a polyhedral cone.

Probably, the chromatic polynomial of a graph is better known than the flow polynomial. Therefore, in this paper we consider the dual case of our NL-flow polynomial, the NL-coflow polynomial which equals the chromatic polynomial for the dichromatic number divided by the number of colors if the digraph is connected. We will present a representation using totally cyclic subdigraphs and decompose them to obtain an even simpler representation. In particular, it will suffice to consider certain subsets of edges of the underlying undirected graph.

Our notation is fairly standard and, if not explicitly defined, should follow the books of Bondy and Murty [5] for digraphs and Beck and Sanyal [3] for polyhedral geometry. Note that all our digraphs may have parallel and antiparallel arcs as well as loops if not explicitly excluded.

2 Definitions and Tools

Let G be a finite Abelian group and $D = (V, A)$ a digraph. Recall that a map $f : A \longrightarrow G$ is a flow in D, if it satisfies Kirchhoff's law of flow conservation

$$\sum_{a \in \partial^+(v)} f(a) = \sum_{a \in \partial^-(v)} f(a) \tag{1}$$

in every vertex $v \in V$, where $\partial^+(v)$ and $\partial^-(v)$ denote the set of outgoing resp. incoming arcs at v.

Analogously, a map $g : A \longrightarrow G$ is a coflow in D, if it satisfies Kirchhoff's law for (weak) cycles $C \subseteq A$

$$\sum_{a \in C^+} g(a) = \sum_{a \in C^-} g(a), \tag{2}$$

where C^+ and C^- denote the set of arcs in C that are traversed in forward resp. in backward direction.

Now let n be the number of vertices, m be the number of arcs and let M denote the totally unimodular $(n \times m)$-incidence matrix of D. While condition (1) is equivalent to the condition that the vector $f = (f(a_1), \ldots, f(a_m))^\top$ is an element of the null space of M, that is $Mf = 0$, condition (2) is equivalent to the condition that the vector $g = (g(a_1), \ldots, g(a_m))$ is an element of the row space of M, that is $g = pM$, for some $(1 \times n)$-vector $p \in G^{|V|}$.

Definition 1 A digraph $D = (V, A)$ is called *totally cyclic*, if every component is strongly connected. A *feedback arc set* of a digraph is a set $S \subseteq A$ such that $D - S$ is acyclic.

Definition 2 Let $D = (V, A)$ be a digraph and G a finite Abelian group. An *NL-G-coflow* in D is a coflow $g : A \longrightarrow G$ in D whose support contains a feedback arc set. For $k \in \mathbb{Z}$ and $G = \mathbb{Z}$, a coflow g is an *NL-k-coflow*, if

$$g(a) \in \{0, \pm 1, \ldots, \pm(k-1)\}, \text{ for all } a \in A,$$

such that its support contains a feedback arc set.

In order to develop a closed formula for the number of NL-G-coflows we use a generalization of the well-known inclusion-exclusion formula, the Möbius inversion.

Definition 3 (See e.g. [1]) Let $(P, \leq)$ be a finite poset, then the *Möbius function* is defined as follows

$$\mu : P \times P \to \mathbb{Z}, \quad \mu(x, y) := \begin{cases} 0 & \text{, if } x \nleq y \\ 1 & \text{, if } x = y \\ -\sum_{x \leq z < y} \mu(x, z) & \text{, otherwise .} \end{cases}$$

Proposition 1 (See [1, 15]) *Let* $(P, \leq)$ *be a finite poset,* $f, g : P \longrightarrow \mathbb{K}$ *functions and* μ *the Möbius function. Then the following equivalence holds*

$$f(x) = \sum_{y \geq x} g(y), \text{ for all } x \in P \iff g(x) = \sum_{y \geq x} \mu(x, y) f(y), \text{ for all } x \in P.$$

With this so called *Möbius inversion from above* it will suffice to compute the number of G-coflows in some given minors B, which is $|G|^{\mathrm{rk}(B)}$, where $\mathrm{rk}(B)$ is the rank of the incidence matrix of $G[B]$ which equals $|V(B)| - c(B)$, i.e. the number of vertices minus the number of connected components of $G[B]$.

3   The NL-Coflow Polynomial

In this chapter we will define the NL-coflow polynomial, which counts the number of NL-G-coflows, using Möbius inversion. Therefor we need a specific partially ordered set. The following poset $(\mathscr{C}, \geq)$ with

$$\mathscr{C} := \Big\{ A/C \mid \exists\, C_1, \ldots, C_r \text{ directed cycles, such that } C = \bigcup_{i=1}^{r} C_i \Big\}$$

and

$$A/\bigcup_{j \in J} C_j \geq A/\bigcup_{i \in I} C_i :\Leftrightarrow \bigcup_{j \in J} C_j \subseteq \bigcup_{i \in I} C_i,$$

will serve our purpose. Note that in case D is strongly connected, A is the unique minimum of this poset.

Definition 4 Let $D = (V, A)$ be a digraph and μ the Möbius function of $\mathscr{C}$. Then the *NL-Coflow Polynomial* of D is defined as

$$\psi_{NL}^{D}(x) := \sum_{Y \in \mathscr{C}} \mu(A, Y) x^{\mathrm{rk}(Y)}.$$

The dual version of Theorem 3.5 in [2] reveals the following.

Theorem 1 *The number of NL-G-coflows of a digraph D depends only on the order k of G and is given by* $\psi_{NL}^{D}(k)$.

Proof Using Proposition 1 with $f_k, g_k : \mathscr{C} \to \mathbb{Z}$, such that $f_k(Y)$ indicates all G-coflows and $g_k(Y)$ all NL-G-coflows in $D[Y]$, it suffices to show that

$$f_k(Z) = \sum_{\substack{Y \leq Z \\ Y \in \mathscr{C}}} g_k(Y) \tag{3}$$

holds for all $Z \in \mathscr{C}$. Then we obtain

$$\psi_{NL}^D(k) = g_k(A) = \sum_{\substack{Y \leq A \\ Y \in \mathscr{C}}} \mu(A, Y) f_k(Y) = \sum_{Y \in \mathscr{C}} \mu(A, Y) k^{\mathrm{rk}(Y)},$$

since the number of G-coflows on $D[Y]$ is given by $k^{\mathrm{rk}(Y)}$.

Concerning (3) let $Z \in \mathscr{C}$ and φ be a G-coflow on $D[Z]$. With d we denote the number of directed cycles in $D[Z]$ and set

$$Y := Z / \bigcup_{i=1}^{d} \{C_i \mid C_i \text{ is a directed cycle in } D[Z] \text{ and } \forall c \in C_i : \varphi(c) = 0\}.$$

Then clearly $Y \in \mathscr{C}$ and $\varphi|_Y$ is an NL-G-coflow on $D[Y]$.

The other direction is obvious since every NL-G-coflow g on $D[Y]$ with $Y \in \mathscr{C}$ can be extended to a G-coflow $\tilde{g}$ on $D[Z]$, setting $\tilde{g}(a) := 0_G$ for all $a \in Z - Y$. $\quad\square$

3.1 Totally Cyclic Subdigraphs

Since many unions of directed cycles determine the same strongly connected subdigraph it suffices to consider all totally cyclic subdigraphs which turn out to form a graded poset.

Lemma 1 *The poset*

$$Q := \{B \subseteq A \mid D[B] \text{ is totally cyclic subdigraph of } D\},$$

ordered by inclusion, is a graded poset with rank function rk_Q *and its Möbius function alternates in the following fashion:*

$$\mu_Q(\emptyset, B) = (-1)^{\mathrm{rk}_Q(B)}.$$

Proof Let M be the totally unimodular $(n \times m)$-incidence matrix of D. We will show that the face lattice of the polyhedral cone PC described by

$$\begin{pmatrix} M \\ -M \\ -I \end{pmatrix} x \leq 0,$$

corresponds to Q.

Since M is totally unimodular all extreme rays of PC are spanned by integral points. It follows that every totally cyclic subdigraph can be represented by a face of PC, where an arc $1 \leq i \leq m$ exists iff for the corresponding entry $x_i > 0$ holds.

Thus the elements of the face lattice of PC coincide with the elements of our poset and so do the Möbius functions. Well-known facts from topological geometry which can be found for instance in Corollary 3.3.3 and Theorem 3.5.1 in [3] yield that Q is a graded poset and

$$\mu_Q(\emptyset, B) = (-1)^{\dim(B)+1}\chi(B) = (-1)^{\mathrm{rk}_Q(B)}\chi(B),$$

where χ denotes the reduced Euler characteristic, which equals one in this case, since the faces of PC build non-empty closed polytopes (see e.g. Thm. 3.4.1 in [3]).   □

Theorem 2 *Let D be a digraph and $(Q, \subseteq)$ the poset defined above. Then the NL-coflow polynomial of D is given by*

$$\psi_{NL}^D(x) = \sum_{B \in Q} (-1)^{\mathrm{rk}_Q(B)} x^{\mathrm{rk}(A/B)}.$$

Proof With Lemma 1 we immediately obtain:

$$\psi_{NL}^D(x) = \sum_{Y \in \mathscr{C}} \mu(A, Y)x^{\mathrm{rk}(Y)} = \sum_{B \in Q} \mu_Q(\emptyset, B)x^{\mathrm{rk}(A/B)} = \sum_{B \in Q} (-1)^{\mathrm{rk}_Q(B)} x^{\mathrm{rk}(A/B)}.$$

□

It is well known that coflows and colorings are in bijection, once the color of some vertex in each connected component has been chosen. As a consequence we have the following corollary, where $c(D)$ denotes the number of connected components in D.

Corollary 1 *The chromatic polynomial of a digraph D is given as*

$$\chi(D, x) = x^{c(D)} \cdot \psi_{NL}^D(x) = \sum_{B \in Q} (-1)^{\mathrm{rk}_Q(B)} x^{\mathrm{rk}(A/B)+c(D)}.$$

4 Decomposing the NL-Coflow Polynomial

In the following we will put our previous results into the setting of polyhedral geometry. There we will find a way to compound some of the objects considered, which will, going back to graph theory, decompose the NL-coflow polynomial such that only certain subsets of edges of the underlying undirected graph need to be considered.

More precisely, fixing the support, implying a fixed exponent in our polynomial, we will show that all existing totally cyclic orientations correlate to the face lattice of some usually unbounded polyhedron. This will yield a relation between the above mentioned poset Q and the maximal faces of a class of polyhedra to be defined in the following. Using the geometric structure of those polyhedra we can contract the corresponding order complex and, by correlating the corresponding Möbius functions, obtain an even simpler representation of the NL-coflow polynomial and therefore of the chromatic polynomial of arbitrary digraphs.

Let $D = (V, A)$ be a digraph, $G = (V, E)$ its underlying undirected graph with $|V| = n$ and $|E| = m$. For $\emptyset \neq B \subseteq E$ a *partial orientation* $\mathcal{O}(B)$ is an orientation of a subset $B' \subseteq B$ of the edges, where the remaining edges in $B \setminus B'$ are considered as pair of antiparallel arcs, called digons. We say a partial orientation is *totally cyclic* if the corresponding induced digraph is. Once the support is fixed, there is a unique inclusionwise maximal partial orientation, denoted with $\bar{\mathcal{O}}(B)$, where we have as many digons as possible.

A flow $x = (\overrightarrow{x}, \overleftarrow{x})^\top \in \mathbb{R}^{2m}$ on D is related to a partial orientation $\mathcal{O}(B)$ by orienting only the edges with $x_i \neq 0$.

Let M be the totally unimodular incidence $(n \times m)$-matrix of the subgraph induced by $\emptyset \neq B \subseteq E$. Then $x \in \mathbb{R}^{2m}$ is a flow iff $(M, -M)x = 0$ holds.

Now, consider the following system

$$
\left.
\begin{aligned}
(M, -M)(\overrightarrow{x}, \overleftarrow{x})^\top &= 0 \\
\overrightarrow{x_i} + \overleftarrow{x_i} &\geq 1 \quad \forall 1 \leq i \leq m \\
\overrightarrow{x_i} &= 0 \quad \text{if } \overrightarrow{i} \notin A \text{ but } \overleftarrow{i} \in A \\
\overleftarrow{x_i} &= 0 \quad \text{if } \overleftarrow{i} \notin A \text{ but } \overrightarrow{i} \in A \\
\overrightarrow{x}, \overleftarrow{x} &\geq 0.
\end{aligned}
\right\} \quad (P)
$$

We denote the polyhedron described above with P and take a look at its vertices, which are the solutions of the program (P), in the first place.

Lemma 2 *Let $x = (\overrightarrow{x}, \overleftarrow{x})^\top$ be a solution of (P). Then a solution $y = (\overrightarrow{y}, \overleftarrow{y})^\top$ of (P) exists with $\mathrm{supp}(y) \subseteq \mathrm{supp}(x)$ and $\overrightarrow{y_a} = \overleftarrow{y_a} = \frac{1}{2}$, if a is a bridge and $\min\{\overrightarrow{y_a}, \overleftarrow{y_a}\} = 0$, otherwise.*

Proof Let y be a solution with minimal support such that the corresponding partial orientation contains a minimum number of directed cycles.

Let $1 \leq \overrightarrow{a} \leq m$. If a is a bridge, then $y_{\overrightarrow{a}} = y_{\overleftarrow{a}}$ has to hold since otherwise the flow condition would be violated. In the other case assume that $y_{\overrightarrow{a}} \geq y_{\overleftarrow{a}} > 0$. Let $\overrightarrow{a} = (v, w)$ and $C := \{\overrightarrow{a}, b_0, b_1, \ldots, b_k\}$ be a directed cycle. After reassigning

$$
\tilde{y}_{\overrightarrow{a}} := 1 + y_{\overrightarrow{a}} - y_{\overleftarrow{a}} \geq 1,
$$

$$
\tilde{y}_{\overleftarrow{a}} := y_{\overleftarrow{a}} - y_{\overleftarrow{a}} = 0,
$$

$$
\tilde{y}_b := y_b + 1, \forall b \in C \setminus \{a\}
$$

$$
\tilde{y}_c := y_c, \quad \text{otherwise,}
$$

the flow condition still holds in v:

$$\sum_{i\in\partial^+(v)} \tilde{y}_i = \sum_{\substack{i\in\partial^+(v)\\ i\neq\overrightarrow{a}}} y_i + 1 + y_{\overrightarrow{a}} - y_{\overleftarrow{a}} = \sum_{i\in\partial^+(v)} y_i + 1 + y_{\overleftarrow{a}} - y_{\overleftarrow{a}} - y_{\overrightarrow{a}}$$

$$= \sum_{i\in\partial^-(v)} y_i + 1 - y_{\overleftarrow{a}} = \sum_{\substack{i\in\partial^-(v)\\ i\neq\overleftarrow{a}, i\neq b_k}} \tilde{y}_i + 1 + y_b = \sum_{i\in\partial^-(v)} \tilde{y}_i,$$

as well as in w:

$$\sum_{i\in\partial^+(w)} \tilde{y}_i = \sum_{\substack{i\in\partial^+(w)\\ i\neq\overleftarrow{a}, i\neq b_0}} y_i + y_b + 1 = \sum_{i\in\partial^+(w)} y_i - y_{\overleftarrow{a}} + 1$$

$$= \sum_{\substack{i\in\partial^-(w)\\ i\neq\overrightarrow{a}}} \tilde{y}_i + y_{\overrightarrow{a}} + 1 - y_{\overleftarrow{a}} = \sum_{i\in\partial^-(w)} \tilde{y}_i.$$

Thus the solution $\tilde{y}$ yields a contradiction to y having minimal support.   □

As a result of the preceding lemma, the vertices $\mathscr{V}$ of P are totally cyclic subdigraphs, where the only remaining digons are bridges.

To describe the polyhedron completely we take a look at the recession cone

$$rec(P) = \{y \in \mathbb{R}^{2m} \mid \forall c \in P, \ \forall\lambda \geq 0 : c + \lambda y \in P\}$$

$$= P(A, 0)$$

$$= Cone\left(\{y \in \mathbb{R}^{2m} \mid y \text{ is directed cycle}\}\right).$$

Thus we have $P = Conv(\mathscr{V}) + Cone\left(\{y \in \mathbb{R}^{2m} \mid y \text{ is directed cycle}\}\right)$.

In the following we would like to correlate the elements of our poset Q to the face lattice of P, where maximal and minimal elements, $\hat{1}$ and $\hat{0}$, are adjoined and the corresponding Möbius function is denoted with μ_P.

Since there may be several faces corresponding to the same element of Q we define a closure operator on the set of faces $cl : \mathscr{F} \to \overline{\mathscr{F}}$ as follows, where $eq(F)$ is the set of constraints in (P) where equality holds:

$$cl(F) = F_{max} := \bigvee\{\tilde{F} \mid \text{supp}(\tilde{F}) = \text{supp}(F)\}$$

$$= \{x \in P \mid \text{supp}(F_{max}) = \text{supp}(F), eq(F_{max}) \text{ is minimal}\},$$

where $\vee$ is the join of all faces with equal support in the face lattice.

This function is well-defined since the dimension of every face is bounded by $2m$ and F_{max} is uniquely determined since the join is. It is also easy to check that cl is indeed a closure operator.

Now we can identify the maximal faces with the elements of Q by either forgetting the values of a flow or by first taking an arbitrary flow $x \in \mathbb{R}_+^{2m}$ satisfying $\overrightarrow{x} + \overleftarrow{x} \geq 1$, that lives on some face F_x and then taking its closure operator $cl(F_x)$.

As a result the Möbius function of $\mathscr{F}$ behaves for $x, y \in P$ as follows (see Prop. 2 on p. 349 in [15]):

$$\sum_{\substack{z \in P \\ cl(F_z) = cl(F_y)}} \mu_P(F_x, F_z) = \begin{cases} \mu_{\tilde{\mathscr{F}}}(cl(F_x), cl(F_y)) & \text{, if } F_x = cl(F_x) \\ 0 & \text{, if } F_x \subset cl(F_x) \end{cases}.$$

This is why we will simply write $\mu_P(B, B')$ instead of $\mu_{\tilde{\mathscr{F}}}(cl(F_x), cl(F_y))$ for flows x, y on $B, B' \in Q$. Also we identify $\hat{0}$ with $\emptyset$ and $\hat{1}$ with $\bar{\mathscr{O}}(B)$, respectively.

Examining the polyhedron P we find three cases which determine the structure and therefore the Möbius function of the face lattice:

1. There is exactly one vertex v in P.

    1.1 There are no further faces in P including v, i.e. $\dim(P) = 0$.
    1.2 There are further faces in P including v, so P is a pointed cone and $\dim(P) \geq 1$.

2. There are at least two vertices in P.

Note that all cases are mutually exclusive and complete since every P has at least one vertex.

Lemma 3 *Let $\emptyset \neq X \in \mathscr{F}$ be a face of P. Then*

$$\mu_P(\emptyset, X) = \begin{cases} -1 & \text{if } \dim(X) = 0, \\ (-1)^{\mathrm{rk}_P(X)} & \text{in cases 1.1 and 2,} \\ 0 & \text{in case 1.2.} \end{cases}$$

Proof If X is a vertex, then $\dim(X) = 0$ and

$$\mu_P(\emptyset, X) = -\mu_P(\emptyset, \emptyset) = -1 = (-1)^{\mathrm{rk}_P(X)}.$$

For the other cases we will use Theorem 3.5.1 and Corollary 3.3.3 in [3]:

$$\mu_P(\emptyset, X) = (-1)^{\dim(X)+1} \chi(X) = (-1)^{\mathrm{rk}_P(X)} \chi(X),$$

where χ denotes the reduced Euler characteristic.

1.2 Since there is only one vertex, every face of dimension greater 0 builds a pointed
     cone. Proposition 3.4.9 in [3] yields that $\chi(X) = 0$.
 2. Since there are at least two vertices, there are also some faces including them.
     Those form non-empty closed polytopes with $\chi(X) = 1$ (see Thm. 3.4.1 in
     [3]).   □

Comparing the Möbius functions of P and Q we find the following relation,
where $cr(B) = |B| - |V(B)| + c(B)$ denotes the corank and $\beta(B)$ the number of
bridges in the graph induced by $B \subseteq E$.

Lemma 4 *Let $\emptyset \neq B \subseteq E$ and $\mathcal{O}(B)$ be a totally cyclic partial orientation of B,
then*

$$\mu_Q(\emptyset, \mathcal{O}(B)) = (-1)^{cr(B)+\beta(B)+1} \mu_P(\emptyset, \mathcal{O}(B))$$

*holds, if $\mu_P(\emptyset, X)$ alternates, i.e. in cases 1.1, 2 and if $\dim(X) = 0$, where $X \in \mathcal{F}$
is the maximal face corresponding to $\mathcal{O}(B)$. Otherwise (in case 1.2) we find*

$$\sum_{\substack{\mathcal{O}(B) \subseteq A \\ tot.cyclic}} \mu_Q(\emptyset, \mathcal{O}(B)) = 0.$$

Proof If both Möbius functions alternate it suffices to consider elements $\mathcal{O}(B) \subseteq A$
where $rk_P(\mathcal{O}(B))$ is minimal. In this case $\mu_P(\emptyset, \mathcal{O}(B)) = -1$ and we are left to
verify

$$\mu_Q(\emptyset, \mathcal{O}(B)) = (-1)^{cr(B)+\beta(B)}.$$

We prove the statement by induction over the number of edges in B. The base cases
can be easily checked. Deleting one edge $d \in B$ yields the following two cases:

1. d is a bridge.
    Then $rk_Q(B-d) = rk_Q(B) - 1$, $cr(B-d) = cr(B)$ and $\beta(B-d) = \beta(B) - 1$.
2. d is not a bridge.
    Then $rk_Q(B-d) = rk_Q(B) - 1$, $cr(B-d) = cr(B) - 1$ and $\beta(B-d) = \beta(B)$.

Using the induction hypothesis we find in both cases

$$(-1)^{rk_Q(B)} = (-1)^{rk_Q(B-d)+1} \overset{IH}{=} (-1)^{cr(B-d)+\beta(B-d)+1} = (-1)^{cr(B)+\beta(B)}.$$

Otherwise, i.e. case 1.2 due to Lemma 3, we have exactly one vertex and some faces
containing it. The number of these faces is determined by the number of digons in

$\bar{\mathscr{O}}(B)$, which we denote with d. Then we have

$$\sum_{\substack{\mathscr{O}(B) \subseteq A \\ tot.cyclic}} \mu_Q(\emptyset, \mathscr{O}(B)) = -\binom{d}{0} + \binom{d}{1} - \binom{d}{2} + \ldots \pm \binom{d}{d} = -\sum_{k=0}^{d} (-1)^k \binom{d}{k} = 0.$$

$\square$

The key point is the following lemma, where the contraction finally takes place.

Lemma 5 *Let $\emptyset \neq B \subseteq E$. Then*

$$\sum_{\emptyset \neq X \subseteq \bar{\mathscr{O}}(B)} \mu_P(\emptyset, X) = -1.$$

Proof Since P is obviously unbounded and has at least one vertex, Corollary 3.4.10 in [3] yields that P has reduced Euler characteristic zero. Consequently the corresponding Möbius function $\mu_P(\emptyset, \bar{\mathscr{O}}(B))$, which is the reduced Euler characteristic (see Prop. 3.8.6 in [16]), equals zero, too. As a result,

$$0 = \mu_P(\emptyset, \bar{\mathscr{O}}(B)) = - \sum_{\emptyset \subseteq X \neq \bar{\mathscr{O}}(B)} \mu_P(\emptyset, X) = -1 - \sum_{\emptyset \neq X \subseteq \bar{\mathscr{O}}(B)} \mu_P(\emptyset, X)$$

holds.

$\square$

Combining the last two lemmas we find two different kinds of compression: In cases 1.1 and 2 it suffices to count the element having minimal support due to Lemma 5 and in case 1.2 all totally cyclic partial orientations sum up to zero due to Lemma 4. The following observation translates these cases from polyhedral language into graph theoretical properties.

Definition 5 Let $D = (V, A)$ be a totally cyclic digraph. A digon $d \subseteq A$ is called *redundant* for cyclicity if $D - d$ is still totally cyclic.

Note that every bridge is redundant for cyclicity. Fig. 1 shows a digon that is redundant but not a bridge.

Lemma 6 *Case 1.2 does not hold true if and only if there exists a digon in $\bar{\mathscr{O}}(B)$ that is redundant for cyclicity but not a bridge, or every digon in $\bar{\mathscr{O}}(B)$ is a bridge.*

Fig. 1 A digon that is redundant for cyclicity

Proof First we proof the following equivalence:

There are at least two vertices in P if and only if there is a digon in $\bar{\mathscr{O}}(B)$ that is redundant for cyclicity but not a bridge.

Let e be a digon in $\bar{\mathscr{O}}(B)$ that is redundant but not a bridge, then $\bar{\mathscr{O}}(B) - \overleftarrow{e}$ and $\bar{\mathscr{O}}(B) - \overrightarrow{e}$ contain vertices including $\overrightarrow{e}$, resp. $\overleftarrow{e}$ which hence are two different vertices in P. For the other direction take vertices $v \neq w$ in P. Then $v \cup w$ is a face in P including a digon e that is no bridge. Assume e is not redundant, then $\bar{\mathscr{O}}(B) - \overleftarrow{e}$ or $\bar{\mathscr{O}}(B) - \overrightarrow{e}$ could not have been totally cyclic and so one of the vertices v or w.

Consequently case 1.2 does not hold true iff there is a digon that is redundant but not a bridge (case 2) or, if there is only one vertex in P, then there are no further faces including it, which means that every digon in $\bar{\mathscr{O}}(B)$ is a bridge (case 1.1). □

This leads to the following main result of this paper, a representation of the NL-coflow polynomial for arbitrary digraphs, where we sum only over certain subsets of the edges of the underlying undirected graph.

Theorem 3 *Let* $D = (V, A)$ *be a digraph and* $G = (V, E)$ *its underlying undirected graph. Then*

$$\psi_{NL}^D(x) = \sum_{B \in TC} (-1)^{|B|} x^{\tilde{c}(B) - c(D)}$$

holds, where $\tilde{c}(B)$ *counts the components in the spanning subgraph of* G *with edge set* B *and* TC *includes all* $B \subseteq E$ *which admit a totally cyclic partial orientation* $\mathscr{O}(B)$ *in* A *such that* $\bar{\mathscr{O}}(B)$ *has no digons but bridges or* $\bar{\mathscr{O}}(B)$ *has a digon that is redundant but not a bridge.*

Proof Instead of counting totally cyclic subdigraphs one can count totally cyclic partial orientations of a fixed underlying subgraph. Thus the preceding lemmas yield

$$\psi_{NL}^D(x) = \sum_{\substack{X \subseteq A \\ tot.cyclic}} \mu_Q(\emptyset, X) x^{\text{rk}(A/X)}$$

$$= \sum_{B \subseteq E} \sum_{\substack{\mathscr{O}(B) \\ tot.cyclic}} \mu_Q(\emptyset, \mathscr{O}(B)) x^{\text{rk}(A/B)}$$

$$= \sum_{\substack{\emptyset \neq B \subseteq E \\ }} \sum_{\substack{\mathscr{O}(B) \\ tot.cyclic}} \mu_Q(\emptyset, \mathscr{O}(B)) x^{\text{rk}(A/B)} + x^{-c(D)}$$

$$\overset{\text{Lemma 4}}{=} \sum_{\substack{\emptyset \neq B \subseteq E \\ (*)}} \sum_{\substack{\mathscr{O}(B) \\ tot.cyclic}} (-1)^{cr(B) + \beta(B) + 1} \mu_P(\emptyset, \mathscr{O}(B)) x^{\text{rk}(A/B)} + x^{-c(D)}$$

Fig. 2 A totally cyclic orientation that is not considered in TC

$$\text{Lemma 5} \atop = \quad \sum_{\emptyset \neq B \subseteq E \atop (*)} (-1)^{cr(B)+\beta(B)} x^{\text{rk}(A/B)} + x^{-c(D)}$$

$$= \sum_{B \subseteq E \atop (*)} (-1)^{cr(B)+\beta(B)} x^{n-|V(B)|+c(B)-c(D)}.$$

Condition $(*)$ means, that we sum over all $B \subseteq E$ having a totally cyclic partial orientation $\mathcal{O}(B) \subseteq A$, where case 1.2 is not true. Due to Lemma 6 this situation occurs if and only if $\bar{\mathcal{O}}(B)$ has no digons but bridges, or there exists a digon that is redundant but not a bridge. Clearly, $n - |V(B)| + c(B) = \tilde{c}(B)$ holds, and we are left to verify

$$(-1)^{cr(B)+\beta(B)} = (-1)^{|B|}.$$

This can be done by induction. Deleting a bridge $d \in B$ yields $cr(B - d) = cr(B)$ and $\beta(B-d) = \beta(B)-1$ while deleting a non-bridge yields $cr(B-d) = cr(B)-1$ and $\beta(B - d) = \beta(B)$. In both cases we find

$$(-1)^{cr(B)+\beta(B)} = (-1)^{cr(B-d)+\beta(B-d)+1} \overset{IH}{=} (-1)^{|B-d|+1} = (-1)^{|B|}. \qquad \square$$

Note that TC includes all $B \subseteq E$ which admit a totally cyclic partial orientation $\mathcal{O}(B)$ in A, but not those, where $\bar{\mathcal{O}}(B)$ includes a digon that is no bridge and no digon is redundant unless it is a bridge in $\mathcal{O}(B)$ (Fig. 2).

5 Symmetric Digraphs

Considering symmetric digraphs $D = (V, A)$, it is obvious that the NL-coflow polynomial equals the chromatic polynomial $\chi(G, x)$ of the underlying undirected graph $G = (V, E)$ divided by the number of colors since both polynomials count the same objects. Using Theorem 3 we find an alternative proof of this fact, where the chromatic polynomial is represented by (see [5])

$$\chi(G, x) = \sum_{B \subseteq E} (-1)^{|B|} x^{\tilde{c}(B)}.$$

Corollary 2 *Let* $D = (V, A)$ *be a symmetric digraph and* $G = (V, E)$ *its underlying undirected graph. Then the following holds*

$$\psi_{NL}^{D}(x) = \chi(G, x) \cdot x^{-c(G)}.$$

Proof In a symmetric digraph every edge is a digon, so for every subset $B \subseteq E$ there exists a totally cyclic partial orientation $\mathcal{O}(B)$. Furthermore, if $cr(D) = 0$, every digon is a bridge and if $cr(D) \geq 1$ there exists a cycle of length ≥ 3 in D where every digon is redundant but no bridge. □

References

1. Aigner, M.: Combinatorial Theory. Springer, Berlin (1980)
2. Altenbokum, B., Hochstättler, W., Wiehe, J.: The NL-flow polynomial. Discrete Appl. Math. (2021 in press). https://doi.org/10.1016/j.dam.2020.02.011
3. Beck, M., Sanyal, R.: Combinatorial Reciprocity Theorems. American Math. Society, Providence (2018)
4. Bokal, D., Fijavz, G., Juvan, M., Kayll, P.M., Mohar, B.: The circular chromatic number of a digraph. J. Graph Theory **46**(3), 227–240 (2004)
5. Bondy, J.A., Murty, U.S.R.: Graph Theory. Springer, London (2008)
6. Ellis, P., Soukup, D.T.: Cycle reversions and dichromatic number in tournaments. Eur. J. Comb. **77**, 31–48 (2019)
7. Erdős, P.: Problems and results in number theory and graph theory. In: Proc. Ninth Manitoba Conf. on Numerical Math. and Computing, pp. 3–21 (1979)
8. Erdős, P., Gimbel, J., Kratsch, D.: Some extremal results in cochromatic and dichromatic theory. J. Graph Theory **15**(6), 579–585 (1991)
9. Hochstättler, W.: A flow theory for the dichromatic number. Eur. J. Comb. **66**, 160–167 (2017)
10. Li, Z., Mohar, B.: Planar digraphs of digirth four are 2-colorable. SIAM J. Discret. Math. **31**, 2201–2205 (2017)
11. Mohar, B., Wu, H.: Dichromatic number and fractional chromatic number. Forum Math. Sigma **4**, e32 (2016)
12. Neumann-Lara, V.: The dichromatic number of a digraph. J. Comb. Theory Ser. B **33**, 265–270 (1982)
13. Neumann-Lara, V.: Vertex colourings in digraphs. Some problems. Tech. rep., University of Waterloo (1985)
14. Neumann-Lara, V.: The 3 and 4-dichromatic tournaments of minimum order. Discrete Math. **135**(1), 233–243 (1994)
15. Rota, G.C.: On the foundations of combinatorial theory. Z. Wahrscheinlichkeitstheorie Verw. Geb. **2**, 340–368 (1964)
16. Stanley, R.P.: Enumerative Combinatorics, vol. 1. Cambridge Studies in Advanced Mathematics. Cambridge University Press, Cambridge (2011)
17. Tutte, W.T.: A contribution to the theory of chromatic polynomials. Canad. J. Math. **6**, 80–91 (1954)

On List k-Coloring Convex Bipartite Graphs

Josep Díaz, Öznur Yaşar Diner, Maria Serna, and Oriol Serra

Abstract List k-Coloring (LI k-COL) is the decision problem asking if a given graph admits a proper coloring compatible with a given list assignment to its vertices with colors in $\{1, 2, \ldots, k\}$. The problem is known to be NP-hard even for $k = 3$ within the class of 3-regular planar bipartite graphs and for $k = 4$ within the class of chordal bipartite graphs. In 2015 Huang, Johnson and Paulusma asked for the complexity of LI 3-COL in the class of chordal bipartite graphs. In this paper, we give a partial answer to this question by showing that LI k-COL is polynomial in the class of convex bipartite graphs. We show first that biconvex bipartite graphs admit a multichain ordering, extending the classes of graphs where a polynomial algorithm of Enright et al. (SIAM J Discrete Math 28(4):1675–1685, 2014) can be applied to the problem. We provide a dynamic programming algorithm to solve the LI k-COL in the class of convex bipartite graphs. Finally, we show how our algorithm can be modified to solve the more general LI H-COL problem on convex bipartite graphs.

Keywords List coloring · Convex bipartite · Biconvex bipartite graphs

J. Díaz (✉) · M. Serna
ALBCOM Research Group, CS Department, Universitat Politècnica de Catalunya, Barcelona, Spain
e-mail: diaz@cs.upc.edu; mjserna@cs.upc.edu

Ö. Y. Diner
Mathematics Department, Universitat Politècnica de Catalunya, Barcelona, Spain

Computer Engineering Department, Kadir Has University, Istanbul, Turkey
e-mail: oznur.yasar@khas.edu.tr

O. Serra
Mathematics Department, Universitat Politècnica de Catalunya, Barcelona, Spain
e-mail: oriol.serra@upc.edu

© The Author(s), under exclusive license to Springer Nature Switzerland AG 2021
C. Gentile et al. (eds.), *Graphs and Combinatorial Optimization: from Theory to Applications*, AIRO Springer Series 5,
https://doi.org/10.1007/978-3-030-63072-0_2

15

1   Introduction

A *coloring* of a graph $G = (V, E)$ is a map $c : V \to \mathbb{N}$. A coloring is *proper* if no two adjacent vertices are assigned the same color. If there is a proper coloring of a graph that uses at most k colors, then we say that G is k-*colorable*, and that c is a k-*coloring* for G. The coloring problem COL asks for a given graph $G = (V, E)$, and a positive integer k, whether there is a k-coloring for G or not. When k is fixed, we have the k-COLORING problem.

A list assignment $L : V \to 2^{\mathbb{N}}$ is a map assigning a set of positive integers to each vertex of G. Given G and L, the List Coloring problem LICOL asks for the existence of a proper coloring c that obeys L, i.e., each vertex receives a color from its own list. If the answer is positive, G is said to be L-*colorable*. Variants of the problem are defined by bounding the total number of available colors or by bounding the list size. In LIST k-COLORING (LI k-COL), $L(v) \subseteq \{1, 2, \ldots, k\}$ for each $v \in V$. Thus, there are k colors in total. On the other hand, in k-LIST COLORING (k-LICOL) each list L has size at most k. In this case, the total number of colors can be larger than k.

Precoloring Extension, PREXT, is a special case of LICOL and a generalization of COL. In PREXT all of the vertices in a subset W of V are previously colored; and the task is to extend this coloring to all of the vertices. If, in addition, the total number of colors is bounded, say by k, then it is called the k-Precoloring Extension, k-PREXT. k-COL is clearly a special case of k-PREXT, which in turn is a special case of LI k-COL. Refer to [16] for a chart summarizing these relationships.

For general graphs COL and its variants LICOL and PREXT are NP-complete; see [14, 24]. Most of their variants are NP-complete even when the parameter k is fixed for small values of k: k-COL, k-LICOL, LI k-COL and k-PREXT are NP-complete when $k \geq 3$ [29]; and they are polynomially solvable when $k \leq 2$ [13, 38].

Concerning the complexity of these problems in graph classes, COL is solvable in polynomial time for perfect graphs [18] whereas LICOL is NP-complete when restricted to perfect graphs and many of its subclasses, such as split graphs, bipartite graphs [28] and interval graphs [2]. On the other hand, LICOL is polynomially solvable for trees, complete graphs and graphs of bounded treewidth [23]. Refer to Tuza [37], and more recently to Paulusma [33] for related surveys.

For small values of k, Jansen and Scheffler [23] have shown that 3-LICOL is NP-complete when restricted to complete bipartite graphs and cographs, as observed in [15]. Kratochvíl and Tuza [27] showed that 3-LICOL is NP-complete even if each color appears in at most three lists, each vertex in the graph has degree at most three and the graph is planar. 3-PREXT is NP-complete even for 3-regular planar bipartite graphs and for planar bipartite graphs with maximum degree 4 [7].

For fixed $k \geq 3$, LI k-COL is polynomially solvable for P_5-free graphs [20]. Note that chordal bipartite graphs contain P_5-free graphs, but P_6 free graphs are incomparable with chordal bipartite graphs [35]. LI 3-COL is polynomial for P_6-free graphs [6] and for P_7-free graphs [3]. Computational complexity of LI 3-COL for P_8-free bipartite graphs is open [3]. Even the restricted case of LI 3-COL for P_8-free chordal bipartite graphs is open. Golovach et al. [16] give a survey that summarizes the results for LI *k*-COL on *H*-free graphs in terms of the structure of *H*.

PREXT problem is solvable in linear time on P_5-free graphs; and it is NP-complete when restricted to P_6-free chordal bipartite graphs [22]. 3-PREXT is NP-complete even for planar bipartite graphs [26], even for those having maximum degree 4 [7]. Recall that PREXT generalizes *k*-PrExt and LI *k*-COL generalizes *k*-PREXT. But there is no direct relation between PREXT and LI *k*-COL [16].

Coloring problems can be placed in the more general class of *H*-coloring problems. Given two graphs *G* and *H*, a function $f : V(G) \rightarrow V(H)$ such that $f(u)$ and $f(v)$ are adjacent in *H* whenever *u* and *v* are adjacent in *G* is called a graph homomorphism from *G* to *H*. For a fixed graph *H* and for an input *G*, the *H*-coloring problem, *H*-COL asks whether there is a *G* to *H* homomorphism. In the list *H*-coloring problem, LI *H*-COL, each vertex of the input graph *G* is associated with a list of vertices of *H*, and the question is whether a *G* to *H* homomorphism exists that maps each vertex to a member of its list. Observe that LI *H*-COL is a generalization of LI *k*-COL. The complexities of the *H*-coloring and list *H*-coloring problems for arbitrary input graphs are completely characterized in terms of the structure of *H*, see Nešetřil and Hell [19].

Although intensive research on this subject has been undertaken in the last two decades, there are still numerous open questions regarding computational complexities on LICOL and its variants when they are restricted to certain graph classes. Huang et al. [21] proved that LI 4-COL is NP-complete for P_8-free chordal bipartite graphs and 4-PREXT is NP-complete for P_{10}-free chordal bipartite graphs. They further pose the problem on the computational complexity of the LI 3-COL and 3-PREXT on chordal bipartite graphs. Here LI *k*-COL and *k*-PREXT on convex bipartite graphs, a proper subclass of chordal bipartite graphs, are studied for fixed *k*, and a partial answer to this question is given. Figure 1 summarizes the related results. Note that, here by LI *k*-COL it is assumed that *k* is fixed.

A bipartite graph $G = (X \cup Y, E)$ is convex if it admits an ordering on one of the parts of the bipartition, say *X*, such that the neighbours of each vertex in *Y* are consecutive in this order. If both color classes admit such an ordering the graph is called biconvex bipartite (see Sect. 2 for formal definitions). Chordal bipartite graphs contain convex bipartite graphs properly. Convex bipartite graphs contain as a proper subclass biconvex bipartite graphs, which contain bipartite permutation graphs properly. More information on these classes can be found in Spinrad [35] and in Brandstädt et al. [4].

Enright et al. [12] have shown that LI *k*-COL is solvable in polynomial time when restricted to graphs with all connected induced subgraphs having a multichain ordering. They apply this result to permutation graphs and interval graphs. Here, we show that connected biconvex graphs also admit a multichain ordering, implying a polynomial time algorithm for LI *k*-COL on this graph class.

From the point of view of parameterized complexity, treewidth can be computed in polynomial time on chordal bipartite graphs [25]. LI *k*-COL can be solved in polynomial time on chordal bipartite graphs with bounded treewidth [9, 23], which includes chordal bipartite graphs of bounded degree [30]. LI *k*-COL is polynomial

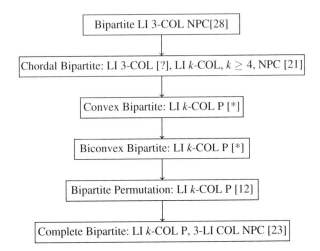

Fig. 1 Chart for known complexities for LICOL and its variants for chordal bipartite graphs and its subclasses, for $k \geq 3$. The complexity results marked with [*] is the topic of this paper, while [?] stands for open cases. Results without reference are trivial. P stands for Polynomial and NPC for NP-complete

for graphs of bounded cliquewidth [8]. Note that convex bipartite graph contains graphs with unbounded treewidth as well as graphs with unbounded cliquewidth.

The paper is organized as follows. In Sect. 2, we give the necessary definitions. In Sect. 3, we show that connected biconvex bipartite graphs admit multichain ordering. In Sect. 4, we show that, for fixed k, LI k-COL is polynomially solvable when it is restricted to convex bipartite graphs. Then, we show how to extend this result to LI H-COL. For an extended version of this paper the reader may refer to [10].

2 Preliminaries

We consider finite simple graphs $G = (V, E)$. For terminology refer to Diestel [11]. An edge joining non adjacent vertices in the cycle, C_n, is called a *chord*. A graph G is *chordal* if every induced cycle of length $n \geq 4$ has a chord. *Chordal bipartite graphs* are bipartite graphs in which every induced $C_n, n \geq 6$ has a chord. This graph class is introduced by Golumbic and Gross [17]. Chordal bipartite graphs may contain induced C_4, so they do not constitute a subclass of chordal graphs but it is a proper subclass of bipartite graphs. Chordal bipartite graphs can be recognized in polynomial time [32].

A bipartite graph is represented by $G = (X \cup Y, E)$, where X and Y form a bipartition of the vertex set into stable sets. An ordering of the vertices X in a bipartite graph $G = (X \cup Y, E)$ has the *adjacency property* (or the ordering is

Fig. 2 A convex bipartite graph which is not biconvex

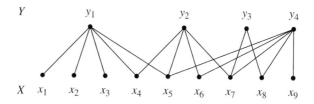

said to be *convex*) and G is said to have *convexity with respect to X* if, for each vertex $v \in Y$, $N(v)$ consists of vertices which are consecutive in the ordering of X. We say that an ordering of the vertices X in a bipartite graph $G = (X \cup Y, E)$ has the *enclosure property* if for every pair of vertices $u, v \in Y$ such that $N(u) \subseteq N(v)$, the vertices in $N(v) \backslash N(u)$ occur consecutively in the ordering of X.

Convex bipartite graphs are bipartite graphs $G = (X \cup Y, E)$ that have the adjacency property on one of the partite sets and *biconvex bipartite graphs* have the adjacency property on both partite sets X and Y. Figure 2 shows a graph that is convex but not biconvex. *Bipartite permutation graphs* are biconvex bipartite graphs in which one of the partite sets obeys both the adjacency and the enclosure properties. There are linear time recognition algorithms for these classes [31, 36].

A *chain graph* [39] is a bipartite graph that contains no induced $2K_2$ (a graph formed by two independent edges). The following characterization from [12] is equivalent: a connected bipartite graph with bipartite sets X and Y is a chain graph if and only if for any two vertices $y_1, y_2 \in Y$ we have $N(y_1) \subseteq N(y_2)$ or $N(y_2) \subseteq N(y_1)$. If the vertices in X are ordered with respect to their degrees starting from the highest degree, then for any $y \in Y$, the vertices in $N(y)$ will be consecutive in the ordering on X and, if the graph is connected, there is always a vertex $y \in Y$ so that $N(y)$ includes the first vertex in X. In particular, chain graphs are a proper subclass of convex bipartite graphs.

3 List k-Coloring on Biconvex Graphs

Enright et al. [12] show that LI k-COL, as well as the general LI H-COL, is solvable in polynomial time when restricted to graphs with all connected induced subgraphs having a multichain ordering. They apply this result to permutation graphs and interval graphs. Here, we show that connected biconvex graphs also admit a multichain ordering.

The *distance layers* of a connected graph $G = (V, E)$ from a vertex v_0 are $L_0, L_1, ..., L_z$, where $L_0 = \{v_0\}$ and, for $i > 0$, L_i consists of the vertices at distance i from v_0 and z is the largest integer for which this set is non-empty (see Fig. 3 for an example). These layers form a *multi-chain ordering* [5] of G if, for every two consecutive layers L_i and L_{i+1}, the edges connecting these two layers form a chain graph (not necessarily the layers themselves). All connected bipartite permutation graphs [5] and interval graphs [12] admit multichain orderings. Observe

Fig. 3 A convex bipartite
graph and its associated
distance layers from x_1

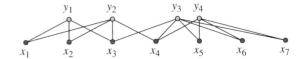

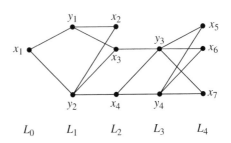

Fig. 4 Subdivision of $K_{1,3}$

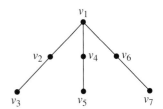

that, for the graph given in Fig. 3, the distance layers from x_1 provide a multichain ordering.

Recall that a subdivision of a graph G is the graph $G' = subd(G)$ obtained from G by replacing each edge by a path of length two. Thus $|E(G')| = 2|E(G)|$ and $|V(G')| = |V(G)| + |E(G)|$.

Lemma 1 *If G is a biconvex graph, then G does not contain $subd(K_{1,3})$ as an induced subgraph.*

Proof Let G be a biconvex graph and let $H = subd(K_{1,3})$. Let v_1 be the vertex of degree 3 in H, v_2, v_4 and v_6 be the vertices in $N(v_1)$ and v_3, v_5 and v_7 the vertices of degree 1 so that v_i is adjacent to v_{i+1} for $i = 2, 4, 6$, see Fig. 4. We observe that there is no ordering of $\{v_1, \ldots, v_7\}$ in which the three sets $N(v_2) = \{v_1, v_3\}$, $N(v_4) = \{v_1, v_5\}$ and $N(v_6) = \{v_1, v_7\}$ become consecutive. Therefore, a bipartite graph which contains H as an induced subgraph does not admit a biconvex ordering. □

Proposition 1 *Every connected biconvex graph admits a multichain ordering.*

Proof To see that biconvex graphs admit a multichain ordering, we use the notion of biconvex straight ordering introduced by Abbas and Stewart [1]. Let $G = (X, Y, E)$ be a bipartite graph with a linear ordering $\leq$ defined on $X \cup Y$. Two edges xy, $x'y' \in E$, where $x, x' \in X$ and $y, y' \in Y$, are said to *cross* if $x < x'$ and $y > y'$. If xy and

$x'y'$ cross, we call (x, y') and (x', y) the *corresponding straight pairs*. An ordering on $X \cup Y$ is a *straight ordering* if, for each pair xy, $x'y'$ of crossing edges, at least one of the corresponding straight pairs, (x, y') or (x', y), is an edge of the graph [1].

Let $G = (X, Y, E)$ be a connected biconvex graph. It follows from [1, Theorem 11] that G admits a biconvex straight ordering, say $v_0, v_1, \ldots, v_n$ of $X \cup Y$. Let $L_0 = \{v_0\}, L_1, \ldots, L_m$ be the distance layers of G from v_0. Since the graph G is connected, $V = L_0 \cup L_1 \cup \cdots \cup L_m$. Let us show that these layers form a multi-chain ordering.

The first layers L_0 and L_1 trivially form a multi-chain ordering. Let $L_1 = \{v_{i_1}, \cdots, v_{i_\ell}\}$, where the vertices are listed according to the ordering. When $\ell = 1$, L_1, L_2 trivially form a chain graph. When $\ell > 1$, since the ordering is straight, all the edges joining v_{i_1} with vertices in L_2 cross with the edge $v_0 v_{i_2}$. As, v_{i_2} is not connected to v_0, the other straight pair (v_{i_1}, v_{i_2}) should be an edge in G. Therefore, $N(v_{i_1}) \subseteq N(v_{i_2})$. By iterating the same argument, we see that $N(v_{i_1}) \subseteq \cdots \subseteq N(v_{i_\ell})$. Thus, the layers L_0, L_1, L_2 form a multi-chain ordering. By a similar discussion, it can be shown that L_0, L_1, L_2, L_3 form a multichain ordering.

Suppose that $m > 3$. Let $i > 3$ be the largest subscript such that $L_0, L_1, \ldots, L_i$ form a multichain ordering. Suppose for a contradiction that $i < m$. Thus, the bipartite graph induced by the layers L_i, L_{i+1} contain an induced copy of $2K_2$, say with edges $uv, u'v', u, u' \in L_i$ and $v, v' \in L_{i+1}$. As the ordering is straight, we may assume $u < u'$ and $v < v'$. We consider two cases:

Case 1 $N(u) \cap N(u') \cap L_{i-1} \neq \emptyset$. Let $w \in N(u) \cap N(u') \cap L_{i-1}$ and consider predecessors $w' \in L_{i-2}$, $w'' \in L_{i-3}$ of w. Then the subgraph induced by w, w', w'', u, u', v, v' is isomorphic to a subdivision H of $K_{1,3}$, contradicting Lemma 1.

Case 2 $N(u) \cap N(u') \cap L_{i-1} = \emptyset$. Let $w \in N(u) \cap L_{i-1}$ and $w' \in N(u') \cap L_{i-1}$ be some predecessors of u and u' in the previous layer. Observe that the two edges wu, $w'u'$ induce a $2K_2$ in the subgraph induced by $L_{i-1} \cup L_i$ contradicting the choice of i.

□

Let us state the main result by Enright et al. in [12] explicitly:

Theorem 1 ([12]) *Let H be a fixed graph.* LI *H-COL is polynomial- time solvable for input graphs G satisfying that every connected induced subgraph of G admits a multichain ordering.*

Proposition 1 and Theorem 1 give us the main result in this section.

Theorem 2 *For any fixed H,* LI *H-COL is solvable in polynomial time when restricted to biconvex graphs.*

As LI *k-COL* is a particular case of LI *H-COL* and LI *k-COL* generalizes k-PREXT, we have the following corollary.

Corollary 1 LI *k-COL and k-PREXT are solvable in polynomial time when restricted to biconvex graphs.*

Concerning the running time of the algorithms, it is shown in Abbas and Stewart [1] that a biconvex straight ordering of a biconvex bipartite graph can be found in linear time on the number of vertices of the graph. On the other hand, the algorithm in [12] is shown to run in time $O(n^{k^2-3k+4})$ time when a multichain ordering in decreasing ordering of degrees is given. Observe that to get such ordering, we only have to reorder the elements in the layers provided by the straight ordering, therefore it can be obtained in linear time. Altogether, it gives an upper bound $O(n^{k^2-3k+4})$ on the complexity of LI k-COL in the class of biconvex graphs.

4   List k-Coloring of Convex Bipartite Graphs

Let $G = (X \cup Y, E)$ be a connected bipartite graph that is convex with respect to X. Let $X = \{x_1, \ldots, x_n\}$ be a convex ordering of X, that is, for each $y \in Y$ there are two positive integers $a_y \le b_y$ such that $N(y) = \{x_i \mid a_y \le i \le b_y\}$.

Consider the set of integers $A = \{a_y \mid y \in Y\}$ and $B = \{b_y \mid y \in Y\}$. For the graph given in Fig. 5, $A = \{1, 4, 5, 7\}$ and $B = \{5, 7, 8, 9\}$.

We use the set B to direct the dynamic programming algorithm and the elements in A to determine the relevant information to be kept for the next step. Assume that $B = \{b_1, \ldots, b_\beta\}$ are sorted so that $b_1 < b_2 < \cdots < b_\beta$. By connectivity of G, we have $b_\beta = n$. For each $1 \le j \le \beta$, let $X_j = \{x_i \in X \mid i \le b_j\}$, $Y_j = \{y \in Y \mid b_y \le b_j\}$, and $Z_j = \{y \in Y \mid a_y \le b_j < b_y\}$. Define $G_j = G[X_j \cup Y_j]$. Observe that $G_\beta = G$, $Z_\beta = \emptyset$ and that Z_j contains those vertices in Y whose neighborhood starts before or at b_j and ends after b_j. For example, for the graph given in Fig. 5, $b_2 = 7$, $X_2 = \{x_1, x_2, \ldots, x_7\}$, $Y_2 = \{y_1, y_2\}$ and $Z_2 = \{y_3, y_4\}$. For sake of simplicity, we assume an initial point $b_0 = 0$, so that G_0 is the empty graph.

Let K be a set of k colors. Assume that each vertex u in G has an associated list $L(u) \subseteq K$. We next define the information that we want to compute for each $1 \le b_j \le b_\beta$. For each $1 \le j \le \beta$, define $A(j) = \{a_y \mid y \in Z_j\} \cup \{b_j\}$. As before,

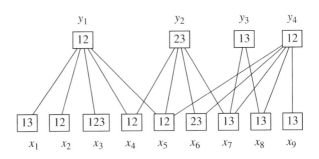

Fig. 5 A list assignment for the convex bipartite graph given in Fig. 2. Labels inside vertex indicate the list of colors, from $\{1, 2, 3\}$, associated to the node

we assume that the elements in $A(j) = \{a_{1,j}, \ldots, a_{\alpha_j,j}\}$ are increasingly ordered, $a_{1,j} < a_{2,j} < \cdots < a_{\alpha_j,j} = b_j$. To simplify notation, set $a_{\alpha_j+1,j} = b_j + 1$ to make sure that a higher value always exists. For the example in Fig. 5, $A(2) = \{5, 6, 7\}$. For the fictitious initial b value $j = 0$, we take $A(0) = \{0\}$.

Fix j, $1 \leq j \leq \beta$. For each $1 \leq i \leq \alpha_j$ and $S \subsetneq K$, $T_j(i, S)$ will hold value true whenever there is a valid list coloring of G_j such that it uses no color in S for the set $X_i^j = \{x_\ell \mid a_{i,j} \leq \ell < a_{i+1,j}\}$. Observe that we are not considering K as a potential set as not using any color is impossible.

The Color Algorithm will compute those values in three steps. In going from $j - 1$ to j, first it computes the values for the $x \in X_j$ that were not in X_{j-1} combining this information with the relevant information computed in the previous step. Next, it incorporates the restriction from $y \in Y_j$ that were not in Y_{j-1}. Finally, it rearranges the information to keep only the values for the index in $A(j)$.

Color Algorithm: Let j, $1 \leq j \leq \beta$. Initially set $A(0) = \{0\}$, $b_0 = 0$ and set $T_0(0, S)$ to TRUE for any S. When $j \geq 1$ assume that the values of T_{j-1} have already been computed.

Step 1   Extending to new parts.

Let $A'(j) = A(j-1) \cup \{a_y \mid b_{j-1} < a_y \leq b_j\} \cup \{b_j\}$. For $j > 1$, by construction, those values lie before b_j and some of them have no corresponding entries in T_{j-1}. Assume that $A'(j) = \{a'_1, \ldots, a'_{\gamma_j}\}$ increasingly ordered. Let $a'_{\gamma_j+1} = b_j + 1$. We set $T_{j-1}(\ell, S)$ for $\alpha_{j-1} < \ell \leq \gamma_j$ and $S \subsetneq K$ to be true whenever there is a valid list coloring of the set $X'(\ell) = \{x_i \mid a'_\ell \leq i < a'_{\ell+1}\}$. For this, the algorithm checks whether $L(x) \setminus S \neq \emptyset$ for each $x \in X'(\ell)$. If this is the case, one can select a color not in S and get a valid coloring. Accordingly we update the value of $T_{j-1}(\alpha_{j-1}, S)$ so that it remains TRUE if it was already set to TRUE and the previous condition holds for the elements in $X'(\alpha_{j-1})$

Step 2   Incorporating Y_j.

For $y \in Y_j$ and $a_i \in [a_y, b_y]$, consider any entry $T_{j-1}(i, S)$ set to TRUE. If $S \cap L(y) = \emptyset$, the corresponding entry is changed to FALSE.

Next, the values on T_{j-1} are processed in increasing order of x_i: any entry (i, S) holding value TRUE will remain TRUE whenever there is an entry $(i - 1, S')$ holding value TRUE with $S \subseteq S'$. By monotonicity, the property holds whenever $T_{j-1}(i - 1, S)$ is TRUE.

After processing y, if $T_{j-1}(l, S)$ holds true, for each piece $[a'_l, a'_{l+1})$ between a_y and b_y, we can pick a common color not in S but in $L(y)$ to color y that is compatible with some list coloring on the X relevant parts that do not use S.

Step 3   Compacting to get T_j.

For each $1 \leq i \leq \alpha_j$ the set X_i^j might contain several subintervals on $X'(j)$, considered either in T_{j-1} that will not be needed later on. We fusion those sets from left to right, adding one at a time, setting $T_j(i, S)$ to true whenever there are corresponding entries holding value true for sets S_1 and S_2 so that $S \subseteq S_1 \cap S_2$.

Lemma 2 *Let $G = (X \cup Y, E)$ be a connected convex bipartite graph, L be a color assignment for G. There is an L-coloring of G if and only if there is $S \subseteq K$ such that at the end of the execution of the Color Algorithm $T_\beta(\alpha_\beta, S) = true$.*

Proof Assume that G admits a list coloring. Let c be an L-coloring of G. For $U \subseteq X$ let $S_U = K \setminus c(X)$. Observe that L does not use any color in S_U on U and furthermore, for any $y \in Y$ so that $N(y) \cap U \neq \emptyset$, $L(y) \cap S_U \neq \emptyset$. Using this fact, it follows that the entries in the tables for the corresponding sets get the value true and at the end of the algorithm $T(\beta, \{c(x_n)\})$ will be true.

Conversely, we can prove that the Color Algorithm correctly computes the values of T_j for $1 \leq j \leq \beta$. The proof is by induction. Observe that for $j = 1$ the table R provides the right indices and the initialization step provides the correct values for the table on an empty graph. By induction hypothesis, we assume that the values of T_j are correctly computed. Step 1 guarantees that the desired coloring exists when adding only the X part on G_j to G_{j-1}. Step 2 has two parts. The first one guarantees that only those entries with sets that are compatible with the list of the vertices in Y_j are still alive. The second one ensures that when combining two consecutive pieces having a common neighborhood on Y_j a common set of colors (a subset) is available to color these vertices. Finally, Step 3 merges tables for pieces that have the same Y neighborhood outside G_j. Again, we need to maintain a common set of colors free for potential use on these neighbors. □

Finally, observe that all the running time of the Color Algorithm is polynomial in $|G|$ and in 2^k. Furthermore, the k-PREXT can be polynomially reduced to LI k-COL. Therefore, we get our main result.

Theorem 3 *For $k \geq 3$, LI k-COL and k-PREXT on convex bipartite graphs can be solved in polynomial time.*

The Color Algorithm can be modified to solve the LI H-COL on convex bipartite graphs. For this, the algorithm keeps track instead of the unused color on the X part of the used ones. For doing that, we have to consider some longer subdivision of the intervals in the X part. Step 2 will check that at least one of the colors in the list of y is connected to all the used colors in the X part. Step 3 is also modified as the global set of used colors will be the union.

Theorem 4 *For fixed H, LI H-COL on convex bipartite graphs can be solved in polynomial time.*

5 Conclusions

In this paper, the problem posed by Huang et al. [21] on the computational complexity of the LI 3-COL and 3-PREXT on chordal bipartite graphs is addressed. A partial answer to a general version of this question is given by increasing the subclasses of chordal bipartite graphs for which polynomial time algorithms for the

LI k-COL are known to biconvex bipartite graphs and convex bipartite graphs. Note that the later class includes convex bipartite graphs with bounded degree, complete bipartite graphs which have unbounded treewidth, as well as graphs with unbounded cliquewidth. Interestingly enough, the second result can also be extended, with a slight modification, to solve LI H-COL for the same graph class. The paper includes another result of independent interest: any connected biconvex bipartite graph admits a multichain ordering.

On the other hand, chordal bipartite graphs form a much larger graph class. Using the terminology of [34], it is a superfactorial graph class whereas convex bipartite graphs is a factorial graph class. Although LI k-COL is hard for $k \geq 4$ when restricted to chordal bipartite graphs, finding the computational complexity of LI 3-COL for chordal bipartite graphs is the next natural open question.

Acknowledgments J. Díaz and M. Serna are partially supported by funds from MINECO and EU FEDER under grant TIN 2017-86727-C2-1-R and AGAUR project ALBCOM 2017-SGR-786. Ö. Y. Diner is partially supported by the Scientific and Technological Research Council Tübitak under project BIDEB 2219-1059B191802095 and by Kadir Has University under project 2018-BAP-08. O. Serra is supported by the Spanish Ministry of Science under project MTM2017-82166-P.

References

1. Abbas, N., Stewart, L.K.: Biconvex graphs: ordering and algorithms. Discrete Appl. Math. **103**(1–3), 1–19 (2000)
2. Biro, M., Hujter, M., Tuza, Zs.: Precoloring extension. I. Interval graphs. Discrete Math. **100**(1–3), 267–279 (1992)
3. Bonomo, F., Chudnovsky, M., Maceli, P., Schaudt, O., Stein, M., Zhong, M.: Three-coloring and list three-coloring of graphs without induced paths on seven vertices. Combinatorica **38**, 779–801 (2018)
4. Brandstädt, A., Le, V.B., Spinrad, J.P.: Graph Classes: A Survey. Society for Industrial and Applied Mathematics, Philadelphia (1999)
5. Brandstädt, A., Lozin, V.V.: On the linear structure and clique-width of bipartite permutation graphs. Ars Comb. **67**, 273–281 (2003)
6. Broersma, H.J., Fomin, F.V., Golovach, P.A., Paulusma, D.: Three complexity results on coloring P_k-freegraphs. Eur. J. Comb. **34**, 609–619 (2013)
7. Chlebik, M., Chlebikova, J.: Hard coloring problems in low degree planar bipartite graphs. Discrete Appl. Math. **154**, 1960–1965 (2006)
8. Courcelle, B., Makowsky, J.A., Rotics, U.: Linear time solvable optimization problems on graphs of bounded clique-width. Theory Comput. Syst. **33**(2), 125–150 (2000)
9. Díaz, J., Serna, M., Thilikos, D.M.: Counting H-colorings of partial k-trees. Theor. Comput. Sci. **281**, 291–309 (2002)
10. Díaz, J., Diner, Ö.Y., Serna, M., Serra, O.: On list k-coloring convex bipartite graphs (2020). arXiv:2002.02729 [cs.CC]
11. Diestel, R.: Graph Theory. Graduate Texts in Mathematics, vol. 173, no. 5. Springer, Heidelberg (2017)
12. Enright, J. Stewart, T., Tardos, G.: On list coloring and list homomorphism of permutation and interval graphs. SIAM J. Discrete Math. **28**(4), 1675–1685 (2014)
13. Erdös, P., Rubin, A.L., Taylor, H.: Choosability in graphs. In: Proceedings of the West Coast Conference on Combinatorics, Graph Theory and Computing, Humboldt State University, Utilitas Mathematica, pp. 125–157 (1979)

14. Garey, M.R., Johnson, D.S.: Computers and Intractability: A Guide to the Theory of NP-Completeness. W. H. Freeman, New York (1979)
15. Golovach, P.A., Paulusma, D.: List coloring in the absence of two subgraphs. Discrete Appl. Math. **166**, 123–130 (2014)
16. Golovach, P.A., Johnson, M., Paulusma, D., Song, J.: A survey on the computational complexity of colouring graphs with forbidden subgraphs. J. Graph Theory **84**(4), 331–363 (2016)
17. Golumbic, M.C., Goss, C.F.: Perfect elimination and chordal bipartite graphs. J. Graph Theory **2**, 155–163 (1978)
18. Gröstchel, M., Lovasz, L., Schrijver, A.: Polynomial algorithms for perfect graphs. Ann. Discret. Math. **21**, 325–356 (1984)
19. Hell, P., Nešetřil, J.: Graphs and Homomorphisms. Oxford University Press, Oxford (2004)
20. Hoàng, C., Kamiński, M., Lozin, V., Sawada, J., Shu, X.: Deciding k-colorability of P_5-free graphs in polynomial time. Algorithmica **57**, 74–81 (2010)
21. Huang, S., Johnson H., Paulusma, D.: Narrowing the complexity gap for coloring (C_s, P_t)–Free Graphs. Comput. J. **58**(11), 3074–3088 (2015)
22. Hujter, M., Tuza, Zs.: Precoloring extension. III. Classes of perfect graphs. Comb. Probab. Comput. **5**, 35–56 (1996)
23. Jansen, K., Scheffler, P.: Generalized coloring for tree-like graphs. Discrete Appl Math. **75**, 135–155 (1997)
24. Karp, R.M.: Reducibility among combinatorial problems. In: Miller, R.E., Thatcher, J.W., Bohlinger J.D. (eds.) Complexity of Computer Computations, pp. 85–103. Plenum Press, New York (1972)
25. Kloks, T., Kratsch, D.: Treewidth of chordal bipartite graphs. J. Algoritm. **19**, 266–281 (1995)
26. Kratochvil, J.: Precoloring extension with fixed color bound. Acta Math. Univ. Comenian. **62**, 139–153 (1993)
27. Kratochvil, J., Tsuza, Z.: Algorithmic complexity of list colorings. Discrete Appl. Math. **50**, 297–302 (1994)
28. Kubale, M.: Some results concerning the complexity of restricted colorings of graphs. Discrete Appl. Math. **36**, 35–46 (1992)
29. Lovasz, L.: Coverings and colorings of hypergraphs. In: Proceedings of the 4th Southeastern Conference on Combinatorics, Graph Theory and Computing, Utilitas Mathematica, pp. 3–12 (1973)
30. Lozin, V., Rautenbach, D.: Chordal bipartite graphs of bounded tree- and clique-width. Discrete Math. **283**, 151–158 (2004)
31. Nussbaum, D., Pu, S., Sack, J.-R., Uno, T., Zarrabi-Zadeh, H.: Finding maximum edge bicliques in convex bipartite graphs. Algorithmica **64**(2), 140–149 (2010)
32. Paige, R., Tarjan, R.E.: Three partition refinement algorithms. SIAM J. Comput. **16**, 973–989 (1987)
33. Paulusma, D.: Open Problems on Graph Coloring for Special Graph Classes. Lecture Notes in Computer Science, vol. 9224, pp. 16–30. Springer, Berlin (2016)
34. Scheinerman, E.R., Zito, J.: On the size of hereditary classes of graphs. J. Combin. Theory Ser. B. **61**, 16–39 (1994)
35. Spinrad, J.: Efficient Graph Representations. Fields Institute Monographs, vol. 19. American Mathematical Society, Providence (2003)
36. Spinrad, J.P., Brandstädt, A., Stewart, L.: Bipartite permutation graphs. Discrete Appl. Math. **18**, 279–292 (1987)
37. Tuza, Zs.: Graph coloring with local constrains - a survey. Discuss. Math. Graph Theory **17**, 161–228 (1997)
38. Vizing, V.G.: Coloring the vertices of a graph in prescribed colors. Metody Diskret. Anal. v Teorii Kodov i Schem **29**, 3–10 (1976) (in Russian)
39. Yannakakis, M.: The complexity of the partial order dimension problem. SIAM J. Algebr. Discrete Methods **3**, 351–358 (1982)

Total Chromatic Sum for Trees

**Ewa Kubicka, Grzegorz Kubicki, Michał Małafiejski,
and Krzysztof M. Ocetkiewicz**

Abstract The total chromatic sum of a graph is the minimum sum of colors (natural numbers) taken over all proper colorings of vertices and edges of a graph. We provide infinite families of trees for which the minimum number of colors to achieve the total chromatic sum is equal to the total chromatic number. We construct infinite families of trees for which these numbers are not equal, disproving the conjecture from 2012.

Keywords Total colorings · Sum of colors · Trees

1 Introduction

Consider a proper coloring ϕ of vertices of a graph G using natural numbers; i.e. $\phi : V(G) \to N$ and $\phi(u) \neq \phi(v)$ whenever uv is an edge of G. The *chromatic sum* of G, denoted $\Sigma(G)$, is the minimum sum $\sum_{v \in V(G)} \phi(v)$ taken over all proper colorings ϕ of G. A coloring is *optimal* if the sum of colors equals $\Sigma(G)$.

This idea was introduced by Kubicka [4] in 1989, and since then much more work has been done with calculating the chromatic sums of graphs, generating algorithms to find chromatic sums and optimal colorings, and calculating the complexity of finding chromatic sums of graphs in certain families. Erdös et al. [2] constructed infinite families of graphs for which the minimum number of colors necessary to get an optimal coloring of G was larger than $\chi(G)$. This graph parameter, the minimum

E. Kubicka (✉) · G. Kubicki
University of Louisville, Louisville, KY, USA
e-mail: ewa@louisville.edu; gkubicki@louisville.edu

M. Małafiejski · K. M. Ocetkiewicz
Gdansk University of Technology, Gdansk, Poland
e-mail: michal@animima.org; krzysztof.ocetkiewicz@eti.pg.edu.pl

© The Author(s), under exclusive license to Springer Nature Switzerland AG 2021
C. Gentile et al. (eds.), *Graphs and Combinatorial Optimization:
from Theory to Applications*, AIRO Springer Series 5,
https://doi.org/10.1007/978-3-030-63072-0_3

number of colors necessary for an optimal coloring, is called the *strength* of G, and denoted by $\sigma(G)$ [3]. In [2], it is shown that even trees can have arbitrarily high strength, even though their chromatic number is 2. In fact, Erdös et al. [2] found for every $k \geq 3$ the smallest tree of strength k. In [3], Jiang and West also constructed trees with strength k but not of minimum order but of minimum maximum degree, $\Delta = 2k - 2$.

We say that a graph G is *strong* if $\chi(G) < \sigma(G)$. The smallest strong graph is the tree on eight vertices given in Fig. 1.

These color-sum concepts can be applied to edge coloring as well. In an analogous way, one can define the *edge chromatic sum* of a graph, its edge strength σ', and ask the question of whether or not $\chi' = \sigma'$. In 1997, Mitchem et al. [7] proved that every graph has a proper edge coloring with minimum sum that uses only Δ or $\Delta + 1$ colors. This implies that the only way for a graph to have $\chi' < \sigma'$ is to have both $\chi' = \Delta$ and $\sigma' = \Delta + 1$. We say that a graph G with this property, namely $\chi'(G) < \sigma'(G)$, is *E-strong*. The smallest known E-strong graph M is presented in Fig. 2.

Fig. 1 The smallest strong tree

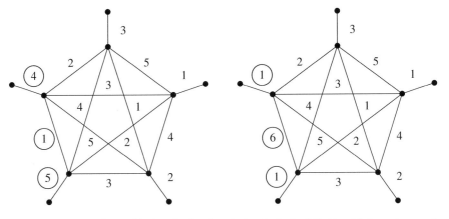

Fig. 2 E-strong graph M with $\Delta = 5$, $\chi' = 5$, and $\sigma' = 6$. On the left side with five colors used, the sum of the colors is 45. If we introduce a sixth color and change the colors of the edges whose color labels are circled, we obtain a coloring with sum 43

2 Total Chromatic Sum and Total Strength

A *total coloring* ϕ of G is an assignment of natural numbers to vertices and edges of a graph. A total coloring of G is *proper* if no pair of adjacent or incident elements (vertices or edges) is assigned the same color. A total-k-coloring ϕ of G is a proper total coloring that uses k colors. The *total chromatic number* $\chi''(G)$ of a graph G is the smallest number k for which G has a total-k-coloring. The famous Total Coloring Conjecture stating that

$$\chi''(G) \le \Delta(G) + 2$$

for every graph G, where $\Delta(G)$ is the maximum degree of G, was posed independently by Vizing [8, 9] and Behzad [1].

The total chromatic sum of a graph is defined in a similar way to the chromatic sum. The *total chromatic sum* of G, denoted $\Sigma''(G)$, is the minimum sum $\sum_{x \in V(G) \cup E(G)} \phi(x)$ taken over all proper total colorings ϕ of G. A total coloring is *optimal* if the sum of colors of vertices and edges of G equals $\Sigma''(G)$. The minimum number of colors necessary for an optimal total coloring is called the *T-strength* of G, and is denoted by $\sigma''(G)$. We say that a graph G is *T-strong* if $\sigma''(G) > \chi''(G)$. The total chromatic sum and the related parameters were introduced by Leidner [6] in his Ph.D. dissertation.

The total chromatic sum was determined for many families of graphs. In [6] and [5], several infinite families of T-strong graphs, it means graphs for which we need more colors for optimal total coloring than the total chromatic number, were constructed. Each graph from the following list is T-strong:

1. Cycles of length $3n$, $n \ge 2$, with one chord joining vertices at distance congruent to 3 along the cycle. Those graphs have $\Delta = 3$, $\chi'' = 4$, and $\sigma'' = 5$.
2. Cycles of length $3n$, $n \ge 2$, with two independent chords with proper distances along the cycle. For those graphs also $\Delta = 3$, $\chi'' = 4$, and $\sigma'' = 5$.
3. Graphs G obtained from M, the graph from Fig. 2, by attaching a copy of K_{2k+1} to each vertex. Here $\Delta(G) = 2k + 5$, $\chi''(G) = 2k + 6$, and $\sigma''(G) = 2k + 7$.

The smallest T-strong graph is a 6-cycle with a diametral chord, or equivalently the graph $P_2 \times P_3$. Two colorings, the first using $\chi''(G)$ colors and the second that is optimal and uses one more color are depicted in Fig. 3. Leidner [6] verified by an exhaustive computer search that $P_2 \times P_3$ is the smallest T-strong graph and the only one of order smaller than 9.

Fig. 3 The grid graph $P_2 \times P_3$ total-colored in two ways

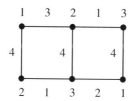
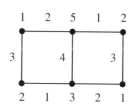

In the next sections, we show new results for the total chromatic sum of trees and the existence of T- strong trees. Two conjectures about the total strength of a graphs were stated in [5].

Conjecture 1 For every graph G, $\chi''(G) \leq \sigma''(G) \leq \chi''(G) + 1$.

Conjecture 2 No tree is T-strong.

In this paper, we prove that no tree requires more than $\Delta + 2$ colors to achieve its total chromatic sum, which proves Conjecture 1 for trees. We construct a polynomial time algorithm to determine the total chromatic sum and the total chromatic strength of a tree. We prove that all trees with no adjacent vertices of maximum degree have the total chromatic number equal to the total chromatic strength. Finally, we disprove Conjecture 2 providing infinite families of T-strong trees.

3 Upper Bound on Total Strength of Trees

Let $G = (V, E)$ be a tree. We define a distance between elements of G.

- If $u, v \in V$, then $d(u, v)$ is the number of edges on the $u - v$ path.
- If $e, f \in E$, then $d(e, f)$ is the number of vertices on the $e - f$ path.
- If $e = uv \in E$ and $w \in V$, then $d(e, w) = \frac{1}{2} + min\{d(u, w), d(v, w)\}$.

It is easy to check that d is a metric on $V \cup E$.

It is well known that every tree is either **unicentral** or **bicentral**, i.e. has the center consisting of one vertex or two adjacent vertices, respectively. For the purpose of this note we call the **strong center** of G:

- the central vertex u if G if unicentral,
- the edge $e = uv$ if G is bicentral with the center $cent(G) = \{u, v\}$.

Theorem 1 *No tree requires $\Delta + 3$ or more colors for an optimal total coloring.*

Proof Let $\Delta = \Delta(G)$. We will show that there is an optimal coloring of G without a color $c = \Delta + 3$. For larger colors, the proof is similar but simpler. Suppose, to the contrary, that the color $c = \Delta + 3$ must occur in any optimal coloring of G. Among all such colorings select a coloring φ in which the color c occurs as far away from the strong center of G as possible. Suppose first that color $c = \Delta + 3$ occurs on the edge $e = xy$, x closer to the center than y (if $\{e\}$ is not a strong center of G). Let $E(x)$ and $E(y)$ denote the set of edges incident to x and y, respectively. Define two sets of elements of G, $L = \{x\} \cup E(x) - \{e\}$ and $R = \{y\} \cup E(y) - \{e\}$. Notice that $|L|, |R| \leq \Delta$; also all elements of L (and of R) must be colored with different colors. If $\varphi(R) \subset \varphi(L)$, then because $|\varphi(L)| \leq \Delta$, we can find a color $c_1 \in \{1, 2, \ldots, \Delta + 1\}$ that was not used on elements from L and from R. Recolor e with c_1 obtaining a total coloring with a smaller sum. If $\varphi(R) \setminus \varphi(L) \neq \emptyset$, say $c_1 \in \varphi(R) \setminus \varphi(L)$, and $c_1 = \varphi(\alpha)$, then recolor the element α by c and the edge e by c_1. This produces the total coloring of G with the same sum but with color

$c = \Delta + 3$ father away from the strong center of G; a contradiction. Notice that the same argument can be used if the color on e is $\Delta + 2$ (not $\Delta + 3$). This observation will be used in the proof of the next theorem.

It remains to consider the case when in the coloring φ of G, the element with color $c = \Delta + 3$ is a vertex v. Of course, v is not a leaf since any leaf can be colored with a color from $\{1, 2, 3\}$. Let u be a neighbor of v closer to the strong center of G than v, or the other vertex of the strong center if the center of G is $\{u, v\}$. Assume that $\varphi(u) = c_1$ and $\varphi(uv) = c_2$. Since $deg(v) \le \Delta$, there are two colors, say c_3 and c_4 from $\{1, 2, \ldots, \Delta + 2\}$ that are not present on the edges incident to v (for illustration see Fig. 4). If $c_1 \in \varphi(E(v) - \{e\})$, then both colors c_3 and c_4 must be used on neighbors of v, say $c_3 = \varphi(x)$, $c_4 = \varphi(y)$, otherwise we could recolor v with c_3 or c_4, obtaining a smaller sum of colors. Therefore, there is a color $c_5(c_5 \ne c_1)$ on an edge f incident to v that is not used on the neighbors of v. We can interchange colors c and c_5 on v and f obtaining a coloring in which c is father away from the strong center of G; a contradiction. Similar argument works if $c_1 \notin \varphi(E(v) - \{e\})$. Without loss of generality, we might assume that $c_1 = c_3$. Then color c_4 must occur on a neighbor of v, say $c_4 = \varphi(x)$; otherwise we could recolor v with c_4. Now, one of the colors from $\varphi(E(v) - \{e\})$ is not used on any neighbor of v, say this color is c_5 and it occurs on the edge f, $c_5 = \varphi(f)$. Similarly, we can interchange colors c and c_5 on v and f obtaining a coloring with the color c farther away from the strong center of G; a contradiction. $\square$

It is well known that, with the exception of K_2, $\chi''(G) = \Delta + 1$ for every tree G with maximum degree Δ. Therefore, as the corollary of Theorem 1, for trees we have the following two possibilities:

1. $\chi''(G) = \Delta + 1 = \sigma''(G)$, which means that G is not T-strong, or
2. $\chi''(G) = \Delta + 1$ and $\sigma''(G) = \Delta + 2$, if a tree G is T-strong.

This observation verifies Conjecture 1 for trees.

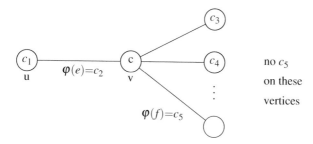

Fig. 4 Illustration of the proof of Theorem 1

4 Total Strength of Trees with No Adjacent Vertices of Maximum Degree

abel3:sec:4

A tree G is called $\triangle\triangle$-**free** if G has no adjacent vertices of maximum degree.

Theorem 2 *No $\triangle\triangle$-free tree is T-strong.*

Proof Assume, to the contrary, that there exists a T-strong tree. Such a tree must have $\triangle \geq 3$. Among all such trees consider those of minimum order. For each tree of this property, select an optimal total coloring φ (the coloring with the sum of colors equal to $\sum''(G)$) in which the color c $(c = \triangle + 2)$ occurs as far away from the strong center of G as possible. Consider a particular tree G with such coloring φ. By the observation in the first part of the proof of Theorem 1, this color cannot occur on any edge of G.

Thus, one can assume that in the coloring $\varphi(G)$, the element colored with c is some vertex, say v. If $deg(v) \leq \triangle - 1$, then we get a contradiction using a similar argument as in the proof of Theorem 1, where numbers $\triangle$ and $\triangle + 3$ denoting the degree of v and the largest color c, respectively, are now replaced by $\triangle - 1$ and $\triangle + 2$. It remains to show that there is no $\triangle\triangle$-free tree with the color $\triangle + 2$ occurring on the vertex v with $deg(v) = \triangle$. So assume that $\varphi(v) = \triangle + 2$, the neighbor of v "toward the center of G" is u (notice that $deg(u) \leq \triangle - 1$ because G is $\triangle\triangle$-free), $\varphi(u) = c_1$, and $\varphi(uv) = c_2$.

Case 1 If $c_1 \notin \varphi(E(v) - \{e\})$ and $\varphi(N(v)) - u) = \varphi(E(v) - \{e\})$ (the same palette of colors is used on $\triangle - 1$ edges and $\triangle - 1$ neighbors of v away from the strong center), then there must be a color among them, say c_3 on the edge f that is not used on $E(u)$. We can modify φ by coloring v by c_2, f by $\triangle + 2$, and e by c_3, obtaining a coloring with the same sum but with color $\triangle + 2$ father away from the strong center of G; a contradiction.

Case 2 If $c_1 \notin \varphi(E(v) - \{e\})$ and $\varphi(N(v)) - u) \neq \varphi(E(v) - \{e\})$, then there is a color $c_3 \in \varphi(E(v) - \{e\})$ that is not present on any neighbor of v, suppose that c_3 occurs on an edge f. By interchanging colors c_3 and $\triangle + 2$ on f and v, f will receive color $\triangle + 2$ which is father away from the strong center of G; a contradiction.

Case 3 If $c_1 \in \varphi(E(v) - \{e\})$, say $\varphi(f) = c_1$, then there is a color $c_3(c_3 \neq c_1, c_3 \neq c_2)$ that is not present on $E(v) - e$ but is present on some neighbor of v, say x.

Case 3a If no other neighbor of v has color c_3, we swap colors of v and x obtaining the coloring with color $\triangle + 2$ farther away from the strong center of G (Fig. 5).

Case 3b If there is another neighbor of v with color c_3, then one of the $\triangle - 2$ colors from the edges $E(v) - e - f$ is not present on the neighbors of v; say color c_4 occurring on an edge h. Modify the coloring φ by swapping colors of v and h; the color $\triangle + 2$ will be on h that is farther away from the strong center of G; a contradiction (Fig. 6). □

Fig. 5 Illustration of Case 3a

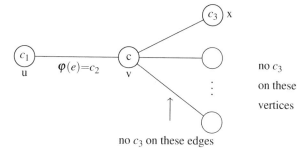

no c_3
on these
vertices

no c_3 on these edges

Fig. 6 Illustration of Case 3b

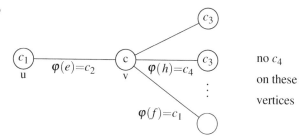

no c_4
on these
vertices

Fig. 7 An example of a tree fragment

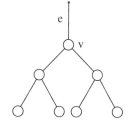

The only case in which we cannot "push" color $\Delta + 2$ away from the strong center is for a tree with two adjacent vertices u and v both of degree Δ such that the pallets of colors on $E(u) - e$, $E(v) - e$, and $N(v) - u$ are identical.

5 Polynomial Time Algorithm for Trees

In this section we propose an $O(n\Delta^4)$ algorithm for finding the total chromatic sum, the minimum sum total coloring and the total strength of an arbitrary tree.

Let G be a tree, and let $v \in V(G)$ and $e \in E(G)$, $e = vu$. We define a **fragment tree** or an **f-tree** with respect to v and e, to be a component of $G \backslash e$ containing v, together with an attached edge e joining v and u, but without the vertex u (see Fig. 7). We denote this fragment tree by $Q(v, e)$. Formally, $Q(v, e)$ is not a graph. By $p(v)$ we mean the vertex u.

Let $F = Q(v, e)$ be an f-tree of some tree. By the **root** of F, denoted by $r(F)$, we mean the vertex v, and by the **extent** of F, denoted by $x(F)$, we mean the edge e. Obviously, $Q(r(F), x(F)) = F$. By $\hat{F}$ we mean F with attached $p(r(F))$. By a total coloring of an f-tree F we mean a partial total coloring of $\hat{F}$, with colors assigned to $V(\hat{F}) \setminus \{p(r(F))\} \cup E(\hat{F})$. By $\Delta(F)$ we mean $\Delta(\hat{F})$.

Let F be an f-tree of some tree G and let $c \geq \Delta(F) + 1$ be an integer (an upper bound for the number of colors). The **cost matrix** C_F^c of the f-tree F is a square $c \times c$ matrix whose (p, q)-entry, for $p \neq q$, denotes the sum of colors of an optimal total coloring of F using at most c colors where the root $r(F)$ has color p and the extend $x(F)$ has assigned color q. The diagonal entries of C_F^c are undefined, since in a proper total coloring we must have $p \neq q$.

If $F = Q(v, e)$ and the neighbors of v are $v_1, v_2, \ldots, v_k$ adjacent to v by the edges $e_1, e_2, \ldots, e_k$ (see Fig. 8), then knowing the cost matrices $C_{F_i}^c$ for the k fragments $F_i = Q(v_i, e_i)$, $1 \leq i \leq k$, we can evaluate the cost matrix for F. Namely, the (p, q)-entry of C_F^c is the minimum of the sums $p + q + C_{F_1}^c[p_1, q_1] + \ldots + C_{F_k}^c[p_k, q_k]$ taken over all colors $p_i, q_i \in \{1, \ldots, c\}$ such that for each $i, 1 \leq i \leq k$, $p_i \neq q_i$ and $p_i \neq p$ and all colors in the set $\{p, q, q_1, \ldots, q_k\}$ are different.

If G is a tree rooted at v and the neighbors of v are $v_1, v_2, \ldots, v_k$, then in a similar manner we can evaluate the cost vector C_G^c for G knowing cost matrices for all k fragments $F_i = Q(v_i, e_i)$, $1 \leq i \leq k$, where $e_i = vv_i$. The p-entry $C_G^c[p]$ of this vector equals the minimum of the sums $p + C_{F_1}^c[p_1, q_1] + \ldots + C_{F_k}^c[p_k, q_k]$ taken over all colors $p_i, q_i \in \{1, \ldots, c\}$ such that for each $i, 1 \leq i \leq k$, $p_i \neq q_i$ and $p_i \neq p$ and all colors in the set $\{p, q_1, \ldots, q_k\}$ are different.

Theorem 3 *There is an algorithm of complexity $O(n\Delta^4)$ for finding the minimum total chromatic sum of a tree in the class of trees of order n and the degree bounded by Δ.*

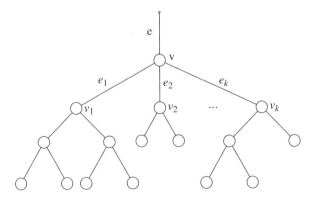

Fig. 8 Decomposing a tree fragment into k smaller fragments

Proof Let G be a tree of order n and maximum degree $\Delta(G) = \Delta$. Select any vertex of G as a root; call it r. Direct all edges of G toward the root. Sort the vertices in a topological order in accordance to the orientation of edges. Select $c = \Delta + 2$, since by Theorem 1 no tree needs more than $\Delta + 2$ colors for a optimal total coloring. Initialize the algorithm by assigning the cost matrix for any fragment F consisting of a leaf with its pendant edge $C_F^c(p, q) = p + q$.

If v is not a root and not a leaf, and has $v_1, v_2, \ldots, v_k$ as predecessors, compute cost matrix C_F^c for the f-fragment $Q(v, e)$, where e is the only edge of v directed towards r. If $v = r$ is the root of G, compute the cost vector C_G^c. Total chromatic sum of G equals

$$\Sigma''(G) = \min_{1 \le p \le \Delta + 2} \{C_G^c[p]\}.$$

It is not difficult to see that the complexity of this algorithm is $O(n\Delta^4)$. □

By running this algorithm with $c = \Delta(G) + 1$ and $c = \Delta(G) + 2$, we can determine the total strength $\sigma''(G)$ of a tree G. If the algorithm returns the same costs for both upper bounds for c, then $\sigma''(G) = \Delta(G) + 1$. If the cost for $c = \Delta(G) + 2$ is smaller than for $c = \Delta(G) + 1$, then $\sigma''(G) = \Delta(G) + 2$ and G is T-strong.

6 Existence of T-strong Trees

From the proof of Theorem 2, we can get some information about an optimal total coloring of any T-strong tree G. The structure of such a tree and an optimal total coloring of G must be as follows:

1. G must have two adjacent vertices, say u and v, both of degree $\Delta(G)$.
2. Color $\Delta + 2$ must occur at one of those vertices, say v.
3. The three palettes of colors occurring on vertices adjacent to v (not counting u), the edges incident to v (not counting uv), and the edges incident to u (without uv as well) must be identical.

Using this observation. we were able to construct a T-strong tree. The smallest in the family of all subcubic trees ($\Delta \le 3$) is the tree T_{50}, of order 50, depicted in Fig. 9. Our algorithm verified that $\sigma''(T_{50}) = 5$. We used the vertex marked black as the root for running our algorithm. Notice that its neighbor $r(L)$ is the root of a subtree isomorphic to the other half of T_{50}.

Theorem 4 *There is an infinite family of T-strong trees.*

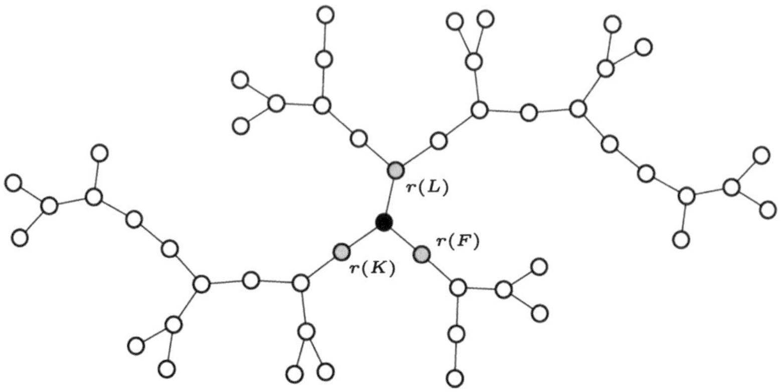

Fig. 9 The smallest subcubic T-strong tree T_{50}

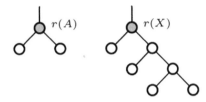

Fig. 10 Fragments A and X used for constructing larger T-strong trees

Proof Two f-trees A and X given in Fig. 10 have the following property. In any optimal coloring of A and X roots $r(A)$ and $r(X)$ must have colors 2, 3, or 4 and the extents must be colored with 1. Moreover, changing colors on those elements increases the cost of both f-trees by the same amount. This means that replacing a fragment A in any tree G by the fragment X does not change the coloring of the rest of G. Notice that X has four more vertices than A and X contains A as a subfragment. Starting with the T-strong tree T_{50}, we can replace any f-tree A in it by a copy of X obtaining a T-strong tree of order 54. Continuing these fragments' replacements, we can construct subcubic T-strong trees of arbitrarily large order.

□

Our algorithm verified that the smallest T-strong tree with vertices of degree 1 and 3 only is the tree T_{122} depicted in Fig. 11. We also found a T-strong tree of order 266 with $\Delta = 4$. Both of these trees can generate infinite families of T-strong trees by similar fragment replacements.

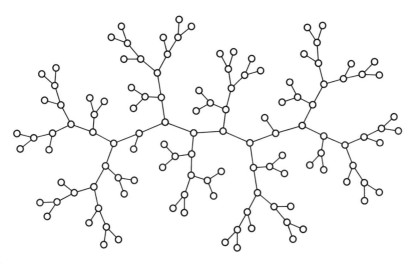

Fig. 11 T-strong tree T_{122} of order 122 with degree set $\{1, 3\}$

References

1. Behzad, M.: Graphs and their chromatic numbers. Ph.D. Thesis, Michigan State University, 69 pp. (1965)
2. Erdős, P., Kubicka, E., Schwenk, A.: Graphs that require many colors to achieve their chromatic sum. Congr. Numer. **71**, 17–28 (1990)
3. Jiang, T., West, D.B.: Coloring of trees with minimum sum of colors. J. Graph Theory **32**, 354–358 (1999)
4. Kubicka, E.: The chromatic sum and efficient tree algorithms. Ph.D. Thesis, Western Michigan University, 149 pp. (1989)
5. Kubicka, E., Kubicki, G., Leidner, M.: Total colorings of graphs with minimum sum of colors. Graphs Comb. **32**, 2515–2524 (2016)
6. Leidner, M.: The study of the total coloring of graphs. Ph.D. Thesis, University of Louisville, 101 pp. (2012)
7. Mitchem, J., Morris, P., Schmeichel, E.: On the cost chromatic number of outerplanar, planar, and line graphs. Discuss. Math. Graph Theory **2**, 229–241 (1997)
8. Vizing, V.G.: On an estimate of the chromatic class of a p-graph. Diskret. Anal. **3**, 29–39 (1964)
9. Vizing, V.G.: The chromatic class of a multigraph. Kibernetika (Kiev) **3**, 25–30 (1965)

An Incremental Search Heuristic for Coloring Vertices of a Graph

Subhankar Ghosal and Sasthi C. Ghosh

Abstract Graph coloring is one of the fundamentally known NP-complete problem. Several heuristics have been developed to solve this problem, among which greedy coloring is the most naturally used one. In greedy coloring, vertices are traversed following an order and hence performance of it highly depends on finding a good order. In this paper, we propose an *incremental search heuristic* (ISH) which considers some ρ_1 random orders and for each of them it calls a *selective search* (SS) procedure with parameter ρ_2. Given an order, SS considers only those orders which produces equal or less number of distinct colors than a given order. We showed that those orders can be partitioned into disjoint subsets of *equivalent orders*. To make effective search, SS selects and evaluates only one order from such a subset. Analytically we have shown that ISH can solve the graph coloring instances on sparse graph in expected polynomial time. Through simulations, we have evaluated ISH on 86 challenging benchmarks and compare results with state of the art existing algorithms. We observed that ISH significantly outperforms existing algorithms specially for sparse graphs and also produces reasonably good results for others.

Keywords Graph coloring · Incremental search · Selective search · Equivalent orders · Benchmarks

1 Introduction

Graph coloring is the problem of assigning positive integers to the vertices of a graph G such that no monochromatic edge exists in G. It is well-known that finding chromatic number of a graph is NP-Complete [1]. It is also known that if unique game conjecture is true providing $n^{1/\epsilon}$ approximation ($\forall \epsilon > 0$) solution of

S. Ghosal (✉) · S. C. Ghosh
Advance Computing and Microelectronics Unit, Indian Statistical Institute, Kolkata, India
e-mail: sasthi@isical.ac.in

© The Author(s), under exclusive license to Springer Nature Switzerland AG 2021
C. Gentile et al. (eds.), *Graphs and Combinatorial Optimization: from Theory to Applications*, AIRO Springer Series 5,
https://doi.org/10.1007/978-3-030-63072-0_4

graph coloring is also NP-hard [2]. Several heuristic algorithms exist for coloring a graph, a good survey of which can be found in [3, 4]. Recently a reinforcement learning based local search algorithm [5], a modified cuckoo algorithm [6], a hybrid evolutionary algorithm [7], a feasible and infeasible search based algorithm [8] and a parallel ordering based algorithm [9] are developed for coloring vertices of a graph.

Greedy coloring is one of the most natural graph coloring technique. Greedy coloring visits vertices of a graph in an order and while visiting a vertex it puts the minimum color absent in all of its neighbors. But performance of greedy coloring entirely depends on the order it visits the vertices. Thus with greedy coloring the problem boils down to the problem of finding an optimum order among the $n!$ possible orders, where n is the number of vertices in G. Several authors have proposed methods to find good orders such as [10, 11] and [12]. One such good order is the order obtained by sorting the vertices of a graph according to non-increasing order of their degrees. Authors of [10] have shown that with this ordering a graph G having degree sequence $d_1 \geq d_2 \geq \cdots \geq d_n$, has total number of colors to be used as $\chi(G) \leq 1 + \max_i \min(d_i, i - 1)$. Another good order is *Kempe order* [13]. The order is generated as following. We find a vertex with minimum degree in the graph and *push* it into a *stack* and *remove* it from the graph. Recursively apply this step on the *residual graph* till the graph become empty. Then *pop* the elements from the stack and this will give the Kempe order. In [12] authors provided an *Dsature order* defined as follow. Choose a random vertex and *enqueue* it into a *queue*. Calculate the *saturation degree* of all the vertices, not in queue, as defined bellow. The saturation degree of a vertex is the number of it's neighbors in the queue. Now choose the vertex with the maximum saturation degree and enqueue it into the queue. Continue this process till the queue contains all the vertices of the graph. *Dequeue* the elements from the queue and consider this order as the Dsature order.

As finding optimum order is a hard problem, several authors have developed different search heuristics to find a near optimum order. Some of them are based on simulated annealing [14, 15], genetic algorithms [16], hill climbing, memetic-algorithm [17] and tabu-search [18, 19] heuristic. Simulated annealing is a probabilistic heuristic where probability of accepting a worse solution is decreased over time. In genetic algorithm, a set of orders is considered initially. They had been subject to mutation, cross over and selection to generate a set of better orders. In hill climbing, starting from an arbitrary order it tries to find a better order by incremental changes to the solution. When it got stuck at local optimum, it jumps to a random location and starts from there. In tabu-search, when stuck in local optimum, it chooses a worsening move. It also keeps record of previous steps to prune the chance to search the same zone again.

It is evident that all of the above mentioned heuristic algorithms may eventually reach to a poorer order than the current best order during the search process. Unlike these search paradigms, we propose an *selective search* algorithm (SS) which considers only those orders which produce at most the same span (no of distinct colors used) than a given order. SS then partitions those orders into disjoint subsets of equivalent orders. An order is equivalent to another if the color vectors produced by applying greedy coloring on them are identical, where colors appear in a color vector in natural order of the vertices. SS evaluates only one order from such a set of equivalent orders. In a call of SS, ρ_2 such orders are evaluated. In this way SS enhances the optimum hitting probability by an exponential factor of n. Note that SS requires an initial order to start its searching. Starting with an initial order, SS may not cover the whole $n!$ orders even for arbitrarily large ρ_2. To complete the search, in ISH, we execute SS with ρ_1 random orders and report the minimum span produced by them. In ISH we essentially call the greedy coloring $\Omega(\rho_1\rho_2)$ times. We then calculate the expected value of $\rho_1\rho_2$ to reach the optimum. We notice that for sparse graphs ISH solves the coloring problem in expected polynomial time. We evaluate ISH on 86 well-known challenging benchmarks and compare the results with state of the art existing algorithms. We observe that ISH significantly outperforms the existing algorithms for all sparse benchmarks and also produces reasonably good results for others. In this paper, we have adopted the concepts of pseudo-vertices and equivalent orders from [20], however, there application and subsequent analysis are significantly different.

2 Key Ideas

Consider a graph $G(V, E)$ with n vertices $v_1, v_2, \cdots, v_n$, where V is the set of vertices and E is the set of edges. Let $S = (v_{l_1}, v_{l_2}, \cdots, v_{l_n})$ be an arbitrary order of the vertices of G, where $1 \leq l_k \leq n$. Assuming colors are positive integers starting from 1, let $c(v_{l_k})$ be the color of vertex v_{l_k} obtained by applying greedy coloring on G following order S. Recall that greedy coloring colors the vertices of a graph following a specific order of the vertices and while coloring a vertex it puts the *minimum* color that is absent in all of its neighboring vertices. Hence $c(v_{l_k})$ depends only on the colors assigned to the vertices $v_{l_1}, v_{l_2}, \cdots, v_{l_{k-1}}$ that appear before v_{l_k} in S. Moreover, greedy coloring always produces a *no-hole* coloring. A coloring of G is a no-hole coloring if it uses all colors between 1 and its maximum color. Given S, let $C = (c(v_1), c(v_2), \cdots, c(v_n))$ be the *color vector* obtained by applying greedy coloring on G following S. Note that in the color vector, colors of the vertices are stored in the ascending order of their vertex indexes. With respect to the resulted color vector obtained by greedy coloring, an order may be considered as equivalent to another order as formally defined in Definition 1.

Definition 1 (Equivalent Order) Let S_1 and S_2 be two orders of the vertices of G. Let C_1 and C_2 be the color vectors generated by greedy coloring while applied upon S_1 and S_2 respectively. Then orders S_1 and S_2 are said to be equivalent to each other, if and only if $C_1 = C_2$.

The span of a color vector C denoted by $span(C)$ is the total number of distinct colors in C. As colors start from 1 and greedy coloring produces no-hole coloring, $span(C)$ is essentially same as the maximum color used in C.

Definition 2 Let S_1 and S_2 be two orders of the vertices of G. Let C_1 and C_2 be the color vectors generated by applying greedy coloring on S_1 and S_2 respectively. Then $S_1 \lhd S_2$ if and only if $span(C_1) \leq span(C_2)$.

Note that C essentially partitions the graph into $k = span(C)$ vertex disjoint independent sets each of which contains all the vertices of a particular color. We call each such independent set as a pseudo-vertex as formally defined in Definition 3.

Definition 3 (Pseudo-Vertex) Let S be an order of the vertices of G and C be the color vector obtained by applying greedy coloring on S. Let $k = span(C)$. A pseudo-vertex V_i is a subset of vertices of G all of which get color i in C, where $1 \leq i \leq k$.

Let S be an order and $V_1, V_2, \cdots, V_k$ be the k pseudo-vertices of C, where $n_1 = |V_1|$, $n_2 = |V_2|$, $\cdots$, $n_k = |V_k|$. Let $\Pi(S)$ be the set of all permutations of $V_1, V_2, \cdots, V_k$, and $\pi = (V_{l_1}, V_{l_2}, \cdots, V_{l_k}) \in \Pi$ be an arbitrary permutation. Let $L(\pi)$ be the set of all orders generated from π by permuting the vertices within each pseudo-vertex while keeping the order of the pseudo-vertices intact. All orders in $L(\pi)$ are said to be the orders represented by π. Let $N_k = |L(\pi)| = n_1! n_2! \cdots n_k!$. Figure 1 shows an example of construction of $L(\pi)$.

In Theorem 1, we will prove that all those N_k orders represented by $L(\pi)$ are equivalent to each other.

Theorem 1 *Let S be an order of the vertices of G and $V_1, V_2, \cdots, V_k$ be the k pseudo-vertices of C, where C is the color vector obtained by applying greedy coloring on S. Let $\pi = (V_{l_1}, V_{l_2}, \cdots, V_{l_k}) \in \Pi(S)$ be an arbitrary permutation. All orders in $L(\pi)$ are equivalent to each other.*

Proof Let $S_1, S_2 \in L(\pi)$ such that $S_1 \neq S_2$. Also assume that $S_1 = (v_{r_1}, v_{r_2}, \cdots v_{r_n})$, where $1 \leq r_i \leq n$ for all i. Let C_1 and C_2 be the color vectors generated by greedy coloring while applied upon S_1 and S_2 respectively. To show S_1 is equivalent to S_2, we have to prove that $c_1(v_{r_i}) = c_2(v_{r_i})$ for all v_{r_i}. We prove this by induction on i. Since V_{l_1} is an independent set, the color of all its vertices must be 1 in both C_1 and C_2. Hence the first vertex of S_1 appears in V_{l_1} corresponding to both orders S_1 and S_2. So, $c_1(v_{r_1}) = c_2(v_{r_1}) = 1$. Hence the base case is done. Our induction hypothesis is, for all vertex v_{r_j} appears before v_{r_i}

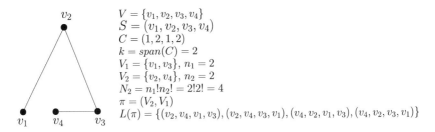

$V = \{v_1, v_2, v_3, v_4\}$
$S = (v_1, v_2, v_3, v_4)$
$C = (1, 2, 1, 2)$
$k = span(C) = 2$
$V_1 = \{v_1, v_3\}, n_1 = 2$
$V_2 = \{v_2, v_4\}, n_2 = 2$
$N_2 = n_1! n_2! = 2! 2! = 4$
$\pi = (V_2, V_1)$
$L(\pi) = \{(v_2, v_4, v_1, v_3), (v_2, v_4, v_3, v_1), (v_4, v_2, v_1, v_3), (v_4, v_2, v_3, v_1)\}$

Fig. 1 Vertices and pseudo-vertices

in S_1, $c_1(v_{r_j}) = c_2(v_{r_j})$. We now left to prove $c_1(v_{r_i}) = c_2(v_{r_i})$. Let $v_{r_i} \in V_{l_m}$. Note that $c_1(v_{r_i})$ and $c_2(v_{r_i})$ depend only on the colors of those vertices which appear before v_{r_i} in S_1 and S_2 respectively. All the vertices of $V_{l_1}, V_{l_2}, \cdots, V_{l_{m-1}}$ and some vertices of V_{l_m} may only appear before v_{r_i} in both S_1 and S_2. Since V_{l_m} is an independent set, eventually $c_1(v_{r_i})$ and $c_2(v_{r_i})$ depend only on the vertices that belong to $\bigcup_{j=1}^{m-1} V_{l_j}$. But according to our induction hypothesis, $c_1(v_{r_j}) = c_2(v_{r_j})$ for all $v_{r_j} \in \bigcup_{j=1}^{m-1} V_{l_j}$. Hence the proof. $\qquad\square$

So far we get that $\forall \pi \in \Pi(S)$, $L(\pi)$ is an equivalent set. We now show that there is a special permutation $\pi_0(S) = (V_1, V_2, \cdots, V_k) \in \Pi(S)$ such that for each $\pi \in \Pi(S)$ if $S \in L(\pi_0(S))$ and $S' \in L(\pi)$, then the span produced by greedy coloring on S is greater than or equals to that of S'. Each order $S \in L(\pi_0(S))$ will be termed as *cardinal order*, as formally defined in Definition 4.

Definition 4 (Cardinal Order) An order $S = (v_{l_1}, v_{l_2}, \cdots, v_{l_n})$ of the vertices of G is said to be a cardinal order if C is the color vector generated by greedy coloring while applied upon S, such that $c(v_{l_1}) \leq c(v_{l_2}) \leq \cdots \leq c(v_{l_n})$, where $c(v_{l_k})$ denotes the color of v_{l_k}.

Theorem 2 *Let S be an order and upon applying greedy coloring on S we get channel vector C with pseudo-vertices $V_1, V_2, \cdots, V_k$ respectively. Then $\forall S' \in L(\pi_0(S))$, S' and S are equivalent orders.*

Proof Let $S' = (v_{r_1}, v_{r_2}, \cdots, v_{r_n})$ and C' be the color vector generated by greedy coloring applied upon S', where $1 \leq v_{r_i} \leq n$ for all i. We are left to show that $c(v_{r_i}) = c'(v_{r_i})$ $\forall i$. We will apply induction on i to prove this. For $i = 1$ the proof is trivial. Hence the base case is done. Our induction hypothesis is, $c(v_{r_j}) = c'(v_{r_j})$ for all vertices v_{r_j} appearing before v_{r_i} in S'. Note that $c'(v_{r_i})$ depends only on the colors of those vertices which appear before v_{r_i} in S'. It is evident that all the vertices which have been colored with less than $c(v_{r_i})$ in C must appear before v_{r_i} in S' according to the construction of S'. Hence $c'(v_{r_i})$ cannot be less than $c(v_{r_i})$. Some vertices which have been colored with $c(v_{r_i})$ in C may also appear before v_{r_i} in S'. But all such vertices belong to an independent set in G. Hence $c(v_{r_i}) = c'(v_{r_i})$. Hence the proof. $\qquad\square$

Theorem 3 *Let S be an order and upon applying greedy coloring on S we get pseudo-vertices $V_1, V_2, \cdots, V_k$ respectively. If $\pi \in \Pi(S)$ then $\forall S' \in L(\pi)$, $S' \triangleleft S$.*

Proof Let $S' = (v_{r_1}, v_{r_2}, \cdots, v_{r_n})$, where $1 \leq v_{r_i} \leq n$ for all $1 \leq i \leq n$. Let C and C' be the color vectors generated by greedy coloring while applied upon S and S' respectively. Let v_{r_i} belongs to the pseudo-vertex which is in m_i-th position in π, where $1 \leq m_i \leq k$. Our claim is that $c'(v_{r_i}) \leq m_i$ $\forall i$. Since $1 \leq m_i \leq k$, if our claim is true, then we can immediately conclude that $span(C') \leq k = span(C)$.

So we are left to prove our claim. We prove this by induction on i. Since all vertices belong to the pseudo-vertex appeared in the first position in π get color 1 in C', our claim is trivially true for all such vertices. Hence base case is done. We now consider in induction hypothesis that for all vertices appear before vertex v_{r_i} in S' our claim is true. This implies that our claim is true for all vertices which belong to the pseudo-vertices of C appeared before m_i-th position in π. Note that $c'(v_{r_i})$ depends only on the colors of those vertices which appear before v_{r_i} in S'. There are $m_i - 1$ pseudo-vertices before m_i-th position in π. Hence according to our induction hypothesis, the colors of the vertices belong to those pseudo-vertices are at most $m_i - 1$. Note that some vertices belong to the pseudo-vertex in m_i-th position may also appear before v_{r_i} in S'. Since each pseudo-vertex is an independent set, $c'(v_{r_i})$ does not depend on those vertices. Hence $c'(v_{r_i}) \leq m_i$. Hence the proof. □

Remark 1 From Theorem 3 we can conclude that for a given order S, there are at least $k! \times N_k$ orders S's such that each $S' \vartriangleleft S$. Also, from Theorem 1 we get that, by visiting an order represented by a particular permutation of the pseudo-vertices, we can essentially find the minimum span generated by N_k orders represented by that permutation. Hence by visiting only one order from each of $k!$ different permutations of the pseudo-vertices, we can essentially find the minimum span generated by $k! \times N_k$ orders.

In Theorem 4 we will show that given a color vector C' we can build an order S and the corresponding C such that $span(C) \leq span(C')$.

Theorem 4 *Given any coloring C' of G we could generate an order S by sorting the vertices of G according to ascending order of their colors in C'. If C is the color vector generated by greedy coloring while applied upon S, then $span(C) \leq span(C')$.*

Proof Let $S = (v_{l_1}, v_{l_2}, \cdots, v_{l_n})$ be an order of vertices of G and C is the channel vector generated by greedy coloring while applied upon S. We claim that $c(v_{l_i}) \leq c'(v_{l_i})$ for all l_i, $1 \leq l_i \leq n$. We prove this by induction on i. For the first vertex, the claim is trivially true. So the base case is done. In induction hypothesis we assume that the claim is true for each vertex which appears prior to v_{l_i} in S. Let $N(v_{l_i})$ be the set of neighbors of v_{l_i} in G which appear before v_{l_i} in S. As S is constructed by sorting the vertices according to ascending order of their colors in C', we get $c'(v_{l_1}) \leq c'(v_{l_2}) \leq \cdots \leq c'(v_{l_i})$. Since $c'(v_{l_i})$ is a valid coloring of v_{l_i} in C', $N(v_{l_i})$ cannot contain $c'(v_{l_i})$. In other words, $c'(v_{l_j}) \leq c'(v_{l_i}) - 1$ for $\forall v_{l_j} \in N(v_{l_i})$. From the induction hypothesis, we get $c(v_{l_j}) \leq c'(v_{l_j})$ for $\forall v_{l_j} \in N(v_{l_i})$. The previous two arguments together imply $c(v_{l_j}) \leq c'(v_{l_i}) - 1$ for $\forall v_{l_j} \in N(v_{l_i})$. Since C is obtained by greedy coloring, $c(v_{l_i})$ must be the minimum color which is not been used in any of the vertices of $N(v_{l_i})$, implying $c(v_{l_i}) \leq c'(v_{l_i})$. Hence the proof. □

3 Incremental Search Heuristic (ISH)

3.1 Selective Search (SS) Algorithm

So far we get that from a given color vector C' of G we can generate an order S such that C is the channel vector generated by greedy coloring while applied upon S where $span(C) \leq span(C')$. Again from C we can construct k pseudo-vertices $V_1, V_2, \cdots, V_k$ where $k = span(C)$. It is also evident that for each order $S' \in L(\pi)$ where $\pi \in \Pi(S)$, $S' \lhd S$. Also all $N_k = |L(\pi)|$ orders are equivalent to S'. From the above discussion we can think of a natural algorithm which can be stated as: Given C', we first generate S and then build C. Next we consider $\pi \in \Pi(S)$, apply greedy coloring on $S' \in L(\pi)$ and check whether the span is improved. If improved, we consider this new order as S and repeat the procedure. If the span is not improved, we consider another permutation of the pseudo-vertices and repeat the process. Since there are $k!$ permutations of the pseudo-vertices of C, it may not be practical to consider all of them. So we introduce a parameter ρ and consider ρ random permutations. If an improvement is found, we further consider ρ permutations. In each iteration, SS evaluates an order using greedy coloring, which takes $O(n^2)$ time and $O(n^2)$ space complexity. It is evident that the span obtained after evaluating an order could be at most $\Delta(G) + 1$, where $\Delta(G)$ is the maximum degree of G. Hence the process could be repeated at most $\Delta(G) + 1$ times. Thus the total time and space complexities of SS are $O(\rho(\Delta(G) + 1)n^2) = O(\rho n^3)$ and $O(n^2)$ respectively. Formally SS is presented in Algorithm 1. Note that in each iteration of SS, by visiting only one order we eventually visit $N_k = \prod_{i=1}^{k} n_i! \geq (\frac{n}{k}!)^k = \Omega((\frac{n}{ek})^n)$ (using Stirling's approximation) orders. Hence we hit the optimum with probability $N_k \times p(k)$, where $p(k)$ is the probability that a random order with span $\leq k$ is optimum.

Algorithm 1: Selective search (SS) algorithm

Input: G, C, ρ
Output: C

1   Find S by sorting the vertices according to ascending order of their colors in C;
2   $C = C'$ be the color vector generated by the greedy coloring while applied upon S;
3   $k = span(C)$;
4   **for** $i = 1, 2, \cdots \rho$ **do**
5   Let $\pi \in \Pi(S)$ and $S' \in L(\pi)$;
6   C' be the color vector generated by the greedy coloring while applied upon S';
7   **if** $span(C') < span(C)$ **then**
8         $C = C'$;
9   Reset $i = 1$;

10   **return** C

3.2   Incremental Search Heuristic (ISH) Algorithm

ISH generates ρ_1 random orders and calls SS with $\rho = \rho_2$ for each of them and finally returns the color vector having the minimum span. ISH is formally presented in Algorithm 2. Similar to SS the time and space complexities of ISH are $O(\rho_1 \rho_2 n^4)$ and $O(n^2)$ respectively.

Algorithm 2: Incremental search heuristic (ISH)

    Input: G, ρ_1, rho_2
    Output: C_{min}
1 Set $C_{min} = (1, 2, \cdots, |V(G)|)$;
2 **for** $i = 1, 2, \cdots, \rho_1$ **do**
3   Generate a random order S of the vertices;
4   C be the color vector generated by the greedy coloring while applied upon S;
5      $C = SS(G, C, \rho_2)$;
6   **if** $span(C) < span(C_{min})$ **then**
7         $C_{min} = C$;
8   Reset $i = 1$;

9 **return** C_{min};

---

4   Expected Value of $\rho_1 \rho_2$

Let's define a *step* as applying greedy coloring on an order and obtaining a color vector. Clearly SS and ISH execute $\Omega(\rho)$ and $\Omega(\rho_1 \rho_2)$ steps.

Theorem 5 *If ISH produces span less than or equals to k in each step, then expected number of steps to find optimum is* $\mathbb{E}[\rho_1 \rho_2] = O(\frac{1}{k!}(\frac{ek^2}{n})^n)$.

Proof Let $A(k)$ be the set of orders which produce span $\leq k$ while greedy coloring is applied upon them. Considering k_o as the optimum span, we get $p(k) = \frac{|A(k_o)|}{|A(k)|}$. Again $|A(k)| \leq n!$ hence $p(k) \geq \frac{|A(k_o)|}{n!}$. If an order with span k_o is optimum, then all the orders represented by all the $k_o!$ different permutations of its pseudo-vertices must also be optimum. Since each permutation represent $N_{k_o} = \Omega((\frac{n}{ek_o})^n)$ (Stirling's approximation) orders, there are $|A(k_o)| = k_o! \times N_{k_o} = \Omega(k_o!(\frac{n}{ek_o})^n))$ optimum orders. Hence using Stirling's approximation we get $p(k) = \Omega(\frac{k_o!}{n!}(\frac{n}{ek_o})^n) = \Omega(\frac{k_o!}{k_o^n}) = \Omega(\frac{k!}{k^n})$, $\forall k \geq k_o$. Since in a step, ISH essentially evaluates N_k orders, expected number of steps to find an optimum order is $\mathbb{E}[\rho_1 \rho_2] = \frac{1}{N_k p(k)} = O(\frac{1}{k!}(\frac{ek^2}{n})^n)$. $\qquad\square$

Corollary 1 *If* $k \leq \sqrt{\dfrac{n}{e}}$ *then* $\mathbb{E}[\rho_1 \rho_2] = O(1)$.

Since for greedy coloring $k \leq \Delta(G) + 1$, Corollary 2 is immediate.

Corollary 2 *If* $\Delta(G) + 1 \leq \sqrt{\dfrac{n}{e}}$, *then* $\mathbb{E}[\rho_1 \rho_2] = O(1)$.

Remark 2 It is a well-known fact that solving the graph coloring problem even for sparse graph is hard. It is known from [21] that even coloring a planer graph with maximum degree just 4 is NP-Complete. Using ISH we can solve these sparse graph instances in expected polynomial time. This is a key feature of ISH algorithm.

It is evident that ISH performs well for graphs whose maximum degree is small. We now consider the performance of ISH for the graphs with small average degree. For this purpose, we consider Erdos-Reyni random graph $G(n, p)$, where n is the number of vertices and each edge is generated independently with probability p. For $G(n, p)$, where $np = O(1)$, if k is the span produced by greedy coloring then $k \sim_n np$ [22]. Here $x(n) \sim_n y(n) \implies \lim\limits_{n \to \infty} \dfrac{x(n)}{y(n)} = 1$. From Corollary 1, the following theorem is immediate, where $x(n) \leq_n y(n) \implies \lim\limits_{n \to \infty} \dfrac{x(n)}{y(n)} \leq 1$.

Theorem 6 *In Erdos-Reyni random graph $G(n, p)$, if average degree $\Delta_a = np \leq_n \sqrt{\dfrac{n}{e}}$ then $\mathbb{E}[\rho_1 \rho_2] = O(1)$.*

5 Simulation

We simulate ISH and compare with 9 state of the art algorithms based on the results obtained on 86 challenging benchmarks taken from [15, 23–31]. Table 1 shows the results. Here n represents number of vertices and e represents number of edges of the corresponding benchmarks. All benchmarks for which $\Delta_a \leq \sqrt{\dfrac{n}{e}}$, we termed them as *sparse* benchmarks and marked in bold. DBG represents a hybrid genetic algorithm reported in [32]. MCOACOL represents a modified cuckoo optimization algorithm reported in [6]. Best of [33] represents the minimum span produced and its corresponding time among the five integer linear programming based algorithms REP, POP, POP2, ASS+(c) and ASS+(e) reported in [33]. DR represents a Doglas-Ranchford algorithm reported in [34]. EBDA represents a Enhanced binary dragonfly algorithm reported in [35]. Note that these algorithms are considered by many authors [36–39]. We run our ISH with $\rho_1 = \rho_2 = \rho = 10{,}000$. For a particular algorithm, χ represents chromatic number and T represents the time taken by the corresponding algorithm to reach the corresponding value of χ. For example, for benchmarks "1-Fullins-5" and "school1-nsh", we get $\chi = 6$, $T = 0.002752s$ and $\chi = 14$, $T = 0.656823s$ by executing only 3 and 1310 solutions respectively.

Table 1 Span and time requirement of different algorithms on benchmark instances

Benchmark instances	n	e	DBG[32] x	DBG[32] T	MCOACOL[6] x	MCOACOL[6] T	Best of [33] x	Best of [33] T	DR[34] x	DR[34] T	EBDA[35] x	EBDA[35] T	ISH x	ISH T
1-FullIns-3	30	100	–	–	4	0.4s	–	–	–	–	–	–	4	0.000077s
1-Insertion-4	67	232	–	–	5	0.5s	–	–	–	–	5	2.63s	5	0.000085s
2-FullIns-3	52	201	–	–	5	0.4s	–	–	–	–	–	–	5	0.000061s
2-Insertions-3	37	72	4	2.23s	4	0.4s	–	–	–	–	4	2.08s	4	0.000093s
3-Insertions-3	56	110	4	3.37s	4	0.5s	–	–	–	–	4	3.49s	4	0.00006s
anna	138	493	11	11.63s	11	0.8s	–	–	11	1.04s	11	2.53s	11	0.000395s
ash331GPIA	662	4181	–	–	5	42.2s	4	3.29s	–	–	–	–	4	21.3364s
david	87	406	11	10.89s	11	0.5s	–	–	11	0.26s	11	2.61s	11	0.000095s
fpsol2.i.1	496	11654	–	–	65	3.6s	65	82.03s	65	463.94s	–	–	65	0.002455s
fpsol2.i.2	451	8691	30	69.79s	30	7.4s	–	–	30	495.94s	–	–	30	0.002805s
fpsol2.i.3	425	8688	30	67.14s	30	6.4s	–	–	30	480.27s	–	–	30	0.002604s
games120	120	638	9	10.77s	9	0.8s	–	–	9	0.24s	9	3.24s	9	0.000176s
homer	561	1628	–	–	13	4.3s	–	–	13	59.01s	13	6.67s	13	0.00333s
huck	74	301	11	11.10s	11	0.5s	–	–	11	0.11s	11	4.78s	11	0.000075s
inithx.i.1	864	18707	–	–	54	19.4s	–	–	54	2443.43s	54	93.14s	54	0.0076s
inithx.i.2	645	13979	–	–	31	9s	–	–	31	1500.45s	31	39.16s	31	0.00429s
inithx.i.3	621	13969	–	–	31	6s	–	–	31	1432.43s	31	42.71s	31	0.000321s
jean	80	254	10	9.92s	10	0.5s	–	–	10	0.13s	10	4.42s	10	0.000082s
le450-25a	450	8260	–	–	25	4.4s	–	–	25	68.93s	25	3.65s	25	0.0028s
le450-25b	450	8263	–	–	25	7s	–	–	25	65.82s	25	5.58s	25	0.0025s
le450-5a	450	5714	–	–	–	–	5	21.17s	5	82.47s	–	–	5	32.7469s
le450-5b	450	5734	–	–	–	–	–	–	5	238.33s	–	–	5	54.597s
le450-5c	450	9803	–	–	–	279s	5	140.16s	5	68.68s	–	–	5	5.04865s
le450-5d	450	9757	–	–	6	80.8s	–	–	5	44.49s	–	–	5	7.0381s
miles1000	128	3216	–	–	42	1s	–	–	42	2.43s	42	5.73s	42	0.007641s

Benchmark Instances	n	e	DBG[32] x	DBG[32] T	MCOACOL[6] x	MCOACOL[6] T	Best of [33] x	Best of [33] T	DR[34] x	DR[34] T	EBDA[35] x	EBDA[35] T	ISH x	ISH T
mug100-1	100	166	4	8.34s	4	0.8s	4	0.13s	4	0.07s	4	2.98s	4	0.00007s
mug100-25	100	166	4	8.21s	4	0.5s	4	0.31s	4	0.06s	4	2.57s	4	0.000391s
school1	385	19095	–	–	14	1934.2s	–	–	–	–	–	–	14	0.644538s
qg.order40	1600	62400	–	–	–	–	40	534.83s	–	–	–	–	40	0.582656s
wap05a	905	43081	–	–	50	12.4s	50	125.45s	–	–	–	–	50	0.022395s
zeroin.i.3	206	3540	30	27.06s	30	1.6s	–	–	30	34.51s	30	9.87s	30	0.001792s
school1-nsh	352	14612	–	–	14	1352.6s	14	12.76s	–	–	–	–	14	0.656823s
DSJR500.5	500	58862	124	728s	–	–	122	572.01s	–	–	122	587s	23	4372.94s
DSJR500.1c	500	121275	85	581s	–	–	85	0.33s	–	–	–	–	85	0.05701s
DSJC125.5	125	3891	17	19.71s	–	–	20	h	–	–	–	–	18	5.58092s
DSJC125.9	125	6961	44	35.87s	–	–	44	h	–	–	–	–	44	3.34038s
DSJC250.9	250	27897	72	75.81s	–	–	–	–	–	–	72	63.28s	73	184.649s
queen10-10	100	2940	–	–	12	9.6s	12	h	–	–	–	–	11	223.45s
queen11-11	121	3960	–	–	14	2.4s	13	h	–	–	–	–	12	5748s
queen12-12	144	5192	–	–	15	2.8s	–	–	–	–	–	–	14	48.5491s
queen13-13	169	6656	–	–	16	9.3s	–	–	–	–	–	–	14	19011s
queen14-14	196	4186	–	–	17	22.2s	–	–	–	–	–	–	16	3933s
queen15-15	225	5180	–	–	18	17.9s	–	–	–	–	–	–	17	1774.38s
le450-15b	450	8169	–	–	–	–	15	700.50s	–	–	15	4.23s	16	0.546667s
le450-15c	450	16680	15	93s	–	–	25	h	–	–	–	–	15	12758.4s
le450-15d	450	16750	15	228s	–	–	26	h	–	–	–	–	15	23125.2s
le450-25c	450	17343	25	740s	–	–	30	h	–	∞	–	–	27	9213.9s
le450-25d	450	17425	25	382s	–	–	30	h	–	∞	–	–	27	52432s
1-FullIns-5	282	3247	–	–	6	1.9s	6	1.54s	–	–	–	–	6	0.0027252s
2-FullIns-4	212	1621	–	–	6	1.2s	6	0.01s	–	–	–	–	6	0.000599s

(continued)

This table spans the full page in landscape (rotated) orientation. It consists of two side‑by‑side halves, each comparing several colouring methods (paired columns: x = colours found, T = time) over a set of instances, followed by Greater / Equal / Smaller summary rows.

Left half

Instance	|V|	|E|	x	T	x	T	x	T	x	T	x	T
miles1500	128	5198	–	–	73	1.2s	73	24.65s	73	9.34s	73	0.0006s
miles250	128	387	8	4.39s	8	1.1s	8	0.4s	8	4.52s	8	0.0007s
miles500	128	1170	20	14.48s	20	1.2s	20	1.07s	20	5.11s	20	0.0008s
miles750	128	2113	–	–	31	1.5s	31	2.54s	31	5.64s	31	0.013746s
mug88-1	88	146	4	4.05s	4	1.1s	4	0.05s	4	2.46s	4	0.0002s
mug88-25	88	146	4	3.67s	4	1.3s	4	0.05s	4	1.65s	4	0.00038s
mulsol.i.1	197	3925	49	25.75s	49	1.2s	49	18.79s	49	9.68s	49	0.0015s
mulsol.i.2	188	3885	31	22.07s	31	1.1s	31	63.18s	31	7.43s	31	0.00028s
mulsol.i.3	184	3916	31	23.49s	31	1.3s	31	55.88s	31	6.49s	31	0.0014s
mulsol.i.4	185	3946	31	25.22s	31	1.7s	31	60.71s	31	6.75s	31	0.00022s
queen8-12	96	1368	–	–	13	1.1s	–	–	10	10.24s	12	0.699735s
queen8-8	64	728	9	9.87s	10	**1.4s**	–	–	–	–	9	4.66952s
queen9-9	81	1056	10	13.96s	11	94s	–	–	–	–	10	9.80493s
r125.1	125	209	–	–	5	1.2s	–	–	–	–	5	0.000265s
r125.1c	125	7501	–	–	46	1.8s	–	–	–	–	46	0.000565s
r125.5	125	3838	–	–	37	6.7s	–	–	–	–	36	0.328828s
r250.1	250	867	–	–	8	2.4s	–	–	–	–	8	0.001849s
r250.1c	250	30227	–	–	64	5s	–	–	–	–	64	0.012621s

Left summary (x | T):

	x	T	x	T	x	T	x	T	x	T
Greater	0	19	7	40	0	1	0	29	1	26
Equal	19	0	34	0	29	0	29	0	25	0
Smaller	0	0	0	1	0	2	0	0	0	0

Right half

Instance	|V|	|E|	x	T	x	T	x	T	x	T	x	T	x	T
2-FullIns-5	852	12201	–	–	7	10.7s	7	5.02s	–	–	–	–	7	0.0043s
3-FullIns-4	405	3524	–	–	7	1.6s	7	0.02s	–	–	–	–	7	0.001365s
4-FullIns-4	690	6650	–	–	8	7.7s	8	0.02s	–	–	–	–	8	0.004098s
5-FullIns-4	1085	11395	–	–	9	28s	9	0.04s	–	–	–	–	9	0.010039s
1-Insertions-5	202	1227	–	–	6	1.2s	–	–	–	–	–	–	6	0.0004s
1-Insertions-6	607	6337	–	–	7	8.1s	–	–	–	–	–	–	7	0.003357s
2-Insertions-4	149	541	–	–	5	1.1s	–	–	5	5.71s	–	–	5	0.00021s
2-Insertions-5	597	3936	–	–	6	6.5s	–	–	–	–	–	–	6	0.003145s
3-Insertions-4	281	1046	–	–	5	2.1s	–	–	–	–	–	–	5	0.000739s
3-Insertions-5	1406	9695	–	–	6	45s	–	–	–	–	–	–	6	0.018598s
4-Insertions-3	79	156	–	–	4	0.6s	–	–	4	3.72s	–	–	4	0.000069s
4-Insertions-4	475	1795	–	–	5	3.7s	–	–	–	–	–	–	5	0.005s
qg.order60	3600	212400	–	–	–	–	62	**h**	–	–	–	–	60	257186s
DSJC250.1	250	3218	**8**	26.07s	8	3s	–	–	**8**	**5.77s**	–	–	9	19.19942s
DSJC250.5	250	15668	**28**	89s	33	600s	–	–	30	**80.12s**	–	–	29	12511s
will199GPIA	701	6772	–	–	8	21.8s	7	6.68s	–	–	–	–	7	1.38738s
zeroin.i.1	211	4100	49	28.24s	49	1.7s	–	–	49	27.1s	49	7.74s	49	0.002171s
zeroin.i.2	211	3541	30	83.39s	30	2.1s	–	–	30	39.08s	30	15.11s	30	0.001804s

Right summary (x | T):

	x	T	x	T	x	T	x	T	x	T	x	T
Greater	8	9	8	23	8	17	2	7	1	9	1	9
Equal	28	0	22	0	14	0	5	0	9	0	7	0
Smaller	6	7	1	8	2	7	0	0	6	7	4	3

The time taken by different algorithms are represented in seconds (s). In case of Best of [33], h represents 1 h. For state of the art algorithms, we summarize the number of instances for which span (χ) and time (T) is greater, equal and smaller than that of ISH at the bottom of the table. We also mark the table entries for both χ and T which are better than those of ISH in bold font.

It is evident from Table 1 that for all sparse benchmarks, ISH produces optimal results very quickly, mostly in the range of microseconds, which is less than several magnitude than other algorithms. This superior performance of ISH on sparse benchmarks is in accordance with our theoretical findings stated in Theorem 6. For other benchmarks also, its performance is reasonably good. From the summary of results presented at the bottom of Table 1, it implies that for a large number of benchmarks, ISH outperforms other algorithms in terms of span or time significantly.

6 Conclusion

In this paper we have proposed an incremental search heuristic for coloring graphs. We simulate ISH on 86 benchmark instances and show that it performs significantly good for sparse graphs and reasonably good for other benchmarks.

References

1. Karp, R.M.: Reducibility among combinatorial problems. In: Complexity of Computer Computations, pp. 85–103 (1972)
2. Zuckerman, D.: Linear degree extractors and the inapproximability of max clique and chromatic number. In: Proceedings of the Thirty-Eighth Annual ACM Symposium on Theory of Computing (STOC '06), pp. 681–690 (2006)
3. Galinie, P., Hertz, A.: A survey of local search methods for graph coloring. Comput. Oper. Res. **33**(9), 2547–2562 (2006)
4. Pardalos, P.M., Mavridou, T., Xue, J.: The graph coloring problem: a bibliographic survey. In: Handbook of Combinatorial Optimization, pp. 1077–1141. Springer, Berlin (1998)
5. Zhou, Y., Hao, J.K., Duval, B.: Reinforcement learning based local search for grouping problems: a case study on graph coloring. Expert Syst. Appl. **64**, 412–422 (2016)
6. Mahmoudi, S., Lotfi, S.: Modified cuckoo optimization algorithm (MCOA) to solve graph coloring problem. Appl. Soft Comput. **33**, 48–64 (2015)
7. Moalic, L., Gondran, A.: Variations on memetic algorithms for graph coloring problems. J. Heuristics **24**(1), 1–24 (2018)
8. Sun, W., Hao, J.-K., Lai, X., Wu, Q.: On feasible and infeasible search for equitable graph coloring. In: Proceedings of the Genetic and Evolutionary Computation Conference, pp. 369–376 (2017)
9. Hasenplaugh, W., Kaler, T., Schardl, T.B., Leiserson, C.E.: Ordering heuristics for parallel graph coloring. In: Proceedings of the 26th ACM Symposium on Parallelism in Algorithms and Architectures, pp. 166–177 (2014)
10. Welsh, D.J.A., Powell, M.B.: An upper bound for the chromatic number of a graph and its application to timetabling problems. Comput. J. **10**(1), 85–86 (1967)

11. Matula, D.W.: A min-max theorem for graphs with application to graph coloring. SIAM Rev. **10**, 481–482 (1968)
12. Brélaz, D.: New methods to color the vertices of a graph. Commun. ACM **22**(4), 251–256 (1979)
13. Kempe, A.B.: On the geographical problem of the four colours. Am. J. Math. **2**(3), 193–200 (1879)
14. Chams, M., Hertz, A., De Werra, D.: Some experiments with simulated annealing for coloring graphs. Eur. J. Oper. Res. **32**(2), 260–266 (1987)
15. Johnson, D.S., Aragon, C.R., McGeoch, L.A., Schevon, C.: Optimization by simulated annealing: an experimental evaluation; part II. Graph coloring and number partitioning. Oper. Res. **39**(3), 378–406 (1991)
16. Holland, J.H.: Adaptation in natural and artificial systems. University of Michigan Press, Ann Arbor, MI (1992)
17. Lü, Z., Hao, J.-K.: A memetic algorithm for graph coloring. Eur. J. Oper. Res. **203**(1), 241–250 (2010)
18. Glover, F.: Tabu search and adaptive memory programming—advances, applications and challenges. In: Interfaces in Computer Science and Operations Research, pp. 1–75. Springer, Berlin (1997)
19. Hertz, A., de Werra, D.: Using tabu search techniques for graph coloring. Computing **39**(4), 345–351 (1987)
20. Ghosal, S., Ghosh, S.C.: A randomized algorithm for joint power and channel allocation in 5g d2d communication. In: 2019 IEEE 18th International Symposium on Network Computing and Applications (NCA), pp. 1–5. IEEE, Piscataway (2019)
21. Garey, M.R., Johnson, D.S., Stockmeyer, L.: Some simplified np-complete problems. In: Proceedings of the Sixth Annual ACM Symposium on Theory of Computing, pp. 47–63. ACM, New York (1974)
22. Łuczak, T.: The chromatic number of random graphs. Combinatorica **11**(1), 45–54 (1991)
23. Mizunoa, K., Nishihara, S.: Constructive generation of very hard 3-colorability instances. Discrete Appl. Math. **156**(2), 218–229 (2008)
24. Zymolka, A., Koster, A., Wessaly, R.: Transparent optical network design with sparse wavelength conversion. In: Proceedings of the 7th IFIP Working Conference on Optical Network Design and Modelling, pp. 61–80 (2003)
25. Gomes, C., Shmoys, D.: Completing quasigroups or Latin squares: a structured graph coloring problem. In: Johnson, D.S., Mehrotra, A., Trick, M. (eds.) Proceedings of the Computational Symposium on Graph Coloring and Its Generalizations, pp. 22–39 (2002)
26. Hossain, S., Steihaug, T.: Graph coloring in the estimation of mathematical derivatives. In: Johnson, D.S., Mehrotra, A., Trick, M. (eds.) Proceedings of the Computational Symposium on Graph Coloring and Its Generalizations, pp. 9–16 (2002)
27. Mizuno, K., Nishihara, S.: Toward ordered generation of exceptionally hard instances for graph 3-colorability. In: Proceedings of the Computational Symposium on Graph Coloring and Its Generalizations, pp. 1–8 (2002)
28. Caramia, M., Dell'Olmo, P.: Coloring graphs by iterated local search traversing feasible and infeasible solutions. Discrete Appl. Math. **156**(2), 201–217 (2008)
29. Lewandowski, G., Condon, A.: Experiments with parallel graph coloring heuristics and applications of graph coloring. DIMACS Ser. Discrete Math. Theoret. Comput. Sci. **26**, 309–334 (1996)
30. Mehrotra, A., Trick, M.A.: A column generation approach for graph coloring. INFORMS J. Comput. **8**(4), 344–354 (1996)
31. Culberson, J., Beacham, A., Papp, D.: Hiding our colors. In: Proceedings of the CP'95 Workshop on Studying and Solving Really Hard Problems, pp. 31–42 (1995)
32. Douiri, S.M., Elbernoussi, S.: Solving the graph coloring problem via hybrid genetic algorithms. J. King Saud Univ. Eng. Sci. **27**(1), 114–118 (2015)

33. Jabrayilov, A., Mutzel, P.: New integer linear programming models for the vertex coloring problem. In: Latin American Symposium on Theoretical Informatics, pp. 640–652. Springer, Berlin (2018)
34. Artacho, F.J.A., Campoy, R., Elser, V.: An enhanced formulation for solving graph coloring problems with the Douglas–Rachford algorithm. J. Glob. Optim. 1–21 (2018). arXiv:1808.01022
35. Baiche, K., Meraihi, Y., Hina, M.D., Ramdane-Cherif, A., Mahseur, M.: Solving graph coloring problem using an enhanced binary dragonfly algorithm. Int. J. Swarm Intell. Res. 10(3), 23–45 (2019)
36. Meraihi, Y., Ramdane-Cherif, A., Mahseur, M., Achelia, D.: A chaotic binary Salp swarm algorithm for solving the graph coloring problem. In: International Symposium on Modelling and Implementation of Complex Systems, pp. 106–118. Springer, Berlin (2018)
37. Mostafaie, T., Khiyabani, F.M., Navimipour, N.J.: A systematic study on meta-heuristic approaches for solving the graph coloring problem. Comput. Oper. Res. 120, 104850 (2019)
38. Zhou, Y., Hao, J.-K., Duval, B.: Reinforcement learning based local search for grouping problems: a case study on graph coloring. Expert Syst. Appl. 64, 412–422 (2016)
39. Zhou, Y., Duval, B., Hao, J.-K.: Improving probability learning based local search for graph coloring. Appl. Soft Comput. 65, 542–553 (2018)

Improved Bounds on the Span of $L(1, 2)$-edge Labeling of Some Infinite Regular Grids

Susobhan Bandopadhyay, Sasthi C. Ghosh, and Subhasis Koley

Abstract For two given non-negative integers h and k, an $L(h, k)$-edge labeling of a graph G is the assignment of labels $\{0, 1, \cdots, n\}$ to the edges so that two edges having a common vertex are labeled with difference at least h and two edges not having any common vertex but having a common edge connecting them are labeled with difference at least k. The span $\lambda'_{h,k}(G)$ is the minimum n such that G admits an $L(h, k)$-edge labeling. Here our main focus is on finding $\lambda'_{h,k}(G)$ for $L(1, 2)$-edge labeling of infinite regular hexagonal (T_3), square (T_4) and triangular (T_6) grids. It was known that $7 \leq \lambda'_{h,k}(T_3) \leq 8$, $10 \leq \lambda'_{h,k}(T_4) \leq 11$ and $16 \leq \lambda'_{h,k}(T_6) \leq 20$. Here we have shown that $\lambda'_{h,k}(T_3) \leq 7$, $\lambda'_{h,k}(T_4) \geq 11$ and $\lambda'_{h,k}(T_6) \geq 19$.

Keywords $L(1, 2)$-edge labelling · Bounds · Minimum span · Infinite regular grids

1 Introduction

Channel assignment problem (CAP) is one of the fundamental problems in wireless communication where frequency channels are assigned to transmitters such that interference can not occur. The objective of the CAP is to minimize the span of frequency spectrum. In 1980, Hale [6] first formulated the CAP as a classical vertex coloring problem. Later on, in 1988 Roberts [9] introduced $L(h, k)$-vertex labeling as defined below:

Definition 1 For two non-negative integers h and k, an $L(h, k)$-vertex labeling of a graph $G(V, E)$ is a function $\mathbf{f} : V \rightarrow \{0, 1, \cdots, n\}$, $\forall v \in V$ such that $|\mathbf{f}(u) - \mathbf{f}(v)| \geq h$ when $d(u, v) = 1$ and $|\mathbf{f}(u) - \mathbf{f}(v)| \geq k$ when $d(u, v) = 2$. For two vertices u and v, the distance, $d(u, v)$ is k' if at least k' edges are required to connect u and v.

S. Bandopadhyay (✉) · S. C. Ghosh · S. Koley
Advanced Computing and Microelectronics Unit, Indian Statistical Institute, Kolkata, India
e-mail: sasthi@isical.ac.in

© The Author(s), under exclusive license to Springer Nature Switzerland AG 2021
C. Gentile et al. (eds.), *Graphs and Combinatorial Optimization:*
from Theory to Applications, AIRO Springer Series 5,
https://doi.org/10.1007/978-3-030-63072-0_5

The *span* $\lambda_{h,k}(G)$ of $L(h, k)$-vertex labeling is the minimum n such that G admits an $L(h, k)$-vertex labeling. In 1992 Griggs and Yeh [5] extended the concept of $L(h, k)$ labeling by introducing $L(k_1, k_2, \cdots, k_l)$-vertex labeling with separation $\{k_1, k_2, \cdots, k_l\}$ for $\{1, 2, \cdots, l\}$ distant vertices and their main focus was on $L(h, k)$-vertex labeling for a special case $h = 2$, $k = 1$. In 2007, Griggs and Jin [4] studied $L(h, k)$-edge labeling, which can be formally defined as:

Definition 2 For two non-negative integers h and k, an $L(h, k)$-edge labeling of a graph $G(V, E)$ is a function $\mathbf{f}' : E \rightarrow \{0, 1, \cdots, n\}, \forall e \in E$ such that $|\mathbf{f}'(e_1) - \mathbf{f}'(e_2)| \geq h$ when $d(e_1, e_2) = 1$ and $|\mathbf{f}'(e_1) - \mathbf{f}'(e_2)| \geq k$ when $d(e_1, e_2) = 2$. Here, for any two edges e_1 and e_2, the distance $d(e_1, e_1)$ is k' if at least $(k' - 1)$ edges are required to connect e_1 and e_2.

Like $L(h, k)$-vertex labeling, the *span* $\lambda'_{h,k}(G)$ of $L(h, k)$-edge labeling is the minimum n such that G admits an $L(h, k)$-edge labeling. In 2011, Calamoneri did a rigorous survey [1] on both vertex and edge labeling problems. Authors in [2, 3, 7, 8] have studied $L(h, k)$-edge labeling of regular infinite hexagonal (T_3), square (T_4) and triangular (T_6) grids for the special case of $h = 1$ and $k = 2$. They obtained some upper and lower bounds on $\lambda'_{1,2}(G)$ for T_3, T_4 and T_6 with a gap between them. In this paper, we improve some of these gaps.

Given a graph $G(V, E)$, its *line graph* $L(G)(V', E')$ is a graph such that each vertex of $L(G)$ represents an edge of G and two vertices of $L(G)$ have an edge if and only if their corresponding edges share a common vertex in G. It is well-known that if G is d-regular then $L(G)$ is $2(d - 1)$-regular. Figure 1 shows $T_3, T_4, L(T_3), L(T_4)$ and T_6. It is also well-known that edge labeling of G is equivalent to vertex labeling of $L(G)$. In our approach, instead of $L(1, 2)$-edge labeling of T_3 and T_4, we use $L(1, 2)$-vertex labeling of $L(T_3)$ and $L(T_4)$. Note that $L(T_6)$ is 10-regular. Because of this high degree, we consider $L(1, 2)$-edge labeling of T_6 directly. Our results on $\lambda'_{1,2}(G)$ for T_3, T_4 and T_6 are stated in Table 1. In this table, $a - b$ represents that $a \leq \lambda'_{1,2}(G) \leq b$. Here, we use the term 'coloring' and 'labeling' interchangeably.

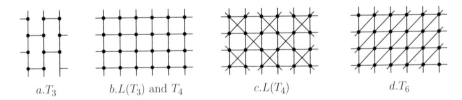

$a.T_3$ $b.L(T_3)$ and T_4 $c.L(T_4)$ $d.T_6$

Fig. 1 T_3, T_4, their line graphs and T_6

Table 1 The main results

	T_3		T_4		T_6	
	Known	Ours	Known	Ours	Known	Ours
$\lambda'_{1,2}(G)$	7-8 [7]	7-7	10-11 [7]	11-11	16-20 [2]	19-20

2 Results

2.1 Hexagonal Grid

Let us consider the induced subgraph G_S of $L(T_3)$ as shown in Fig. 2a, where all vertices are at mutual distance at most three. It is clear that $\lambda_{1,2}(L(T_3)) \geq \lambda_{1,2}(G_S)$. A color can be reused at a pair of vertices at mutual distance three apart in G_S. But we observe that if any color is reused at distance three in G_S, then there exists a color which remains unused in G_S. Thus there is no such benefit of reusing a color over using all different colors in G_S. This motivates us to consider reusing a color at distance four only keeping all colors distinct at G_S. We show in Theorem 1 that such a coloring of $L(T_3)$ exists which uses colors from 0 to 7 only.

Theorem 1 $\lambda'_{1,2}(T_3) = 7$.

Proof Consider the coloring function g of vertices $v = (x, y)$ as $g(v)_{(x,y)} = (x + 5y) \bmod 8$. Here coordinates (x, y) of a vertex v can be computed from the origin $O(0, 0)$ as shown in Fig. 2b. The minimum and maximum color used here are 0 and 7 respectively. It can also be verified that g satisfies the $L(1, 2)$-vertex labeling requirements of $L(T_3)$. Hence $\lambda_{1,2}(L(T_3)) \leq 7$. It has been shown in [7] that $\lambda_{1,2}(L(T_3)) \geq 7$. Hence $\lambda'_{1,2}(T_3) = \lambda_{1,2}(L(T_3)) = 7$. In Fig. 2b, an $L(1, 2)$-vertex labeling of $L(T_3)$ has been shown.   □

It is evident that $\lambda'_{1,2,1}(T_3) \geq \lambda'_{1,2}(T_3) = 7$. In the coloring function g stated above, observe that no vertices at distance three have the same color in $L(T_3)$. Hence

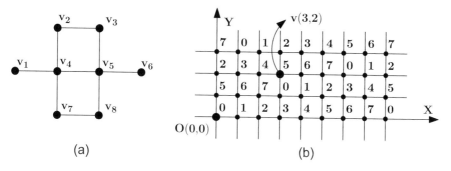

Fig. 2 (a) Sub graph G_s of $L(T_3)$. (b) A feasible $L(1, 2)$-labeling of $L(T_3)$

g also satisfies the $L(1, 2, 1)$-edge labeling requirements for T_3. So, $\lambda'_{1,2,1}(T_3) \leq 7$. Hence we have the following result.

Corollary 1 $\lambda'_{1,2,1}(T_3) = 7$.

2.2 Square Grid

Let us consider the induced subgraph G of $L(T_4)$ as shown in Fig. 3 where all vertices are at mutual distance at most three. Let $S_1 = \{a, b\}$, $S_2 = \{k, l\}$, $S_3 = \{c, g\}$, $S_4 = \{f, j\}$ and $S_5 = \{d, e, h, i\}$.

Definition 3 The set of vertices in S_5 are termed as **central vertices** in G.

Definition 4 The set of vertices in $S_1 \cup S_2 \cup S_3 \cup S_4$ are termed as **peripheral vertices** in G.

Now we have the following observations in G. Here the color of vertex a is denoted by $\mathbf{f}(a)$.

Observation 1 *If colors of vertices of G are all distinct then* $\lambda_{1,2}(G) \geq 11$.

Proof As G has 12 vertices, if all of them get distinct colors then $\lambda_{1,2}(G) \geq 11$. ☐

Observation 2 *No color can be used thrice in G. Colors used at the central vertices in S_5 cannot be reused in G. Colors used at the peripheral vertices in S_1 can be reused only at the peripheral vertices in S_2. Similarly, colors used at the peripheral vertices in S_3 can be reused only at the peripheral vertices in S_4.*

Proof No three vertices are mutually distant three apart. Hence no color can be used thrice in G. For any central vertex in S_5 there does not exist any vertex in G which is distance three apart from it. So colors used in the central vertices in S_5 cannot be reused in G. For all peripheral vertices in $S_1 \cup S_2$, $d(x, y) = 3$ only when $x \in S_1$ and $y \in S_2$. Hence color used at peripheral vertex in S_1 can only be reused in S_2. Similarly, color used at peripheral vertex in S_3 can only be reused in S_4. ☐

Observation 3 *If $\mathbf{f}(x) = \mathbf{f}(y) = \mathbf{c}$ where $x \in S_1$ and $y \in S_2$ then either $\mathbf{c} \pm 1$ is to be used in $(S_1 \cup S_2) \setminus \{x, y\}$ or it should remain unused in G. Similarly, if $\mathbf{f}(x) = \mathbf{f}(y) = \mathbf{c}$ where $x \in S_3$ and $y \in S_4$ then either $\mathbf{c} \pm 1$ is to be used in $(S_3 \cup S_4) \setminus \{x, y\}$ or it should remain unused in G.*

Fig. 3 Sub graph G of $L(T_4)$

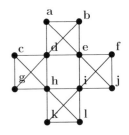

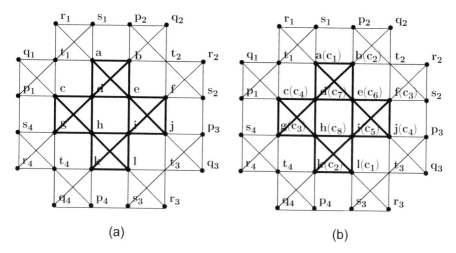

Fig. 4 (a) The subgraph G_1. (b) A feasible $L(1, 2)$-labeling of G

Proof Note that for all vertices $z \in V(G) \setminus (S_1 \cup S_2)$, either $d(z, x) = 2$ or $d(z, y) = 2$, where $x \in S_1$ and $y \in S_2$. Hence $\mathbf{c} \pm 1$ cannot be used in $V(G) \setminus (S_1 \cup S_2)$. So $\mathbf{c} \pm 1$ can only be used in $(S_1 \cup S_2) \setminus \{x, y\}$ or it should remain unused in G. Similarly, if $\mathbf{f}(x) = \mathbf{f}(y) = \mathbf{c}$, where $x \in S_3$ and $y \in S_4$, then $\mathbf{c} \pm 1$ can only be used in $(S_3 \cup S_4) \setminus \{x, y\}$ or it should remain unused in G.   □

Observation 4 *Let* $\mathbf{f}(x) = \mathbf{f}(y) = \mathbf{c}$ *where* $x \in S_1$ *and* $y \in S_2$. *If* $|\mathbf{f}(x) - \mathbf{f}(x')| \geq 2$, *where* $x' \in S_1 \setminus \{x\}$, *then one of* $\mathbf{c} \pm 1$ *must remain unused in* G. *Similarly if* $|\mathbf{f}(y) - \mathbf{f}(y')| \geq 2$, *where* $y' \in S_2 \setminus \{y\}$, *then one of* $\mathbf{c} \pm 1$ *must remain unused in* G. *Similar facts hold when* $x \in S_3$, $x' \in S_3 \setminus \{x\}$, $y \in S_4$ *and* $y' \in S_4 \setminus \{y\}$.

Proof Since $|\mathbf{f}(x) - \mathbf{f}(x')| \geq 2$, $\mathbf{f}(x') \neq \mathbf{c} \pm 1$. Hence from Observation 3, one of $\mathbf{c} \pm 1$ must remain unused in G.   □

If no color is reused in G, then $\lambda_{1,2}(G) \geq 11$ from Observation 1. To make $\lambda_{1,2}(G) < 11$, at least one color must be reused in G. From Observation 2, there are at most 4 distinct pairs of peripheral vertices in G where a pair can have the same color. Now consider the subgraph G_1 of $L(T_4)$ as shown in Fig. 4a. Note that G_1 consists of 5 subgraphs G', G'_1, G'_2, G'_3 and G'_4 which all are isomorphic to G having central vertices $\{d, h, i, e\}$, $\{t_1, c, d, a\}$, $\{b, e, f, t_2\}$, $\{i, l, t_3, j\}$ and $\{g, t_4, k, h\}$ respectively. Based on the span requirements of coloring G_1, we derive the following theorem.

Theorem 2 $\lambda_{1,2}(L(T_4)) \geq \lambda_{1,2}(G_1) \geq 11$.

Proof

Case 1 When at most one pair of peripheral vertices use the same color in any sub graph of $L(T_4)$ isomorphic to G.

If no color is reused in G', then $\lambda_{1,2}(G') \geq 11$ from Observation 1. We now consider the case when exactly one pair reuse a color in G'. Without loss of generality, consider $\mathbf{f}(a) = \mathbf{f}(l) = c_1$. From Observation 3, $c_1 \pm 1$ can only be put in $\{b, k\}$. Let $\mathbf{f}(k) = c_1 - 1$ and $\mathbf{f}(b) = c_1 + 1$. We assume that $c_1 - 1$ is the minimum color. Let us consider $\mathbf{f}(d) = c_1 + n$ where $n \in \mathbb{N}$ and $n \geq 2$. From Observation 4, $x \in \{c_1, c_1 + n\}$ can be reused in G'_2 only if one of $x \pm 1$ remains unused in G'_2. In either case, $\lambda_{1,2}(G'_2) \geq 11$. So x cannot be reused in G'_2. Since $\mathbf{f}(a) = \mathbf{f}(l) = c_1, c_1 - 1$ can only be put in $\{r_2, s_2\}$ as vertex b is already colored and for all other vertices $z \in V(G'_2) \setminus \{r_2, s_2\}$, either $d(z, a) = 2$ or $d(z, l) = 2$. Without loss of generality, let $\mathbf{f}(r_2) = c_1 - 1$. In that case, $c_1 + n \pm 1$ can only be put in $\{e, s_2\}$. Without loss of generality, let $\mathbf{f}(e) = c_1 + n - 1$ and $\mathbf{f}(s_2) = c_1 + n + 1$. Since $\mathbf{f}(a) = \mathbf{f}(l) = c_1, \mathbf{f}(i) \neq c_1 \pm 1$ and hence $|\mathbf{f}(l) - \mathbf{f}(i)| \geq 2$. Now if $|\mathbf{f}(d) - \mathbf{f}(c)| \geq 2$, then from Observation 4, one of $\mathbf{f}(c) \pm 1$, $\mathbf{f}(d) \pm 1$ and $\mathbf{f}(i) \pm 1$ remains unused in G'_4 if $\mathbf{f}(c)$ or $\mathbf{f}(d)$ or $\mathbf{f}(i)$ is reused in G'_4 respectively. In either case, this implies $\lambda_{1,2}(G'_4) \geq 11$. So $|\mathbf{f}(d) - \mathbf{f}(c)| = 1$ and $\mathbf{f}(c) = c_1 + n + 1$. There are 5 more vertices $\{g, h, i, j, f\}$ in G' which are to be colored with 5 distinct colors. Hence at least color $c_1 + n + 6$ must be used. Observe that if $\mathbf{f}(f) = c_1 + n + 2$ then $|\mathbf{f}(e) - \mathbf{f}(f)| = 3$ and $|\mathbf{f}(k) - \mathbf{f}(h)| \geq 3$ implying $\lambda_{1,2}(G'_3) \geq 11$ from Observation 4. As $d(s_2, i) = d(s_2, j) = 2$ and $\mathbf{f}(s_2) = c_1 + n + 1$, we get $\mathbf{f}(i) \neq c_1 + n + 2$ and $\mathbf{f}(j) \neq c_1 + n + 2$. Therefore, either $\mathbf{f}(g) = c_1 + n + 2$ or $\mathbf{f}(h) = c_1 + n + 2$. So, $\mathbf{f}(p_4) \neq c_1 + n + 1$ and $\mathbf{f}(q_4) \neq c_1 + n + 1$. In that case, $\mathbf{f}(p_4)$ and $\mathbf{f}(q_4)$ must be in $\{c_1 + n, c_1 + n - 1\}$ if color $c_1 + n$ is to be reused in G'_4, otherwise, $\lambda_{1,2}(G_1) \geq 11$. As c_1 cannot be reused in G'_4, either $\mathbf{f}(r_4) = c_1 + 1$ or $\mathbf{f}(s_4) = c_1 + 1$. Let $\mathbf{f}(r_4) = c_1 + 1$. When $n = 2$, $c_1 + n - 1 = c_1 + 1$ and when $n = 3$, $c_1 + n - 1 = c_1 + 2$. As $d(p_4, l) = d(p_4, r_4) = d(q_4, l) = d(q_4, r_4) = 2$, $\mathbf{f}(p_4), \mathbf{f}(q_4) \notin \{c_1 + 1, c_1 + 2\}$. So, $n \geq 4$ and hence $c_1 + n + 6 \geq c_1 + 10$. So at least 12 color are required in G_1 including $c_1 - 1$ and $c_1 + 10$. Hence $\lambda_{1,2}(G_2) \geq 11$.

Case 2 There exists at least one subgraph of $L(T_4)$ isomorphic to G where two pairs of peripheral vertices use a color each.

There are two different ways of reusing two colors in G'.

Case 2.1 First consider the case when $\mathbf{f}(a) = \mathbf{f}(l) = c_1$ and $\mathbf{f}(c) = \mathbf{f}(j) = c_2$. From Observation 3, $c_1 \pm 1$ and $c_2 \pm 1$ must be used in $\{b, k\}$ and $\{g, f\}$ respectively. From Observation 2, c_1 can only be reused in $\{r_2, s_2\}$ in G'_2. But $\mathbf{f}(r_2) \neq c_1$ and $\mathbf{f}(s_2) \neq c_1$ as $|\mathbf{f}(b) - c_1| = 1$ and $d(b, r_2) = d(b, s_2) = 2$. Again, from Observation 2, c_2 can only be reused in $\{p_2, q_2\}$. But $\mathbf{f}(p_2) \neq c_2$ and $\mathbf{f}(q_2) \neq c_2$ as $|\mathbf{f}(f) - c_2| = 1$ and $d(f, p_2) = d(f, q_2) = 2$. From Observation 3, if $\mathbf{f}(i)$ is to be reused in G'_2, then $|\mathbf{f}(i) - c_2| = 1$. But $\mathbf{f}(i) \neq c_2 \pm 1$ as $d(c, i) = 2$ and $\mathbf{f}(c) = c_2$. If $\mathbf{f}(d)$ is to be reused in G'_2, then $|\mathbf{f}(d) - c_1| = 1$. But $\mathbf{f}(d) \neq c_1 \pm 1$ as $d(d, l) = 2$ and $\mathbf{f}(l) = c_1$. Therefore, no color can be reused in G'_2 and hence $\lambda_{1,2}(G_1) \geq 11$.

Case 2.2 Consider the case when $\mathbf{f}(a) = \mathbf{f}(l) = c_1$ and $\mathbf{f}(b) = \mathbf{f}(k) = c_2$. Without loss of generality, assume $c_2 > c_1$. From Observation 3, $c_1 \pm 1$ and $c_2 \pm 1$ must be used in $\{b, k\}$ and $\{a, l\}$ respectively. Even if we set $c_2 = c_1 + 1$, at least one of $c_1 - 1$ and $c_2 + 1$ must remain unused in G'. So the 8 vertices in $V(G') \setminus (\{a, l\} \cup \{b, k\})$ must get 8 distinct colors other than c_1 and c_2. So, $\lambda_{1,2}(G') \geq 10$. Note that $\lambda_{1,2}(G') = 10$ only if $c_2 = c_1 + 1$, c_1 is minimum color ($c_1 - 1$ does not exists) or c_2 is maximum color ($c_2 + 1$ does not exists). If both c_1 and c_2 are non-extreme color, then $\lambda_{1,2}(G') \geq 11$ and we are done. So, we consider $c_1 = 0$, $c_2 = c_1 + 1 = 1$ and $c_2 + 1 = 2$ as unused in G'. In that case, $\mathbf{f}(d) = x \geq 3$ and hence $|\mathbf{f}(d) - \mathbf{f}(a)| \geq 3$. From Observation 4, if x is reused in G'_2, then one of $x \pm 1$ cannot be used in G'_2. If only x is reused in G'_2, then $\lambda_{1,2}(G'_2) \geq 11$. If x and one of $\{\mathbf{f}(i), \mathbf{f}(j)\}$ are reused in G'_2, then from Case 2.1 above, $\lambda_{1,2}(G_1) \geq 11$. If x and both of $\{\mathbf{f}(i), \mathbf{f}(j)\}$ are reused in G'_2, from Case 3 below, we will see that $\lambda_{1,2}(G_1) \geq 11$. So, to keep $\lambda_{1,2}(G_1) < 11$, x should not be reused in G'_2. In that case, $x - 1$ must be used at one of $\{c, g, h, e\}$ in G'. Now arguing similarly as stated in case **1**, we can conclude that $x + 7$ must be used in G'_1 or G'_2. If $x = 3$, then $x - 1 = 2$ must be used in G' which is a contradiction, as 2 must remain unused in G'. Hence $x \geq 4$ implying $x + 7 = 11$. Hence $\lambda_{1,2}(G_1) \geq 11$.

Case 3 The exists at least one sub graph of $L(T_4)$ isomorphic to G where three pairs of peripheral vertices use a color each.

Without loss of generality, let us consider $\mathbf{f}(a) = \mathbf{f}(l) = c_1$, $\mathbf{f}(b) = \mathbf{f}(k) = c_2$ and $\mathbf{f}(c) = \mathbf{f}(j) = c_3$. From Observation 3, $c_1 \pm 1$ and $c_2 \pm 1$ must be used in $\{b, k\}$ and $\{a, l\}$ respectively. It can be observed that $\lambda_{1,2}(G') = 9$ only if $|c_1 - c_2| = 1$, $|c_3 - \mathbf{f}(g)| = 1$, $|c_3 - \mathbf{f}(f)| = 1$ and any one of $\{c_1, c_2\}$ is one extreme color. Without loss of generality consider $\mathbf{f}(g) = c_3 + 1$, $\mathbf{f}(f) = c_3 - 1$, c_1 is minimum color and $c_2 = c_1 + 1$. From Observation 2, c_3 can only be reused in $\{p_2, q_2\}$. But $\mathbf{f}(p_2) \neq c_3$ and $\mathbf{f}(q_2) \neq c_3$ as $\mathbf{f}(f) = c_3 - 1$ and $d(f, p_2) = d(f, q_2) = 2$. From Observation 3, if $\mathbf{f}(i)$ is to be reused in G'_2, then $|\mathbf{f}(i) - c_3| = 1$. But $\mathbf{f}(i) \neq c_3 \pm 1$ as $d(c, i) = 2$ and $\mathbf{f}(c) = c_3$. From Observation 2, c_1 can only be reused in $\{r_2, s_2\}$. But $\mathbf{f}(r_2) \neq c_1$ and $\mathbf{f}(s_2) \neq c_1$ as $\mathbf{f}(b) = c_2 = c_1 + 1$ and $d(b, r_2) = d(b, s_2) = 2$. Now arguing similarly as stated in case **2.2** above, we can conclude that $c_2 + 1$ must remain unused in G'. So, $(c_1 - \mathbf{f}(d)) \geq 3$. Now from Observation 4, if $\mathbf{f}(d)$ is reused in G'_2 then any one of $\mathbf{f}(d) \pm 1$ must remain unused in G'_2. Thus in G'_2, only $\mathbf{f}(d)$ can be reused by keeping one of $\mathbf{f}(d) \pm 1$ as unused. Hence $\lambda_{1,2}(G_1) \geq 11$. If we consider $\lambda_{1,2}(G') = 10$, the same result can be obtained by considering the corresponding G'_i, $1 \leq i \leq 4$.

Case 4 The exists at least one subgraph of $L(T_4)$ isomorphic to G where all four pairs of peripheral vertices use a color each.

Let us consider $\mathbf{f}(a) = \mathbf{f}(l) = c_1$, $\mathbf{f}(b) = \mathbf{f}(k) = c_2$, $\mathbf{f}(g) = \mathbf{f}(f) = c_3$ and $\mathbf{f}(c) = \mathbf{f}(j) = c_4$. From Observation 3, $c_1 \pm 1$, $c_2 \pm 1$, $c_3 \pm 1$ and $c_4 \pm 1$ must be used in $\{b, k\}$, $\{a, l\}$, $\{c, j\}$ and $\{g, f\}$ respectively. It can be observed that $\lambda_{1,2}(G') = 9$ only if $|c_1 - c_2| = 1$, $|c_3 - c_4| = 1$, one of $\{c_1, c_2\}$ is an extreme color and one of $\{c_3, c_4\}$ is the other extreme color. Without loss of generality, consider $c_1 = 0$, $c_4 = 9$, $c_2 = c_1 + 1 = 1$ and $c_3 = c_4 - 1 = 8$. So $c_2 + 1 = 2$ and $c_3 - 1 = 7$ are two distinct unused colors. Without loss of generality, consider $c_8 = c_2 + 2$, $c_5 = c_8 + 1$, $c_6 = c_5 + 1$ and $c_7 = c_6 + 1$. Since $|c_3 - c_4| = 1$ and $d(g, p_4) = d(g, q_4) = 2$, we get $\mathbf{f}(p_4) \neq c_4$ and $\mathbf{f}(q_4) \neq c_4$. Similarly, $\mathbf{f}(r_4) \neq c_1$ and $\mathbf{f}(s_4) \neq c_1$. From

Observation 2, c_5 can only be reused at $\{s_4, r_4\}$ in G'_4 but $\mathbf{f}(s_4) \neq c_5$ and $\mathbf{f}(r_4) \neq c_5$ as $d(h, s_4) = d(h, r_4) = 2$ and $\mathbf{f}(h) = c_8 = c_5 - 1$. Therefore, only c_7 can be reused in $\{p_4, q_4\}$. From Observation 4, one of $c_7 \pm 1$ must remain unused in G'_4 as $(c_4 - c_7) = 3$. Hence $\lambda_{1,2}(G_1) \geq 11$. For other assignment of central vertices and for the case when $\lambda_{1,2}(G') = 10$, we can obtain the same result by considering the corresponding G'_i, $1 \leq i \leq 4$.

□

2.3 Triangular Grid

Here we first define some notations.

For any vertex u, the set of vertices which are adjacent to u is called $N(u)$. Let us define $N(S) = \{\cup_{u \in S} N(u) : u \in S\}$. Let v be any vertex in T_6. Consider the subgraph $G_v(V, E)$ of T_6 centering v as shown in Fig. 5, where $V = N(v) \cup N(N(v))$ and E is set of all the edges which are incident to u where $u \in N(v)$. Observe that in G_v, for any two edges e_1 and e_2, $d(e_1, e_2) \leq 3$. Now we define the following three sets of edges S_1, S_2 and S_3:

S_1: Edges of G_v incident to v.
S_2: Edges of G_v whose both end points incident to e_1 and e_2 where e_1, $e_2 \in S_1$.
S_3: $E \setminus (S_1 \cup S_2)$.

Consider the 6-cycle, H_v formed with the edges of S_2 in G_v. We say e and e_1 as a pair of *opposite edges* in H_v if and only if $d(e, e_1) = 3$. This implies that the same color can be used at a pair of opposite edges in $L(1, 2)$-edge labeling. An edge $e(v, w)$ *covers* the set of edges E' if for every $e' \in E'$, $d(e, e') \leq 2$. This implies that a color used at e cannot be used at any edge $e' \in E'$ in $L(1, 2)$-edge labeling. Now we have the following lemmas.

Lemma 1 *If c be a color used to color an edge e in S_1, then c cannot be used in $E \setminus e$.*

Proof Since e is incident to v, for any other edge $e_1 \in E$, $d(e, e_1) \leq 2$. Hence $f'(e_1) \neq c$ for $L(1, 2)$-edge labeling, where $f'(e_1)$ denotes the color of e_1.   □

Fig. 5 A subgraph G_v of T_6

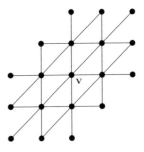

Lemma 2 *If c be a color used to color an edge in S_1, then $c + 1$ and $c - 1$ both can be used at most once in G_v.*

Proof Let e be an edge in S_1 such that $f'(e) = c$. Since e is incident to v, for any other edge $e_1 \in E$, $d(e, e_1) \leq 2$. Let $S_e = \{e_1 : d(e, e_1) = 1\}$. For $L(1, 2)$-edge labeling, $c + 1$ can only be used in an edge e_1 in S_e. It can be noted that for any two edges $e_1, e_2 \in S_e$, $d(e_1, e_2) \leq 2$. Hence $c + 1$ can be used at most once. Proof for $c - 1$ can be done in similar manner.   □

Lemma 3 *If c be a color used to color an edge e in S_2, then c can be used at most one edge in $E \setminus e$ in G_v.*

Proof Note that c cannot be used at any edge in S_1. Here c can be used at the *opposite edge* e_1 of e in S_2 or at an edge e_2 in S_3, which is adjacent to e_1. When c is used at e and e_1, then c cannot be used again in G_v as e and e_1 together cover all the edges of G_v. When c is used at e and e_2, c cannot be used again in G_v as e and e_2 together also cover all the edges of G_v.   □

Lemma 4 *If c be a color used to color an edge e in S_2, then $c + 1$ and $c - 1$ both can be used at most twice in G_v.*

Proof Suppose e_1 be an edge colored with $c + 1$. If e_1 is not adjacent to e then $d(e_1, e) = 3$. From statement of Lemma 3, it follows that there does not exist two edges along with e in G_v which are mutually distance 3 apart, otherwise c would have been used for three times. Hence $c + 1$ can be used at most once.

When e_1 is adjacent to e, e_2 can be colored with $c + 1$ if e_2 is at distance 3 apart from both e_1 and e. Again from the statement of Lemma 3, it follows that there does not exist two edges along with e in G_v which are mutually distance 3 apart, otherwise c would have been used for three times. So, $c + 1$ can be used at most twice, one in one of the edges adjacent to e and other in one of the edges which are at distance 3 apart from e. Proof for $c - 1$ can be done in similar manner.   □

Lemma 5 *If c be a color used to color an edge e in S_3, then c can be used at most twice in $E \setminus e$.*

Proof It follows from Fig. 5 that exactly one end point of e is incident to a vertex in H_v. Note that for any walk through H_v, every third vertex is distance 2 apart. So edges incident to those vertices are distance 3 apart. Since the order of H_v is 6, there can be at most $6/2 = 3$ vertices which are mutually distance 2 apart. Hence c can be used thrice.   □

Lemma 6 *If c be a color used to color an edge e in S_3, then $c + 1$ and $c - 1$ both can be used at most thrice in G_v.*

Proof We know that $c + 1$ can be used at an edge adjacent to e. From Lemma 5 it is clear that c can be used at most thrice. So, $c + 1$ can also be used at most thrice, where each such edge is adjacent to one of the three edges colored with c. It can be proved similarly for $c - 1$.   □

Lemma 7

   i. *To color the edges of S_1, at least 6 colors are required.*
  ii. *To color the edges of S_2, at least 3 colors are required.*
 iii. *To color the edges of S_3, at least 6 colors are required.*

Proof

   i. From Lemma 1, every edge of S_1 has an unique color. As there are 6 edges in S_1, 6 distinct colors are required here.
  ii. In S_2, there are 3 pairs of *opposite edges*. Each pair of opposite edges requires at least one unique color. So at least 3 colors are required.
 iii. A color can be used thrice in S_3 by Lemma 5. In S_3, there are 18 edges. So, at least 6 colors are required.

$\square$

Theorem 3 *For any optimal labeling of G_v, 6 consecutive colors including either the minimum color or the maximum color must be used in S_1.*

Proof It is clear from Lemma 7.i that S_1 needs at least 6 colors to color its edges. From Lemma 2, note that if c be a color used in an edge of S_1 then both $c + 1$ and $c - 1$ can be used at most once in G_v. Whereas a color can be used twice in S_2 and thrice in S_3. Thus our aim should be to minimize the number of colors which can be used only once in G_v. This implies that consecutive colors should be used in S_1 for optimal coloring. If the minimum color (min) or the maximum color (max) is used in S_1 then further benefit can be achieve as $min - 1$ or $max + 1$ does not exist. Therefore, optimal span can be achieved only when the colors of S_1 are consecutive including either min or max. $\square$

Theorem 4 *For any optimal labeling of G_v, 3 colors like $\{c, c + 2, c + 4\}$ have to be used twice each in S_2.*

Proof Let c be a color used in S_2. From Lemma 3, observe that c can be used at most twice in G_v. Also, no matter how many times c is used in S_2, it follows from Lemma 4 that both $c + 1$ and $c - 1$ can be used at most twice in G_v. Let $C_{S_2} = \{c, c + 1, c - 1| \forall c \text{ used at } S_2\}$. Note that a color can be used at most thrice in G_v. So our goal is to minimize $|C_{S_2}|$, where $|C_{S_2}|$ is the cardinality of set C_{S_2}. Observe that minimum 3 colors are required and maximum 6 colors can be used to color S_2. If 3 colors $\{c, c + 2, c + 4\}$ are used then $|C_{S_2}| \geq 6$, assuming one of them is an extreme color. If 6 consecutive colors are used then $|C_{S_2}| \geq 7$, assuming one of them is an extreme color. One can follow that in all the other cases $|C_{S_2}| > 7$. So for optimal coloring of G_v, 3 colors such as $\{c, c + 2, c + 4\}$ have to be used twice each in S_2. $\square$

Lemma 8 *If three consecutive colors c, $c + 1$, $c + 2$ are used thrice each in S_3 then neither $c - 1$ nor $c + 3$ can be used in S_3.*

Proof Observe that there are exactly 2 sets of three alternating vertices in H_v where a color can be used thrice at edges incident to any set of alternating vertices. If $c - 1$

would have been used in S_3 then either it was used at an edge adjacent to the edges colored with c or at an edge distance 3 apart from the edge colored with c. Now observe that c and $c - 1$ are used at two edges of S_3 which form a triangle with one edge of S_2. Suppose $c, c - 1$ be the colors used at those two edges $e, e_1 \in S_3$ respectively, where e is incident to u and e_1 is incident to w where $uw \in S_2$. Note that c is used thrice in S_3. Then c must be reused at an edge incident to x, and $xw \in S_2$. So c and $c - 1$ are used at two edges at distance 2 apart, which violets the condition of $L(1, 2)$-edge labeling. Hence $c - 1$ cannot be used in G_v. Similarly it can be shown that $c+3$ can also not be used in G_v. This implies that no 4 consecutive colors can be used thrice each in G_v. $\qquad \square$

Lemma 9 *If all colors in $\{c, c + 2, c + 4\}$ are used twice each in S_2 then at least 6 colors are required which all are either higher than $c + 4$ or lower than c.*

Proof From Theorem 4, it follows that $|C_{S_2}| = 6$ for optimal coloring of G_v. So $c+1, c+3$ and one of $c+5$ and $c-1$ must be used in S_3. Without loss of generality, assume that colors $\{c + 1, c + 3, c + 5\}$ are used in S_3. Using Lemma 4 it can be verified that colors $c + 1, c + 3$ and $c + 5$ can be used at most twice in S_3. That means using these three colors at most 6 edges can be colored in S_3. So 12 edges remain uncolored till now. By Lemma 6 a color can be used thrice in S_3. Again, it follows from Lemma 8 that no 4 consecutive colors can be used thrice each in S_3. Hence the maximum color used in S_3 will be at least $(c + 5) + 5$. Similarly it can be shown that minimum color used in S_3 will be at most $(c - 1) - 5$ for the case when $\{c - 1, c + 1, c + 3\}$ are used at S_3. $\qquad \square$

Theorem 5 $\lambda'_{1,2}(G_v) \geq 17$.

Proof By Theorem 3, 6 consecutive colors must be used to color the edges of S_1. Recall that, we assume the minimum color is used at S_1. Let c' be the maximum color used at S_1 and c'' be the minimum color used at S_2. From Theorem 4, 3 colors must be used to color the edges of S_2 and in that case by Lemma 1, $(c'' - c') \geq 2$. Note that, $c'+1$ and $c''-1$ can be used at most once and twice respectively. However a color can be used thrice in S_3. Therefore, it is beneficial if $c' + 1 = c'' - 1$. Now if $\{c, c + 2, c + 4\}$ are used at S_2 then $\{c - 2, c - 3, \cdots c - 7\}$ are used at S_1. Now if $\{c+1, c+3, c+5\}$ are used in S_3 then from Lemma 9, it follows that $c+10$ must be used at S_3. So, $\lambda'_{1,2}(G_v) \geq ((c+10)-(c-7)) = 17$. Similarly, if $\{c-1, c+1, c+3\}$ are used at S_3, $\lambda'_{1,2}(G_v) \geq ((c + 11) - (c - 6)) = 17$. Hence the proof. $\qquad \square$

We assume that the minimum color is used in S_1. The maximum color can be used at most thrice in S_3 and at most twice in S_2. In all cases, there exists a vertex say v' in H_v such that color of any edge incident to v' is neither minimum nor maximum. Now we consider the subgraph $G_{v'}$ of T_6 centering v' and isomorphic to G_v.

Let min_1 and max_1 be the minimum and maximum colors used to color the edges of S'_1 in $G_{v'}$.

Lemma 10 *If $max_1 - min_1 \geq 7$, i.e., there exists at least two intermediate colors between min_1 and max_1 which are not used in S_1', then $\lambda_{1,2}'(G_{v'}) \geq 19$.*

Proof There must be at least two unused colors say, $\{c_1, c_2\}$ such that for each $c \in \{c_1, c_2\}$ either $c + 1$ or $c - 1$ is used in S_1'. From Lemma 2, it can be said that each of $c_1, c_2, min_1 - 1$ and $max_1 + 1$ can be used at most once in $G_{v'}$. Observe from discussion of Theorem 5, that for any optimal coloring of $G_{v'}$, each color must be used at least twice in $G_{v'} \setminus S_1'$. Note that at most 4 edges can be colored by c_1, c_2, $min_1 - 1$ and $max_1 + 1$. But for optimal coloring, $c_1, c_2, min_1 - 1$ and $max_1 + 1$ should have been colored at least 8 edges. For those four uncolored edges, at least two additional colors must be required as a color can be used at most thrice in $G_{v'}$. Hence the proof. □

Theorem 6 $\lambda_{1,2}'(T_6) \geq 19$.

Proof Assume that x be such a vertex which is not adjacent to edges colored with any of min or max in G_x. Let us consider G_x is not colored and u, w be two vertices of H_x in G_x. Let us define S_{x1} as the set of edges adjacent to x. Now we consider the following two cases.

- When $w \in N(u)$: u and w are connected by an edge e. Let $\{c_1, \cdots, c_6\}$ and $\{c_1', \cdots, c_6'\}$ be two sequences consisting of consecutive colors are used at the edges incident to u and w respectively. It is possible to assign consecutive colors at those edges when e is colored with either $c_6 = c_1'$ or $c_1 = c_6'$. Now observe two edges e' and e_1' of S_{x1} are already colored and those are not consecutive. Note that $|f'(e') - f'(e_1')| \geq 2$. If $|f'(e') - f'(e_1')| = 2$ then $f'(e')$ and $f'(e_1')$ is neither minimum nor maximum color used in u and w. Then any color of any other edge in S_{x1} is neither consecutive to $f'(e')$ nor $f'(e_1')$. So $max - min \geq 7$ where min and max be the minimum and maximum colors used to color the edges of S_{x1}. If $|f'(e') - f'(e_1')| > 2$, then also $max - min \geq 7$. Therefore from Lemma 10, at least 20 colors are required for G_x. Hence $\lambda_{1,2}'(T_6) \geq \lambda_{1,2}'(G_x) \geq 19$.
- When $w \notin N(u)$: Note that $x \in \{N(u) \cap N(w)\}$. Let two sequences $\{c_1, \cdots, c_6\}$ and $\{c_1', \cdots, c_6'\}$ consisting of consecutive colors are used at the edges incident to u and w respectively. Let uv and wv are e' and e_1' respectively. If $f'(e')$ and $f'(e_1')$ are consecutive then either $f'(e') = c_6$, $f'(e_1') = c_1'$ or $f'(e') = c_1$, $f'(e_1') = c_6'$. Now observe that for any other edge e in S_{x1}, $|f'(e) - f'(e')| > 2$ implying $max - min \geq 7$ where min and max be the minimum and maximum colors used to color the edges of S_{x1}. If $f'(e')$ and $f'(e_1')$ are not consecutive then $|f'(e') - f'(e_1')| \geq 2$. If $|f'(e') - f'(e_1')| = 2$ then the intermediate color must be used at an edge $e \in S_{x1}$. There are still 4 edges remain uncolored. It can be checked that for any coloring of the rest of the graph, there exists a vertex y in $H_x \cup x$, for which $max - min \geq 7$ where min and max be the minimum and maximum colors used to color the edges incident to u. Hence from Lemma 10, at least 20 colors are required for G_x. Hence $\lambda_{1,2}'(T_6) \geq \lambda_{1,2}'(G_x) \geq 19$.

Hence the proof. □

3 Conclusions

In this article we improved some lower and upper bounds for infinite regular hexagonal, square and triangular grids using structural properties of those graphs. An interesting problem will be to improve or introduce new bounds on those graphs for other values of h and k. It would also be interesting to examine similar bounds for other infinite regular grids.

References

1. Calamoneri, T.: The $L(h, k)$-labelling problem: an updated survey and annotated bibliography. Comput. J. **54**(8), 1344–1371 (2011). https://doi.org/10.1093/comjnl/bxr037
2. Calamoneri, T.: Optimal l(j, k)-edge-labeling of regular grids. Int. J. Found. Comput. Sci. **26**(4), 523–535 (2015). https://doi.org/10.1142/S012905411550029X
3. Chen, Q., Lin, W.: L(j, k)-labelings and l(j, k)-edge-labelings of graphs. Ars Comb. **106**, 161–172 (2012)
4. Griggs, J.R., Jin, X.T.: Real number labelings for paths and cycles. Internet Math. **4**(1), 65–86 (2007). https://doi.org/10.1080/15427951.2007.10129140
5. Griggs, J.R., Yeh, R.K.: Labelling graphs with a condition at distance 2. SIAM J. Discrete Math. **5**(4), 586–595 (1992). https://doi.org/10.1137/0405048
6. Hale, W.K.: Frequency assignment: theory and applications. Proc. IEEE **68**(12), 1497–1514 (1980). https://doi.org/10.1109/PROC.1980.11899
7. He, D., Lin, W.: L(1, 2)-edge-labelings for lattices. Appl. Math. J. Chin. Univ. **29**, 230–240 (2014). https://doi.org/10.1007/s11766-014-3176-4
8. Lin, W., Wu, J.: Distance two edge labelings of lattices. J. Comb. Optim. **25**(4), 661–679 (2013). https://doi.org/10.1007/s10878-012-9508-5
9. Roberts, F.: Working group agenda. In: DIMACS/DIMATIA/Renyi Working Group on Graph Colorings and Their Generalizations (2003)

Optimal Tree Decompositions Revisited: A Simpler Linear-Time FPT Algorithm

Ernst Althaus and Sarah Ziegler

Abstract In 1996, Bodlaender showed the celebrated result that an optimal tree decomposition of a graph of bounded treewidth can be found in linear time. The algorithm is based on an algorithm of Bodlaender and Kloks that computes an optimal tree decomposition given a non-optimal tree decomposition of bounded width. Both algorithms, in particular the second, are hardly accessible. We present the second algorithm in a much simpler way in this paper and refer to an extended version for the first. In our description of the second algorithm, we start by explaining how all tree decompositions of subtrees defined by the nodes of the given tree decomposition can be enumerated. We group tree decompositions into equivalence classes depending on the current node of the given tree decomposition, such that it suffices to enumerate one tree decomposition per equivalence class and, for each node of the given tree decomposition, there are only a constant number of classes which can be represented in constant space.

Keywords Tree decompositions · Parametrized complexity · Simplified description

1 Introduction

Tree decompositions and treewidth are important concepts in parameterized complexity and are therefore introduced in many textbooks on graph-theory, graph algorithms, or parameterized complexity, e.g., [4–6, 8]. They are even introduced in the basic algorithms book of Kleinberg and Tardos [7]. Many NP-hard problems are fixed-parameter tractable in the treewidth—i.e., for a graph $G = (V, E)$ of treewidth tw, they can be solved in time $\mathcal{O}(f(tw) \cdot \text{poly}(|V|,|E|))$ for a computable function f and a polynomial poly. A necessary condition for these algorithms is

E. Althaus (✉) · S. Ziegler
Johannes Gutenberg-Universität Mainz, Mainz, Germany
e-mail: ernst.althaus@uni-mainz.de

© The Author(s), under exclusive license to Springer Nature Switzerland AG 2021
C. Gentile et al. (eds.), *Graphs and Combinatorial Optimization:*
from Theory to Applications, AIRO Springer Series 5,
https://doi.org/10.1007/978-3-030-63072-0_6

that a tree decomposition with a similar width as the treewidth of the graph can be computed. In most textbooks, a (rather complicated) algorithm that computes a tree decomposition of width at most $4tw$ is shown.

Bodlaender [2] proved that an optimal tree decomposition of a graph G with fixed treewidth can be computed in linear time. Since its publication, this paper has been cited more than 1500 times. The algorithm, however, does not appear in textbooks, as its presentation is way too complicated. We aim to give a simpler description in order to make the algorithm accessible to a wider audience.

The algorithm is based on an algorithm by Bodlaender and Kloks [3] that computes an optimal tree decomposition in linear time, if a (non-optimal) tree decomposition of fixed width is given. Bodlaender [2] shows how, given a graph G, one can find a graph G' of at most the same treewidth that is a constant factor smaller and it is easy to construct a tree decomposition for G of width $2tw$ from an optimal tree decomposition of G'. The algorithm has linear running time. Together with the algorithm of Bodlaender and Kloks, this gives an exact linear time algorithm:

- Compute G'.
- Compute an optimal tree decomposition of G' recursively.
- Construct a tree decomposition of G of width at most $2tw$ from the tree decomposition of G'.
- Use the algorithm of Bodlaender and Kloks to find the optimal tree decomposition.

Several attempts were made to simplify the construction of an appropriate graph G' (i.e., with the properties mentioned above) from G (see e.g., [5, 9]). Hence, we only show the main ideas of our work on the algorithm of Bodlaender and Klocks in this extended abstract and refer to [1] for more details the overall algorithm.

2 Definitions and Basic Properties

In this Section, we give sketch some definitions of some basic properties, that we will use later in the paper. All definitions are standard and the properties are well known (see, e.g., [4]).

A tree decomposition $(T, (X_t)_{t \in T})$ for a graph G is a tuple of a tree T over some set of vertices $V(T)$ and subsets of vertices $X_t \subseteq V(G)$ of G, one for each vertex in T, such that the following three properties hold:

(Node coverage)   For each $v \in V(G)$ there is at least one $t \in V(T)$ such that $v \in X_t$.
(Edge coverage)   For each $uv \in E(G)$ there is at least one $t \in V(T)$ such that u and v are in X_t.
(Coherence)   If $v \in X_{t_1}$ and $v \in X_{t_2}$ for $v \in V(G)$ and $t_1, t_2 \in V(T)$ then $v \in X_{t_3}$ for all vertices t_3 on the unique path in T from t_1 to t_2.

The sets X_t are called the bags. The width of a tree decomposition $(T, (X_t)_{v \in V(T)})$ is the maximal cardinality of one of the sets X_t minus one, i.e., $\max_{t \in V(T)} |X_t| - 1$. The treewidth of a graph G is the minimal width of a tree decomposition of G.

To make a clearer distinction between the vertices $V(G)$ of the graph and those of the tree of the tree decomposition, we call the elements of $V(T)$ nodes and the elements of $V(G)$ vertices in the following.

If there are bags $t \neq t'$ in a tree decomposition with $X_t \subseteq X_{t'}$, we call the tree decomposition redundant and otherwise non-redundant. There is always a non-redundant tree decomposition of minimal width. The number of nodes of a non-redundant tree decomposition is at most $|V(G)|$.

Choosing an arbitrary node $r \in V(T)$ of $(T, (X_t)_{t \in T})$ as root, we get a rooted tree decomposition with natural parent-child and ancestor-descendant relations. For a node $t \neq r$ let $p(t)$ be the parent of t. For a rooted tree decomposition $(T, (X_t)_{t \in V(t)})$ and $t \in V(T)$, let X_t^+ be the set of all vertices in descendants of t and $G_t^+ = G[X_t^+]$ the induced graph of these vertices.

A rooted tree decomposition $(T, (X_t)_{t \in T})$ with root r is called nice, if the following properties hold:

- $X_r = \emptyset$ and $X_t = \emptyset$ for all leaves t of T, i.e., the bags of all leaves and the root are empty
- Every non-leaf node $t \in V(t)$ is of one of the following three types:

> Join-node: t has exactly two children t_1, t_2 and $X_t = X_{t_1} = X_{t_2}$.
> Introduce-node: t has exactly one child t' and $X_t = X_{t'} \cup \{v\}$ for some vertex $v \in V \setminus X_{t'}$. We say that v is introduced at t.
> Forget-node: t has exactly one child t' and $X_{t'} = X_t \cup \{v\}$ for some vertex $v \in V \setminus X_t$. We say that v is forgotten at t.

Given a tree decomposition $(T, (X_t)_{t \in T})$ of width tw of G, we can compute in time $\mathcal{O}(tw^2(|V(T)| + |V(G)|))$ a nice tree decomposition of G of width tw with at most $\mathcal{O}(tw|V(G)|)$ nodes.

3   Computing an Optimal Tree Decomposition from an Arbitrary One

We want to compute an optimal tree decomposition of a graph $G = (V, E)$. Notice that there is an optimal tree decomposition consisting of at most n nodes. There are only a finite number of tree topologies with at most n nodes. Hence we can enumerate all such topologies and all assignments of subsets of V with a size of at most tw to the nodes and check the three properties of a tree decomposition to compute the optimal tree decomposition. We now want to improve upon this simple algorithm by using a given tree decomposition of G.

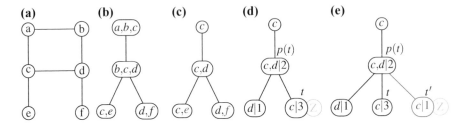

Fig. 1 Consider the graph on the left (**a**). (**b**) Shows an optimal tree decomposition for it. If we remove the vertices a and b from all bags as shown in (**c**), we obtain a tree decomposition for $G[\{c, d, e, f\}]$. Assume that the given tree decomposition has a node $\hat{t}$ with $X_{\hat{t}} = \{c, d\}$ and $X_{\hat{t}}^+ = \{c, d, e, f\}$. Then the tree decompositions enumerated for $\hat{t}$ already covered all edges incident to e and f and hence, it does not matter whether a bag contains these vertices, only the current size of the bag matters. Formally, this is captured by the restricted bags depicted in (**d**) ignoring the green part. We can assume that no node is added to the bottom-right leaf in (**d**) later in the construction, as its restricted bag is contained in the restricted bag of its parent and contains additional nodes. This is as we could add an additional leave only containing the nodes of the restricted bag and the additional nodes as depicted in (**e**)

3.1 Enumerating Tree Decompositions with a Detour

Let $G = (V, E)$ be the given graph and $\hat{\mathcal{T}} = (\hat{T}, (Y_{\hat{t}})_{\hat{t} \in V(\hat{T})})$ be a nice (and hence rooted) tree decomposition of G of width $\hat{tw}$. Both the graph G and the tree decomposition $\hat{\mathcal{T}}$ are given to the algorithm and we fix this notation for the remainder of this section. We assume that G is connected and hence edge coverage implies node coverage.

In order to reduce the number of enumerated tree decompositions, we want to make use of the given (non-optimal) tree decomposition. Notice that given a tree decomposition $\mathcal{T}$ for G, we obtain a tree decomposition for $G_{\hat{t}}^+$ if we intersect all bags of $\mathcal{T}$ with $Y_{\hat{t}}^+$ (see Fig. 1a–c). For all $\hat{t} \in V(\hat{T})$, we enumerate all tree decompositions of $G_{\hat{t}}^+$ of width at most $\hat{tw}$ with at most n nodes bottom up, denoted as $\mathcal{TD}_{\hat{t}}$. Hence, for the root $\hat{r}$ of the given tree decomposition, we enumerate all tree decompositions of width at most $\hat{tw}$ of G and the problem is solved. Moreover, it is easy to enumerate $\mathcal{TD}_{\hat{t}}$ from its children as follows:

Construction 1 (Base Case)

Leaves: For a leaf $\hat{t}$ of $\hat{T}$, the set $\mathcal{TD}_{\hat{t}}$ contains all tree decompositions $(T, (\emptyset)_{t \in V(T)})$ for an arbitrary tree T of at most n nodes.

Join-Node: Consider an arbitrary tree decomposition $\mathcal{T} = (T, (X_t)_{t \in V(T)})$ in $\mathcal{TD}_{\hat{t}}$ for a join-node $\hat{t}$ with children $\hat{t}^1$ and $\hat{t}^2$. Notice that $\mathcal{T}^i = (T, (X_t \cap Y_{\hat{t}^i}^+)_{t \in V(t)}) \in \mathcal{TD}_{\hat{t}^i}$ for $i \in \{1, 2\}$ hence enumerated in $\mathcal{TD}_{\hat{t}^i}$. Furthermore the tree T and $X_t \cap Y_{\hat{t}}$ for all $t \in T$ are the same on all three tree decompositions. Therefore all tree decompositions in $\mathcal{TD}_{\hat{t}}$ can be constructed by choosing tree decompositions $\mathcal{T}^1 = (T^1, (X_t^1)_{t \in V(T^1)}) \in \mathcal{TD}_{\hat{t}^1}$ and $\mathcal{T}^2 =$

$(T^2, (X_t^2)_{t \in V(T^2)}) \in \mathcal{TD}_{\hat{i}^2}$ with the same tree, i.e., $T^1 = T^2$ and such that $X_t^1 \cap Y_{\hat{i}} = X_t^2 \cap Y_{\hat{i}}$ for all $t \in T^1$ and constructing $\mathcal{T} = (T^1, (X_t^1 \cup X_t^2)_{t \in V(T^1)})$. For a join-node $\hat{i}$, we define $J_{\hat{i}} : \mathcal{TD}_{\hat{i}^1} \times \mathcal{TD}_{\hat{i}^2} \to 2^{\mathcal{TD}_{\hat{i}}}$ by $J_{\hat{i}}(\mathcal{T}^1, \mathcal{T}^2) = \{\mathcal{T}\}$ for $\mathcal{T}^1, \mathcal{T}^2$ and $\mathcal{T}$ as above, i.e., $J_{\hat{i}}(\mathcal{T}^1, \mathcal{T}^2)$ is the set containing the tree decomposition constructed from $\mathcal{T}^1$ and $\mathcal{T}^2$ as explained above as the single element. $J_{\hat{i}}$ maps to the power set of all tree decompositions as this is consistent with the corresponding definition for introduce- and forget-nodes.

Introduce-Node: Consider an tree decomposition $\mathcal{T} = (T, (X_t)_{t \in V(T)})$ in $\mathcal{TD}_{\hat{i}}$ for an introduce-node $\hat{i}$ with child $\hat{i}'$ for which the vertex $v \in V$ is introduced. Notice that $\mathcal{T}' = (T, (X_t \setminus \{v\})_{t \in V(T)})$ is a tree decomposition in $\mathcal{TD}_{\hat{i}'}$. Furthermore the nodes t of $V(T)$ such that the bag X_t contains v form a subtree of T. For each edge $vw \in E(G_{\hat{i}}^+)$ adjacent to v, there is at least one bag containing v and w in T. Hence, we can enumerate all tree decompositions in $\mathcal{TD}_{\hat{i}}$ as follows: We choose a tree decomposition $\mathcal{T}'$ in $\mathcal{TD}_{\hat{i}'}$ and a subtree of the tree of $\mathcal{T}'$ with the properties mentioned earlier. For these choices, we construct the tree decomposition in $\mathcal{TD}_{\hat{i}}$ from $\mathcal{T}'$ by adding v into the bags of the selected subtree. We refer to the set that contains all these tree decompositions as $I_{\hat{i}}(\mathcal{T}')$.

Forget-Node: Consider an arbitrary tree decomposition $\mathcal{T} = (T, (X_t)_{t \in V(T)})$ in $\mathcal{TD}_{\hat{i}}$ for a forget-node $\hat{i}$ with child $\hat{i}'$ for which the vertex $v \in V$ is forgotten. Notice that $\mathcal{T}$ is also a tree decomposition in $\mathcal{TD}_{\hat{i}'}$ (as $G_{\hat{i}}^+ = G_{\hat{i}'}^+$) and hence the tree decomposition in $\mathcal{TD}_{\hat{i}}$ are the same as the ones in $\mathcal{TD}_{\hat{i}'}$. We define $F_{\hat{i}} : \mathcal{TD}_{\hat{i}'} \to 2^{\mathcal{TD}_{\hat{i}}}$ by $F_{\hat{i}}(\mathcal{T}) = \{\mathcal{T}\}$.

We do not argue about the running time to enumerate these tree decompositions (since we will be computing something different anyway), but we notice that the number of enumerated tree decompositions is at most $T_n \cdot n^{n \cdot t\hat{w}}$, where T_n is the number of tree topologies with at most n nodes.

This estimate of the number of enumerated tree decompositions is a function in n and $t\hat{w}$ and not only in $t\hat{w}$. We now reduce the number of enumerated tree decompositions to make it a function in $t\hat{w}$ only. We show that it suffices to store some limited information for a tree decomposition $\mathcal{T}$ that is enumerated at a node $\hat{i}$ of the given tree decomposition $\hat{\mathcal{T}}$: We store only a part of the tree of $\mathcal{T}$ and only those vertices in the bags of $\mathcal{T}$ that are in $Y_{\hat{i}}$ of $\hat{\mathcal{T}}$. Hence, several tree decompositions have the same limited information. We write $\mathcal{T}^1 \equiv_s^{\hat{i}} \mathcal{T}^2$ if two tree decompositions have the same limited information and call them equivalent. Before defining $\equiv_s^{\hat{i}}$, we will define finer classes $\equiv_b^{\hat{i}}$ and $\equiv_c^{\hat{i}}$, i.e., we iteratively show that we can ignore some information in the construction of the tree decompositions.

The classes will be such that equivalent tree decompositions have the same treewidth and if one can use one tree decomposition to construct a certain tree decomposition in the parent node of the given tree decomposition, we can use each equivalently to construct an equivalent one. In order to compensate for the removal of nodes of the tree, we allow to add additional leaves whose bags are subsets of the parental bags and copy certain nodes of degree two. The set of all tree decompositions that can be constructed from the limited information of $\mathcal{T}$ is denoted by $\mathcal{E}^{\hat{i}}(\mathcal{L}^{\hat{i}}(\mathcal{T}))$. More formally, the following will hold for the equivalence class $\equiv_s^{\hat{i}}$ and similarly for the classes $\equiv_b^{\hat{i}}$ without possible modifications of the tree decomposition $\mathcal{T}$ and $\equiv_c^{\hat{i}}$ by allowing to add additional leaves to $\mathcal{T}$.

Lemma 1

Join-node: Let $\hat{i}$ be a join-node with child nodes $\hat{i}^1$ and $\hat{i}^2$, $A, A' \in \mathcal{TD}_{\hat{i}^1}$ with $A \equiv_b^{\hat{i}^1} A'$ and $B, B' \in \mathcal{TD}_{\hat{i}^2}$ with $B \equiv_b^{\hat{i}^2} B'$. If $C \in J_{\hat{i}}(E^{\hat{i}}(\mathcal{L}^{\hat{i}}(A)), E^{\hat{i}}(\mathcal{L}^{\hat{i}}(B)))$ then there is $C' \in J_{\hat{i}}(E^{\hat{i}}(\mathcal{L}^{\hat{i}}(A')), E^{\hat{i}}(\mathcal{L}^{\hat{i}}(B')))$ with $C' \equiv_b^{\hat{i}} C$.

Introduce-node: Let $\hat{i}$ be an introduce-node with child node $\hat{i}'$ and $A, A' \in \mathcal{TD}_{\hat{i}'}$ with $A \equiv_b^{\hat{i}'} A'$. If $C \in I_{\hat{i}}(E^{\hat{i}}(\mathcal{L}^{\hat{i}}(A)))$ then there is $C' \in I_{\hat{i}}(E^{\hat{i}}(\mathcal{L}^{\hat{i}}(A')))$ with $C' \equiv_b^{\hat{i}} C$.

Forget-node: Let $\hat{i}$ be a forget-node with child node $\hat{i}'$ and $A, A' \in \mathcal{TD}_{\hat{i}'}$ with $A \equiv_b^{\hat{i}'} A'$. If $C \in F_{\hat{i}}(E^{\hat{i}}(\mathcal{L}^{\hat{i}}(A)))$ then there is $C' \in F_{\hat{i}}(E^{\hat{i}}(\mathcal{L}^{\hat{i}}(A')))$ with $C' \equiv_b^{\hat{i}} C$.

Proof As we will see, the limited information of the tree decompositions in $J_{\hat{i}}(E^{\hat{i}}(\mathcal{L}^{\hat{i}}(A)), E^{\hat{i}}(\mathcal{L}^{\hat{i}}(B)))$, $I_{\hat{i}}(E^{\hat{i}}(\mathcal{L}^{\hat{i}}(A)))$ and $F_{\hat{i}}(E^{\hat{i}}(\mathcal{L}^{\hat{i}}(A)))$ can be computed from the limited information of A (and B). Therefore, if we replace A by A' (and B by B') in the construction, the construction will lead to tree decompositions with the same limited information. □

3.2 Equivalence Classes on Bags

Given a tree decomposition $\mathcal{T} = (T, \{X_t\}_{t \in V(T^1)})$, we define the restricted bags with respect to $\hat{i}$ to be the tuples $(X_t \cap Y_{\hat{i}}, |X_t|)_{t \in V(T)}$, i.e., the vertices of the bag are restricted to the vertices of the current bag of the given tree decomposition and we store the sizes of the bags. The restricted bag representation of a tree decomposition $\mathcal{T} = (T, (X_t)_{t \in V(T)})$ consists of the tree T and of the restricted bags for all $t \in V(T)$. We call tree decompositions $\mathcal{T}^1 = (T^1, \{X_t^1\}_{t \in V(T^1)})$ and $\mathcal{T}^2 = (T^2, \{X_t^2\}_{t \in V(T^2)})$ in $\mathcal{TD}_{\hat{i}}$ bag-equivalent, if their restricted bag representations are the same and write $\mathcal{T}^1 \equiv_b^{\hat{i}} \mathcal{T}^2$ in his case.

We will not show the construction of $J^{\hat{i}}$, $I^{\hat{i}}$ and $F^{\hat{i}}$ but give the main points. For a join-node, the vertices of the restricted bags have to be the same and the size can easily be computed out of the sizes of the children. For an introduce-node, we note that the validity of a chosen subtree for the new vertex only depends on the vertices that are contained in the restricted bags of the given tree decomposition. For a forget-node, we simply remove the forgotten vertex from all restricted bags.

3.3 The Core of a Tree Decomposition

In this section, we will use the following fact about tree decompositions: If t is a node with bag X_t, we can add a new node t', the edge tt' and chose $X_{t'} \subseteq X_t$ to obtain another tree decomposition of the same width. We call this an addition of a leaf to the tree decomposition.

In the following, we will show that we can assume some structure of the constructed optimal tree decomposition which depends on the given tree decomposition $\hat{\mathcal{T}} = (\hat{T}, (Y_{\hat{t}})_{\hat{t} \in V(\hat{T})})$.

Lemma 2 *Assume t is a leaf with adjacent node $p(t)$ in the constructed tree decomposition $\mathcal{T} = (T, (X_t)_{t \in V(t)})$ such that $X_t \setminus Y_{\hat{t}} \neq \emptyset$ and $X_t \cap Y_{\hat{t}} \subseteq X_{p(t)} \cap Y_{\hat{t}}$ for some node $\hat{t} \in V(\hat{T})$. Then there is an optimal tree decomposition such that no node from $V \setminus Y_{\hat{t}}^+$ is contained in X_t.*

Proof Let the notation be given as stated in the lemma. Assume $Z := X_t \cap (V \setminus Y_{\hat{t}}^+)$ is non-empty. We can remove all vertices in Z from X_t and create a new child of $p(t)$ with bag $Z \cup (X_t \cap Y_{\hat{t}}^+)$ to get another tree decomposition. This transformation cannot increase with the width of the tree decomposition and all edges are still covered. □

Informally, a leaf with a bag whose vertices are a subset of the vertices of its parent gets only additional vertices later on in the construction if it does not contain vertices that are already forgotten (see Fig. 1d, e). Notice that if a node becomes a leaf by the removal of a node, the rule can be applied to this node too.

Given a tree decomposition $\mathcal{T} = (T, \{X_t\}_{t \in V(T)}) \in \mathcal{TD}_{\hat{t}}$ we define its $\hat{t}$-core as follows: As long as there is a leaf t in T such that $X_t \cap Y_{\hat{t}} \subseteq X_{p(t)} \cap Y_{\hat{t}}$ remove the leaf from t where $p(t)$ is the node adjacent to t (see Fig. 2a and b). Since it is possible to remove the largest bag in this way, we store the size of the largest bag with the core. We call two tree decompositions $\mathcal{T}^1$ and $\mathcal{T}^2$ $\hat{t}$-core-equivalent if their cores are identical (including the restricted bags of the nodes that were not removed from the core and the size of the largest bag), denoted by $\mathcal{T}^1 \equiv_c^{\hat{t}} \mathcal{T}^2$. In the following, it is easier if we can assume that $Y_{\hat{t}} \neq \emptyset$ and hence, we start the construction with the parents $p(\hat{t})$ of the leaves $\hat{t}$. Their cores consist of a single node whose bag contains the single vertex in $Y_{p(\hat{t})}$.

In order to compensate this removal of bags, we have to allow the adding of additional leaves in the construction. By Lemma 2, we can assume that the bag of a leaf connected to a node t that contains a node from $V \setminus Y_{\hat{t}}^+$ does not contain nodes from $Y_{\hat{t}}^+ \setminus X_t$. For a tree decomposition $\mathcal{T}$ let $\mathcal{L}^{\hat{t}}(\mathcal{T})$ be the (infinite) set of all tree decompositions that can be obtained by iteratively adding leaves to nodes of the $\hat{t}$-core of $\mathcal{T}$, whose bags are subsets of $Y_{\hat{t}}$. $\mathcal{L}^{\hat{t}}$ depends on the node $\hat{t}$ of the given tree decomposition as we have to start the addition on a node of the core. Notice that all elements of $\mathcal{L}^{\hat{t}}(\mathcal{T})$ have the same core as $\mathcal{T}$.

Notice that each leaf of the core has a unique vertex as otherwise each vertex of the leaf would also be contained in its adjacent node and hence the leaf can

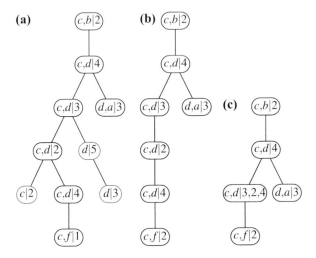

Fig. 2 Assume we consider a tree with its restricted bags as shown in (**a**). The nodes shown in red are removed from the core (depicted in (**b**)) as they are leaves and their restricted bags are subsets of the bags of their adjacent nodes. Notice that we remove the largest bag and hence we store its size with the core. In (**c**) we show the compact representation of the core, i.e., the path of nodes of degree 2 having equal vertices in the restricted bag are replaced by a single node and the sizes of the original nodes are represented as an integer sequence in the replacement

be removed. Hence, the number of leaves is at most $t\hat{w}$. Furthermore, the number of nodes of the core of degree at least three is at most $t\hat{w}$ as the tree has at most $t\hat{w}$ leaves. In the following, we give up assumption that the constructed tree decomposition has at most n nodes, but we assume that each path of nodes of degree 2 in the core has a length of at most n.

We will not show the construction for $J^{\hat{i}}(\mathcal{L}^{\hat{i}}(A), \mathcal{L}^{\hat{i}}(B))$, $I^{\hat{i}}(\mathcal{L}^{\hat{i}}(A))$ and $F^{\hat{i}}(\mathcal{L}^{\hat{i}}(A))$, but give the main points. For a join-node, we note that if we can join two tree decompositions, their cores must be the same. For an introduce-node we have to distinguish two cases. If the vertex is added to at least one bag of the core, the core does not change. Otherwise the new vertex will be in exactly one bag of the new core, which is connected to the old core by a single path which we removed then constructing the core. Hence, we have to try all possibilities of adding path of nested subsets of length at most n. For a forget-node, we potentially have to remove further nodes from the core.

3.4 The Compressed Core of a Tree Decomposition

In the following, it is easier to consider a different representation of the cores. We replace each maximal path of nodes in the tree of degree 2 such that the vertices in the restricted bags are the same by a single node, and assign the sequence of

integers of the sizes of the bags of the path with the new node (see Fig. 2c). We
call this the compact representation of the core. Notice that in this representation,
the number of nodes becomes bounded in $t\hat{w}$ since between two nodes t and t' of
degree at least three, we have at most $|X_t \setminus X_{t'}| + |X_{t'} \setminus X_t| \leq 2t\hat{w}$ nodes of degree
two. On the other hand, each node has an integer sequence assigned to it of length
up to n. We define equivalence classes on integer sequences such that the number
of equivalence classes becomes bounded in $t\hat{w}$. We call two tree decompositions
$\mathcal{T}^1$ and $\mathcal{T}^2$ sequence-equivalent, if the compact representations of their cores are
the same and each pair of integer sequences assigned to a node in the compact
representation of the core are equivalent, denoted as $\mathcal{T}^1 \equiv_s^{\hat{i}} \mathcal{T}^2$.

In this section, we use the following fact about tree decompositions: Assume that
$\mathcal{T} = (T, (X_t)_{t \in V(T)})$ is a tree decomposition of G and tt' is an edge of T. If we
add a bag t'' with $X_{t''} = X_t$ to T, remove the edge tt' and add edges tt'' and $t''t'$,
we get another tree decomposition of G of the same width. We call the resulting
tree decomposition an extension of $\mathcal{T}$. In the following, we restrict this operation to
edges tt' such that the degree of t is two in the core of $\mathcal{T}$. For a tree decomposition $\mathcal{T}$
let $\mathcal{E}^{\hat{i}}(\mathcal{T})$ be the set of all tree decomposition that can be obtained from $\mathcal{T}$ by a series
of such extensions. Notice that although $\mathcal{E}^{\hat{i}}(\mathcal{T})$ is infinite, we will define a finite
number of equivalence classes. Notice that the tree of the compact representation of
the core is the same for each element of $\mathcal{E}^{\hat{i}}(\mathcal{L}^{\hat{i}}(\mathcal{T}))$, but the integer sequences differ.

In the next lemma we show that we may assume that a vertex added to a subpath
in $\mathcal{T}$ of $t_1, \ldots, t_k$ with integer sequence $(a_1, \ldots, a_k)$ is added up to a node t_ℓ such
that there are no $i \leq \ell \leq j$ with $\min(a_i, a_j) < a_\ell < \max(a_i, a_j)$.

Lemma 3 *Consider a node $\hat{i}$ of the given tree decomposition and the compact
representation of the core of a tree decomposition in $\mathcal{TD}_i$. Assume $(a_1, \ldots, a_k)$
is the integer sequence of a node and $i < j$ are such that $a_i \geq a_k$ and $a_j \leq a_k$ for
all $i \leq k \leq j$. Let ℓ be the smallest index such that $a_\ell > a_i$ (if such an ℓ exists,
set $\ell = j$ otherwise). Let $Z_i, \ldots, Z_j$ be the vertices of $V \setminus Y_{\hat{i}}^+$ of bags of the nodes
corresponding to $a_i, \ldots, a_j$ in the final tree decomposition. There is an optimal tree
decomposition such that $Z_{\ell'} = Z_j$ for all $\ell \leq \ell' \leq j$ and no edges adjacent to $a_{\ell'}$
are added to the core later in the construction. Similarly, if $a_i \leq a_k$ and $a_j \geq a_k$,
we can assume $Z_{\ell'} = Z_i$ and no edges adjacent to $a_{\ell'}$ are added to core for all
$i \leq \ell' \leq \ell$, for ℓ being the largest index such that $a_\ell > a_j$.*

Proof Consider an arbitrary tree decomposition $\mathcal{T}$ and assume it does not have the
property. Let t be a node of the compact representation of the core for which the
property does not hold for the subsequence $(a_i, \ldots, a_j)$ of its integer sequence and
let $t_i, \ldots, t_j$ the nodes of the tree of $\mathcal{T}$ corresponding to $a_i, \ldots, a_j$. Notice that we
can assume that the edges adjacent to the nodes corresponding to $t_i, \ldots, t_j$ have
bags that are either subsets of $Y_{\hat{i}}^+ \cap X_t$ or subsets of $(V \setminus Y_{\hat{i}}^+) \cup X_t$ by Lemma 2.

We extend t_i, i.e., the node with the smallest bag, $j - i + 1$ times, i.e., instead of a
single node t_i for size a_i, we now have nodes $t_i^1, \ldots, t_i^{j-i+1}$, all of size a_i. We move
the sets $Z_i, \ldots, Z_j$ to the nodes $t_i^1, \ldots, t_i^{j-i+1}$ and with them all edges whose other

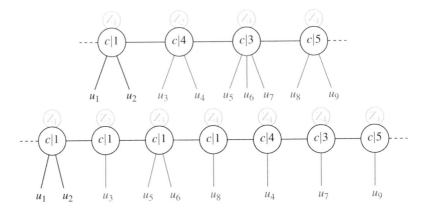

Fig. 3 Assume we have a path of four nodes in the core such that the sets of vertices of the restricted bags are equal (and hence these and possible further nodes are replaced by a single node in the restricted core), say X. Assume furthermore that the first node has the smallest number of vertices and the last has the largest number, and that the sets of vertices $Z_1, \ldots, Z_4$ are added to these nodes later on in the construction. We proved that we may assume that $Z_2 = Z_3 = Z_4$. Otherwise, we could add three copies of the first node in the beginning and give the first three nodes the additional vertices Z_1, Z_2 and Z_3 and all remaining nodes the additional vertices Z_4. By Lemma 2, the bags of nodes incident to the nodes may be assumed to be either subsets of $X \cup Y_i^+$ (shown in red) or subsets of $X \cup (V \setminus Y_i^+)$ (shown in blue). The first kind of edges stay at their original positions whereas the second kind are moved to the front

incident nodes have bags are subsets of $(V \setminus Y_i^+) \cup X_t$. The bags of nodes $t_2, \ldots, t_j$ are all set to Z_j (see Fig. 3 for an illustration).

This results in a tree decomposition of at most the same width (The bags of t_i and t_j are not changed. For the new nodes t_i^k, we have that $|X_{t_i^k}|$ at most the size of bag of the node that contained the assigned Z_ℓ in the originating tree decomposition. The bags of the nodes $t_{i+1}, \ldots, t_{j-1}$ have at most the size of that of t_j.) that has the property for $a_i, \ldots, a_j$. We can repeat this construction until all such conditions are satisfied.   □

Notice furthermore that if we add a vertex v up to the node t_ℓ, we can first extend t_ℓ and add only up to the first appearance of t_ℓ such that after the insertion, the first node not containing v is the copy of t_ℓ and still has size a_ℓ.

For an integer sequence $A = (a_1, \ldots, a_k)$, we define its typical sequence $\tau(A)$ as follows: Apply one of the following two operations until no further such operation is possible (we will show that the resulting sequence is unique and hence it is indeed a definition)

- Remove duplicates: if $a_i = a_{i+1}$ for $1 \leq i \leq k - 1$, replace $(a_1, \ldots, a_k)$ by $(a_1, \ldots, a_i, a_{i+2}, \ldots, a_k)$
- Delete dominated parts: for $1 \leq i < j \leq k$ with $min(a_i, a_j) \leq a_\ell \leq max(a_i, a_j)$ for all $i \leq \ell \leq j$, replace $(a_1, \ldots, a_k)$ by $(a_1, \ldots, a_i, a_j, \ldots, a_k)$.

In [1] we show that $\tau(A)$ is a unique subsequence of A (Property P1) and there are at most $2 \cdot 2^{2t\hat{w}}$ different typical sequences (Property P2). Hence, there are only a bounded number of typical sequences if their integers are bounded by $t\hat{w}$. We will call two integer sequences equivalent, if their typical sequences are the same. An extension of an integer sequence $(a_1, \ldots, a_n)$ is a sequence of the form $(a_1, \ldots, a_i, a_i, \ldots, a_n)$ for $1 \leq i \leq n$, i.e., one integer is repeated. The set of all extensions of an integer sequence A is denoted as $\mathcal{E}(A)$. Notice that an extension of a tree decomposition corresponds to an extension of the integer sequence assigned to one of the nodes of the compact core. Notice that all extensions $A^{\mathcal{E}} \in \mathcal{E}(A)$ of A have the same typical sequence.

Let $\mathcal{T}, \mathcal{T}' \in TD_{\hat{i}}$ with $\mathcal{T}' \in \mathcal{E}^{\hat{i}}(\mathcal{T})$ and assume that we can use $\mathcal{T}$ in the construction of a tree decomposition of G of width tw. Notice that we can also use $\mathcal{T}'$ to construct such a tree decomposition $\mathcal{T}''$ (if we do not restrict to tree decompositions of at most n nodes) since we can extend all tree decompositions used in the construction of $\mathcal{T}''$ at the same nodes as we extended $\mathcal{T}$ to obtain $\mathcal{T}'$. We call an integer sequence A superior to B if there are $A' \in \mathcal{E}(A)$ and $B' \in \mathcal{E}(B)$ such that $A' \leq B'$. In [1] we show that A is superior to $\tau(A)$ (Property (P3)) and that $\tau(A)$ is superior to A (Property (P4)). Hence, if we constructed a tree decomposition for which a node of the compact representation has the integer sequence A, we can assume that we have a tree decomposition with integer sequence $\tau(A)$ (even if no such tree decomposition exists) since for all tree decompositions that can be constructed from the latter, we can construct a tree decomposition of the same width as the first.

Finally, we show in [1] that we can compute all typical sequences that can arise when adding up two extensions of two typical sequences (Property (P5)).

The compressed core of a tree decomposition is the compact representation of its core, where each node is assigned its typical sequence together with the size of the largest bag. We call two tree decompositions sequence-equivalent, denoted as $\equiv_s^{\hat{i}}$, if their compressed cores are the same.

In order to compensate the replacement of an integer sequence by its typical sequence, we allow arbitrary extensions of the compressed core.

We will not show how the sets $J^{\hat{i}}(\mathcal{E}^{\hat{i}}(\mathcal{L}^{\hat{i}}(A)), \mathcal{E}^{\hat{i}}(\mathcal{L}^{\hat{i}}(B)))$, $I^{\hat{i}}(\mathcal{E}^{\hat{i}}(\mathcal{L}^{\hat{i}}(A)))$ and $F^{\hat{i}}(\mathcal{E}^{\hat{i}}(\mathcal{L}^{\hat{i}}(A)))$ are constructed but give the main points. First, we note that we can consider the integer sequences assigned to the nodes of the compact core independently. For a join-node, we have to compute all typical sequences of integer sequences obtained by adding two extensions of the given typical sequences which can be done by (P5). For an introduce-node, use the Lemma 3 by which we can assume that the vertex is either added to all nodes of the integer sequence or it is added up to a node which is contained in the typical sequence and this node can be assumed to be extended before. Hence, we can perform the necessary operations directly on the typical sequences. In a forget-node, we potentially have to join typical sequences if restricted bags become equal. Notice that the compressed cores of the elements in $J^{\hat{i}}(\mathcal{E}^{\hat{i}}(\mathcal{L}^{\hat{i}}(A)), \mathcal{E}^{\hat{i}}(\mathcal{L}^{\hat{i}}(B)))$, $I^{\hat{i}}(\mathcal{E}^{\hat{i}}(\mathcal{L}^{\hat{i}}(A)))$ and $F^{\hat{i}}(\mathcal{E}^{\hat{i}}(\mathcal{L}^{\hat{i}}(A)))$

can be directly computed from the compressed cores of A (and B) and hence Lemma 1 holds for $\equiv_s^l$.

For proof of the necessary properties of typical sequences, we refer to the extended version of this paper[1]. Furthermore, we estimate the number of compact cores enumerated for each bag and the running time in that paper. Here, we only remark that it is not hard to see that this number and the running time can be bounded by a function computable from $\hat{tw}$.

4　Conclusion

We gave simpler descriptions of two algorithms: The algorithm of Bodlaender and Kloks [3] that computes an optimal tree decomposition for a graph G given a (non-optimal) tree decomposition of bounded width in linear time, and the algorithm of Bodlaender [2] that uses the first algorithm to compute an optimal tree decomposition of a graph with bounded treewidth in linear time.

Although we were able to shorten the text significantly, the description is still too long to become part of textbooks. We hope that even simpler descriptions of the algorithms can be found in the future, which will finally allow these algorithms to be shown in textbooks.

References

1. Althaus, E., Ziegler, S.: Optimal tree decompositions revisited: a simpler linear-time FPT algorithm. CoRR 1912.09144 (2019)
2. Bodlaender, H.L.: A linear-time algorithm for finding tree-decompositions of small treewidth. SIAM J. Comput. **25**(6), 1305–1317 (1996)
3. Bodlaender, H.L., Kloks, T.: Efficient and constructive algorithms for the pathwidth and treewidth of graphs. J. Algorithms **21**(2), 358–402 (1996)
4. Cygan, M., Fomin, F.V., Kowalik, L., Lokshtanov, D., Marx, D., Pilipczuk, M., Pilipczuk, M., Saurabh, S.: Parameterized Algorithms. Springer, Berlin (2015)
5. Downey, R.G., Fellows, M.R.: Fundamentals of Parameterized Complexity. Texts in Computer Science. Springer, Berlin (2013)
6. Flum, J., Grohe, M.: Parameterized Complexity Theory. Texts in Theoretical Computer Science. An EATCS Series. Springer, Berlin (2006)
7. Kleinberg, J.M., Tardos, É.: Algorithm Design. Addison-Wesley, Boston (2006)
8. Niedermeier, R.: Invitation to Fixed-Parameter Algorithms. Oxford Lecture Series in Mathematics and Its Applications. Oxford University Press, Oxford (2006)
9. Perkovic, L., Reed, B.A.: An improved algorithm for finding tree decompositions of small width. Int. J. Found. Comput. Sci. **11**(3), 365–371 (2000)

On Superperfection of Edge Intersection Graphs of Paths

Hervé Kerivin and Annegret Wagler

Abstract The routing and spectrum assignment problem in flexgrid elastic optical networks can be modeled in two phases: a selection of paths in the network and an interval coloring problem in the edge intersection graph of these paths. The interval chromatic number equals the smallest size of a spectrum such that a proper interval coloring is possible, the weighted clique number is a natural lower bound. Graphs where both parameters coincide for all possible non-negative integral weights are called superperfect. We examine the question which minimal non-superperfect graphs can occur in the edge intersection graphs of paths in different underlying networks. We show that for any possible network (even if it is restricted to a path) the resulting edge intersection graphs are not necessarily superperfect and discuss some consequences.

Keywords Routing and spectrum assignment problem · Edge intersection graph of paths · Interval coloring · Superperfection

1 Introduction

Flexgrid elastic optical networks constitute a new generation of optical networks in response to the sustained growth of data traffic volumes and demands in communication networks. In optical networks, light is used as communication medium between sender and receiver nodes, and the frequency spectrum of an optical fiber is divided into narrow frequency slots of fixed spectrum width. Any sequence of consecutive slots can form a channel that can be switched in the network to create a lightpath (i.e., an optical connection represented by a route and a channel). The *routing and spectrum assignment (RSA) problem* consists of establishing the lightpaths for a set of end-to-end traffic demands, that is, finding a

H. Kerivin · A. Wagler (✉)
LIMOS (UMR 6158 CNRS), Université Clermont Auvergne, Clermont-Ferrand, France
e-mail: herve.kerivin@uca.fr; annegret.wagler@uca.fr

© The Author(s), under exclusive license to Springer Nature Switzerland AG 2021
C. Gentile et al. (eds.), *Graphs and Combinatorial Optimization:
from Theory to Applications*, AIRO Springer Series 5,
https://doi.org/10.1007/978-3-030-63072-0_7

route and assigning an interval of consecutive frequency slots for each demand such that the intervals of lightpaths using a same edge in the network are disjoint, see e.g. [17]. Thereby, the following constraints need to be respected when dealing with the RSA problem:

1. *spectrum continuity*: the frequency slots allocated to a demand remain the same on all the edges of a route;
2. *spectrum contiguity*: the frequency slots allocated to a demand must be contiguous;
3. *non-overlapping spectrum*: a frequency slot can be allocated to at most one demand.

The RSA problem has started to receive a lot of attention over the last few years. It has been shown to be NP-hard [3, 18]. In fact, if for each demand the route is already known, the RSA problem reduces to the so-called *spectrum assignment (SA) problem* and only consists of determining the demands' channels. Even the SA problem has been shown to be NP-hard on paths [16].

More formally, for the RSA problem, we are given an optical network G and a set $\mathscr{D}$ of end-to-end traffic demands where each demand is specified by a pair u, v of distinct nodes in G and the number d_{uv} of required frequency slots. The routing part of the RSA problem consists of selecting a route through G from u to v, i.e. a (u, v)-path P_{uv} in G, for each such traffic demand. The spectrum assignment can then be interpreted as an *interval coloring* of the *edge intersection graph* $I(\mathscr{P})$ of the set $\mathscr{P}$ of selected paths:

- Each path $P_{uv} \in \mathscr{P}$ becomes a node of $I(\mathscr{P})$ and two nodes are joined by an edge if the corresponding paths in G are in conflict as they share an edge (notice that we do not care whether they share nodes).
- Any interval coloring in this graph $I(\mathscr{P})$ weighted with the demands d_{uv} correctly solves the spectrum assignment: we assign a frequency interval of d_{uv} consecutive frequency slots (*spectrum contiguity*) to every node of $I(\mathscr{P})$ (and, thus, to every path $P_{uv} \in \mathscr{P}$ (*spectrum continuity*)) in such a way that the intervals of adjacent nodes are disjoint (*non-overlapping spectrum*).

Let $\mathbf{d} \in \mathbb{Z}_{+}^{|\mathscr{D}|}$ be the vector whose entries d_{uv} are the slot requirements associated with the demands between pairs u, v of nodes in $\mathscr{D}$. The *interval chromatic number* $\chi_I(I(\mathscr{P}), \mathbf{d})$ is the minimum spectrum width such that $I(\mathscr{P})$ weighted with the vector $\mathbf{d}$ of traffic demands d_{uv} for each path P_{uv} has a proper interval coloring. Given G and $\mathscr{D}$, the minimum spectrum width of any solution of the RSA problem, thus, equals

$$\chi_I(G, \mathscr{D}) = \min\{\chi_I(I(\mathscr{P}), \mathbf{d}) : \mathscr{P} \text{ possible routing of demands } \mathscr{D} \text{ in } G\}.$$

For each routing $\mathscr{P}$, the *weighted clique number* $\omega(I(\mathscr{P}), \mathbf{d})$, also taking the traffic demands d_{uv} as weights, equals the weight of a heaviest clique in $I(\mathscr{P})$ and is a natural lower bound for $\chi_I(I(\mathscr{P}), \mathbf{d})$ (as clearly the intervals of all nodes in a clique in $I(\mathscr{P})$ have to be disjoint by construction of $I(\mathscr{P})$). However, it is not always possible to find a solution with this lower bound as spectrum width, as weighted clique number and interval chromatic number are not always equal.

Graphs where weighted clique number and interval chromatic number coincide for all possible non-negative integral weights are called *superperfect*.

A graph is *perfect* if and only if this holds for every $(0, 1)$-weighting **d** of its nodes. According to a characterization achieved by Chudnovsky et al. [4], perfect graphs are precisely the graphs without chordless cycles C_{2k+1} with $k \geq 2$, termed *odd holes*, or their complements, the *odd antiholes* $\overline{C}_{2k+1}$ (the complement $\overline{G}$ has the same nodes as G, but two nodes are adjacent in $\overline{G}$ if and only if they are non-adjacent in G).

In particular, every superperfect graph is perfect.

On the other hand, comparability graphs form a subclass of superperfect graphs. A graph $G = (V, E)$ is *comparability* if and only if there exists a partial order $\mathscr{O}$ on $V \times V$ such that $uv \in E$ if and only if u and v are comparable w.r.t. $\mathscr{O}$. Hoffman [12] proved that every comparability graph is superperfect. Gallai [6] characterized comparability graphs by giving a complete list of minimal non-comparability graphs, that are

- odd holes C_{2k+1} for $k \geq 2$ and antiholes $\overline{C}_n$ for $n \geq 6$,
- the graphs J_k and J'_k for $k \geq 2$ and the graphs J''_k for $k \geq 3$ (see Fig. 1),
- the complements of D_k for $k \geq 2$ and of E_k, F_k for $k \geq 1$ (see Fig. 2),
- the complements of $A_1, \ldots, A_{10}$ (see Fig. 3).

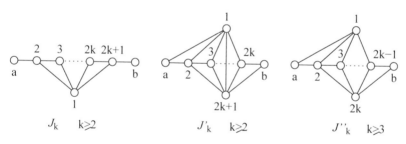

Fig. 1 Minimal non-comparability graphs: J_k, J'_k for $k \geq 2$ and J''_k for $k \geq 3$

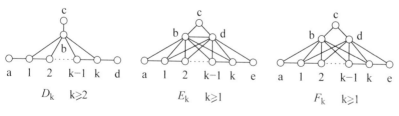

Fig. 2 Minimal non-comparability graphs: the complements of D_k, E_k, F_k

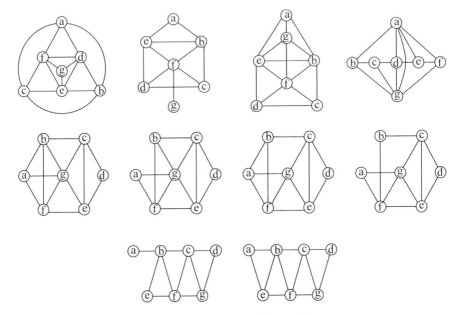

Fig. 3 Minimal non-comparability graphs: the graphs $\overline{A}_1, \ldots, \overline{A}_{10}$

As comparability graphs form a subclass of superperfect graphs, we have that every non-superperfect graph is in particular non-comparability, which raises the question which minimal non-comparability graphs are also minimal non-superperfect. Clearly, odd holes and odd antiholes are minimal non-superperfect (as they are minimal non-perfect). It has been shown by Golumbic [7] that $\overline{A}_1$, $\overline{D}_2$, $\overline{E}_1$, $\overline{E}_2$ and J_2 are non-superperfect, but that there are also superperfect non-comparability graphs such as e.g. even antiholes $\overline{C}_{2k}$ for $k \geq 3$.

Furthermore, Andreae showed in [1], that the graphs J_k'' for $k \geq 3$ and the complements of $A_3, \ldots, A_{10}$ are superperfect, but that the graphs J_k for $k \geq 2$ and J_k' for $k \geq 3$ as well as the complements of D_k for $k \geq 2$ and of E_k, F_k for $k \geq 1$ are non-superperfect.

Note that Andreae wrongly determined $\overline{A}_2$ as superperfect which is, in fact, not the case (see Fig. 4 for a weight vector $\mathbf{d}$ and an optimal interval coloring showing that $\omega(\overline{A}_2, \mathbf{d}) = 5 < 6 = \chi_I(\overline{A}_2, \mathbf{d})$ holds). Moreover, Andreae wrongly determined J_2' as non-superperfect which is, in fact, not the case:

Lemma 1 J_2' *is a superperfect graph.*

Hence, all the previous results together imply the following:

Corollary 1 *The following minimal non-comparability graphs are also minimal non-superperfect:*

- $\overline{A}_1$ *and* $\overline{A}_2$,
- *odd holes* C_{2k+1} *and odd antiholes* $\overline{C}_{2k+1}$ *for* $k \geq 2$,

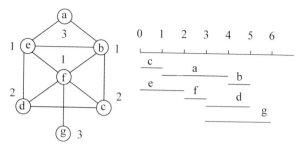

Fig. 4 The graph $\overline{A}_2$ together with node weights $\mathbf{d}$ and an optimal interval coloring showing $\omega(\overline{A}_2, \mathbf{d}) = 5 < 6 = \chi_I(\overline{A}_2, \mathbf{d})$

- *the graphs J_k for $k \geq 2$ and J'_k for $k \geq 3$ as well as*
- *the complements of D_k for $k \geq 2$ and of E_k, F_k for $k \geq 1$.*

Note that we have $\omega(G, \mathbf{1}) < \chi_I(G, \mathbf{1})$ with $\mathbf{1} = (1, \ldots, 1)$ if G is an odd hole or an odd antihole (as they are not perfect), whereas the other minimal non-comparability non-superperfect graphs are perfect and, thus, $\omega(G, \mathbf{d}) < \chi_I(G, \mathbf{d})$ is attained for some $\mathbf{d} \neq \mathbf{1}$ (see Fig. 4).

We examine, for different underlying networks G, the question whether or not there is a solution of the RSA problem with

$$\omega(G, \mathscr{D}) = \min\{\omega(I(\mathscr{P}), \mathbf{d}) : \mathscr{P} \text{ possible routing of demands } \mathscr{D} \text{ in } G\}$$

as spectrum width which depends on the occurrence of (minimal) non-superperfect graphs in the edge intersection graphs $I(\mathscr{P})$.

Note that for some networks G, the edge intersection graphs form well-studied graph classes: if G is a path (resp. tree, resp. cycle), then $I(\mathscr{P})$ is an *interval graph* (resp. *EPT graph*, resp. *circular-arc graph*). However, if G is a sufficiently large grid, then it is known by Golumbic et al. [9] that $I(\mathscr{P})$ can be *any* graph. Modern optical networks do not fall in any of these classes, but are 2-connected, sparse planar graphs with small maximum degree with a grid-like structure.

We first study the cases when the underlying network G is a path, a tree or a cycle (see Sects. 2–4). We recall results on interval graphs, EPT graphs and circular-arc graphs from [5, 8, 14] and then discuss which minimal non-comparability non-superperfect graphs can occur. In addition, we exhibit new examples of minimal non-superperfect graphs within these classes.

All of these non-superperfect graphs are inherited for the case when G is an optical network, and we give also representations as edge intersection graphs for the remaining minimal non-comparability non-superperfect graphs. In view of the result on edge intersection graphs of paths in a sufficiently large grid [9], we expect that any further minimal non-superperfect graph has such a representation and give some further new examples of such graphs.

To find new examples, we make use of the complete list of minimal non-comparability graphs found by Gallai [6] and the fact that any candidate for a new minimal non-superperfect graph can neither be imperfect nor a comparability graph. Thus, among the graphs with n nodes, the candidates of new minimal non-superperfect graphs are all graphs that are

- perfect (i.e. do not contain odd holes or odd antiholes),
- do not contain any minimal non-superperfect graph with $\leq n$ nodes,
- contain a minimal non-comparability superperfect graph with $< n$ nodes.

We close with some concluding remarks and open problems.

2   If the Network Is a Path

If the underlying optical network is a path P, then there exists exactly one (u, v)-path P_{uv} in P for every traffic demand between a pair u, v of nodes. Hence, if P is a path, then $\mathscr{P}$ and $I(\mathscr{P})$ are uniquely determined for any set of end-to-end traffic demands, and the RSA problem reduces to the spectrum assignment part. The edge intersection graph $I(\mathscr{P})$ of the (unique) routing $\mathscr{P}$ of the demands is an *interval graph* (i.e. the intersection graph of intervals in a line, here represented as subpaths of a path).

Interval graphs are known to be perfect by Berge [2]. In order to examine which minimal non-comparability non-superperfect graphs are interval graphs, we rely on a characterization of minimal non-interval graphs from [14].

A graph is *triangulated* if it does not have holes C_k with $k \geq 4$ as induced subgraph. Interval graphs are triangulated [11] hence all holes are in particular minimal non-interval graphs.

Theorem 1 *If $\mathscr{P}$ is a set of paths in a path, then $I(\mathscr{P})$ is an* interval graph *and can contain the graphs J_k for all $k \geq 2$, J_k' for all $k \geq 3$ and $\overline{E}_2$, but none of the other minimal non-comparability non-superperfect graphs.*

This implies that edge intersection graphs of paths in a path are not necessarily superperfect.

We next briefly discuss which further minimal non-superperfect graphs can be interval graphs. Recall that all of them have to contain a minimal non-comparability superperfect graph as proper induced subgraph. We observe that any further minimal non-superperfect interval graph can contain

- no even antihole $\overline{C}_{2k}$ for $k \geq 3$ (as they all contain a C_4 induced by 1, 2, 4, 5),
- none of the graphs J_k'' for all $k \geq 3$ (as they all contain a C_4 induced by 1, 2, 2k, 2k - 1),
- none of the graphs $\overline{A}_3, \ldots \overline{A}_8$ (as they all contain a C_4, see Fig. 3),

but only $\overline{A}_9$, $\overline{A}_{10}$ and J_2'. However, there is no example of a minimal non-superperfect interval graph containing $\overline{A}_9$, $\overline{A}_{10}$ or J_2' known yet.

3 If the Network Is a Tree

If the underlying network G is a tree, then there exists also exactly one (u, v)-path P_{uv} in G for every traffic demand between a pair u, v of nodes. Hence, if G is a tree, then $\mathscr{P}$ and $I(\mathscr{P})$ are uniquely determined for any set $\mathscr{D}$ of end-to-end traffic demands, and the RSA problem again reduces to the spectrum assignment part. The resulting edge intersection graph $I(\mathscr{P})$ belongs to the class of EPT graphs studied in [8]. We recall results from [8] on holes in EPT graphs and examine which minimal non-superperfect graphs can occur in such graphs.

It is known from [8] that EPT graphs are not necessarily perfect as they can contain odd holes. More precisely, Golumbic and Jamison showed the following:

Theorem 2 (Golumbic and Jamison [8]) *If the edge intersection graph $I(\mathscr{P})$ of a collection $\mathscr{P}$ of paths in a tree T contains a hole C_k with $k \geq 4$, then T contains a star $K_{1,k}$ with nodes $b, a_1, \ldots, a_k$ and there are k paths $P_1, \ldots, P_k$ in $\mathscr{P}$ such that P_i precisely contains the edges ba_i and ba_{i+1} of this star (where indices are taken modulo k).*

Figure 5 illustrates the case of $C_5 = I(\mathscr{P})$. From the above result, Golumbic and Jamison deduced the possible adjacencies of a hole which further implies that several graphs cannot occur as induced subgraphs of EPT graphs, including the complement of the P_6 and the two graphs G_1 and G_2 shown in Fig. 6.

That $\overline{P}_6$ is a non-EPT graph shows particularly that no antihole $\overline{C}_k$ for $k \geq 7$ can occur in such graphs. This implies:

Theorem 3 (Golumbic and Jamison [8]) *An EPT graph is perfect if and only if it does not contain an odd hole.*

With view on Theorem 2, this is clearly the case when the underlying tree has maximum degree 4, as noted in [8].

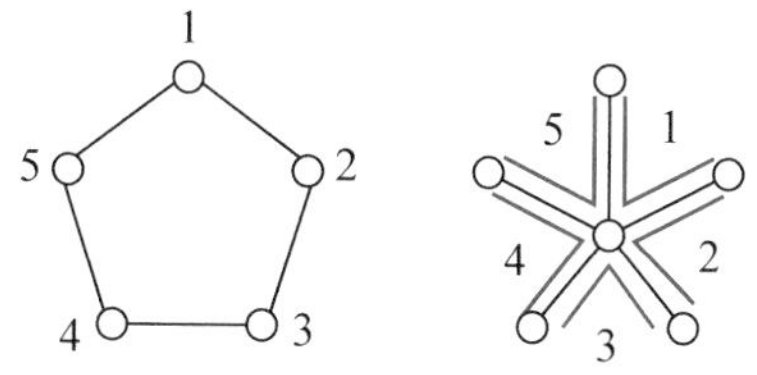

Fig. 5 The odd hole
$C_5 = I(\mathscr{P})$ with $\mathscr{P}$ in a star

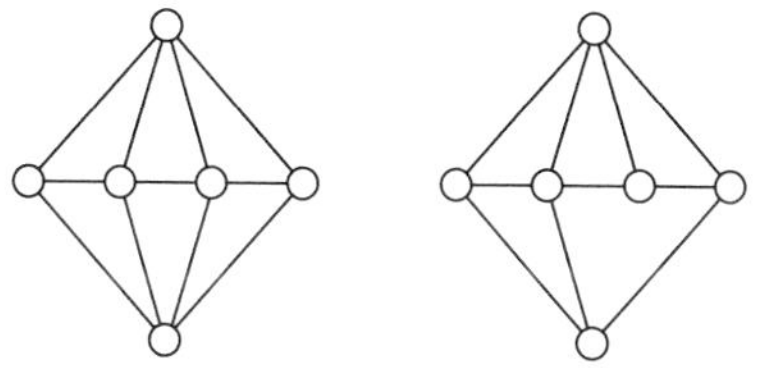

Fig. 6 The non-EPT graphs
G_1 and G_2

Based on the above results, we further examine which minimal non-comparability non-superperfect graphs can occur in edge intersection graphs of paths in a tree:

Theorem 4 *If $\mathscr{P}$ is a set of paths in a* tree, *then the EPT graph $I(\mathscr{P})$ can contain $\overline{A}_1, \overline{A}_2$ and*

- *odd holes C_{2k+1} for $k \geq 2$, but no odd antiholes $\overline{C}_{2k+1}$ for $k \geq 3$,*
- *the graphs J_k for all $k \geq 2$ and J'_k for all $k \geq 3$,*
- *$\overline{D}_2, \overline{D}_3, \overline{E}_1, \overline{E}_2, \overline{E}_3, \overline{F}_1, \overline{F}_2, \overline{F}_3$, but none of $\overline{D}_k, \overline{E}_k, \overline{F}_k$ for $k \geq 4$.*

This implies that perfect EPT graphs are not necessarily superperfect.

We next briefly discuss which further minimal non-superperfect graphs can be EPT graphs. Recall that all of them have to be perfect and have to contain a minimal non-comparability superperfect graph as proper induced subgraph. Among the minimal non-comparability superperfect graphs, the following are EPT graphs: J'_2 and

- $\overline{C}_6$ but no even antihole $\overline{C}_{2k}$ for $k \geq 4$ (as they all contain $\overline{P}_6$) by Golumbic and Jamison [8],
- none of the graphs J''_k for all $k \geq 3$ (as they all contain G_1 induced by the nodes $1, 2, 3, 4, 5, 2k$),
- the graphs $\overline{A}_3, \dots \overline{A}_6, \overline{A}_8, \dots \overline{A}_{10}$ (but not $\overline{A}_7$ as it has a G_2).

Hence, any minimal non-superperfect EPT graph not being minimal non-comparability has to contain one of $\overline{C}_6, \overline{A}_3, \dots \overline{A}_6, \overline{A}_8, \dots \overline{A}_{10}$ or J'_2 as proper induced subgraph. Figure 7 shows one example containing $\overline{A}_{10}$: it is non-superperfect (due to the indicated weight vector **d** causing a gap between weighted clique and interval chromatic number), it is minimal (as it does not have a non-comparability subgraph different from $\overline{A}_{10}$), it is an EPT graph (see the according path representation). However, note that the graph is not an interval graph (as it contains a C_4 induced by a, e, f, h).

Fig. 7 A minimal non-superperfect EPT graph containing $\overline{A}_{10}$

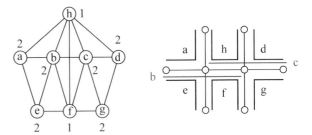

4 If the Network Is a Cycle

If the underlying optical network is a cycle C, then there exist exactly two (u, v)-paths P_{uv} in C for every traffic demand between a pair u, v of nodes. Hence, if C is a cycle, then the number of possible routings $\mathscr{P}$ (and their edge intersection graphs $I(\mathscr{P})$) is exponential in the number $|\mathscr{D}|$ of end-to-end traffic demands, namely $2^{|\mathscr{D}|}$.

Moreover, the edge intersection graphs of paths in a cycle are clearly *circular-arc graphs* (that are the intersection graphs of arcs in a cycle, here represented as paths in a hole C_n). It is well-known that circular-arc graphs are not necessarily perfect as they can contain both odd holes and odd antiholes, see e.g. [5] and Fig. 8 for illustration.

In order to address the question which of the studied perfect minimal non-comparability, non-superperfect graphs can occur in circular-arc graphs, we either present according path collections for the affirmative cases or exhibit a minimal non-circular-arc graph otherwise. For that, we first show the following:

Lemma 2 $\overline{E}_3$ is a minimal non-circular-arc graph.

Making use of the above facts, we can prove:

Theorem 5 *If $\mathscr{P}$ is a set of paths in a cycle, then the circular-arc graph $I(\mathscr{P})$ can contain $\overline{A}_1$ but not $\overline{A}_2$,*

- *all odd holes C_{2k+1} and odd antiholes $\overline{C}_{2k+1}$ for $k \geq 2$,*
- *the graphs J_k for all $k \geq 2$ and J'_k for all $k \geq 3$,*
- *$\overline{D}_2, \overline{D}_3, \overline{D}_4$, but not the graphs $\overline{D}_k$ for $k \geq 5$,*
- *$\overline{E}_1$ and $\overline{E}_2$, but not the graphs $\overline{E}_k$ for $k \geq 3$,*
- *$\overline{F}_2$, but not $\overline{F}_1$ neither the graphs $\overline{F}_k$ for $k \geq 3$.*

We next discuss which further minimal non-superperfect graphs can be circular-arc graphs. For that, we first show the following:

Lemma 3 J''_3 is a minimal non-circular-arc graph.

Fig. 8 The odd antihole $\overline{C}_7 = I(\mathscr{P})$ with $\mathscr{P}$ in a cycle

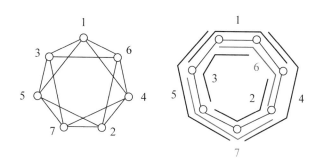

Fig. 9 A minimal
non-superperfect circular-arc
graph containing $\overline{A}_6$

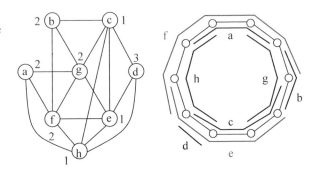

Remark 1 Note that $\overline{E}_3$ and J_3'' are, to the best of our knowledge, *new* examples of minimal non-circular-arc graphs (see e.g. the results on circular-arc graphs surveyed in [5]).

Recall that every further minimal non-superperfect graph has to be perfect and has to contain a minimal non-comparability superperfect proper induced subgraph. Among the perfect minimal non-comparability superperfect graphs, the following are circular-arc graphs: J_2' but

- no even antihole $\overline{C}_{2k}$ for $k \geq 3$ ("folklore"),
- neither J_3'' (by Lemma 3) nor the graphs J_k'' for all $k \geq 4$ (as they all contain the well-known minimal non-circular-arc graph $K_{2,3}$ induced by the nodes $1, 2, 4, 6, 2k$),
- all of the graphs $\overline{A}_3, \ldots \overline{A}_{10}$.

Hence, any minimal non-superperfect circular-arc graph not being minimal non-comparability has to contain one of $\overline{A}_3, \ldots \overline{A}_{10}$ or J_2' as proper induced subgraph. Figure 9 shows one example containing $\overline{A}_6$: it is non-superperfect (due to the indicated weight vector **d** causing a gap between weighted clique and interval chromatic number), it is minimal (as it does not have a non-comparability subgraph different from $\overline{A}_6$), it is a circular-arc graph (see the according path representation). However, note that the graph is not an interval graph (as $\overline{A}_6$ is not).

5 The General Case

Modern optical networks have clearly not a tree-like structure neither are just cycles due to survivability aspects concerning node or edge failures in the network G, see e.g. [13]. Instead, today's optical networks are 2-connected, sparse planar graphs with small maximum degree and have more a grid-like structure, see as example Fig. 10 showing the Telefónica network of Spain taken from [15].

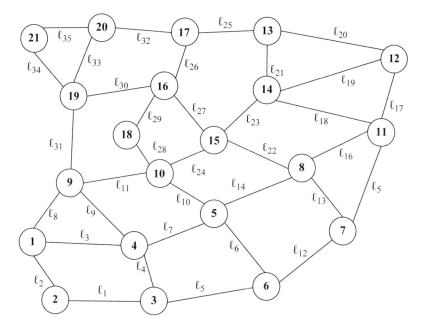

Fig. 10 The Telefónica network of Spain from [15]

We first wonder which minimal non-comparability non-superperfect graphs can occur in edge intersection graphs of paths in such networks G and can show:

Theorem 6 *All minimal non-comparability non-superperfect graphs can occur in edge intersection graphs $I(\mathscr{P})$ of sets $\mathscr{P}$ of paths in optical networks G.*

In addition, there are further minimal non-superperfect graphs in edge intersection graphs of paths in networks.

Figure 11 shows one example containing $\overline{A}_7$: it is non-superperfect (due to the indicated weight vector **d** causing a gap between weighted clique and interval chromatic number), it is minimal (as removing node g or h yields $\overline{A}_7$, and removing any other node results in a comparability graph), and it has a path representation in

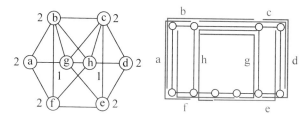

Fig. 11 A minimal non-superperfect graph containing $\overline{A}_7$ and a path representation in a sparse planar graph

a sparse planar graph. However, note that the graph is neither an EPT graph (as $\overline{A}_7$ is not), nor a circular-arc graph (as nodes a, e, f, g, h induce a $K_{2,3}$).

We expect that *all* minimal non-superperfect graphs can occur in edge intersection graphs of paths in networks, as soon as the networks G satisfy minimal survivability conditions concerning edge or node failures.

6   Concluding Remarks

From the fact that both, EPT graphs and circular-arc graphs, are not necessarily perfect, we notice that also edge intersection graphs of paths in networks are not necessarily perfect and, thus, also not necessarily superperfect. If we restrict the networks to paths, then $I(\mathscr{P})$ is an interval graph, but still not necessarily superperfect (as the minimal non-superperfect graphs J_k for all $k \geq 2$, J'_k for all $k \geq 3$ and $\overline{E}_1$ can occur). This is in accordance with the fact that the SA problem has been showed to be NP-hard on paths [16].

Hence, in all networks, it depends on the weights $\mathbf{d}$ induced by the traffic demands whether there is a gap between the weighted clique number $\omega(I(\mathscr{P}), \mathbf{d})$ and the interval chromatic number $\chi_I(I(\mathscr{P}), \mathbf{d})$. To determine the size of this gap, we propose to extend the concept of χ-binding functions introduced in [10] for usual coloring to interval coloring in weighted graphs, that is, to χ_I-binding functions f with

$$\chi_I(I(\mathscr{P}), \mathbf{d}) \leq f(\omega(I(\mathscr{P}), \mathbf{d}))$$

for edge intersection graphs $I(\mathscr{P})$ in a certain class of networks and all possible non-negative integral weights $\mathbf{d}$.

It is clearly of interest to study such χ_I-binding functions for different families of minimal non-superperfect graphs and to identify a hierarchy of graph classes between trees respectively cycles and sparse planar graphs resembling the structure of modern optical networks in terms of the gap between $\omega_I(I(\mathscr{P}), \mathbf{d})$ and $\chi_I(I(\mathscr{P}), \mathbf{d})$.

Furthermore, in networks different from trees, the routing part of the RSA problem is crucial and raises the question whether it is possible to select the routes in $\mathscr{P}$ in such a way that neither non-superperfect subgraphs nor unnecessarily large weighted cliques occur in $I(\mathscr{P})$.

Finally, giving a complete list of minimal non-superperfect graphs is an open problem, so that our future work comprises to find more minimal non-superperfect graphs and to examine the here addressed questions for them.

Acknowledgments We would like to thank Martin C. Golumbic and Martin Safe for interesting discussions on the topic, in particular concerning EPT graphs and circular-arc graphs. This work was supported by the French National Research Agency grant ANR-17-CE25-0006, project FLEXOPTIM.

References

1. Andreae, T.: On superperfect noncomparability graphs. J. Graph Theory **9**, 523–532 (1985)
2. Berge, C.: Les problèmes de coloration en théorie des graphes. Publ. Inst. Statist. Univ. Paris **9**, 123–160 (1960)
3. Christodoulopoulos, K., Tomkos, I., Varvarigos, E.: Elastic bandwidth allocation in flexible OFDM based optical networks. IEEE J. Lightwave Technol. **29**, 1354–1366 (2011)
4. Chudnovsky, M., Robertson, N., Seymour, P., Thomas, R.: The strong perfect graph theorem. Ann. Math. **164**, 51–229 (2006)
5. Durán, G., Grippo, L.N., Safe, M.D.: Structural results on circular-arc graphs and circle graphs: a survey and the main open problems. Discrete Appl. Math. **164**, 427–443 (2014)
6. Gallai, T.: Transitiv orientierbare Graphen. Acta Math. Acad. Sci. Hungar. **18**, 25–66 (1967)
7. Golumbic, M.: Algorithmic Graph Theory and Perfect Graphs, 2nd edn. North Holland, Amsterdam (2004)
8. Golumbic, M., Jamison, R.: The edge intersection graphs of paths in a tree. J. Comb. Theory B **38**, 8–22 (1985)
9. Golumbic, M., Lipshteyn, M., Stern, M.: Edge intersection graphs of single bend paths in a grid. Networks **54**, 130–138 (2009)
10. Gyárfás, A.: Problems from the world surrounding perfect graphs. Zastos. Mat. **19**, 413–431 (1987)
11. Hajös, G.: Über eine Art von Graphen. Int. Math. Nachr. **11**, Problem 65 (1957)
12. Hoffman, A.: A generalization of max flow-min cut. Math. Prog. **6**, 352–359 (1974)
13. Kerivin, H., Mahjoub, A.R.: Design of survivable networks: a survey. Networks **46**, 1–21 (2005)
14. Lekkerkerker, C., Boland, D.: Representation of finite graphs by a set of intervals on the real line. Fundam. Math. **51**, 45–64 (1962)
15. Ruiz, M., Pioro, M., Zotkiewicz, M., Klinkowski, M., Velasco, L.: Column generation algorithm for RSA problems in flexgrid optical networks. Photonic Netw. Commun. **26**, 53–64 (2013)
16. Shirazipourazad, S., Zhou, Ch., Derakhshandeh, Z., Sen, A.: On routing and spectrum allocation in spectrum sliced optical networks. In: Proceedings of IEEE INFOCOM, pp. 385–389 (2013)
17. Talebi, S., Alam, F., Katib, I., Khamis, M., Salama, R., Rouskas, G.N.: Spectrum management techniques for elastic optical networks: a survey. Opt. Switch. Netw. **13**, 34–48 (2014)
18. Wang, Y., Cao, X., Pan, Y.: A study of the routing and spectrum allocation in spectrum-sliced elastic optical path networks. In: Proc. of IEEE INFOCOM (2011)

A Cycle-Based Formulation for the Distance Geometry Problem

Leo Liberti, Gabriele Iommazzo, Carlile Lavor, and Nelson Maculan

Abstract The distance geometry problem consists in finding a realization of a weighted graph in a Euclidean space of given dimension, where the edges are realized as straight segments of length equal to the edge weight. We propose and test a new mathematical programming formulation based on the incidence between cycles and edges in the given graph.

Keywords Mathematical programming · Cycle basis · Protein conformation

1 Introduction

The DISTANCE GEOMETRY PROBLEM (DGP), also known as the *realization problem* in geometric rigidity, belongs to a more general class of metric completion and embedding problems.

L. Liberti (✉)
LIX CNRS Ecole Polytechnique, Institut Polytechnique de Paris, Palaiseau, France
e-mail: liberti@lix.polytechnique.fr

G. Iommazzo
LIX CNRS Ecole Polytechnique, Institut Polytechnique de Paris, Palaiseau, France

DI, Università di Pisa, Pisa, Italy
e-mail: giommazz@lix.polytechnique.fr

C. Lavor
IMECC, University of Campinas, Campinas, Brazil
e-mail: clavor@ime.unicamp.br

N. Maculan
COPPE, Federal University of Rio de Janeiro (UFRJ), Rio de Janeiro, Brazil
e-mail: maculan@cos.ufrj.br

C. Gentile et al. (eds.), *Graphs and Combinatorial Optimization:
from Theory to Applications*, AIRO Springer Series 5,
https://doi.org/10.1007/978-3-030-63072-0_8

DGP. Given a positive integer K and a simple undirected graph $G = (V, E)$ with an edge weight function $d : E \rightarrow \mathbb{R}_{\geq 0}$, establish whether there exists a *realization* $x : V \rightarrow \mathbb{R}^K$ of the vertices such that Eq. (1) below is satisfied:

$$\forall \{i, j\} \in E \qquad \|x_i - x_j\| = d_{ij}, \tag{1}$$

where $x_i \in \mathbb{R}^K$ for each $i \in V$ and d_{ij} is the weight on edge $\{i, j\} \in E$.

In its most general form, the DGP might be parametrized over any norm. In practice, the ℓ_2 norm is the most usual choice. The DGP with the ℓ_2 norm is sometimes called the EUCLIDEAN DGP (EDGP). For the EDGP, Eq. (1) is often reformulated to:

$$\forall \{i, j\} \in E \qquad \|x_i - x_j\|_2^2 = d_{ij}^2, \tag{2}$$

which is a system of quadratic polynomial equations with no linear terms.

The EDGP is motivated by many scientific and technological applications. The clock synchronization problem, for example, aims at establishing the absolute time of a set of clocks when only the time difference between subsets of clocks can be exchanged [29]. The sensor network localization problem aims at finding the positions of a moving wireless sensor on a 2D manifold given an estimation of some of the pairwise Euclidean distances [2]. The MOLECULAR DGP (MDGP) aims at finding the positions of atoms in a protein, given some of the pairwise Euclidean distances [15, 16]. In general, the DGP is an inverse problem which occurs every time one can measure some of the pairwise distances in a set of entities, and needs to establish their position.

The DGP is weakly **NP**-hard even when restricted to simple cycle graphs and strongly **NP**-hard even when restricted to integer edge weights in $\{1, 2\}$ in general graphs [27]. It is in **NP** if $K = 1$ but not known to be in **NP** if $K > 1$ for general graphs [4], which is an interesting open question [19]. More information about the DGP can be found in [22].

There are many approaches to solving the DGP. Generally speaking, application-specific solution algorithms exploit some of the graph structure, if induced by the application. For example, a condition often asked when reconstructing the positions of sensor networks is that the realization should be unique (as one would not know how to choose between multiple realizations), a condition called *global rigidity* [7] which can, at least generically, be imposed directly on the unweighted input graph. For protein structures, on the other hand, which are found in nature in several isomers, one is often interested in finding all (incongruent) realizations of the given protein graph [20]. Since such graphs are rigid, one can devise an algorithm (called Branch-and-Prune) which, following a given vertex order, branches on reflections of the position of the next vertex, which is computed using trilateration [18, 21]. In absence of any information on the graph structure, however, one can resort to Mathematical Programming (MP) formulations and corresponding solvers [8, 23].

The MP formulation which is most often used reformulates Eq. (2) to the minimization of the sum of squared error terms:

$$\min_{x} \sum_{\{i,j\}\in E} (\|x_i - x_j\|_2^2 - d_{ij}^2)^2. \tag{3}$$

This formulation describes an unconstrained polynomial minimization problem. The polynomial in question has degree 4, is always nonnegative, and generally nonconvex and multimodal. Each solution x^* having global minimum value equal to zero is a realization of the given graph.

As far as we know, all existing MP formulations for the EDGP are based on the incidence of edges and vertices. In this paper we discuss a new MP formulation for the EDGP based on the incidence of cycles and edges instead, some variants, and a computational comparison with a well-known edge-based formulation.

2   A New Formulation Based on Cycles

In this section we propose a new formulation for the EDGP. The basic idea stems from the fact that the quantities $x_{ik} - x_{jk}$ sum up to zero over all edges of any cycle in the given graph for each dimensional index $k \leq K$. This idea was used in [27] for proving weak **NP**-hardness of the DGP on cycle graphs, by reduction from PARTITION. For a subgraph H of a graph $G = (V, E)$, we use $V(H)$ and $E(H)$ to denote vertex and edge set of H explicitly; given a set F of edges we use $V(F)$ to denote the set of incident vertices. Let $m = |E|$ and $n = |V|$. For a mapping $x : V \to \mathbb{R}^K$ we denote by $x[U]$ the restriction of x to a subset $U \subseteq V$.

Lemma 1 *Given an integer $K > 0$, a simple undirected weighted graph $G = (V, E, d)$ and a mapping $x : V \to \mathbb{R}^K$, then for each cycle C in G, each orientation of the edges in C given by a closed trail $W(C)$ in the cycle, and each $k \leq K$ we have:*

$$\sum_{(i,j)\in W(C)} (x_{ik} - x_{jk}) = 0. \tag{4}$$

Proof We renumber the vertices in $V(C)$ to $1, 2, \ldots, \gamma = |V(C)|$ following the walk order in $W(C)$. Then Eq. (4) can be explicitly written as:

$$(x_{1k} - x_{2k}) + (x_{2k} - x_{3k}) + \cdots + (x_{\gamma k} - x_{1k}) =$$
$$= x_{1k} - (x_{2k} - x_{2k}) - \cdots - (x_{\gamma k} - x_{\gamma k}) - x_{1k} = 0,$$

as claimed.   □

We introduce new decision variables y_{ijk} replacing the terms $x_{ik} - x_{jk}$ for each $\{i, j\} \in E$ and $k \leq K$. Equation (2) then becomes:

$$\forall \{i, j\} \in E \qquad \sum_{k \leq K} y_{ijk}^2 = d_{ij}^2. \tag{5}$$

We remark that for the DGP with other norms this constraint changes. For the ℓ_1 or ℓ_∞ norms, for example, we would have:

$$\forall \{i, j\} \in E \quad \sum_{k \leq K} |y_{ijk}| = d_{ij} \quad \text{or} \quad \max_{k \leq K} |y_{ijk}| = d_{ij}. \tag{6}$$

Next, we adjoin the constraints on cycles:

$$\forall k \leq K, C \subset G \quad \left(C \text{ is a cycle} \Rightarrow \sum_{\{i,j\} \in E(C)} y_{ijk} = 0 \right). \tag{7}$$

We also note that the feasible value of a y_{ijk} variable is the (oriented) length of the segment representing the edge $\{i, j\}$ projected on the k-th coordinate. We can therefore infer bounds for y as follows:

$$\forall k \leq K, \{i, j\} \in E \quad -d_{ij} \leq y_{ijk} \leq d_{ij}. \tag{8}$$

We now prove our main result, i.e. that Eqs. (5) and (7) are a valid MP formulation for the EDGP.

Theorem 1 *There exists a vector $y^* \in \mathbb{R}^{Km}$ which satisfies Eqs. (5) and (7), parametrized on K, G, if and only if (K, G) is a YES instance of the EDGP.*

Proof ($\Leftarrow$) Assume that (K, G) is a YES instance of the EDGP. Then G has a realization $x^* \in \mathbb{R}^{nK}$ in $\mathbb{R}^K$. We define $y_{ijk}^* = x_{ik}^* - x_{jk}^*$ for all $\{i, j\} \in E$ and $k \leq K$. Since x^* is a realization of G, by definition it satisfies Eq. (2), and, by substitution, Eq. (5). Moreover, any realization of G satisfies Eq. (4) over each cycle by Lemma 1. Hence, by replacement, it also satisfies Eq. (7).

($\Rightarrow$) Assume next that (K, G) is a NO instance of the EDGP, and suppose that Eqs. (5) and (7) have a non-empty feasible set Y. For every $y \in Y$ we consider the K linear systems

$$\forall \{i, j\} \in E \quad x_{ik} - x_{jk} = y_{ijk}, \tag{9}$$

for each $k \leq K$, each with n variables and m equations. We square both sides then sum over $k \leq K$ to obtain $\sum_{k \leq K} (x_{ik} - x_{jk})^2 = \sum_{k \leq K} y_{ijk}^2$ for all $\{i, j\} \in E$. By Eq. (5) we have $\sum_{k \leq K} y_{ijk}^2 = d_{ij}^2$ whence follows Eq. (2), contradicting the assumption that the EDGP is NO. So we need only show that there is a solution x to Eq. (9) for any given $y \in Y$.

We first consider the case where G is not biconnected: let G be a union of two graphs G', G'' with a single common vertex v. Assume recursively that the claim holds for both G' and G'', so we get realizations x', x'' for G', G'': simply translate x'' so that $x'_v = x''_v$, and let x be the concatenation of x', x''. Then x satisfies Eq. (9) by translation invariance (given by $x_i - x_j = (x_i - z) - (x_j - z)$ for any translation vector z).

Next, we consider the case where G is a tree: by the above reasoning, we can assume that G is a path (since a tree is a connected union of pendant paths, dealt with above) with n vertices and $n - 1$ edges. Then for each fixed $k \leq K$ Eq. (9) has n variables and $n - 1$ equations. Let A be the set of vertices incident to a single edge and B the set of vertices incident to two edges (clearly $A \cup B = V$). If $i \in A$ then x_i occurs in a single equation; if $i \in B$ then x_i occurs in exactly two equations. Thus the linear dependence condition $\sum_{\{i,j\} \in E} \lambda_{ijk}(x_{ik} - x_{jk}) = 0$ (†) requires all of the λ_{ijk} involving $i \in A$ to be zero, which implies $j \in A$ too (if $j \in B$ there would be an x_j term left in (†)): this implies $\lambda = 0$, showing that the system has rank $n - 1$. Hence Eq. (9) has uncountably many solutions. This is repeated for every $k \leq K$ to yield a realization of the tree in $\mathbb{R}^K$.

Now we assume WLOG that G is biconnected, since any pendant trees can be easily treated separately as shown above, and proceed by induction on the simple cycles of G. For the base case, we consider a cycle C with corresponding y satisfying Eqs. (5) and (7). Since C is a cycle, it has the same number of vertices and edges, say q. This implies that, for any fixed $k \leq K$, Eq. (9) is a linear system of equations $Mx = y$ with a $q \times q$ matrix M as shown below:

$$M = \begin{pmatrix} 1 & -1 & & & \\ & 1 & -1 & & \\ & & 1 & \ddots & \\ & & & \ddots & -1 \\ -1 & & & & 1 \end{pmatrix}.$$

By Eq. (4) and by inspection it is clear that the rank of M is exactly $q - 1$: then Eq. (7) ensures that Eq. (9) has uncountably many solutions. Repeating this for every $k \leq K$ we obtain a realization x of C with K degrees of freedom.

Since any cycle basis generates the set of all cycles in a graph, for the induction step we consider a cycle basis $\mathscr{B}$ of G that is fundamental (see Sect. 3). We assume that G' is a union of fundamental cycles in $\mathscr{B}$ with realization x' satisfying Eq. (9), and that C is another fundamental cycle in $\mathscr{B}$ with realization x^C. We aim at proving that Eq. (9) has a solution for $G' \cup C$. Since G is biconnected, the induction can proceed by ear decomposition [25], which means that G' is also biconnected, and that C is such that $E(G') \cap E(C) = F$ is a non-empty path in G'. We want to show that C can be realized so the edges in F are realized according to x': we argue that there is $\tilde{x} : V(C) \setminus V(F) \to \mathbb{R}^K$ such that $\bar{x}^C = (x'[V(F)], \tilde{x})$ is a realization of C. It suffices to assume that $E(C) \setminus F$ consists of a single edge, say $\{u, v\}$, since any more edges can be considered as a pendant path attached to G' (easily dealt with as we saw above since paths are trees) and a single edge. This means that $u, v \in V(F)$, i.e. x' already maps $V(G') \cup V(C)$ to $\mathbb{R}^K$. Thus we only need to check that $x'_{uk} - x'_{vk} = y_{uvk}$ for each $k \leq K$.

By Eq. (4) applied to $C = F \cup \{\{u, v\}\}$ and the facts that (a) C is a cycle in G and (b) x' realizes G', which contains F, we have

$$\forall k \leq K \quad \sum_{\{i,j\} \in C} (x'_{ik} - x'_{jk}) = 0. \tag{10}$$

By Eq. (7) applied to C and the fact that $Y \neq \varnothing$ we have

$$\forall k \leq K \quad \sum_{\{i,j\} \in C} y_{ijk} = 0. \tag{11}$$

By induction hypothesis x' satisfies Eq. (9), whence

$$\forall k \leq K, \{i, j\} \in F \quad x'_{ik} - x'_{jk} = y_{ijk}. \tag{12}$$

We replace Eq. (12) in Eq. (11), obtaining

$$\forall k \leq K \quad \sum_{\{i,j\} \in F} (x'_{ik} - x'_{jk}) = -y_{uvk}. \tag{13}$$

Subtracting Eq. (13) from Eq. (10) finally yields $x'_{uk} - x'_{vk} = y_{uvk}$ for all $k \leq K$, which concludes the proof. □

The issue with Theorem (1) is that it relies on the exponentially large family of constraints Eq. (7). While this is sometimes addressed by algorithmic techniques such as row generation, we shall see in the following that it suffices to consider a polynomial set of cycles (which, moreover, can be found in polynomial time) in the quantifier of Eq. (7).

3   The Cycle Vector Space and its Bases

We recall that incidence vectors of cycles (in a Euclidean space having $|E|$ dimensions) form a vector space over a field $\mathbb{F}$, which means that every cycle can be expressed as a weighted sum of cycles in a basis. In this interpretation, a *cycle* in G is simply a subgraph of G where each vertex has even degree: we denote their set by $\mathscr{C}$. This means that Eq. (7) is actually quantified over a subset of $\mathscr{C}$, namely the simple connected cycles. Every basis has cardinality $m - n + a$, where a is the number of connected components of G. If G is connected, cycle bases have cardinality $m - n + 1$ [28].

Our interest in introducing cycle bases is that we would like to quantify Eq. (7) polynomially rather than exponentially in the size of G. Our goal is to replace "C a simple connected cycle in $\mathscr{C}$" by "C in a cycle basis of G". In order to show that this limited quantification is enough to imply every constraint in Eq. (7), we have to show that, for each simple connected cycle $C \in \mathscr{C}$, the corresponding constraint in Eq. (7) can be obtained as a weighted sum of constraints corresponding to the basis elements.

Another feature of Eq. (7) to keep in mind is that edges are implicitly given a direction: for each cycle, the term for the *undirected* edge $\{i, j\}$ in Eq. (7) is $(x_{ik} - x_{jk})$. Note that while $\{i, j\}$ is exactly the same vertex set as $\{j, i\}$, the corresponding term is either positive or not, depending on the direction (i, j) or (j, i). We deal with this issue by arbitrarily directing the edges in E to obtain a set A of arcs, and considering *directed* cycles in the directed graph $\bar{G} = (V, A)$. In this interpretation, the incidence vector of a directed cycle C of $\bar{G}$ is a vector $c^C \in \mathbb{R}^m$ satisfying [14, §2, p. 201]:

$$\forall j \in V(C) \quad \sum_{(i,j)\in A} c^C_{ij} = \sum_{(j,\ell)\in A} c^C_{j\ell}. \tag{14}$$

A directed circuit D of $\bar{G}$ is obtained by applying the edge directions from $\bar{G}$ to a connected subgraph of G where each vertex has degree exactly 2 (note that a directed circuit need not be strongly connected, although its undirected version is connected). Its incidence vector $c^D \in \{-1, 0, 1\}^m$ is defined as follows:

$$\forall (i, j) \in A \quad c^D_{ij} \triangleq \begin{cases} 1 & \text{if} \quad (i, j) \in A(D) \\ -1 & \text{if} \quad (j, i) \in A(D) \\ 0 & \text{otherwise} \end{cases}$$

where we have used $A(D)$ to mean the arcs in the subgraph D. In other words, whenever we walk over an arc (i, j) in the natural direction $i \to j$ we let the (i, j)-th component of c^D be 1; if we walk over (i, j) in the direction $j \to i$ we assign a -1, and otherwise a zero.

3.1 Constraints Over Cycle Bases

The properties of undirected and directed cycle bases have been investigated in a sequence of papers by many authors, culminating with [14]. We now prove that it suffices to quantify Eq. (7) over a directed cycle basis.

Proposition 1 *Let $\mathscr{B}$ be a directed cycle basis of $\bar{G}$ over $\mathbb{Q}$. Then Eq. (7) holds if and only if:*

$$\forall k \le K, B \in \mathscr{B} \qquad \sum_{(i,j) \in A(B)} c_{ij}^B y_{ijk} = 0. \tag{15}$$

Proof Necessity (7) $\Rightarrow$ (15) follows because Eq. (7) is quantified over all cycles: in particular, it follows for any undirected cycle in any undirected cycle basis. Moreover, the signs of all terms in the sum of Eq. (15) are consistent, by definition, with the arbitrary edge direction chosen for $\bar{G}$.

Next, we claim sufficiency (15) $\Rightarrow$ (7). Let $C \in \mathscr{C}$ be a simple cycle, and $\bar{C}$ be its directed version with the directions inherited from $\bar{G}$. Since $\mathscr{B}$ is a cycle basis, we know that there is a coefficient vector $(\gamma_B \mid B \in \mathscr{B}) \in \mathbb{R}^{|\mathscr{B}|}$ such that:

$$c^{\bar{C}} = \sum_{B \in \mathscr{B}} \gamma_B c^B. \tag{16}$$

We now consider the expression:

$$\forall k \le K \qquad \sum_{B \in \mathscr{B}} \gamma_B \sum_{(i,j) \in A(B)} c_{ij}^B y_{ijk}. \tag{17}$$

On the one hand, by Eq. (16), Eq. (17) is identically equal to $\sum_{(i,j) \in A(\bar{C})} c_{ij}^{\bar{C}} y_{ijk}$ for each $k \le K$; on the other hand, each inner sum in Eq. (17) is equal to zero by Eq. (15). This implies $\sum_{(i,j) \in A(\bar{C})} c_{ij}^{\bar{C}} y_{ijk} = 0$ for each $k \le K$. Since C is simple and connected $\bar{C}$ is a directed circuit, which implies that $c^{\bar{C}} \in \{-1, 0, 1\}$. Now it suffices to replace $-y_{ijk}$ with y_{jik} to obtain

$$\forall k \le K \qquad \sum_{\{i,j\} \in E(C)} y_{ijk} = 0,$$

where the edges on C are indexed in such a way as to ensure they appear in order of consecutive adjacency. $\qquad \square$

Obviously, if $\mathscr{B}$ has minimum (or just small) cardinality, Eq. (15) will be sparsest (or just sparse), which is often a desirable property of linear constraints occurring in MP formulations. Hence we should attempt to find short cycle bases $\mathscr{B}$.

In summary, given a basis $\mathscr{B}$ of the directed cycle space of $\bar{G}$ where c^B is the incidence vector of a cycle $B \in \mathscr{B}$, the following:

$$\left. \begin{array}{ll} \min\limits_{s \ge 0, y} \sum\limits_{\{i,j\} \in E} (s_{ij}^+ + s_{ij}^-) & \\ \forall (i,j) \in A(\bar{G}) \quad \sum\limits_{k \le K} y_{ijk}^2 - d_{ij}^2 = s_{ij}^+ - s_{ij}^- & \\ \forall k \le K, B \in \mathscr{B} \quad \sum\limits_{(i,j) \in A(B)} c_{ij}^B y_{ijk} = 0 & \end{array} \right\} \tag{18}$$

is a valid formulation for the EDGP. The solution of Eq. (18) yields a feasible vector y^*. We must then exploit Eq. (9) to obtain a realization x^* for G.

3.2 How to Find Directed Cycle Bases

We require directed cycle bases over $\mathbb{Q}$. By [14, Thm. 2.4], each undirected cycle basis gives rise to a directed cycle basis (so it suffices to find a cycle basis of G and then direct the cycles using the directions in $\bar{G}$). Horton's algorithm [12] and its variants [11, 24] find a minimum cost cycle basis in polynomial time. The most efficient deterministic variant is $O(m^3 n)$ [24], and the most efficient randomized variant has the complexity of matrix multiplication. Existing approximation algorithms have marginally better complexity.

It is not clear, however, that the provably sparsest constraint system will make the DGP actually easier to solve. We therefore consider a much simpler algorithm: starting from a spanning tree, we pick the $m - n + 1$ circuits that each *chord* (i.e., non-tree) edge defines with the rest of the tree. This algorithm [26] yields a *fundamental cycle basis* (FCB). Finding the minimum FCB is known to be **NP**-hard [9], but heuristics based on spanning trees prove to be very easy to implement and work reasonably well [9] (optionally, their cost can be improved by an edge-swapping phase [1, 17]).

4 Computational Results

The aim of this section is to compare the computational performance of the new "cycle formulation" Eqns. (18) and (9) with the standard "edge formulation" Eq. (3). We note that both formulations are nonconvex Nonlinear Programs (NLP), which are generally hard to solve. We therefore used a very simple 3-iteration multi-start heuristic based on calling a local NLP solver from a random initial starting point at each iteration, and updating the best solution found so far as needed.

We remark that we added the centroid constraints:

$$\forall k \leq K \quad \sum_{i \leq n} x_{ik} = 0$$

to the edge formulation Eq. (3). In our experience, these constraints (which simply remove the degrees of translation freedom) give a slight stability advantage to the edge formulation when solved with most local NLP solvers.

We evaluate the quality of a realization x of a graph G according to mean (MDE) and largest distance error (LDE), defined this way:

$$\mathsf{MDE}(x, G) = \frac{1}{|E|} \sum_{\{i,j\} \in E} \left| \|x_i - x_j\|_2 - d_{ij} \right|$$

$$\mathsf{LDE}(x, G) = \max_{\{i,j\} \in E} \left| \|x_i - x_j\|_2 - d_{ij} \right|.$$

We remark that these realization quality measures are formally different from the objective functions of the formulations we benchmarked.

The CPU time taken to find the solution may also be important, depending on the application. In real-time control of underwater vehicles [3], for example, DGP instances might need to be solved every second. In other applications, such as finding protein structure from distance data [5], the CPU time is not so important.

Our tests were carried out on a single CPU of a 2.1 GHz 4-CPU 8-core-per-CPU machine with 64 GB RAM running Linux. We used AMPL [10] to implement our formulations and solution algorithms, and the local NLP IpOpt solver [6] to solve each formulation locally (Tables 1–2).

Our first benchmark contains a diverse collection of randomly generated weighted graphs of small size and many different types (Table 2), realized in $\mathbb{R}^2$. The cycle formulation finds better MDE values, while the edge formulation generally finds better LDE values and is faster. The instance names in Table 2 label the graph type and some random generation parameters: almostreg-k-n are almost k-regular graphs on n vertices, bipartite-n-p are bipartite graphs on $2n$ vertices with edge density p, cluster-n-k-p-q are k-clustered n-graphs

Table 1 Cycle formulation vs. edge formulation performance on protein graphs (realizations in $K = 3$ dimensions). Boldface figures denote best results

Instance	m	n	mdeC	mdeE	ldeC	ldeE	cpuC	cpuE
1guu	955	150	**0.057**	0.061	1.913	**1.884**	**18.18**	37.14
1guu-1	959	150	**0.035**	0.038	2.025	**1.824**	24.27	**5.48**
1guu-4000	968	150	0.061	**0.060**	2.324	**2.121**	24.24	**6.97**
pept	999	107	**0.104**	0.161	3.367	**2.963**	34.67	**10.89**
2kxa	2711	177	**0.053**	0.155	**3.613**	3.936	169.95	**35.44**
res_2kxa	2627	177	0.131	**0.045**	**3.197**	3.442	153.00	**32.40**
C0030pkl	3247	198	**0.009**	0.059	**2.761**	3.965	156.09	**76.58**
cassioli-130731	4871	281	**0.005**	0.060	**3.447**	3.963	376.33	**143.31**
100d	5741	488	**0.146**	0.246	4.295	**4.090**	3024.67	**253.56**
helix_amber	6265	392	**0.038**	0.059	**3.528**	4.578	1573.10	**212.68**
water	11939	648	**0.222**	0.422	4.557	**4.322**	9384.08	**3836.23**
3all	17417	678	**0.084**	0.124	4.165	**4.087**	4785.91	**1467.74**
1hpv	18512	1629	**0.334**	0.338	**4.256**	4.619	53848.33	**6620.70**
il2	45251	2084	1.481	**0.248**	9.510	**4.415**	**2323.90**	24321.25

Table 2 Cycle formulation vs. edge formulation performance on various small sized graphs (realizations in $K = 2$ dimensions). Boldface figures denote best results

Instance	m	n	mdeC	mdeE	ldeC	ldeE	cpuC	cpuE
almostreg-3-100	298	100	**0**	**0**	0.048	**0.041**	0.88	**0.23**
almostreg-3-150	448	150	**0**	**0**	0.330	**0.282**	1.29	**0.30**
almostreg-3-200	598	200	**0**	**0**	0.030	**0.020**	2.15	**0.44**
almostreg-3-50	146	50	**0**	**0**	**0**	**0**	0.31	**0.11**
almostreg-6-100	591	100	**0.077**	0.093	0.740	**0.410**	6.85	**0.35**
almostreg-6-150	893	150	**0.085**	0.099	1.030	**0.485**	16.52	**0.68**
almostreg-6-200	1192	200	**0.076**	0.098	0.729	**0.501**	34.07	**1.35**
almostreg-6-50	292	50	**0.082**	0.099	0.648	**0.471**	1.80	**0.13**
almostreg-8-100	777	100	**0.105**	0.131	0.846	**0.577**	8.89	**0.42**
almostreg-8-150	1189	150	**0.104**	0.121	0.805	**0.528**	34.84	**0.83**
almostreg-8-200	1581	200	**0.104**	0.125	0.974	**0.654**	48.10	**1.79**
almostreg-8-50	387	50	**0.104**	0.113	0.670	**0.520**	2.46	**0.13**
bipartite-100-03	3044	200	**0.206**	0.218	0.931	**0.790**	209.15	**7.86**
bipartite-100-06	6024	200	**0.225**	0.234	0.978	**0.753**	439.74	**8.00**
bipartite-150-03	6708	300	**0.220**	0.232	0.951	**0.724**	582.71	**14.37**
bipartite-150-06	13466	300	**0.231**	0.240	0.852	**0.808**	1904.18	**30.79**
bipartite-200-03	11906	400	**0.223**	0.235	0.936	**0.812**	3183.43	**33.06**
bipartite-200-06	23963	400	**0.235**	0.244	0.888	**0.741**	4885.52	**64.03**
bipartite-50-03	744	100	**0.166**	0.185	0.936	**0.787**	29.27	**1.11**
bipartite-50-06	1468	100	**0.201**	0.217	1.011	**0.754**	80.80	**1.38**
cluster-120-4-05-01	1495	120	**0.191**	0.206	0.873	**0.838**	98.67	**1.69**
cluster-120-8-05-01	1149	120	**0.181**	0.196	0.892	**0.740**	62.29	**1.04**
cluster-150-2-05-01	3337	150	**0.218**	0.230	**0.901**	0.936	605.00	**3.66**
cluster-150-8-05-01	1750	150	**0.190**	0.205	0.886	**0.831**	70.66	**2.44**
cluster-200-2-05-01	5957	200	**0.231**	0.241	**0.931**	0.952	612.82	**8.01**
cluster-200-4-05-01	4155	200	**0.221**	0.233	0.924	**0.906**	397.45	**7.67**
cluster-200-8-05-01	3046	200	**0.206**	0.220	0.988	**0.851**	462.46	**5.61**
cluster-50-2-05-01	361	50	**0.159**	0.171	0.742	**0.679**	7.52	**0.20**
cluster-50-4-05-01	242	50	**0.145**	0.167	0.899	**0.588**	3.63	**0.18**
cluster-50-8-05-01	187	50	**0.113**	0.133	0.716	**0.500**	2.73	**0.16**
euclid-150-02	2341	150	**0**	**0**	**0**	**0**	286.09	**2.69**
euclid-150-05	5678	150	**0**	**0**	**0**	**0**	991.87	**2.86**
euclid-150-08	8915	150	**0**	**0**	**0**	**0**	1507.94	**3.88**
euclid-200-05	10037	200	**0**	**0**	**0**	**0**	1881.40	**5.47**
euclid-200-08	15877	200	**0**	**0**	**0**	**0**	3114.95	**7.96**
flowersnark120	720	480	**0**	**0**	0.151	**0.109**	**7.86**	8.21
flowersnark-150	900	600	**0**	**0**	0.101	**0.086**	36.53	**15.50**
flowersnark-200	1200	800	**0**	**0**	0.141	**0.123**	**18.02**	31.04
flowersnark40	240	160	**0**	**0**	0.016	**0.005**	1.92	**0.35**
flowersnark80	480	320	**0**	**0**	0.068	**0.059**	3.18	**1.08**
hypercube-10	5120	1024	**0.128**	0.152	1.004	**0.653**	4965.30	**133.93**

(continued)

Table 2 (continued)

Instance	m	n	mdeC	mdeE	ldeC	ldeE	cpuC	cpuE
hypercube-5	80	32	**0.054**	0.058	0.401	**0.321**	0.95	**0.10**
hypercube-6	192	64	**0.075**	0.087	0.774	**0.426**	4.20	**0.20**
hypercube-8	1024	256	**0.104**	0.127	0.876	**0.631**	81.68	**2.59**
powerlaw-100-2-05	148	100	**0.024**	0.025	0.338	**0.309**	1.24	**0.38**
powerlaw-100-2-08	178	100	**0.042**	0.042	0.464	**0.398**	1.64	**0.59**
powerlaw-150-2-05	223	150	**0.034**	0.035	0.404	**0.360**	**1.37**	1.94
powerlaw-150-2-08	268	150	**0.047**	0.047	0.471	**0.404**	2.44	**1.73**
powerlaw-200-2-05	298	200	**0.025**	0.026	0.581	**0.443**	2.64	**1.27**
powerlaw-200-2-08	358	200	**0.037**	0.038	0.454	**0.376**	3.75	**1.78**
random-100-02	1093	100	**0.193**	0.203	0.874	**0.742**	48.43	**0.67**
random-100-05	2479	100	**0.224**	0.234	0.938	**0.855**	168.40	**1.48**
random-150-02	2394	150	**0.209**	0.223	0.932	**0.809**	226.60	**3.98**
random-150-05	5675	150	**0.241**	0.250	0.965	**0.953**	580.59	**6.10**
random-200-02	4097	200	**0.218**	0.228	0.930	**0.887**	271.94	**7.68**
random-200-05	10023	200	**0.248**	0.255	**0.949**	0.952	1024.32	**11.43**
random-50-02	291	50	**0.143**	0.161	0.922	**0.638**	7.03	**0.17**
random-50-05	665	50	**0.195**	0.212	**0.836**	0.953	16.20	**0.23**
rnddegdist-100	2252	100	**0.223**	0.235	**0.929**	0.963	136.74	**1.48**
rnddegdist-150	5293	150	**0.240**	0.249	**0.939**	0.955	819.86	**3.91**
rnddegdist-30	174	30	**0.156**	0.179	0.767	**0.667**	2.26	**0.11**
rnddegdist-40	221	40	**0.156**	0.175	0.672	**0.628**	2.93	**0.17**
tripartite-100-02	4038	300	**0.198**	0.213	0.968	**0.737**	369.77	**10.39**
tripartite-100-05	10003	300	**0.227**	0.238	0.917	**0.729**	1150.35	**21.37**
tripartite-150-02	9061	450	**0.213**	0.227	0.956	**0.765**	2005.30	**32.43**
tripartite-150-05	22431	450	**0.235**	0.245	0.876	**0.751**	4687.28	**45.27**
tripartite-30-02	359	90	**0.106**	0.118	0.736	**0.547**	10.31	**0.37**
tripartite-50-02	995	150	**0.153**	0.173	0.958	**0.722**	38.55	**1.00**
tripartite-50-05	2519	150	**0.208**	0.220	0.849	**0.736**	160.43	**2.39**

with intercluster density p and intracluster density q, euclid-n-p are graphs on n random points in the plane with density p, flowersnark-n are flower snark graphs [13] of order n, hypercube-n are graphs on 2^n vertices connected with a hypercube topology, powerlaw-n-t-a are $\deg_i = ani^{-t}$ power law graphs on n vertices with biconnectedness guaranteed by the addition of a Hamiltonian cycle, random-n-p are Erdős-Renyi graphs on n vertices with density p, rnddegdist-n are biconnected random graphs on n vertices with a randomly generated degree distribution, tripartite-n-p are tripartite graphs on $3n$ vertices with edge density p.

Our second benchmark contains medium to large scale protein graph instances (Table 1), realized in $\mathbb{R}^3$. It turns out that the cycle formulation gives generally better quality solutions (the MDE is better on all instances but two, the LDE is better a little less than half of the times), but takes more time in order to find them. In our largest tested instance (il2) the trend is reversed, meaning that the cycle formulation found a bad quality solution but in a tenth of the time.

In all cases, finding the cycle basis and solving the auxiliary retrieval problem Eq. (9) takes a tiny fraction of the total solution time.

Acknowledgments While the seminal idea for considering DGPs over cycles dates from Saxe's NP-hardness proof [27], the "cycle formulation" concept occurred to us as one of the authors (LL) attended a talk by Matteo Gallet given at the Erwin Schrödinger Institute (ESI), Vienna, during the Geometric Rigidity workshop 2018. LL has received funding from the European Union's Horizon 2020 research and innovation programme under the Marie Sklodowska-Curie grant agreement n. 764759 "MINOA". CL is grateful to the Brazilian research agencies FAPESP and CNPq for support.

References

1. Amaldi, E., Liberti, L., Maffioli, F., Maculan, N.: Edge-swapping algorithms for the minimum fundamental cycle basis problem. Math. Methods Oper. Res. **69**, 205–223 (2009)
2. Aspnes, J., Eren, T., Goldenberg, D., Morse, S., Whiteley, W., Yang, R., Anderson, B., Belhumeur, P.: A theory of network localization. IEEE Trans. Mobile Comput. **5**(12), 1663–1678 (2006)
3. Bahr, A., Leonard, J., Fallon, M.: Cooperative localization for autonomous underwater vehicles. Int. J. Robot. Res. **28**(6), 714–728 (2009)
4. Beeker, N., Gaubert, S., Glusa, C., Liberti, L.: Is the distance geometry problem in NP? In: Mucherino, A., Lavor, C., Liberti, L., Maculan, N. (eds.) Distance Geometry: Theory, Methods, and Applications, pp. 85–94. Springer, New York (2013)
5. Cassioli, A., Bordeaux, B., Bouvier, G., Mucherino, A., Alves, R., Liberti, L., Nilges, M., Lavor, C., Malliavin, T.: An algorithm to enumerate all possible protein conformations verifying a set of distance constraints. BMC Bioinf. **16**, 23–38 (2015)
6. COIN-OR: Introduction to IPOPT: a tutorial for downloading, installing, and using IPOPT (2006)
7. Connelly, R.: Generic global rigidity. Discret. Comput. Geom. **33**, 549–563 (2005)
8. D'Ambrosio, C., Vu, K., Lavor, C., Liberti, L., Maculan, N.: New error measures and methods for realizing protein graphs from distance data. Discrete Comput. Geom. **57**(2), 371–418 (2017)
9. Deo, N., Prabhu, G., Krishnamoorthy, M.: Algorithms for generating fundamental cycles in a graph. ACM Trans. Math. Softw. **8**(1), 26–42 (1982)
10. Fourer, R., Gay, D.: The AMPL Book. Duxbury Press, Pacific Grove (2002)
11. Golynski, A., Horton, J.: A polynomial time algorithm to find the minimum cycle basis of a regular matroid. In: 8th Scandinavian Workshop on Algorithm Theory (2002)
12. Horton, J.: A polynomial-time algorithm to find the shortest cycle basis of a graph. SIAM J. Comput. **16**(2), 358–366 (1987)
13. Isaacs, R.: Infinite families of nontrivial trivalent graphs which are not Tait colorable. Am. Math. Month. **82**(3), 221–239 (1975)
14. Kavitha, T., Liebchen, C., Mehlhorn, K., Michail, D., Rizzi, R., Ueckerdt, T., Zweig, K.: Cycle bases in graphs: characterization, algorithms, complexity, and applications. Comput. Sci. Rev. **3**, 199–243 (2009)

15. Lavor, C., Liberti, L., Maculan, N.: Molecular distance geometry problem. In: Floudas, C., Pardalos, P. (eds.) Encyclopedia of Optimization, 2nd edn., pp. 2305–2311. Springer, New York (2009)
16. Lavor, C., Liberti, L., Maculan, N., Mucherino, A.: Recent advances on the discretizable molecular distance geometry problem. Eur. J. Oper. Res. **219**, 698–706 (2012)
17. Lee, J., Liberti, L.: A matroid view of key theorems for edge-swapping algorithms. Math. Methods Oper. Res. **76**, 125–127 (2012)
18. Liberti, L., Lavor, C.: Euclidean Distance Geometry: An Introduction. Springer, New York (2017)
19. Liberti, L., Lavor, C.: Open research areas in distance geometry. In: Migalas, A., Pardalos, P. (eds.) Open Problems in Optimization and Data Analysis. Springer Optimization and Its Applications, vol. 141, pp. 183–223. Springer, New York (2018)
20. Liberti, L., Lavor, C., Alencar, J., Abud, G.: Counting the number of solutions of ^{k}DMDGP instances. In: Nielsen, F., Barbaresco, F. (eds.) Geometric Science of Information. Lecture Notes in Computer Science, vol. 8085, pp. 224–230. Springer, New York (2013)
21. Liberti, L., Lavor, C., Maculan, N.: A branch-and-prune algorithm for the molecular distance geometry problem. Int. Trans. Oper. Res. **15**, 1–17 (2008)
22. Liberti, L., Lavor, C., Maculan, N., Mucherino, A.: Euclidean distance geometry and applications. SIAM Rev. **56**(1), 3–69 (2014)
23. Liberti, L., Lavor, C., Mucherino, A., Maculan, N.: Molecular distance geometry methods: from continuous to discrete. Int. Trans. Oper. Res. **18**, 33–51 (2010)
24. Liebchen, C., Rizzi, R.: A greedy approach to compute a minimum cycle basis of a directed graph. Inf. Process. Lett. **94**, 107–112 (2005)
25. Lovász, L., Plummer, M.: On minimal elementary bipartite graphs. J. Combin. Theory B **23**, 127–138 (1977)
26. Paton, K.: An algorithm for finding a fundamental set of cycles of a graph. Commun. ACM **12**(9), 514–518 (1969)
27. Saxe, J.: Embeddability of weighted graphs in k-space is strongly NP-hard. In: Proceedings of 17th Allerton Conference in Communications, Control and Computing, pp. 480–489 (1979)
28. Seshu, S., Reed, M.: Linear Graphs and Electrical Networks. Addison-Wesley, Reading (1961)
29. Singer, A.: Angular synchronization by eigenvectors and semidefinite programming. Appl. Comput. Harmonic Anal. **30**, 20–36 (2011)

The Unsuitable Neighbourhood Inequalities for the Fixed Cardinality Stable Set Polytope

Phillippe Samer and Dag Haugland

Abstract Given an undirected graph $G = (V, E)$ and an integer $k \in \{1, \ldots, |V|\}$, we initiate the combinatorial study of stable sets of cardinality exactly k in G. Our aim is to instigate the polyhedral investigation of the convex hull of fixed cardinality stable sets, and we begin by introducing a large class of valid inequalities to the natural integer programming formulation of the problem.

Keywords Stable sets · Independent sets · Cardinality constraints · Valid inequalities · Integer programming · Combinatorial optimization

1 From Conflict-Free Trees to Fixed Cardinality Stable Sets

We investigate a problem that is appealing to different research directions around algorithms, combinatorics and optimization. Let $G = (V, E)$ be a finite, simple, undirected graph, and denote $n = |V|$, and $m = |E|$. A stable set (or independent set, or co-clique) in G is a subset of pairwise non-adjacent vertices. Given $k \in \{1, \ldots, n\}$ and a vertex-weighting function $w : V \to \mathbb{Q}_+$, the *k stable set problem* consists in finding a minimum weight stable set of cardinality k in G, or deciding that none exists. Note that k is also part of the input to this problem; if it were an arbitrary fixed integer, the enumeration and optimization problems over stable sets of that cardinality could be solved in time polynomially bounded by a function of n.

Our original motivation for considering fixed cardinality stable sets stems from the NP-hard problem of determining *minimum spanning trees under conflict constraints* (MSTCC). Given a graph $G = (V, E)$ and a set of conflicting edge pairs $C \subseteq E \times E$, a *conflict-free spanning tree* in G is a set of edges $T \subseteq E$ inducing a spanning tree in G, such that for each $(e, f) \in C$, at most one of the edges e and

P. Samer (✉) · D. Haugland
Institutt for Informatikk, Universitetet i Bergen, Bergen, Norway
e-mail: samer@uib.no; dag.haugland@uib.no

© The Author(s), under exclusive license to Springer Nature Switzerland AG 2021
C. Gentile et al. (eds.), *Graphs and Combinatorial Optimization:*
from Theory to Applications, AIRO Springer Series 5,
https://doi.org/10.1007/978-3-030-63072-0_9

f is in T. The MSTCC problem, introduced by Darmann et al. [5], asks for such a conflict-free spanning tree of minimum weight.

Different combinatorial and algorithmic results about the MSTCC problem explore the associated conflict graph $H = (E, C)$, which has a vertex corresponding to each edge in the original graph G, and we represent each conflict constraint by an (undirected) edge connecting the corresponding vertices in H. Note that each conflict-free spanning tree in G is a subset of E which corresponds both to a spanning tree in G and to a stable set in H. Therefore, one can equivalently search for stable sets in H of cardinality exactly $|V| - 1$ which do not induce cycles in the original graph G.

It is not hard to devise different approaches for studying the MSTCC problem exploring the connection with fixed cardinality stable sets. Therefore, results of different nature from research on the k stable set problem (e.g. integer programming formulations and valid inequalities, well-solved particular cases, primal and dual bounds) could provide fundamental components to advance knowledge on the MSTCC problem as well.

It is surprising that the combinatorics and optimization literature has not addressed the k *stable set problem* problem in depth before. The convex hull of stable sets of cardinality *at most* k was studied by Janssen and Kilakos [8], but only for $k \in \{2, 3\}$. Apart from that article, it has also appeared as part of an algorithm for a variant of the survivable network design problem [3, Chapter 2], where only an alternative proof of one of the original results on [8] is given. We remark that the thorough survey on fixed cardinality versions of combinatorial optimization problems by Bruglieri et al. [4] does not mention stable sets, in spite of the major role played by that structure throughout the development of polyhedral combinatorics.

Our contribution with this work is twofold. First, we draw attention to the fixed cardinality version of a classical structure in combinatorial optimization and graph theory, motivated by its application in the MSTCC problem. Second, we introduce an exponential class of valid inequalities to the fixed cardinality stable set polytope, whose separation problem is interesting in its own right.

2 Polyhedral Results

For any graph G, we denote by $V(G)$ and $E(G)$ the sets of vertices and edges of G, respectively. For conciseness, we abbreviate 'stable set of cardinality k' as k-stab. The family of all k-stabs in G is denoted $\mathscr{F}(G, k)$. Recall that the incidence vector of any $S \subset V$ is $\chi^S \in \{0, 1\}^V$ defined by $\chi_i^S = 1$ if $i \in S$, and $\chi_i^S = 0$ if $i \in V \setminus S$; so the central object of our interest is $\mathfrak{C}(G, k) = \mathbf{conv}\left\{\chi^S : S \in \mathscr{F}(G, k)\right\}$, i.e. the convex hull of incidence vectors of all the k-stabs in G.

The natural integer programming (IP) formulation for minimum-weight k-stabs in G is

$$\min\left\{\sum_{v\in V} w(v)x_v : \mathbf{x} \in \mathscr{P}(G, k) \cap \{0, 1\}^n\right\}, \tag{1}$$

where $\mathscr{P}(G, k)$ denotes the polyhedral region defined by:

$$\sum_{v\in V} x_v = k \tag{2}$$

$$x_u + x_v \leq 1 \qquad\qquad \forall\, \{u, v\} \in E \tag{3}$$

$$0 \leq x_v \leq 1 \qquad\qquad \forall v \in V \tag{4}$$

Constraints (3) are known as *edge inequalities*, imposing that no two adjacent vertices belong to the selection in $\mathbf{x}$. Together with bounds (4), they determine the *fractional stable set polytope* [14, Section 64.5].

Remark 1 Recall that a vector z is *half-integer* if $2z$ is integer. A classical result of Nemhauser and Trotter [11] shows that the *fractional stable set polytope* is half-integer, i.e. all its vertices are $\left\{0, \frac{1}{2}, 1\right\}$-valued. Since that is the starting point for a series of both polyhedral and algorithmic advances, one could be interested in extending that result for $\mathscr{P}(G, k)$ as well. Unfortunately, we could verify that is not the case. While no small counterexample is found, we report a computational finding using benchmark instances from the minimum spanning tree under conflict constraints problem [13]. When $\hat{G}$ corresponds to the conflict graph associated with instance z100-300-1344 in that paper, which has 300 vertices and 1344 edges, and $k = 60$, the primal simplex method implemented in Gurobi Optimizer 8.1 (with all presolve, heuristics and cut options disabled) terminates with a solution corresponding to a vertex of $\mathscr{P}(\hat{G}, k)$ which is not half-integer.

We introduce next a class of valid inequalities for $\mathfrak{C}(G, k)$, exploring the relationship between k, the size of the neighbourhood

$$N(S) = \{u \in V\backslash S : \exists\, \{u, v\} \in E \text{ for some } v \in S\}$$

of any set $S \subset V$, and how many vertices from S can appear in any k-stab. First, denoting the set of neighbours of a vertex $v \in V$ by $\delta(v)$, that is $\delta(v) = N(\{v\})$, one can immediately observe that no vertex which has too many neighbours to still build a k-stab can be chosen. This gives the following simple reduction rule.

Proposition 1 *If $\mathbf{x}$ is the incidence vector of any k-stab, and $v \in V$ is such that $|\delta(v)| > n - k$, then $x_v = 0$.*

In an attempt to enforce an algebraic expression that enough vertices are left upon choosing a set $S \subset V$ towards building a k-stab, we introduce a class of exponentially-many constraints, which we refer to as *unsuitable neighbourhood inequalities* (UNI).

Theorem 1 *The inequality $\sum_{v \in S} x_v \leq |S| - 1$ is valid for $\mathfrak{C}(G, k)$, for each $S \subset V$ such that $1 \leq |S| < k$ and $|N(S)| > n - k$.*

Proof From $|S| < k$, it follows that S is not a k-stab in itself. If S were a subset of any k-stab, there should be at least $k - |S|$ vertices left to choose from, while no neighbour in $N(S)$ can be selected towards building a stable set. That is

$$n - |S| - |N(S)| \geq k - |S| \,, \qquad \forall S \subset V, 1 \leq |S| < k,$$

$$\Leftrightarrow \quad |N(S)| \leq n - k \,, \qquad \forall S \subset V, 1 \leq |S| < k.$$

Since $|N(S)| > n - k$ by hypothesis, S cannot be part of a k-stab. Therefore no incidence vector **x** of a k-stab induces the selection of all the vertices in S, and the result follows.   □

While Proposition 1 is clearly a special case of Theorem 1, one could ask whether the UNI indeed give a stronger condition. The positive answer follows next.

Theorem 2 *For any graph G and $k > 1$, the UNI imply the condition enforced by Proposition 1 in the description of $\mathfrak{C}(G, k)$, but the converse does not hold.*

Proof Let **x** be a vector satisfying all UNI. The inequalities in Proposition 1 are implied by the UNI with $|S| = 1$. Suppose that $S = \{u\}$ and $|N(S)| = |\delta(u)| > n - k$. Then u cannot be extended to a k-stab and the UNI include $x_u = \sum_{v \in S} x_v \leq |S| - 1 = 0$, which is the condition on the former proposition.

Now the converse does not hold, i.e. even if $|\delta(v)| \leq n - k$ for each $v \in V$, the UNI need not be automatically satisfied, as the following counterexample shows (see Fig. 1). Consider the graph $G = 2P_3$, which consists of two copies of the path graph on 3 vertices put together, so that $n = 6$, and suppose that $k = 3$. Since all vertices have degree 1 or 2, it follows that $|\delta(u)| \leq n - k = 3$ for each vertex u. On the other hand, with a test set S consisting of the two vertices of degree 2 in the middle of the paths, we have $1 \leq |S| < k$ and $|N(S)| = 4 > n - k$, thus yielding the unsuitable neighbourhood inequality given by $\sum_{v \in S} x_v \leq |S| - 1 = 1$ which separates from the convex hull $\mathfrak{C}(G, k)$ any vector selecting those two vertices.   □

Proposition 2 *In either of the following two conditions, the corresponding unsuitable neighbourhood inequality is redundant in $\mathfrak{C}(G, k)$: (i) if $S \subset V$ is not*

Fig. 1 The graph $2P_3$ and the selection of its two central vertices

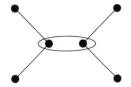

independent, or (ii) if $S \subset V$ is not minimal with respect to the condition $|N(S)| > n - k$.

Proof If $u, v \in S$ are adjacent vertices, the edge inequality $x_u + x_v \le 1$ implies $\sum_{v \in S} x_v \le |S| - 1$.

Otherwise, let $S \subset V$ with $1 \le |S| < k$ and $N(S) > n - k$ be a given independent set, and suppose that $T \subsetneq S$ is such that $|N(T)| > n - k$. The UNI corresponding to T is $\sum_{v \in T} x_v \le |T| - 1$. Combined with $x_v \le 1$ for each $v \in S \backslash T$, it implies the UNI corresponding to S, i.e. $\sum_{v \in S} x_v \le |S| - 1$, which is thus redundant in the description of $\mathfrak{C}(G, k)$. □

Recall that the *domination number* $\gamma(G)$ gives the least cardinality of a dominating set in $G = (V, E)$, i.e. a subset $D \subset V$ such that every vertex $u \in V \backslash D$ has a neighbour in D. If a lower bound on the domination number of G is known, the following result might be useful.

Proposition 3 *If $\gamma(G) \ge k$, then there exists no UNI for $\mathfrak{C}(G, k)$.*

Proof Suppose there were $S \subset V$ with $1 \le |S| < k$ and $|N(S)| > n - k$, and denote $T = V \backslash \{S \cup N(S)\}$. Note that any vertex belongs to exactly one among S, $N(S)$, or T; then

$$|S| + |N(S)| + |T| = n \implies |S| + |T| = n - |N(S)| \implies |S| + |T| < n - [n - k] = k,$$

since $|N(S)| > n - k$. Now, $S \cup T$ would be a dominating set of cardinality strictly less than k, contradicting the hypothesis that $\gamma(G) \ge k$. □

On the algorithmic side, it is in general impractical to include *a priori* all minimal UNI in an IP formulation for a black-box solver, since the number of those inequalities may grow exponentially with the size of the input (n, k). The natural approach in this case is to try to cut off successive solutions x^* to a linear programming (LP) relaxation, by finding cutting planes corresponding to UNI violated at x^*, i.e. separating x^* from $\mathfrak{C}(G, k)$, or deciding that none exists. Answering that question is known as the *separation problem* for a class of valid inequalities.

Definition 1 (Separation Problem for UNI) Given a graph $G = (V, E)$, with $n = |V|$, $k \in \{2, \ldots, n - 1\}$, and $x^* \in [0, 1]^n$ satisfying the conditions that $\sum_{v \in V} x_v^* = k$ and that $x_u^* + x_v^* \le 1$ for each $\{u, v\} \in E$, determine

i. either a set $S \subset V$, with $1 \le |S| \le k - 1$ and $|N(S)| \ge n - (k - 1)$, such that $\sum_{v \in S} x_v^* > |S| - 1$, in which case the unsuitable neighbourhood inequality corresponding to S separates x^* from $\mathfrak{C}(G, k)$,

ii. or that no such set exists, in which case all UNI are satisfied at x^*.

We give next a slight reformulation of the separation problem which might be useful in future work. Given the input $[G, k, x^*]$ corresponding to Definition 1, define $y^* \in [0, 1]^n$ such that $y_v^* = 1 - x_v^*$. Note now that $\sum_{v \in S} x_v^* > |S| - 1$

if and only if $\sum_{v \in S} y_v^* < 1$. We thus have the following equivalent statement of the problem.

Definition 2 (Equivalent Formulation of the Separation Problem for UNI)
Given a graph $G = (V, E)$, with $n = |V|$, $k \in \{2, \ldots, n - 1\}$, and $y^* \in [0, 1]^n$ satisfying the conditions that $\sum_{v \in V} y_v^* = n - k$ and that $y_u^* + y_v^* \geq 1$ for each $\{u, v\} \in E$, determine

i. either a set $S \subset V$, with $|N(S)| \geq n - (k-1)$ and $\sum_{v \in S} y_v^* < 1$, in which case the unsuitable neighbourhood inequality corresponding to S separates $x^* = \mathbf{1} - y^*$ from $\mathfrak{C}(G, k)$,
ii. or that no such set exists, in which case all UNI are satisfied at $x^* = \mathbf{1} - y^*$.

We consider this statement of the problem to be particularly appealing. Note that if S has size exactly $k - 1$, then $|N(S)| \geq n - (k - 1)$ implies that it would be a dominating set. Given the condition that adjacent vertices have y^* values summing up to at least 1, and that we require $\sum_{v \in S} y_v^* < 1$, we would actually have an *independent dominating set* if $|S| = k - 1$, i.e. a subset of vertices which is both dominating and independent (stable). Now, allowing $|S| \leq k - 1$ means that there might be $q \in \{0, 1, \ldots, k - 2\}$ vertices neither in S nor dominated by it. If we define a *q-quasi dominating set* in a graph $G = (V, E)$ to be a subset of vertices which is dominating in $G[V \backslash X]$, for some $X \subset V$, $|X| \leq q$, our separation problem corresponds to finding an independent $(k - 2)$-quasi dominating set of weight at most 1, or deciding that none exists. (Recall that, for any graph G and $U \subset V(G)$, the *induced subgraph* $G[U]$ is a graph with vertex set U and all of the edges in $E(G)$ which have both endpoints in U.)

We leave the open question of establishing the complexity of that problem.

Conjecture 1 The separation problem for UNI is NP-hard.

3   Concluding Remarks and Directions Towards a Branch-and-Cut Algorithm

We investigate in this work the fixed cardinality version of the classical stable set problem, highlighting an interesting gap in the combinatorial optimization literature. Generalizing the remark that vertices of too high degree cannot be in a stable set of cardinality k, we derive a large class of valid inequalities for the *k-stab polytope*. The corresponding separation problem asks for optimizing over subgraphs with a domination-like property and an additional budget constraint.

We are interested in a deeper polyhedral investigation, the starting point of which is to shed light on the relevance of the inequalities we introduce here, and how they relate to other families of valid inequalities for the classical stable set polytope. Moreover, progress in this direction could lead to an interesting algorithm for solving the MSTCC problem, as we indicate in Sect. 1. The remaining ingredient to find conflict-free spanning trees in the original graph G, from a cardinality $k = |V(G)| - 1$ stable set in the conflict-graph H, is to enforce an acyclic solution in the original graph. That could be attained by using a *relax-and-cut* approach (see [6, 9], for instance), separating subtour elimination constraints and immediately dualizing them in a Lagrangean fashion. In fact, Lucena [9] introduced an effective relax-and-cut algorithm for the fixed cardinality set partitioning problem.

We conclude by indicating selected insights on the practical issue of leveraging a modern branch-and-cut solver for the classical stable set problem (referring the reader to the eminently readable tutorial of Rebennack et al. [12]) towards one for the fixed cardinality version.

3.1 UNI Separation with MIP Heuristics

Besides the natural strategies of designing separation heuristics or including *a priori* some UNI corresponding to sets S of small cardinality, it might prove useful to explore an IP formulation of the separation problem. One can actually use good but not necessarily optimal solutions to that auxiliary IP, which give very effective cutting planes, for instance, in the context of an example of optimizing over the first Chvátal closure [2, Section 5.4]. Most MIP solvers include a collection of general purpose heuristics to accelerate the availability of integer feasible solutions, like local branching, feasibility pump and neighbourhood diving methods; see [7] for a recent survey.

The following is described in light of Definition 2, with input $[G, k, \mathbf{y}^*]$. We suppose further that the input is preprocessed by the reduction rules:

(i) Remove any vertex v such that $y_v^* = 1$;
(ii) Remove isolated vertices

Those operations do not change the problem answer, since a UNI is automatically satisfied if it contains a vertex with $y_v^* = 1$, and since isolated vertices are not contained in a minimal set S corresponding to a UNI.

For each $v \in V$, let variables $z_v \in \{0, 1\}$ be such that $z_v = 1$ if and only if $v \in S$, and $w_v \in \{0, 1\}$ be such that $w_v = 1$ if and only if $v \in N[S] = S \cup N(S)$, the *closed neighbourhood* of $S \subset V$. Then, we have to determine

$$\rho = \min \left\{ \sum_{v \in V} y_v^* \cdot z_v : (\mathbf{z}, \mathbf{w}) \in \mathscr{P}_{\text{UNI}}(G, \mathbf{y}^*) \cap \{0, 1\}^{2n} \right\}, \tag{5}$$

where $\mathscr{P}_{\mathrm{UNI}}(G, \mathbf{y}^*)$ denotes the polyhedral region:

$$\sum_{v \in V} (w_v - z_v) \geq n - (k - 1) \tag{6}$$

$$z_u \leq w_v \qquad\qquad \forall u \in V, \forall v \in N[u] \tag{7}$$

$$\sum_{u \in N[v]} z_u \geq w_v \qquad\qquad \forall v \in V \tag{8}$$

$$z_u + z_v \leq 1 \qquad\qquad \forall \{u, v\} \in E \tag{9}$$

$$0 \leq z_v \leq 1 \qquad\qquad \forall v \in V \tag{10}$$

$$0 \leq w_v \leq 1 \qquad\qquad \forall v \in V \tag{11}$$

The objective function in (5) accounts for the used $\mathbf{y}^*$ budget, as prescribed in Definition 2. Inequality (6) guarantees the minimum number of vertices dominated by S (excluding those which are in S). Inequalities (7) and (8) bind the binary variables $\mathbf{w}$ and $\mathbf{z}$, to enforce the domination condition that $w_v = 1$ if and only if $z_u = 1$ for some $u \in N[v]$.

Inequalities (9) are redundant, being implied at integer points in $\mathscr{P}_{\mathrm{UNI}}(G, \mathbf{y}^*)$ by (6) and the fact the input parameter satisfies $y_u^* + y_v^* \geq 1$ for each $\{u, v\} \in E$. Still, adding those inequalities is likely to tighten the LP relaxation bounds, and hence speed up the overall optimization procedure.

The exact separation problem thus reduces to deciding if $\rho < 1$. The MIP heuristic, on the other hand, consists of searching (e.g. allowing a MIP solver to run with a prescribed time limit) for any integer feasible solution $(\mathbf{z}', \mathbf{w}')$ with an objective value less than 1, which determines the UNI $\sum_{v \in S'} x_v \leq |S'| - 1$, with $S' = \{v \in V : z_v' = 1\}$, violated at $x^* = \mathbf{1} - \mathbf{y}^*$.

3.2 Balanced Branching

A fundamental component for the performance of branch-and-cut algorithms for the classical stable set problem is the balanced branching rule of Balas and Yu [1]; see also [12] and [10]. Its original motivation also applies to the fixed cardinality setting: avoiding unbalanced branch-and-bound trees when branching on a fractional variable x_v, since fixing $x_v = 1$ has the larger impact of implying $x_u = 0$ for each $u \in N(v)$, while fixing $x_v = 0$ has no impact on the neighbourhood.

The general branching scheme can be adapted to find minimum weight k-stabs with little effort. Suppose that, on a given node of the enumeration tree, $G' = (V', E')$ denotes the subgraph induced by vertices not fixed in this subproblem, and that $\bar{z}$ is the best primal bound available. Let $W \subseteq V'$ be such that we can determine *efficiently* that the minimum weight of a k-stab in the subgraph induced by W, denoted $z(W)$, is such that $z(W) \geq \bar{z}$. Note that, if $W = V'$, the subproblem is fathomed and the whole subtree rooted on this node can be pruned. Otherwise, if the search on this subtree is to eventually find that $z(V') < \bar{z}$, any bound-improving solution must intersect $V' \setminus W = \{v_1, \ldots, v_p\}$. That is, we can partition the search space into the sets

$$V'_i = \{v_i\} \bigcup V' \setminus (N(v_i) \cup \{v_{i+1}, \ldots, v_p\})$$

for $1 \leq i \leq p$. The enumeration can therefore branch on p subproblems, each fixing $x_{v_i} = 1$, and fixing at 0 those variables corresponding to $N(v_i) \cup \{v_{i+1}, \ldots, v_p\}$.

Now, there are different strategies to determine subgraph W. The standard one is to find a collection of cliques in G', e.g. with as many cliques as the currently available lower bound, when searching for maximum cardinality stable sets. For minimum-weight k-stabs, the natural idea would be to greedily find k cliques, such that the combined weight of the *cheapest* vertices in each exceed $\bar{z}$. We describe next an alternative approach tailored for optimizing over k-stabs, leaving for future work the task of comparing those two strategies, whether theoretically or according to computational experience.

Recall that a *matching* in a graph is a subset of pairwise non-adjacent edges, that is, a subset of edges without common vertices. Since each k-stab contains at most one vertex from each edge in a matching, a lower bound on $z(W)$ can be derived by simply picking the k vertices of lowest weight among: (i) the cheapest vertex in each matched edge, and (ii) the remaining vertices not covered by the matching. We thus have the following result, which we state in the general form of a combinatorial dual bound for the minimum weight of k-stab in an arbitrary graph.

Theorem 3 *Suppose that $\mathscr{P}(G, k) \cap \{0, 1\}^n \neq \varnothing$, so that problem (1) is well-defined. Let $M \subset E$ be any matching in G. Define $c_e = \min\{w(v_i), w(v_j)\}$ for each edge $e = \{v_i, v_j\} \in M$. Also define $c_u = w(v_u)$ for any vertex v_u not covered by the matching M. Then, the sum of the k lowest values in the image of $c(\cdot)$ is a lower bound on (1). That is, given an order $c_1 \leq c_2 \cdots \leq c_{(n-|M|)}$ on $\{c_e\}_{e \in M} \cup \{c_u\}_{u \in V \setminus V_M}$, where V_M corresponds to the set of vertices covered by M, we have that $\sum_{i=1}^k c_i$ is a lower bound on the weight of a k-stab in G.*

Therefore, using the weight function corresponding to $c(\cdot)$ in the above theorem, we can determine candidate subgraphs W by inspecting, for each $l \in \{1, \ldots, k\}$:

1. A minimum-weight matching in G' with cardinality l
2. A suitable choice of $k - l$ vertices not covered by the matching

Finally, note that finding a minimum-weight matching of a specified cardinality in a graph is a well-solved problem. More generally, for any $l, u \in \mathbb{Z}_+$, $l \leq u$, the convex hull of incidence vectors of matchings $M \subset E(G)$ such that $l \leq |M| \leq u$ is equal to the set of those vectors in the matching polytope of G satisfying $l \leq \mathbf{1}^\top x \leq u$, that is, $l \leq \sum_{e \in E(G)} x(e) \leq u$; see [14, Section 18.5f].

References

1. Balas, E., Yu, C.S.: Finding a maximum clique in an arbitrary graph. SIAM J. Comput. **15**(4), 1054–1068 (1986). https://doi.org/10.1137/0215075
2. Bertsimas, D., Weismantel, R.: Optimization Over Integers. Dynamic Ideas, Belmont (2005)
3. Botton, Q.: Survivable network design with quality of service constraints: extended formulations and benders decomposition. Ph.D. thesis, Université Catholique de Louvain, Louvain-la-Neuve (2010). http://hdl.handle.net/2078.1/33300
4. Bruglieri, M., Ehrgott, M., Hamacher, H.W., Maffioli, F.: An annotated bibliography of combinatorial optimization problems with fixed cardinality constraints. Discrete Appl. Math. **154**(9), 1344–1357 (2006). https://doi.org/10.1016/j.dam.2005.05.036
5. Darmann, A., Pferschy, U., Schauer, J., Woeginger, G.J.: Paths, trees and matchings under disjunctive constraints. Discrete Appl. Math. **159**(16), 1726–1735 (2011). http://dx.doi.org/10.1016/j.dam.2010.12.016
6. Guignard, M.: Efficient cuts in Lagrangean 'relax-and-cut' schemes. Eur. J. Oper. Res. **105**(1), 216–223 (1998). https://doi.org/10.1016/S0377-2217(97)00034-9
7. Hanafi, S., Todosijević, R.: Mathematical programming based heuristics for the 0–1 mip: a survey. J. Heuristics **23**(4), 165–206 (2017). https://doi.org/10.1007/s10732-017-9336-y
8. Janssen, J., Kilakos, K.: Bounded stable sets: polytopes and colorings. SIAM J. Discrete Math. **12**(2), 262–275 (1999). https://doi.org/10.1137/S089548019630978X
9. Lucena, A.: Non delayed relax-and-cut algorithms. Ann. Oper. Res. **140**(1), 375–410 (2005). https://doi.org/10.1007/s10479-005-3977-1
10. Mannino, C., Sassano, A.: Edge projection and the maximum cardinality stable set problem. In: Johnson, D.S., Trick, M. (eds.) Cliques, Coloring, and Satisfiability: Second DIMACS Implementation Challenge, 11–13 Oct 1993, vol. 26, pp. 205–219. American Mathematical Society, Providence (1996). DIMACS Series in Discrete Mathematics and Theoretical Computer Science
11. Nemhauser, G.L., Trotter, L.E.: Properties of vertex packing and independence system polyhedra. Math. Program. **6**(1), 48–61 (1974). https://doi.org/10.1007/BF01580222
12. Rebennack, S., Reinelt, G., Pardalos, P.M.: A tutorial on branch and cut algorithms for the maximum stable set problem. Int. Trans. Oper. Res. **19**(1–2), 161–199 (2012). https://doi.org/10.1111/j.1475-3995.2011.00805.x
13. Samer, P., Urrutia, S.: A branch and cut algorithm for minimum spanning trees under conflict constraints. Optim. Lett. **9**(1), 41–55 (2015). https://doi.org/10.1007/s11590-014-0750-x
14. Schrijver, A.: Combinatorial Optimization: Polyhedra and Efficiency. Algorithms and Combinatorics, vol. 24. Springer, Berlin (2003)

Relating Hypergraph Parameters of Generalized Power Graphs

Lucas L. S. Portugal, Renata Del Vecchio, and Simone Dantas

Abstract Graph parameters like the chromatic number, independence number, clique number and many others alongside with their corresponding adjacency matrix have been broadly studied and extended to hypergraphs classes. A generalized power graph G_s^k of a graph G is a k-uniform hypergraph constructed by blowing up each vertex of G into a s-set of vertices and then adding $k - 2s$ vertices of degree one to each edge, where $k \geq 2s$. A natural question is whether there exists any relation between structural parameters and spectral parameters of G_s^k with the respective parameters of the original graph G. In this paper we positively answer this question and investigate the parameters behavior.

Keywords Hypergraph · Generalized power graph · Strong chromatic number · Adjacency matrix of hypergraph · Spectral parameters

1 Introduction

A *hypergraph* $H = (V, E)$ is given by a vertex set V and a set $E = \{e : e \subseteq V\}$, whose elements are called (hyper) edges. A *graph* $G = (V, E)$ is a hypergraph such that $|e| \leq 2$ for every $e \in E$.

Different aspects of a graph like clique number, vertex or edge coloring, matching, connectivity, have been widely studied in many areas and can be generalized to hypergraph theory, for example hypergraph coloring and strong hypergraph coloring, weak and strong vertex connectivity [4, 9]. In [1], the authors stated that strong hypergraph coloring captures many previously studied graph coloring properties. These different ways of expanding a graph parameter have attracted the attention of researchers: [9] studied the difference between weak and strong vertex connectivity; and [2, 7, 11] exclusively focused their work on a single parameter.

L. L. S. Portugal (✉) · R. D. Vecchio · S. Dantas
IME, Universidade Federal Fluminense, Rio de Janeiro, Brazil
e-mail: lucasportugal@id.uff.br; rrdelvecchio@id.uff.br; sdantas@id.uff.br

© The Author(s), under exclusive license to Springer Nature Switzerland AG 2021
C. Gentile et al. (eds.), *Graphs and Combinatorial Optimization:*
from Theory to Applications, AIRO Springer Series 5,
https://doi.org/10.1007/978-3-030-63072-0_10

Spectral graph theory is another area that can be extended to hypergraphs. The goal of spectral graph theory is to study eigenvalues and eigenvectors of matrices associated with graphs finding information of structural properties of these graphs. Many graph matrices are studied in spectral graph theory which can also be extended to hypergraphs in different ways. The study of hypergraph matrices started in the 1990s with a generalization of the graph adjacency matrix [10], and new matrices are still being defined. In 2019, [3], a similar but different adjacency matrix of a hypergraph is defined, allowing a generalization of important spectral graph theory results to hypergraphs. Another approach to spectral hypergraph theory was given in 2012, [8], when it is proposed the study of hypergraphs through tensors.

This work aims to investigate the relation between hypergraph structural parameters and spectral parameters of a class of uniform hypergraphs, called generalized power graph, that was first considered in [15]. Recently, this class was studied by considering its tensor spectra [13–15].

Since these hypergraphs are constructed from a base graph, we discuss four main topics: the relation between hypergraph parameters with their respective graph parameters; the behavior of distinct variations of generalized graph parameters on this hypergraph class; the relation between the adjacency matrix of this hypergraph with matrices of the base graph; and new relations of hypergraph parameters and the adjacency matrix eigenvalues.

2 Preliminaries

A *hypergraph* $H = (V, E)$ is *k-uniform* if $|e| = k$ for every edge $e \in E(H)$. A *simple graph* $G = (V, E)$ is a 2-uniform hypergraph. In this work we consider only *simple hypergraphs*, i.e. it contains no loops (edges with $|e| = 1$) and no repeated edges. A *null hypergraph* contains no vertices (or no edges) and a hypergraph with only one vertex is called *trivial*. Two vertices in a hypergraph are *adjacent* if there is an edge which contains both vertices, and the *degree of a vertex* $v \in V$ is $d(v) = |\{e : v \in e\}|$, the number of edges that contain v.

A *path* P in a hypergraph H is a vertex-edge alternating sequence: $P = v_0, e_1, v_1, e_2, \ldots, v_{r-1}, e_r, v_r$ such that $v_0, v_1, \ldots, v_r$ are distinct vertices; $e_1, e_2, \ldots, e_r$ are distinct edges; and $v_{i-1}, v_i \in e_i$, $i = 1, 2, \ldots, r$. The *length* of a path P is the number of distinct edges. A hypergraph is *connected* if for any pair of vertices, there is a path which connects these vertices; it is *not connected* otherwise.

Let G be a graph and $s \geq 1$ an integer. The *s-extension* G_s of G is a $2s$-uniform hypergraph obtained from G by replacing each vertex $v_i \in V$ by a set $S_{v_i} = \{v_{i1}, \ldots, v_{is}\}$, where $S_{v_i} \cap S_{v_j} = \emptyset$ for every $v_i \neq v_j$. These s new vertices are called *copies* of v_i. More precisely, $V(G_S) = \{v_{11}, \ldots, v_{1s}, \ldots, v_{n1}, \ldots, v_{ns}\}$ and $E(G_s) = \{S_{v_i} \cup S_{v_j} : \{v_i, v_j\} \in E\}$. Note that $|V(G_s)| = s \cdot |V(G)|$ and $|E(G_s)| = |E(G)|$.

For a graph $G = (V, E)$ and an integer $k \geq 2$, the *k-expansion* G^k of G (also called the kth *power graph of* G) is a k-uniform hypergraph obtained from G by

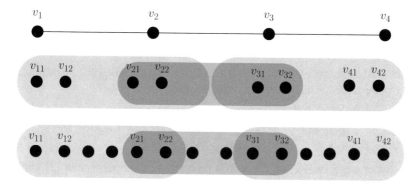

Fig. 1 Graph $G = P_4$ an its respective G_2 and G_2^6

adding $k - 2$ new vertices of degree one to each edge of G. Note that $|V(G^k)| = |V(G)| + (k - 2) \cdot |E(G)|$ and $|E(G^k)| = |E(G)|$.

Let $s \geq 1$ and $k \geq 2s$ be two integers and consider a graph G. The *generalized power graph* G_s^k is the k-uniform hypergraph $(G_s)^k$, obtained by adding $k - 2s$ new vertices to each edge of G_s. These $(k - 2s) \cdot |E(G)|$ new vertices of degree one are called *additional vertices* of G_s^k. Note that $|V(G_s^k)| = s \cdot |V(G)| + (k - 2s) \cdot |E(G)|$ and $|E(G_s^k)| = |E(G)|$. See an example in Fig. 1.

Let G be a simple graph with n vertices. The *adjacency matrix* of G, denoted by $A(G)$, is the $n \times n$ symmetric matrix with entries $a_{ij} = 1$ if there is an edge joining vertices v_i and v_j; and $a_{ij} = 0$ otherwise. The *degree matrix* of G, denoted by $D(G)$, is the $n \times n$ diagonal matrix defined as $D(G) = Diag(d(v_1), \ldots, d(v_n))$ where $d(v_i)$ is the degree of the vertex v_i. The *signless Laplacian matrix* for G, denoted by $Q(G)$, is the $n \times n$ symmetric matrix given by $Q(G) = D(G) + A(G)$. We denote the eigenvalues of $A(G)$ as $\lambda_1(G) \geq \ldots \geq \lambda_n(G)$ and the eigenvalues of $Q(G)$ as $q_1(G) \geq \ldots \geq q_n(G)$.

Let H be a hypergraph with n vertices. The *adjacency matrix* of H, denoted by $A(H)$ is the $n \times n$ symmetric matrix with entries $a_{ij} = |\{e \in E(H) : v_i, v_j \in e\}|$. We also denote the eigenvalues of $A(H)$ as $\lambda_1(H) \geq \ldots \geq \lambda_n(H)$.

Note that all previously defined matrices are real and symmetric, so they are Hermitian (a square matrix that is equal to its own conjugate transpose).

Now, we recall some matrix theory results that we use latter. Let X be a $m \times n$ matrix and let Y be a $p \times q$ matrix. The *kronecker product* $X \otimes Y$ is the $mp \times nq$ matrix:

$$X \otimes Y = \begin{bmatrix} x_{11}Y & \ldots & x_{1n}Y \\ \vdots & \ddots & \vdots \\ x_{m1}Y & \ldots & x_{mn}Y \end{bmatrix}.$$

Theorem 1 ([16]) *Let X be a $n \times n$ matrix and Y a $m \times m$ matrix. If $x_1 \geq \ldots \geq x_n$ are the eigenvalues of X and $y_1 \geq \ldots \geq y_m$ the eigenvalues of Y, then the nm eigenvalues of $X \otimes Y$ are: $x_1 y_1, \ldots, x_1 y_m, x_2 y_1, \ldots, x_2 y_m, \ldots, x_n y_1, \ldots, x_n y_m$.*

The next theorem, by Weyl [12], is a well known inequality that gives lower and upper bounds for the eigenvalues of a matrix sum.

Theorem 2 ([12]) *Let X and Y be square $n \times n$ Hermitian matrices with eigenvalues $x_1 \geq \ldots \geq x_n$ and $y_1 \geq \ldots \geq y_n$ respectively. If the eigenvalues of the sum $Z = X + Y$ are $z_1 \geq \ldots \geq z_n$, then $x_k + y_n \leq z_k \leq x_k + y_1$.*

A hypergraph version of the Wilf's theorem was established in [3] stating a relation between the chromatic number and the largest eigenvalue of its adjacency matrix. This generalization can be restricted to uniform hypergraphs as follows:

Theorem 3 ([3]) *Let H be a k-uniform hypergraph, then $\chi_S(H) \leq 1 + \lambda_1(H)$.*

3 Structural Parameters

Graph parameters can be extended to hypergraphs and most of them in more than one way. In this section we investigate how these parameters behave on the class G_s^k and their relation with the respective parameters of the original graph G.

Let H be a k-uniform hypergraph. A set $U \subseteq V(H)$ is a *clique* if every subset of U with k elements is an edge of H. The *clique number* is $\omega(H) = max\{|U| : U \subseteq V(H)$ is a clique$\}$.

Proposition 1 *Given a graph G with at least one edge, $s \geq 1$ and $k \geq 2s$ (except the case where $s = 1$ and $k = 2$, i.e. $G_s^k = G$), we have that $\omega(G_s^k) = k$. Moreover, every clique in G_s^k is composed by the k vertices of an edge.*

Proof First, observe that the intersection between two edges of G_s^k is formed by a set of s vertices or is empty. Choose any set of $k + 1$ vertices of G_s^k and suppose it is a clique. This means that there exist two edges in G_s^k which share $k - 1$ common vertices. This is a contradiction since $k \geq 2s$, $s \neq 1$ and $k \neq 2$. Clearly any set of k vertices of an edge is a clique.   □

A *matching* of a hypergraph $H = (V, E)$ is a set $M \subset E$ of pairwise disjoint hyperedges of H. The *matching number* $v(H)$ is the cardinality of a maximum matching.

Proposition 2 *If G is a graph with $s \geq 1$ and $k \geq 2s$, then $v(G_s^k) = v(G)$.*

A *perfect matching* of a hypergraph H is a matching M such that each vertex in $V(H)$ is covered by exactly one edge in M. It is easy to see that for $s \geq 1$, G_s has a perfect matching if and only if G has a perfect matching.

Proposition 3 *Let G be a graph that is not the union of disjoint edges. For $s \geq 1$ and $k > 2s$ the hypergraph G_s^k does not have a perfect matching.*

Proof Since $k > 2s$, each edge of G_s^k have $k - 2s$ vertices of degree one. The only way that all those vertices are covered by a matching M is when $M = E(G_s^k)$, and that can happen only when G is the union of disjoint edges. $\qquad\square$

Given a hypergraph $H = (V, E)$ we construct new hypergraphs by deleting vertices in the following ways. The *strong vertex deletion* of a vertex $v \in V$ creates the hypergraph $H' = (V', E')$ where $V' = V - v$ and $E' = \{e \in E : v \notin e\}$. That is, the strong deletion of v removes v from the vertex set and removes all edges that contain v from the hypergraph. For any subset X of V, we use $H -_{(S)} X$ to denote the hypergraph formed by strongly deleting all the vertices of X from H. A vertex $v \in V$ is called a *strong cut vertex* of H if $H -_{(S)} v$ has more connected components than H, and a set $X \subseteq V$ is called a *strong vertex cut* of H if $H -_{(S)} X$ is disconnected. We define the *strong vertex connectivity* of H, denoted $\kappa_S(H)$ as follows: if H has at least one strong vertex cut, then $\kappa_S(H)$ is the cardinality of a minimum strong vertex cut of H; otherwise, $\kappa_S(H) = |V| - 1$. By convention, the strong vertex connectivity of a null or trivial hypergraph is 1. Observe that $\kappa_S(H) \leq \delta(H)$.

Proposition 4 *Given a connected graph G, $s \geq 1$ and $k \geq 2s$ integers such that $G_s^k \neq G$ then $\kappa_S(G_s^k) = 1$.*

Proof If $k > 2s$ removing a vertex that is originally from G_s disconnects G_s^k, since its deletion removes at least one edge and hence the $k - 2s$ additional vertices of this edge become isolated. Similarly, if $k = 2s$ then $s > 1$ and $G_s^k = G_s$. Removing any vertex leaves the $s - 1$ vertices that are its copies isolated. $\qquad\square$

The *weak vertex deletion* of a vertex $v \in V$ creates the hypergraph $H' = (V', E')$ where $V' = V - v$ and $E' = \{e - \{v\} : e \in E\}$. That is, the weak deletion of v removes v from the vertex set, and all occurrences of v from the edges of the hypergraph H. For any subset X of V, we use $H -_{(W)} X$ to denote the hypergraph formed by weakly deleting all the vertices of X from H. Since we are only considering simple hypergraphs, we remove edges with only one vertex. A vertex $v \in V$ is called a *weak cut vertex* of H if $H -_{(W)} v$ has more connected components than H, and a set $X \subseteq V$ is called a *weak vertex cut* of H if $H -_{(W)} X$ is disconnected. We define the *weak vertex connectivity* of H, denote $\kappa_W(H)$ as follows: if H has at least one weak vertex cut, then $\kappa_W(H)$ is the cardinality of a minimum weak vertex cut of H; otherwise, $\kappa_W(H) = |V| - 1$. By convention, the weak vertex connectivity of a null or trivial hypergraph is 1.

Proposition 5 *Given a connected graph G that is not the complete graph and an integer $s \geq 1$, then $\kappa_W(G_s) = s.\kappa(G)$.*

Proof Note that by the construction of G_s, we have that if $X \subset V(G_s)$ is a weak vertex cut of G_s then $X = S_{v_1} \cup S_{v_2} \ldots \cup S_{v_r}$ and $\{v_1, \ldots, v_r\} \subseteq V(G)$ is a vertex cut of G. Now, let $\{v_1, \ldots, v_r\}$ be a minimum vertex cut of G. So, $S_{v_1} \cup S_{v_2} \ldots \cup S_{v_r}$ is a minimum weak vertex cut in G_s with $s.\kappa(G)$ elements, otherwise $v_1, \ldots v_r$ would not be a minimum vertex cut of G, a contradiction. $\qquad\square$

Proposition 6 *Let G be a connected graph that is not the complete graph, $s \geq 1$ and $k > 2s$ be two integers. Then:*

(i) If $\kappa_W(G_s) = s$, then $\kappa_W(G_s^k) = s$.
(ii) If $\kappa_W(G_s) \geq 2s$, then $\kappa_W(G_s^k) = 2s$.

Proof First we observe that a vertex cut of G_s is a vertex cut of G_s^k. Also, after the k-expansion G_s^k of G_s, the only new minimum vertex cut is the one where we isolate the additional $k - 2s$ new vertices of an edge by removing the $2s$ already existing vertices (since the new vertices of G_s^k make no difference in a vertex cut). Hence:

(i) if X is a minimum weak vertex cut of G_s with less than $2s$ elements, then it is a minimum weak vertex cut of G_s^k.
(ii) if $\kappa_W(G_s) \geq 2s$, a minimum weak vertex cut of G_s has more than $2s$ elements. For each edge of G_s^k, the set of the $2s$ vertices that came from G^s is a minimum weak vertex cut of G_s^k since their removal leaves the additional $k - 2s$ remaining vertices isolated. □

Next result follows from the fact that if $s = 1$ then $G_s = G$ and $G_s^k = G^k$.

Corollary 1 *Let G be a connected graph. For any $k > 2$ we have that:*

(i) if $\kappa(G) = 1$, then $\kappa_W(G^k) = 1$;
(ii) if $\kappa(G) \geq 2$, then $\kappa_W(G^k) = 2$.

We observe from the previous results that the difference between weak and strong vertex connectivity of hypergraphs can be arbitrarily large, since $\kappa_S(G_s^k) = 1$ and $\kappa_W(G_s^k) \geq s$, with s as large as desired. Finally, we also remark that the inequality $\kappa_W(H) \leq \delta(H)$ is not valid: if G is a connected graph with $\kappa(G) \geq 2$, we have for $k > 2s$ that $\delta(G_s^k) = 1 < 2s = \kappa_W(G_s^k)$.

The *distance* $d(v, u)$ between two vertices v and u is the minimum length of a path that connects v and u. The *diameter* $d(H)$ of H is defined by $d(H) = max\{d(v, u) : v, u \in V\}$. It is easy to see that given a graph G and $s \geq 1$, then $d(G_s) = d(G)$. But this is not always true for the k-expansion.

Proposition 7 $d(G) \leq d(G_s^k) \leq d(G) + 2$, *for any graph G, $s \geq 1$ and $k \geq 2s$,* .

Proof Suppose $d(G_s) = r$ and $P = v_1, e_1, v_2, e_2, \ldots, v_r, e_r, v_{r+1}$ be a maximum path of G_s. If $k > 2s$, we add $k - 2s$ vertices on each edge to obtain G_s^k. After that, if there is an additional vertex u such that $\{u, v_1\}$ belongs to an edge $e \neq e_1$ and another additional vertex w such that $\{w, v_{r+1}\}$ belongs to an edge $f \neq e_r$, the path $P = u, e, v_1, e_1, v_2, e_2, \ldots, v_r, e_r, v_{r+1}, f, w$ have length $d(G) + 2$. Moreover, $d(G_s^k) = d(G) + 2$ since we have at most 2 additional vertices on a path and the path must start and end on them, otherwise we would have to repeat edges. □

A *hypergraph coloring* is an assigning of colors $\{1, 2, \ldots, c\}$ to each vertex of $V(H)$ in such a way that each edge contains at least two vertices of distinct colors. A coloring using at most c colors is called a *c-coloring*. The *chromatic number* $\chi(H)$ of a hypergraph H is the least integer c such that H has a c-coloring.

It is easy to see that given a graph G we have that $\chi(G^k) = \chi(G_s) = \chi(G_s^k) = 2$ (except when $s = 1$ and $k = 2$). Another type of coloring, that is also a generalization of graph coloring, is the *strong hypergraph coloring*: is an assigning of colors $\{1, 2, \ldots, c\}$ to each vertex of $V(H)$ in such a way that every vertex of an edge has distinct colors. The *strong chromatic number* $\chi_S(H)$ of a hypergraph H is the least integer c such that H has a strongly c-coloring. Given a hypergraph H, note that:

1. $\chi_S(H) \geq |e|$ for every $e \in E(H)$;
2. $\chi(H) \leq \chi_S(H)$, since a strong hypergraph coloring is also a hypergraph coloring;
3. $\omega(H) \leq \chi_S(H)$ (similarly to graphs);
4. for the class G_s^k, the inequality $\omega(H) \leq \chi(H)$ is not valid, since $\chi(G_s^k) = 2$ but we can have edges (cliques) arbitrarily large.

We do not consider $\chi(G) = \chi_S(G_s^k) = 1$, since G has at least one edge. The following results establish relations between $\chi(G)$, $\chi_S(G_s)$ and $\chi_S(G_s^k)$.

Proposition 8 *If G is a graph and $s \geq 1$ is an integer, then $\chi_S(G_s) \leq s.\chi(G)$.*

Proof Let $\chi(G) = c$, we obtain a sc-strong coloring of G_s as follows: if $v \in V(G)$ has color $c(v) \in \{1, \ldots, c\}$ then, in G_s, assign colors $\{1 + (c(v) - 1)s, 2 + (c(v) - 1)s, \ldots, s + (c(v) - 1)s\}$ to S_v. □

Note that this bound is tight in the sense that the equality holds for any s-extension of the complete graph and does not hold for the 2-extension of C_5.

Proposition 9 *Let $s \geq 1$, $k > 2s$ be two integers and let G be a graph. We have that:*

(i) if $\chi_S(G_s) < k$ then $\chi_S(G_s^k) = k$;
(ii) if $\chi_S(G_s) \geq k$ then $\chi_S(G_s^k) = \chi_S(G_s)$.

Proof

(i) Let $\chi_S(G_s) = c < k$, we obtain a k-strong coloring of G_s^k as follows: we color the vertices of G_s^k that came from G_s with the same c-colors used in G_s. Hence, for each edge of G_s^k, we already used $2s$ colors from the set $\{1, 2, .., k\}$, $k > 2s$. Again, for each edge, we color the $k - 2s$ new additional vertices with the remaining $k - 2s$ distinct colors. This k-strong coloring of G_s^k is minimum, since k is the size of each edge of G_s^k.

(ii) Let $\chi_S(G_s) = c \geq k$ and consider a c-strong coloring of G_s. We color the vertices of G_s^k that came from G_s with the same c-colors used in G_s. For each edge, we color the $k - 2s$ additional vertices with any $k - 2s$ distinct colors from $\{1, 2, .., c\}$ different from the $2s$ colors already used in the vertices that came from G_s (since $c \geq k \geq 2s$ such colors exist). Suppose that it is possible to use less than c-colors in G_s^k. This implies that we can color all the vertices of G_s^k that came from G_s with less than c-colors and hence G_s with less than c colors, a contradiction. □

Corollary 2 *Let G be a graph and $k \geq 2$ an integer. Thus:*

(i) if $\chi(G) < k$ then $\chi_S(G^k) = k$;
(ii) if $\chi(G) \geq k$ then $\chi_S(G^k) = \chi(G)$.

A set $U \subseteq V$ is a *strong independent set* if no two vertices of U are adjacent. The *strong independence number* is $\alpha'(H) = max\{|U| : U \subseteq V(H)$ is a strong independent set of H }. Let G be a graph and $s \geq 1$. From the construction of G_s we have that $\alpha'(G_s) = \alpha(G)$.

Proposition 10 *If G is a graph, $s \geq 1$ and $k > 2s$, then $\alpha'(G_s^k) = |E(G)|$.*

Proof Since $k > 2s$, every edge of G_s^k has at least one additional vertex. A set formed by choosing, for each edge, one of these additional vertices is a strong independent set of size $|E(G_s^k)| = |E(G)|$. This set is maximum since $\alpha'(H) \leq |E(H)|$, for any hypergraph H.   □

Another generalization of a graph independent set is as follows: a set $U \subseteq V$ is an independent set if no edge of H is contained in U. As before, the *independence number* is $\alpha(H) = max\{|U| : U \subseteq V(H)$ is an independent set of H }. Observe that if U is a strong independent set of a hypergraph H then U is also an independent set of H, since if U contains no two adjacent vertices then U does not contain an edge of H. So we have that $\alpha'(H) \leq \alpha(H)$.

Proposition 11 *If G is a graph and $s \geq 1$, then $\alpha(G_s) \geq (s-1) \cdot |V(G)| + \alpha(G)$.*

Proof Let $V(G) = \{v_1, \ldots, v_n\}$ and $V(G_s) = S_{v_1} \cup \ldots \cup S_{v_n}$. We obtain an independent set with $(s-1) \cdot n$ elements by choosing $s-1$ vertices of S_{v_i}, for each $i = 1, \ldots, n$. Now, adding a maximum stable set of G to the previous set produces a stable set of G_s with $(s-1) \cdot n + \alpha(G)$ vertices.   □

Proposition 12 *If G be a graph, $s \geq 1$ and $k \geq 2s$, then $\alpha(G_s^k) \geq (s-1) \cdot |V(G)| + \alpha(G) + (k-2s) \cdot |E(G)|$.*

Proof By the construction of G_s^k and Proposition 11, a stable set of G_s is also a stable set of G_s^k with $(s-1) \cdot |V(G)| + \alpha(G)$ vertices. Adding to this stable set every $k - 2s$ additional vertices of each edge of G_s^k produces a stable set with $(s-1) \cdot |V(G)| + \alpha(G) + (k-2) \cdot |E(G)|$ elements.   □

Corollary 3 *Let G be a graph and $k \geq 2$, then $\alpha(G^k) \geq \alpha(G) + (k-2) \cdot |E(G)|$.*

4   Spectral Parameters

In this section we investigate spectral properties of hypergraphs and establish relations with structural parameters. The following result relates the adjacency matrix of G_s with the matrices $A(G)$ and $Q(G)$.

Proposition 13 *Let G be a graph with n vertices and s > 1. The adjacency matrix $A(G_s)$ is given on $s \times s$ blocks of size $n \times n$ by:*

$$
A(G_s) = \begin{bmatrix} A(G) & Q(G) & Q(G) & \cdots & Q(G) \\ Q(G) & A(G) & Q(G) & \cdots & Q(G) \\ Q(G) & Q(G) & A(G) & \cdots & Q(G) \\ \vdots & \vdots & \vdots & \ddots & \vdots \\ Q(G) & Q(G) & Q(G) & \cdots & A(G) \end{bmatrix} = (J_s \otimes Q(G)) + (I_s \otimes -D(G)),
$$

where J_s is the $s \times s$ matrix with 1 on all entries and I_s is the $s \times s$ identity matrix.

Proof First we show that $A(G_s)$ can be written in blocks like above. Let G be a graph on n vertices, then $|V(G_s)| = sn$. So, we order the vertices of the matrix as follows: $V(G_s) = \{v_{11}, v_{21}, \ldots, v_{n1}, v_{12}, v_{22}, \ldots, v_{n2}, v_{13}, v_{23}, \ldots, v_{n3}, \ldots, v_{1s}, v_{2s}, \ldots, v_{sn}\}$, where $S_{v_1} = \{v_{11}, v_{12}, \ldots, v_{1s}\}$, $S_{v_2} = \{v_{21}, v_{22}, \ldots, v_{2s}\}$, $\ldots, S_{v_n} = \{v_{n1}, v_{n2}, \ldots, v_{ns}\}$. We suppose that the vertices $v_{11}, v_{21}, \ldots, v_{n1}$ are the vertices that come from G. So the $n \times n$ block formed by these is $A(G)$, since two vertices that are not copies from each other, share an edge in G_s if and only if they share an edge in G. Hence, we can see that all the diagonal blocks, formed by the vertices $\{v_{1i}, v_{2i}, \ldots, v_{ni}\} \times \{v_{1i}, v_{2i}, \ldots, v_{ni}\}$, $i = 1, \ldots, s$, also correspond to $A(G)$.

For the other blocks we observe that, for every $i \neq j$, the blocks formed by $\{v_{1i}, v_{2i}, \ldots, v_{ni}\} \times \{v_{1j}, v_{2j}, \ldots, v_{nj}\}$ are always the same, since the vertices are copies from one another.

The block where $i = 1$ and $j = 2$ have the following structure: the vertices v_{11} and v_{12} are copies so they are in the same edges; and the number of edges they belong is exactly $d_G(v_1)$. So, their entry is equal $d_G(v_1)$, the degree of v_1 in G. The same works for the entries $v_{21} \times v_{22}, v_{31} \times v_{32}, \ldots, v_{n1} \times v_{n2}$. So, the diagonal of the block is made of the degrees in G. The entries that are not in the diagonal, for example, the entry $v_{11} \times v22$ is the same entry as $v_{11} \times v_{21}$, since v_{22} is a copy of the vertex v_{21}. So, these blocks are equal $D(G) + A(G) = Q(G)$. □

Proposition 14 *Let G be a graph on n vertices, $s > 1$ an integer and $d_1, \ldots, d_n$ the vertices degree of G. Then $-d_1, \ldots, -d_n$ are eigenvalues of $A(G_s)$. Moreover, each $-d_i$ has multiplicity at least $s - 1$.*

Proof Consider the vector $(-1, 0, \ldots, 0|, 1, 0, \ldots, 0|, 0, \ldots, 0|, \ldots, |0, \ldots, 0) \in R^{sn}$, formed of s "blocks" with n entries each (ie, $|-1, 0, \ldots, 0|$ has n entries, $|1, 0, \ldots, 0|$ has n entries, $|0, \ldots, 0|$ has n entries). This vector is an eigenvector of

$A(G_s)$ associated to the eigenvalue $-d_1$. Indeed:

$$
\begin{bmatrix}
A(G) & Q(G) & Q(G) & \cdots & Q(G) \\
Q(G) & A(G) & Q(G) & \cdots & Q(G) \\
Q(G) & Q(G) & A(G) & \cdots & Q(G) \\
\vdots & \vdots & \vdots & \ddots & \vdots \\
Q(G) & Q(G) & Q(G) & \cdots & A(G)
\end{bmatrix}
\cdot
\begin{bmatrix}
-1 \\
0 \\
\vdots \\
0 \\
-- \\
1 \\
0 \\
\vdots \\
0 \\
-- \\
0 \\
\vdots \\
0
\end{bmatrix}
=
\begin{bmatrix}
-1.0 + 1.d_1 \\
-a_{2,1} + a_{2,1} \\
\vdots \\
-a_{n,1} + a_{n,1} \\
-- \\
-1.d_1 + 1.0 \\
-a_{2,1} + a_{2,1} \\
\vdots \\
-a_{n,1} + a_{n,1} \\
-- \\
0 \\
\vdots \\
0
\end{bmatrix}
=
\begin{bmatrix}
d_1 \\
0 \\
\vdots \\
0 \\
-- \\
-d_1 \\
0 \\
\vdots \\
0 \\
-- \\
0 \\
\vdots \\
0
\end{bmatrix}
$$

Note that the vectors
$(-1, 0, \ldots, 0|, 0, \ldots, 0|, 1, 0, \ldots, 0|, \ldots, |0, \ldots, 0),$
$(-1, 0, \ldots, 0|, 0, \ldots, 0|, 0, \ldots, 0|, 1, 0, \ldots, 0|, \ldots, |0, \ldots, 0), \ldots,$
$(-1, 0, \ldots, 0|, 0, \ldots, 0|, 0, \ldots, 0|, \ldots, |1, \ldots, 0)$
are also eigenvectors of $A(G_s)$ associated to the eigenvalue $-d_1$. Since we have s blocks, the multiplicity of $-d_1$ is at least $s - 1$. Similarly to $-d_2$, starting with the eigenvector:
$(0, -1, \ldots, 0|, 0, 1, \ldots, 0|, 0, \ldots, 0|, \ldots, |0, \ldots, 0)$
up to $-d_n$, when starting with the eigenvector:
$(0, 0, \ldots, -1|, 0, 0, \ldots, 1|, 0, \ldots, 0|, \ldots, |0, \ldots, 0).$ □

Next result immediately follows from the previous proposition observing that if G is connected then every vertex degree is positive.

Corollary 4 *If G is a graph on n vertices and $s > 1$ an integer, then $A(G_s)$ has at least $n \cdot (s - 1)$ non positive eigenvalues. Moreover, if G is connected then $A(G_s)$ has at least $n \cdot (s-1)$ negative eigenvalues (hence, $A(G_s)$ has at most n non negative eigenvalues).*

Next proposition provides bounds for the greatest eigenvalue of $A(G_s)$.

Proposition 15 *If G be a graph with n vertices and $s > 1$ integer, then*

$$
s.q_1(G) - \Delta(G) \leq \lambda_1(G_s) \leq s.q_1(G) - \delta(G).
$$

Proof For the left inequality, we observe that is known that the largest eigenvalue of J_s is s. Thus, by Theorem 1, the largest eigenvalue of $J_s \otimes Q(G)$ is $s \cdot q_1(G)$. Also, the smallest eigenvalue of $-D(G)$ is $-\Delta(G)$. So, by Theorem 1, the smallest eigenvalue of $I_s \otimes -D(G)$ is $-\Delta(G)$. Since $A(G_s) = (J_s \otimes Q(G)) + (I_s \otimes -D(G))$, from Theorem 2, we have that $s \cdot q_1(G) - \Delta(G) \leq \lambda_1(G_s)$.

For the right inequality, we observe again that the largest eigenvalue of $J_s \otimes Q(G)$ is $s.q_1(G)$. Also, the largest eigenvalue of $-D(G)$ is $-\delta(G)$. So, by Theorem 1, the largest eigenvalue of $I_s \otimes -D(G)$ is $-\delta(G)$. Since $A(G_s) = (J_s \otimes Q(G)) + (I_s \otimes -D(G))$, we have from theorem 2 that $\lambda_1(G_s) \leq s.q_1(G) - \delta(G)$. $\qquad \square$

We observe that the bound given by Proposition 15 is tight in the sense that the equality holds for any regular graph G and for any $s > 1$. In what follows we obtain some results relating structural and spectral parameters.

A well known spectral graph theory result is: given a connected graph G the number of distinct eigenvalues of $A(G)$ is at least $d(G) + 1$ (this is also true for the number of distinct eigenvalues of $Q(G)$). This result is still true on hypergraphs, and the proof is basically the same. In [5] this bound is proved for the signless Laplacian matrix of a hypergraphs. We prove this result for hypergraphs adjacency matrix but first we present the following lemma.

Lemma 1 *Let H be a hypergraph and $A = A(H)$ its adjacency matrix. $(A^l)_{i,j} > 0$ if there is a path with length l connecting two distinct vertices i and j, and $(A^l)_{i,j} = 0$ otherwise (where $(A^l)_{i,j}$ denotes the entry i, j of $A(H)^l$).*

Proof The proof is by induction on l. If $l = 1$ the property clearly holds. Suppose the statement is true for $l \geq 1$ and now we check for $l + 1$. Note that $(A^{l+1})_{i,j} = \sum_{k=1}^{n} (A^l)_{i,k}(A)_{k,j}$. If there is a path with length $l + 1$ joining i and j then there must exist a path with length l joining i to a neighbor u of j. So $(A)_{u,j} = 1$ and by induction hypothesis $(A^l)_{i,u} > 0$. Therefore $(A^{l+1})_{i,j} > 0$. If there is no path with length $l + 1$ joining i and j then there does exist no path with length l joining i to any neighbor of j. So, if u is a neighbor of j we have that $(A^l)_{i,u} = 0$. When u is not a neighbor of j, we have that $(A)_{u,j} = 0$. Therefore $(A^{l+1})_{i,j} = 0$. $\qquad \square$

Proposition 16 *If H is a connected hypergraph then $|\{\text{distinct eigenvalues of } A(H)\}| \geq d(H) + 1$.*

Proof Let $\lambda_1, \ldots, \lambda_t$ be all the distinct eigenvalues of $A = A(H)$. Then $(A - \lambda_1 I) \ldots (A - \lambda_t I) = 0$. So, we have that A^t is a linear combination of $A^{t-1}, \ldots, A, I$. Suppose by contradiction that $t \leq d(H)$. Hence there exist vertices i and j such that $d(i, j) = t$ and from our previous lemma, we have that $(A^t)_{i,j} > 0$. since there exists no path with length shorter than t joining i and j, $(A^{t-1})_{i,j} = 0, \ldots, (A)_{i,j} = 0, (I)_{i,j} = 0$. This is a contradiction, since $(A^t)_{i,j} = c_1(A^{t-1})_{i,j} + \ldots + c_{t-1}(A)_{i,j} + c_t(I)_{i,j}$. $\qquad \square$

Previous proposition together with Proposition 7 result this simple corollary.

Corollary 5 *If G is connected then $|\{\text{distinct eigenvalues of } A(G_s^k)\}| \geq d(G) + 1$.*

In other words, to find connected hypergraphs of the class G_s^k with few distinct adjacency eigenvalues, we have to look for graphs G with small diameter.

The next proposition gives us a different bound for $\chi_S(G_s^k)$, in terms of the largest eigenvalue of $Q(G)$ and the minimum degree of the graph G.

Proposition 17 *Given a graph G, $s > 1$ and $k \geq 2s$ we have that $\chi_S(G_s^k) = k$ or $\chi(G_s^k) \leq 1 + s.q_1(G) - \delta(G)$.*

Proof If $\chi_S(G_s^k) \neq k$ then by Proposition 9 $\chi_S(G_s^k) = \chi_S(G_s)$. Where by Theorem 3 and Proposition 15 we have: $\chi_S(G_s) \leq 1+\lambda_1(G_s) \leq 1+s.q_1(G)-\delta(G)$. $\square$

A result from spectral graph theory states that if G is a graph, then $\alpha(G) \leq min\{\lambda(G)^-, \lambda(G)^+\}$, where $\lambda(G)^-$ is the number of non positive eigenvalues of $A(G)$ and $\lambda(G)^+$ is the number of non negative eigenvalues of $A(G)$". We show that this is not valid for the independence number of the class G_s.

Proposition 18 *If $s > 1$ and G is a connected graph on n vertices, then $\alpha(G_s) > min\{\lambda(G_s)^-, \lambda(G_s)^+\}$.*

Proof From Corollary 4, we have that $A(G_s)$ has at most n non negative eigenvalues. Hence, by Proposition 11: $\alpha(G_s) \geq (s - 1)n + \alpha(G) > n \geq \lambda(G_s)^+ \geq min\{\lambda(G_s)^-, \lambda(G_s)^+\}$. $\square$

Another result states that, for any graph G, $\frac{|V(G)|}{\alpha(G)} \leq \lambda_1(G) + 1$". This fact has not yet been generalized for hypergraphs and we prove its validity for connected hypergraphs in the class G_s.

Proposition 19 *If G is connected on n vertices and $s > 1$ then $\frac{|V(G_s)|}{\alpha(G_s)} \leq \lambda_1(G_s) + 1$.*

Proof By Proposition 11, we have that $\frac{|V(G_s)|}{\alpha(G_s)} = \frac{sn}{\alpha(G_s)} \leq \frac{sn}{(s-1)n+\alpha(G)} \leq \frac{sn}{(s-1)n} = \frac{s}{s-1}$. From Proposition 15, we have that $s \cdot q_1(G) - \Delta(G) \leq \lambda_1(G_s)$. Thus, it suffices to show that $\frac{s}{(s-1)} \leq s.q_1(G) - \Delta(G) + 1$ or, in other words, that $s \leq (s - 1)(s.q_1(G) - \Delta(G) + 1)$. Since $s > 1$, if $s.q_1(G) - \Delta(G) + 1 \geq 2$ then the above inequality is valid, indeed: $s.q_1(G) - \Delta(G) + 1 \geq s(\Delta(G)+1) - \Delta(G) + 1 = (s - 1)\Delta(G) + s + 1 \geq 2$. Where the first inequality holds because: [6] If G is a connected graph then $q_1(G) \geq \Delta(G) + 1$. $\square$

Acknowledgments This study was financed in part by the Coordenação de Aperfeiçoamento de Pessoal de Nível Superior—Brasil (CAPES)—Finance Code 001, CAPES-PrInt project number 88881.310248/2018-01, CNPq and FAPERJ.

References

1. Agnarsson, G., Halldórsson, M.: Strong colorings of hypergraphs. In: G. Persiano (ed.) Approximation and Online Algorithms, pp. 253–266. Springer, Heidelberg (2005)
2. Annamalai, C.: Finding perfect matchings in bipartite hypergraphs (2016). http://arxiv.org/pdf/1509.07007.pdf

3. Banerjee, A.: On the spectrum of hypergraphs. arXiv:1711.09356v3 [math.CO] (2019). http://arxiv.org/pdf/1711.09356.pdf
4. Bretto, A.: Hypergraph Theory: An Introduction. Springer, Heidelberg (2013)
5. Cardoso, K., Trevisan, V.: The signless laplacian matrix of hypergraphs. arXiv:1909.00246v2 [math.SP] (2019). http://arxiv.org/pdf/1909.00246
6. Chen, Y., Wang, L.: Sharp bounds for the largest eigenvalue of the signless laplacian of a graph. Linear Algebra Appl. **433**, 908–913 (2010)
7. Chishti, T., Zhou, G., Pirzada, S., Ivanyi, A.: On vertex independence number of uniform hypergraphs. Acta Univ. Sapientiae Inf. **6**, 132–158 (2014)
8. Cooper, J., Dutle, A.: Spectra of uniform hypergraphs. Linear Algebra Appl. **436**, 3268–3292 (2012)
9. Dewar, M., Pike, D., Proos, J.: Connectivity in hypergraphs. arXiv:1611.07087v3 [math.CO] (2018). http://arxiv.org/pdf/1611.07087.pdf
10. Feng, K., Ching, W., Li, W.: Spectra of hypergraphs and applications. J. Number Theory **60**, 1–22 (1996)
11. Friezea, A., Mubayib, D.: Coloring simple hypergraphs. J. Combin. Theory **103**, 767–794 (2013)
12. Horn, R., Johnson, C.: Topics in Matrix Analysis. Cambridge University Press, Cambridge (1991)
13. Jin, Y., Zhang, J., Zhang, X.: Equitable partition theorem of tensors and spectrum of generalized power hypergraphs. Linear Algebra Appl. **555**, 21–38 (2018)
14. Kang, L., Liu, L., Qi, L., Yuan, X.: Spectral radii of two kinds of uniform hypergraphs. Appl. Math. Comput. **338**, 661–668 (2018)
15. Khan, M., Fan, Y.: On the spectral radius of a class of non-odd-bipartite even uniform hypergraphs. Linear Algebra Appl. **480**, 93–106 (2015)
16. Schacke, K.: On the kronecker product. Ph.D. Thesis, University of Waterloo, Waterloo (2004). Masters Thesis

Assur Decompositions
of Direction-Length Frameworks

Anthony Nixon

Abstract A bar-joint framework is a realisation of a graph consisting of stiff bars linked by universal joints. The framework is rigid if the only bar-length preserving continuous motions of the joints arise from isometries. A rigid framework is isostatic if deleting any single edge results in a flexible framework. Generically, rigidity depends only on the graph and we say an Assur graph is a pinned isostatic graph with no proper pinned isostatic subgraphs. Any pinned isostatic graph can be decomposed into Assur components which may be of use for mechanical engineers in decomposing mechanisms for simpler analysis and synthesis. A direction-length framework is a generalisation of bar-joint framework where some distance constraints are replaced by direction constraints. We initiate a theory of Assur graphs and Assur decompositions for direction-length frameworks using graph orientations and spanning trees and then analyse choices of pinning set.

Keywords Assur decomposition · Assur graph · Bar-joint framework · Direction-length framework · Pinned framework · Rigid graph

1 Introduction

A (bar-joint) framework (G, p) in $\mathbb{R}^d$ is the combination of a finite graph $G = (V, E)$ and a map $p : V \to \mathbb{R}^d$. (G, p) is *rigid* if the only edge-length-preserving continuous motions of the vertices arise from isometries of $\mathbb{R}^d$ and *flexible* if it is not rigid. It is typically of interest to characterise minimal rigidity, or *isostaticity*, which is when (G, p) is rigid but $(G - e, p)$ is flexible for any $e \in E$.

In this article we will work with *pinned* frameworks where the locations of some subset of the vertex set are fixed in the framework; hence these points are completely immobilised. An *Assur decomposition* of an isostatic framework (G, p)

A. Nixon (✉)
Mathematics and Statistics, Lancaster University, Lancaster, UK
e-mail: a.nixon@lancaster.ac.uk

© The Author(s), under exclusive license to Springer Nature Switzerland AG 2021
C. Gentile et al. (eds.), *Graphs and Combinatorial Optimization:*
from Theory to Applications, AIRO Springer Series 5,
https://doi.org/10.1007/978-3-030-63072-0_11

is a decomposition of the edge set of G such that each component is rigid as a pinned framework and no subframework of any component has that property.

In mechanical engineering, analysis of isostatic graphs, often through Assur decompositions, is used in the design, synthesis and control of mechanisms [2, 16, 18, 19]. Mathematically, Assur decompositions of frameworks in $\mathbb{R}^d$ have been studied from combinatorial and geometric perspectives [13, 14, 17]. Of most relevance to us is the main result of [17] which shows that the Assur decomposition of a pinned isostatic graph is exactly equivalent, on the one hand, to a block decomposition of the pinned rigidity matrix and, on the other hand, to a strongly connected component decomposition of a d-orientation of the graph.

We extend these techniques to allow direction constraints. A DL-graph $G = (V; D, L)$ consists of a graph G in which the edge set E is partitioned into two parts D and L. We refer to edges in D as direction edges and edges in L as length edges. A d-dimensional *direction-length framework* (G, p), abbreviated henceforth to *DL-framework*, consists of a DL-graph $G = (V; D, L)$ and a map $p : V \to \mathbb{R}^d$. (Throughout we will assume $d \geq 2$.) The framework has two types of constraint for the edges: each $e \in L$ will correspond to a length constraint; and each $f \in D$ to a direction constraint. We will say that a DL-framework is specifically one in which D and L are non-empty. We will use the terms: *pure* if either D or L is empty; *length pure* if $D = \varnothing$; and *direction pure* if $L = \varnothing$. Frameworks with direction constraints were first considered in [23]. Subsequently, in the 2-dimensional case [15], it was proved that for a graph $G = (V, E)$, a generic direction pure framework (G, p) is rigid if and only if the corresponding length pure framework (G, p) is rigid.

A DL-framework (G, p) is generic if the set containing the coordinates of the vertices is algebraically independent over $\mathbb{Q}$. The following characterisation of rigidity for generic DL-frameworks in $\mathbb{R}^2$ was proved by Servatius and Whiteley.

Theorem 1 ([15]) *A generic DL-framework (G, p) is isostatic in $\mathbb{R}^2$ if and only if*

1. $|D \cup L| = 2|V| - 2$;
2. $|E'| \leq 2|V'| - 2$ *for all* $(V', E') \subset G$;
3. $|E'| \leq 2|V'| - 3$ *for all* $(V', E') \subset G$ *with* $|E'| > 0$ *and* $E' \subset D$ *or* $E' \subset L$.

In Sect. 2 we provide further background on DL-frameworks and then we develop some basic results on pinned DL-frameworks. In Sect. 3 we define and characterise the Assur decomposition of a DL-framework. Section 4 discusses drivers. In particular we show which components of the Assur decomposition are in motion when a single edge is removed from a particular component. In Sect. 5 we look in more detail at the special case when the DL-framework is pinned with exactly 1 pinned vertex. In this case we describe how the Assur decomposition changes when we vary the choice of pinned vertex. We conclude, in Sect. 6, by discussing further avenues for exploration.

We expect Assur decompositions of pinned isostatic DL-frameworks will complement the existing uses of Assur decompositions in mechanical engineering [18, 19]. Further applications may be possible in wireless sensor networks [3] or in the control of robotic formations [22, 24] where direction frameworks are often used under the name bearing rigidity. In particular, it may already be interesting to the bearing rigidity community to be aware that the Assur decomposition results of [13, 14, 17] immediately adapt to the 2-dimensional direction pure case. In what follows we will work with DL-frameworks and develop analogous results.

2 Pinned Direction-Length Frameworks

Two DL-frameworks (G, p) and (G, q) in $\mathbb{R}^d$ are said to be *equivalent* if $q(u) - q(v)$ is a scalar multiple of $p(u) - p(v)$ for all $uv \in D$ with $p(u) \neq p(v)$ and $\|p(u) - p(v)\| = \|q(u) - q(v)\|$ for all $uv \in L$. They are *congruent* if (G, q) can be obtained from (G, p) by a translation and a dilation by ± 1. We say (G, p) is *rigid* if there exists an $\epsilon > 0$ such that if a DL-framework (G, q) is equivalent to (G, p) and satisfies $\|p(v) - q(v)\| < \epsilon$ for all $v \in V$ then (G, q) is congruent to (G, p). Equivalently, every continuous motion of the points $p(v)$, $v \in V$ respecting the length and direction constraints results in a DL-framework which is congruent to (G, p).

To introduce the rigidity matrix for a DL-framework, as in [5], take a DL-framework (G, p) in $\mathbb{R}^d$ where p is injective. For any direction edge $e = uv$ we let B_e be a $(d - 1) \times d$ matrix whose rows are a basis for the subspace of $\mathbb{R}^d$ orthogonal to $\langle p(u) - p(v) \rangle$. A rigidity matrix $R_{DL}(G, p)$ for (G, p) is a $((d-1)|D| + |L|) \times d|V|$ matrix constructed as follows. We first choose an arbitrary reference orientation for the edges of D, and use the notation $e = uv$ to mean that e has been oriented from u to v. Each edge in D corresponds to $d - 1$ consecutive rows of $R_{DL}(G, p)$, each edge in L to one row of $R_{DL}(G, p)$, and each vertex in V to d consecutive columns of $R_{DL}(G, p)$. The submatrix of $R_{DL}(G, p)$ with rows labelled by $e = uv \in D$ and columns labelled by $x \in V$ is B_e if $x = u$, is $-B_e$ if $x = v$, and is the $(d - 1) \times d$ zero matrix otherwise. The submatrix of $R_{DL}(G, p)$ with row labelled by $e = uv \in L$ and columns labelled by $x \in V$ is $p(u) - p(v)$ if $x = u$, is $p(v) - p(u)$ if $x = v$, and is zero otherwise. It is easy to see that the kernel of $R_{DL}(G, p)$ always contains at least d linearly independent vectors, corresponding to translations.

Next we introduce pinned frameworks with direction and length constraints. Let $G = (P, I; D, L)$ consist of a graph G on a vertex set V which is partitioned into two parts P and I, and an edge set E which is also partitioned into two parts D and L. We will consider rigidity where vertices in P are *pinned* and vertices in I are known as *inner* vertices.

Let $G = (P, I; D, L)$ and $p : V \rightarrow \mathbb{R}^d$. In the DL-framework (G, p) we have length and direction constraints as described above and each $v \in P$ is immobilised by any continuous motion. We say that (G, p) is *pinned rigid* if every continuous motion of the points $p(v)$, $v \in I$ respecting the length and direction constraints results in a DL-framework which is congruent to (G, p). The rigidity matrix $R_{DL}^{pin}(G, p)$ for a pinned DL-framework (G, p) in $\mathbb{R}^d$ arises from $R_{DL}(G, p)$ by deleting the d-tuple of columns corresponding to each pinned vertex $v \in P$. We define (G, p) to be: *pinned infinitesimally rigid* if rank $R_{DL}(G, p) = d|I|$; *pinned independent* if $R_{DL}^{pin}(G, p)$ has linearly independent rows; *pinned isostatic* if it is pinned infinitesimally rigid and pinned independent; and *generic* if the set of coordinates of the inner and pinned vertices is algebraically independent over $\mathbb{Q}$. We also say that (non-zero) vectors in ker $R_{DL}^{pin}(G, p)$ are *infinitesimal motions* of (G, p). Hence infinitesimal motions only apply at inner vertices.

Remark 1 Note that in dimension greater than two each direction constraint provides more than 1 row in the rigidity matrix. Hence it is possible, simply for parity reasons, that a (pinned) direction pure framework is "minimally rigid" in the sense that it is rigid but deleting any edge results in a flexible framework, while at the same time having linearly dependent rows in its rigidity matrix. Since length edges provide precisely one row we can avoid this problem in the general case.

Example 1 Consider the graph $G = (P, I; D, L)$ on 5 vertices where $P = \{v\}$, $G[I]$ induces the complete graph on 4 vertices, the edge set of $G[I]$ has been partitioned into two paths of length 3 one in D and one in L, and the final two edges are incident to v and to distinct points of I, again one each in D and L. (See Fig. 1.) The rigidity matrix, $R_{DL}^{pin}(G, p)$, of (G, p) in $\mathbb{R}^2$ is as follows:

$$
\begin{pmatrix}
[p(v_1) - p(v_2)]^\perp & [p(v_2) - p(v_1)]^\perp & 0 & 0 \\
p(v_1) - p(v_3) & 0 & p(v_3) - p(v_1) & 0 \\
[p(v_1) - p(v_4)]^\perp & 0 & 0 & [p(v_4) - p(v_1)]^\perp \\
0 & p(v_2) - p(v_3) & p(v_3) - p(v_2) & 0 \\
0 & p(v_2) - p(v_4) & 0 & p(v_4) - p(v_2) \\
0 & 0 & [p(v_3) - p(v_4)]^\perp & p(v_4) - p(v_3)]^\perp \\
p(v_1) - p(v) & 0 & 0 & 0 \\
0 & [p(v_2) - p(v)]^\perp & 0 & 0
\end{pmatrix},
$$

where you notice that the submatrix obtained by deleting the last two rows is precisely the matrix $R_{DL}(K_4, p|_{K_4})$. It is not hard to check, for any generic $p \in \mathbb{R}^{2|P \cup I|}$, that rank $R_{DL}^{pin}(G, p) = 8 = |D| + |L|$ and hence (G, p) is pinned isostatic in $\mathbb{R}^2$.

Any generic realisation of the graph H is also pinned isostatic which is easy to deduce from the fact that (G, p) is pinned isostatic as H is formed from G by a sequence of degree 2 vertex additions.

Note that it is easy to see that both examples are rigid, though not isostatic, when re-interpreted as pure frameworks. To obtain an example that is pinned iostatic as a DL-framework but flexible as a pure framework is also easy. One example would be to add four new vertices to G, add a K_4 on these new vertices and attach them to G by two edges, say incident to v_3 and v_4. That the result will be pinned-isostatic

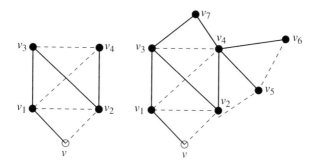

Fig. 1 Two pinned isostatic DL-frameworks G, H. We adopt the convention throughout that vertices in P will be represented by unfilled circles and vertices in I by filled circles. Furthermore, edges in D will be represented by dashed lines and edges in L by unbroken lines

follows from Theorem 1 (provided the new K_4 contains edges of each type), see also Proposition 1 below, but the framework has an obvious motion as a pure framework.

Jackson and Keevash [5] proved that, generically, rigidity and infinitesimal rigidity coincide for unpinned DL-frameworks and hence rigidity depends only on the underlying graph. Their techniques extend to pinned DL-frameworks giving us the following lemmas.

Lemma 1 *Let (G, p) be a generic pinned DL-framework. Then (G, p) is rigid if and only if it is infinitesimally rigid.*

Lemma 2 *Let (G, p) be a generic pinned isostatic DL-framework in $\mathbb{R}^d$. Then G satisfies*

(1) $(d-1)|D| + |L| = d|I|$;
(2) $(d-1)|D'| + |L'| \leq d|I'|$ *for all* $(V', E') \subset G$ *with* $V' = P' \cup I'$ *and* $E' = D' \cup L'$.

In fact, one may easily derive more precise necessary counts by considering the number of pinned vertices in a subgraph and whether it is pure or not. Indeed in $\mathbb{R}^2$ there are the following 3 additional conditions:

(3) $|D'| + |L'| \leq 2|I'| - 1$ for all pure subgraphs with $|P'| = 1$;
(4) $|D'| + |L'| \leq 2|I'| - 2$ for all subgraphs with $|P'| = 0$;
(5) $|D'| + |L'| \leq 2|I'| - 3$ for all pure subgraphs with $|P'| = 0$.

We note the following converse to Lemma 2 in dimension 2.

Proposition 1 *Let $G = (V, E)$. A DL-framework (G, p) is pinned isostatic in $\mathbb{R}^2$ if and only if G satisfies conditions (1)–(5).*

Proof Necessity was discussed above. We prove that if G satisfies (1)–(5) then (G, p) is a pinned isostatic DL-framework. To do this we add a (non-pinned) isostatic DL-graph $G_P = (P, E_P)$ on the pinned vertices and then replace each pinned vertex with an inner vertex to get a graph $G^* = (V^*, E^*)$ where $V^* = V = I \cup P$ and $E^* = D^* \cup L^* = E \cup E_P$. Observe that, by (1) and by applying Theorem 1 to G_P, G^* satisfies

$$|E^*| = |E| + |E_P| = 2|I| + 2|P| - 2 = 2|V^*| - 2,$$

and similarly, using (2)–(5) we have $|E'| \leq 2|V'| - 2$ for all $(V', E') \subset G^*$ and $|E'| \leq 2|V'| - 3$ for all $(V', E') \subset G^*$ with $E' \subset D^*$ or $E' \subset L^*$. Hence Theorem 1 implies that (G^*, p) is (unpinned) isostatic. It follows that (G, p) is isostatic.   □

In a DL-framework (G, p) in $\mathbb{R}^d$, each direction constraint produces $d-1$ rows in $R_{DL}^{\text{pin}}(G, p)$. Hence it will be convenient to consider the graph $G^+ = (V^+, E^+) = (P^+, I^+; D^+, L^+)$ which arises from G by replacing each $e \in D$ with $d-1$ copies of the edge e (and setting $P^+ = P, I^+ = I$ and $L^+ = L$). Note in dimension 2, $G = G^+$. A *DL-orientation* of G^+ is an orientation such that: for each edge of D all parallel copies in D^+ have the same orientation; all inner vertices have out-degree d; and all pinned vertices have out-degree 0.

Lemma 3 *Let (G, p) be a pinned isostatic DL-framework in $\mathbb{R}^d$. Then there is a DL-orientation of G^+. Moreover let $\mathcal{O}$ and $\mathcal{O}'$ be two DL-orientations of G^+. Then the strongly connected components are the same in both DL-orientations.*

Proof For any subgraph $(P', I'; D', L')$ of G we have, By Lemma 2, $|(D')^+| + |(L')^+| = (d-1)|D'| + |L'| \leq d|I'|$. The first assertion now follows from a standard result on orientations of sparse graphs first proved by Hakimi [4, Theorem 2]. The second conclusion is a consequence of the fact that $\mathcal{O}$ may be obtained from $\mathcal{O}'$ by reversing directions on the edges in some set of cycles [17, Corollary 2.2].   □

3   Assur Graphs and Assur Decompositions

For a pinned DL-framework we can consider the minimal pinned isostatic subframework. This corresponds to the smallest subframework of (G, p) which is pinned isostatic (necessarily this subframework contains at least one pinned vertex v_1). The edge set of such a subgraph is the first Assur component C_1. With C_1 chosen we consider a new graph in which the entire subgraph induced by C_1 is pinned. We then find the smallest pinned isostatic subframework and call it's edge set C_2. By repeating until $C_1, C_2, \ldots, C_t$ partitions the edge set of G we obtain the *Assur decomposition* of G. It would be equivalent (see Proposition 2) to, at each stage, contract C_i to a single pinned vertex. We can decompose the (square) pinned rigidity matrix into indecomposable blocks by permuting rows and columns until the blocks are in lower triangular form. Given a DL-orientation we can read off the strongly connected components of G^+. We augment the definition of strongly connected component by including edges directed out of the component.

Lastly, we use *d-tree decomposition* for the decomposition of G^+ into compo-nents, each of which is the edge-disjoint union of d spanning trees and no proper subgraph has that property. We insist that the first component contains some number of pinned vertices (and for the purpose of the tree decomposition, the pinned vertices are considered as a single vertex), and in subsequent components the earlier components are considered as a single pinned vertex. Given a graph that is the edge-disjoint union of d spanning trees, a DL-orientation can be assigned. In particular we can choose P as the sink (with out-degree 0) and direct the edges in each spanning tree towards v. Hence one may think of this decomposition as into edge-disjoint spanning trees along with edges directed out of the component.

Our first main result shows that these four decompositions are equivalent provid-ing multiple ways of understanding, testing and computing Assur decompositions of generically isostatic DL-frameworks.

Theorem 2 *Given a generic pinned isostatic DL-framework (G, p) and any DL-orientation of G^+, the following are equivalent:*

(1) the Assur decomposition of (G, p);
(2) the strongly connected decomposition of G^+;
(3) the block decomposition of $R_{DL}^{pin}(G, p)$;
(4) the d-tree decomposition of G^+.

Proof Since the equivalence of (1), (2) and (3) can be proven by adapting the technique used in [11, Theorem 3] (or alternatively in [17, Theorem 3.5]) we are brief. Observe first that, since (G, p) is pinned isostatic, every square submatrix is invertible and hence (3) $\Rightarrow$ (1) is immediate.

For (1) $\Rightarrow$ (2) let G_1 be the graph of the first Assur component. By Lemma 3 we may choose a DL-orientation of G^+. Suppose G_1^+ contains a proper subgraph H_1^+ which is a strongly connected component of G^+ containing some pinned vertex. If $|E(H_1^+)| < d|I(H_1^+)|$ then counting edges in $G_1^+ - H_1^+$ (including edges between them) contradicts the fact that $(G_1, p|_{G_1})$ is isostatic. Hence Lemma 2 implies that $|E(H_1^+)| = d|I(H_1^+)|$, contradicting the assumption that G_1 is an Assur graph.

To see (2) $\Rightarrow$ (3) suppose there are two or more strongly connected components. Then take the bottom component with its edges to the pinned vertices. In $R_{DL}^{pin}(G, p)$ apply a permutation of rows and a permutation of column vertices to place these rows and columns at the top left of the matrix. The remaining matrix forms a second block to which we iterate this process giving the desired lower block triangular form.

(1) $\Rightarrow$ (4) follows from Lemma 2 using a classical theorem of Nash-Williams [9] and similarly we can deduce (4) $\Rightarrow$ (3) noting that (G, p) is isostatic so $R_{DL}^{pin}(G, p)$ has no linearly dependent submatrices. $\square$

The equivalence also holds at the level of components. That is, the first Assur component is exactly the first strongly connected component, the first component of the block decomposition of the pinned direction-length rigidity matrix and the first component of the *d*-tree decomposition.

Example 2 In Fig. 1 we gave examples of pinned isostatic DL-frameworks (G, p) and (H, q) in $\mathbb{R}^2$. G itself is an Assur graph so the Assur decomposition is trivial. G occurs as a subgraph of H and hence is the first component C_1 in the Assur decomposition C_1, C_2, C_3, C_4 of (H, p). The remainder of the decomposition consists of single vertices attached to the below components. Note that C_2 could either be v_5 or v_7; they are incomparable in the induced partial order, however v_6 must come after v_5. In general the partial order of the Assur decomposition is unique, but there can be multiple different linear extensions.

For pure isostatic frameworks in 2-dimensions, it was proved in [13] that contracting the set of pinned vertices to a single vertex takes us from a pinned isostatic framework to a framework whose graph is a *generic circuit*: that is the graph induced by a circuit in the generic 2-dimensional rigidity matroid (see [1] for details on this class of graphs). We can give the following analogue for DL-frameworks in our next result.

Proposition 2 *Let (G, p) be a pinned DL-framework in $\mathbb{R}^2$ with $|D'| + |L'| \leq 2|I'| - 1$ for all pure subgraphs. Suppose that $P = \{v_1, \ldots, v_k\}$. Let $G' = G/P$ denote the graph formed by contracting $v_1, \ldots, v_k$ to a single pinned vertex v. Then (G, p) is pinned isostatic if and only if (G', p') is pinned isostatic, for any generic p'.*

Proof It is easy to verify that G satisfies conditions (1), (2), (4) and (5) of Proposition 1 if and only if G' does. That G satisfies (3) if and only if G' does follows from the hypothesis that $|D'| + |L'| \leq 2|I'| - 1$ for all pure subgraphs of G.   □

For example, suppose G is the pinned graph consisting of a cycle C_k, for $k \geq 3$, of inner vertices, P consists of a set of k vertices and the remaining edges form a perfect matching between C_k and P (with any pattern of direction and length edges that respects the hypotheses of the proposition). Note that Proposition 1 implies that generic realisations of G and the wheel graph W_k obtained from G by contracting the k pinned vertices are pinned isostatic in $\mathbb{R}^2$.

We conclude this section with an algorithmic remark. The pebble game [6], as extended in [8], can be used to efficiently assign a DL-orientation to G. It is not hard to extend this to check the pure subgraph conditions and hence determine whether a pinned DL-framework is pinned isostatic in $\mathbb{R}^2$. Moreover finding strongly connected components, and hence Assur components, can be done in linear time [20].

4   Drivers and Strongly Assur Graphs

Key to applications of Assur graphs in mechanical engineering is the control and synthesis of mechanisms [16, 19]. Thus, in this section, we derive several results showing how knowledge of the Assur decomposition allows us to control the 1 degree of freedom motion which results from deleting a single edge from a pinned isostatic DL-framework.

Lemma 4 *Let (G, p) be a generic pinned isostatic DL-framework in $\mathbb{R}^d$. Suppose C_k is an Assur component containing the edge $e \in D \cup L$. Then $(G-e, p)$ has a non-trivial continuous motion which is necessarily zero on all vertices in components below or incomparable to C_k.*

Proof Since (G, p) is pinned isostatic, $(G - e, p)$ is not infinitesimally rigid and hence Lemma 1 implies that $(G - e, p)$ has a non-trivial continuous motion. By the definition of Assur decomposition all components below or incomparable to C_k are pinned isostatic and hence are fixed by the motion. □

An Assur graph is *strongly Assur* if the infinitesimal motion created by removing any edge has a non-trivial velocity at every inner vertex.

Lemma 5 *Let (G, p) be a generic pinned isostatic DL-framework in $\mathbb{R}^2$. Suppose each component in the Assur decomposition is strongly Assur and let C_k be an Assur component containing the edge e. Then $(G - e, p)$ has a non-trivial continuous motion which is non-zero on all inner vertices in C_k and all inner vertices in components above C_k in the Assur decomposition.*

Proof As in Lemma 4 there is a non-trivial continuous motion of $(G - e, p)$. That this motion is non-zero on all vertices in C_k and all vertices in components above C_k in the Assur decomposition follows from Proposition 1 and the definitions of Assur decomposition and strongly Assur graph. □

The following lemma will put a strong condition on the nature of continuous motions for DL-frameworks in $\mathbb{R}^2$.

Lemma 6 *Let (G, p) be a generic pinned isostatic DL-framework in $\mathbb{R}^2$. Then (G, p) is Assur if and only if (G, p) is strongly Assur.*

Proof One direction is obvious. For the converse, assume G is an Assur graph. Delete an edge $e = ab$ and we have $|E(G - e)| = 2|I(G - e)| - 1$ by Lemma 2. This implies there is a non-trivial infinitesimal motion u of $(G - e, p)$. Suppose $u(v) = 0$ for some $v \in I$. Then v is rigidly connected to the pinned vertices. Hence Proposition 1 implies that v must be contained in a pinned subgraph H with $|E(H)| = 2|I(H)|$. Since H contains at most one of a, b we have $|I(H)| < |I(G)|$ contradicting the minimality of G. □

We remark that the corresponding result with $d \geq 3$ is already false in the length pure case (see [17]) and similar examples can be constructed for DL-frameworks.

5 Grounding Isostatic DL-Frameworks

In this section we consider how to pin isostatic DL-frameworks. Note that Proposition 2 motivates us to focus our attention on pinning a single vertex. We consider how the choice of this pinned vertex affects the resulting Assur decomposition.

Lemma 7 *Let $\mathcal{T}$ be a tree on n vertices. Let G be the DL-graph on $V(\mathcal{T})$ formed from doubling every edge in $E(\mathcal{T})$ and assigning one copy of each edge to D and one to L. Then, after pinning any single vertex, (G, p) is pinned isostatic in $\mathbb{R}^d$, for any generic p, and the Assur decomposition of (G, p) has $n - 1$ components.*

Proof Let (G, p) be generic in $\mathbb{R}^d$. Since $\mathcal{T}$ is a tree it is easy to see that G can be constructed by a sequence of degree 2 vertex additions starting at any vertex. Let $K_1 \to G_1 \to G_2 \to \cdots \to G_n = G$ be any such sequence. By analysing each $R_{DL}(G_i, p|_{G_i})$ we see that $(G_i, p|_{G_i})$ is isostatic for each $1 \leq i \leq n$. The first conclusion is now clear and the second follows from the fact that any subtree of $\mathcal{T}$ induces an isostatic subframework.   □

At the other extreme we have the following lemma.

Lemma 8 *Let (G, p) be a generic pinned isostatic DL-framework in $\mathbb{R}^d$ and suppose that any proper subgraph H of G^+ satisfies $|E(H)| \leq d|V(H)| - (d + 1)$. Then the Assur decomposition of (G, p) has precisely one component.*

Proof The hypothesis on G^+ ensures that (G, p) has no proper pinned rigid subframework. The result follows.   □

A special case, when $d = 2$, is to take $G = (V, E)$ to be a generic circuit, with any non-trivial partition of E (into D and L), and identify some $v \in V$ as pinned.

Next, given an arbitrary pinned isostatic DL-framework, we consider how to determine which vertex is the optimal choice to pin in order to maximise, or minimise, the number of components. To answer this question, we introduce the following directed acyclic graph. Let $G = (V, E)$ and (G, p) be a generic (unpinned) isostatic DL-framework in $\mathbb{R}^d$. We form a directed graph $\overrightarrow{D} = (U, F)$, which we shall call the *pinning digraph for G* as follows. The set U is the set of subsets of V which induce unpinned isostatic graphs in G (including V itself and each single vertex). There is an edge directed from $X \in U$ to $Y \in U$ if and only if $X \subsetneq Y$ and there is no $Z \in U$ such that $X \subsetneq Z \subsetneq Y$.

Example 3 Let (H, p) be the pinned isostatic DL-framework in Fig. 1. Consider the (unpinned) isostatic framework $(H - v, p|_{H-v})$. In Fig. 2 we construct the pinning digraph $\overrightarrow{D}$ for $(H - v, p|_{H-v})$. If we take v_6 as the pinned vertex then we obtain the Assur decomposition $C_1 = K_4 + \{v_5, v_6\}$ and $C_2 = K_4 + \{v_5, v_6, v_7\}$, whereas if we take v_3 as the pinned vertex then the Assur decomposition has alternative linear extensions of the partial order, one such choice being $C_1 = K_4$, $C_2 = K_4 + v_5$, $C_3 = K_4 + \{v_5, v_7\}$, $C_4 = K_4 + \{v_5, v_6, v_7\}$.

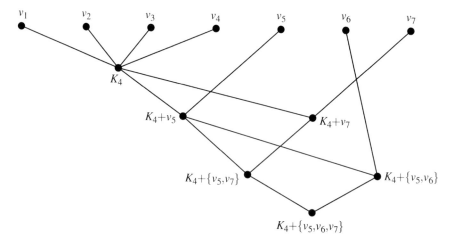

Fig. 2 A pinning digraph of an unpinned isostatic DL-framework. All edges are directed vertically downwards in the figure, arrows omitted. For brevity K_4 is used to denote its vertex set

The following lemma records some basic properties of pinning digraphs; each property follows quickly from the definition.

Lemma 9 *Let (G, p) be a generic isostatic DL-framework in $\mathbb{R}^d$ and let $\overrightarrow{D}$ be the pinning digraph for G. Then: $\overrightarrow{D}$ is acyclic and triangle-free; V is the unique sink of $\overrightarrow{D}$; and each vertex $v \in V$ is a source of $\overrightarrow{D}$.*

Theorem 3 *Let $G = (V, E)$ and let (G, p) be a generic isostatic DL-framework in $\mathbb{R}^d$. Then the Assur decomposition of (G, p) with $x \in V$ pinned is in one-one correspondence with the set of directed paths from x to V in $\overrightarrow{D}$.*

Proof It follows from the construction of $\overrightarrow{D}$ that each path from x to V corresponds to a linear extension of the partial order of the Assur decomposition of (G, p) with x pinned. The theorem follows from the uniqueness of the partial order associated to an Assur decomposition of a pinned isostatic framework. $\square$

In particular, this implies that every directed path from x to V has the same length. Thus we can use $\overrightarrow{D}$ to choose a pinned vertex which will minimise, or maximise, the number of components in an Assur decomposition.

Corollary 1 *Let $G = (V, E)$ and let (G, p) be a generic isostatic DL-framework in $\mathbb{R}^d$ with pinning digraph $\overrightarrow{D}$. Let $x \in V$ be the source of $\overrightarrow{D}$ whose distance to V is minimal (resp. maximal). Then pinning x results in a pinned isostatic graph G whose Assur decomposition has the minimum (resp. maximum) number of Assur components.*

6 Concluding Remarks

There are an array of open questions and potential extensions. We mention just three.

Complexity of the Assur decomposition. How may we minimise the complexity of the Assur decomposition? This could be in terms of the number of components, how close to linear the partial order can be or the complexity of individual components. We pose the question, given a random generic isostatic DL-framework in $\mathbb{R}^2$, what structure does the associated pinning digraph have?

Special positions of Assur graphs. For length pure frameworks, it was proved in [14] that any Assur graph has a special position in which there is a nowhere zero equilibrium stress[1] and a special position in which there is a nowhere zero infinitesimal motion. It is not clear how to extend these results to DL-frameworks as their proof technique breaks down (see Proposition 2).

Alternative constraint systems. There are a number of other rigidity contexts where the count $|E| = k|V| - k$ is fundamental including: frameworks on the cylinder [10], in ℓ_q spaces [7], fixed lattice periodic frameworks [12] and body-bar frameworks [21]. We expect our techniques can be adapted to each of these contexts.

References

1. Berg, A.R., Jordán, T.: A proof of Connelly's conjecture on 3-connected circuits of the rigidity matroid. J. Combin. Theory Ser. B **88**(1), 77–97 (2003)
2. Durango, S., Correa, J., Ruiz, O.E.: Graph-based structural analysis of planar mechanisms. Meccanica **52**(1–2), 441–455 (2017)
3. Eren, T.: Cooperative localization in wireless ad hoc and sensor networks using hybrid distance and bearing (angle of arrival) measurements. EURASIP J. Wireless Commun. Netw. **2011**(1), 72 (2011)
4. Hakimi, S.L.: On the degrees of the vertices of a directed graph. J. Franklin Inst. **279**, 290–308 (1965)
5. Jackson, B., Keevash, P.: Bounded direction-length frameworks. Discrete Comput. Geom. **46**(1), 48–71 (2011)
6. Jacobs, D.J., Hendrickson, B.: An algorithm for two-dimensional rigidity percolation: the pebble game. J. Comput. Phys. **137**(2), 346–365 (1997)
7. Kitson, D., Power, S.C.: Infinitesimal rigidity for non-Euclidean bar-joint frameworks. Bull. Lond. Math. Soc. **46**(4), 685–697 (2014)
8. Lee, A., Streinu, I.: Pebble game algorithms and sparse graphs. Discrete Math. **308**(8), 1425–1437 (2008)
9. Nash-Williams, C.S.J.A.: Edge-disjoint spanning trees of finite graphs. J. Lond. Math. Soc. **36**, 445–450 (1961)
10. Nixon, A., Owen, J.C., Power, S.C.: Rigidity of frameworks supported on surfaces. SIAM J. Discrete Math. **26**(4), 1733–1757 (2012)

---

[1]An equilibrium stress is a vector in the cokernel of the rigidity matrix.

11. Nixon, A., Schulze, B., Sljoka, A., Whiteley, W.: Symmetry adapted Assur decompositions. Symmetry **6**(3), 516–550 (2014)
12. Ross, E.: Inductive constructions for frameworks on a two-dimensional fixed torus. Discrete Comput. Geom. **54**(1), 78–109 (2015)
13. Servatius, B., Shai, O., Whiteley, W.: Combinatorial characterization of the Assur graphs from engineering. Eur. J. Combin. **31**(4), 1091–1104 (2010)
14. Servatius, B., Shai, O., Whiteley, W.: Geometric properties of Assur graphs. Eur. J. Combin. **31**(4), 1105–1120 (2010)
15. Servatius, B., Whiteley, W.: Constraining plane configurations in computer-aided design: combinatorics of directions and lengths. SIAM J. Discrete Math. **12**(1), 136–153 (1999)
16. Shai, O.: Topological Synthesis of All 2D Mechanisms Through Assur Graphs. ASME (2010). Paper No. DETC2010-28926.10.1115/DETC2010-28926
17. Shai, O., Sljoka, A., Whiteley, W.: Directed graphs, decompositions, and spatial linkages. Discrete Appl. Math. **161**(18), 3028–3047 (2013)
18. Sljoka, A., Shai, O., Whiteley, W.: Checking mobility and decomposition of linkages via pebble game algorithm. In Proceedings of the American Society of Mechanical Engineers 2011 International Design Engineering Technical Conferences and Computers and Information in Engineering Conference, Washington, DC, USA, 28–31, 493–502 (2011)
19. Sun, Y., Ge, W., Zheng, J., Dong, D.: Solving the kinematics of the planar mechanism using data structures of assur groups. J. Mech. Robot. **8**(6), 061002 (2016)
20. Tarjan, R.: Depth-first search and linear graph algorithms. SIAM J. Comput. **1**(2), 146–160 (1972)
21. Tay, T.S.: La méthode de Henneberg appliquée aux charpentes de barres et de corps rigides. Struct. Topol. **17**, 53–58 (1991). Dual French-English text
22. Trinh, M.H., Zhao, S., Sun, Z., Zelazo, D., Anderson, B.D.O., Ahn, H.S.: Bearing-based formation control of a group of agents with leader-first follower structure. IEEE Trans. Autom. Control **64**(2), 598–613 (2019)
23. Whiteley, W.: Some matroids from discrete applied geometry. In: Matroid Theory (Seattle, WA, 1995). Contemporary Mathematics, vol. 197, pp. 171–311. American Mathematical Society, Providence (1996)
24. Zhao, S., Zelazo, D.: Bearing rigidity theory and its applications for control and estimation of network systems: life beyond distance rigidity. IEEE Control Syst. **39**(2), 66–83 (2019)

On the Burning Number
of p-Caterpillars

Michaela Hiller, Arie M. C. A. Koster, and Eberhard Triesch

Abstract The burning number is a recently introduced graph parameter indicating the spreading speed of content in a graph through its edges. While the conjectured upper bound on the necessary number of time steps until all vertices are reached is proven for some specific graph classes, it remains open for trees in general. We present two different proofs for ordinary caterpillars and prove the conjecture for a generalised version of caterpillars and for trees with a sufficient number of legs. Furthermore, determining the burning number for spider graphs, trees with maximum degree three and path-forests is known to be $\mathcal{NP}$-complete; however, we show that the complexity is already inherent in caterpillars with maximum degree three.

Keywords Burning number · Computational complexity · Caterpillar graphs

1 Introduction

Given an undirected graph $G = (V, E)$, the burning number $b(G)$ indicates the minimum number of steps to inflame the whole graph while in each time step the fire spreads from all burning vertices to their neighbours and one additional vertex can be lit. This concept was introduced as a possible representation of the spread of content in an online social network in [2], but also other issues, e.g. the contagion of illnesses, can be modelled.

A sequence of vertices $B = (b_1, \ldots, b_m)$ is said to be a burning sequence or burning strategy if the vertices burn off the whole graph in m steps when lit successively. For $m = b(G)$, we say B is an optimum burning sequence or an optimum burning strategy. The set of all vertices which receive the fire from a vertex

M. Hiller (✉) · A. M. C. A. Koster · E. Triesch
Lehrstuhl II für Mathematik, RWTH Aachen, Aachen, Germany
e-mail: hiller@math2.rwth-aachen.de; koster@math2.rwth-aachen.de;
triesch@math2.rwth-aachen.de

© The Author(s), under exclusive license to Springer Nature Switzerland AG 2021
C. Gentile et al. (eds.), *Graphs and Combinatorial Optimization:
from Theory to Applications*, AIRO Springer Series 5,
https://doi.org/10.1007/978-3-030-63072-0_12

145

b_i (or theoretically would, if they were not already burning) together with b_i itself is called a burning circle and is denoted by V_i. Thus, the problem of finding a burning strategy can be reformulated into a covering problem $V = V_1 \cup \cdots \cup V_m$. The extent of a burning circle is given by $\text{diam}(V_i) + 1 = 2i - 1$. We denote the problem of determining the burning number for a graph by BURNING NUMBER.

In 2014, an upper bound for the burning number was conjectured for all connected graphs [2].

Burning Number Conjecture If G is a connected graph of order n, then

$$b(G) \leq \lceil \sqrt{n} \rceil .$$

The conjecture is proven for paths, cycles, Hamiltonian graphs and spiders [3]. Further, it can be easily checked that graphs with a small vertex number fulfil the conjecture. For paths whose length is a square number, the conjecture holds with equality and, as shown in [2], the conjecture is true for all connected graphs if it holds for trees in general.

Firstly, in Sect. 2 the Burning Number Conjecture is proven for caterpillars in two different ways: once by using the principle of infinite descent and alternatively, by determining a burning strategy complying with the conjectured bound. Subsequently, in Sect. 3, we show that BURNING NUMBER is $\mathcal{NP}$-complete for caterpillars. In Sect. 4, we focus on the validity of the conjecture for 2-caterpillars and p-caterpillars with a sufficient number of leaves relative to the order of the graph.

2 The Burning Number Conjecture for Caterpillars

In this section, we investigate the Burning Number Conjecture for caterpillars, trees in which all vertices are within the distance one of a central spine or more vivid:

> A caterpillar is a tree which metamorphoses into a path when its cocoon of endpoints is removed. [4]

Consequently, the graph class of caterpillars can also be described by forbidden minors C_3 and $S_{2,2,2}$ as in Fig. 1.

Let $G = (V, E)$ denote a caterpillar with $n := |V|$ vertices, a spine $P_\ell = \{v_1, \ldots, v_\ell\}$ of length ℓ and $n - \ell$ vertices adjacent to $P_\ell \setminus \{v_1, v_\ell\}$, which we call legs. We assume $\ell \geq 4$ and $n \geq \ell + 2$; otherwise G is a spider graph and the conjecture holds. Further, it can easily be seen that the conjecture is true for all graphs with $n \leq 9$.

Applying the (proven) conjecture for paths to the Spine P_ℓ, we clearly get the following upper bound for the caterpillar.

Fig. 1 Forbidden minors C_3
and $S_{2,2,2}$ in a caterpillar

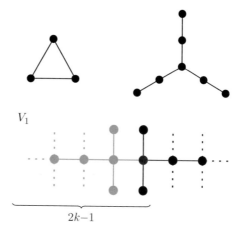

Fig. 2 In the proof of
Theorem 1, we generate G'
by removing the grey vertices
of V_1 in the minimum
counterexample G

V_1

$2k-1$

Proposition 1 *If G is a caterpillar, then $b(G) \leq \left\lceil \sqrt{\ell} \right\rceil + 1$. Thus, the conjecture is proven to be true for $\left\lceil \sqrt{n} \right\rceil \geq \left\lceil \sqrt{\ell} \right\rceil + 1$.*

In fact, the conjecture holds for all caterpillars, which can be shown using the principle of infinite descent.

Theorem 1 (Burning Number Conjecture for Caterpillars) *The burning number of a caterpillar G satisfies $b(G) \leq \left\lceil \sqrt{n} \right\rceil$.*

Proof Let the graph G be a caterpillar and a minimum counterexample regarding n with $b(G) > \left\lceil \sqrt{n} \right\rceil =: k$. We distinguish two cases:

- If either the spine vertex v_{2k-1} has no legs or v_{2k-1} has a leg, but at least one of the vertices $v_1, \ldots, v_{2k-2}$ has an adjacent leg as well, we remove the largest burning circle V_1 with extent $\mathrm{diam}(V_1) + 1 = 2k - 1$ without loss of generality at the end of the spine P_l as shown in Fig. 2. Depending on whether v_{2k-1} is legless or not, we shorten the spine by $2k - 1$ or $(2k - 1) - 1$ vertices, respectively, to maintain the connectivity.
  In both sub-cases, we obtain a new caterpillar G' with

$$\ell' \leq \ell - (2k - 1) + 2 = \ell - 2\left\lceil \sqrt{n} \right\rceil + 3,$$
$$n' \leq n - (2k - 1) \quad = n - 2\left\lceil \sqrt{n} \right\rceil + 1,$$

and for the burning number of G' it follows that $b(G') > \left\lceil \sqrt{n} \right\rceil - 1$; otherwise G would not be a counterexample. Since G is minimum by assumption, we further have $b(G') \leq \left\lceil \sqrt{n'} \right\rceil$.

This yields

$$\left\lceil \sqrt{n'} \right\rceil \geq b(G') > \left\lceil \sqrt{n} \right\rceil - 1,$$

and thus $\left\lceil \sqrt{n'} \right\rceil = \left\lceil \sqrt{n} \right\rceil$. With the estimate from above $\left\lceil \sqrt{n - 2\left\lceil \sqrt{n} \right\rceil + 1} \right\rceil = \left\lceil \sqrt{n} \right\rceil$, and therefore the two radicands lie between the same square numbers $\left\lceil \sqrt{n} \right\rceil^2$ and $\left(\left\lceil \sqrt{n} \right\rceil - 1 \right)^2$. As a consequence,

$$n - \left(\left\lceil \sqrt{n} \right\rceil - 1 \right)^2 \geq 2\left\lceil \sqrt{n} \right\rceil - 1 + 1,$$

or equivalently, $n \geq \left(\left\lceil \sqrt{n} \right\rceil - 1 \right)^2 + 2\left\lceil \sqrt{n} \right\rceil = \left\lceil \sqrt{n} \right\rceil^2 + 1$. This is a contradiction.

- If otherwise $v_1, \ldots, v_{2k-2}$ are legless, but v_{2k-1} is not, we remove the two largest burning circles V_1 with extent $\mathrm{diam}(V_1) + 1 = 2k - 1$ and V_2 with extent $\mathrm{diam}(V_2) + 1 = 2k - 3$ without loss of generality at the end of the spine P_ℓ. We shorten the spine by $(2k - 3) + (2k - 1) - 1$ vertices as shown in Fig. 3. Analogously to the first case, for the remaining caterpillar G'' it follows that

$$\ell'' \leq \ell - (2k - 3) - (2k - 1) + 2,$$
$$n'' \leq n - (2k - 3) - (2k - 1) + 1 - 1 \leq \left(\left\lceil \sqrt{n} \right\rceil - 2 \right)^2,$$

and $b(G'') > \left\lceil \sqrt{n} \right\rceil - 2$; otherwise, G would not be a counterexample. Since G is minimum, we further have $b(G') \leq \left\lceil \sqrt{n''} \right\rceil$. This yields the contradiction

$$\left\lceil \sqrt{n} \right\rceil - 2 < b(G'') \leq \left\lceil \sqrt{n''} \right\rceil \leq \left\lceil \sqrt{n} \right\rceil - 2.$$

Therefore, the minimum counterexample cannot exist. □

The following alternative proof works without the principle of infinite descent and provides a burning strategy in $\left\lceil \sqrt{n} \right\rceil$ steps for all caterpillars.

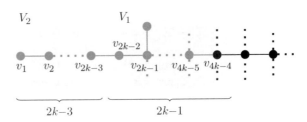

Fig. 3 Let v_{2k-1} have adjacent legs and $v_1, \ldots, v_{2k-2}$ be legless. We remove the grey vertices

Fig. 4 In the first case, v_{2k-1} is legless and we delete the grey vertices

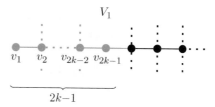

Fig. 5 We assume v_{2k-1} and at least one of $v_1, \ldots, v_{2k-2}$ to have legs and delete the grey vertices

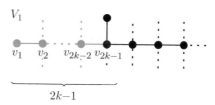

Proof Let again $k := \lceil \sqrt{n} \rceil$ denote the maximum number of steps such that the conjecture still holds. Recursively removing burning circles to reduce the vertex number at least down to the next smaller square number, we consider two cases:

- In the first case, $v_{2k-1} \in P_\ell$ has no legs. After deleting $v_1, \ldots, v_{2k-1}$ with all adjacent legs, the remaining graph has at most $n - (2k-1) \leq \lceil \sqrt{n} \rceil^2 - 2 \lceil \sqrt{n} \rceil + 1 = (\lceil \sqrt{n} \rceil - 1)^2$ vertices as depicted in Fig. 4.
- In the other case, we distinguish whether any of the vertices $v_1, \ldots, v_{2k-2}$ has an adjacent leg or not. If not all of these spine vertices are legless, we remove $v_1, \ldots, v_{2k-2}$ together with their legs as outlined in Fig. 5. Again, the vertex set of the remaining graph contains—just as in the first case—at most $(\lceil \sqrt{n} \rceil - 1)^2$ vertices.

  Otherwise, if $v_1, \ldots, v_{2k-2}$ are legless and v_{2k-1} has an adjacent leg as shwon in Fig. 3, we delete $v_1, \ldots, v_{2k-3}$ and further $v_{(2k-3)+1}, \ldots, v_{(2k-3)+(2k-2)}$ with all their legs (at least the leg adjacent to v_{2k-1}) such that the new graph consists of at most $n - (2k-3) - (2k-2) - 1 \leq \lceil \sqrt{n} \rceil^2 - (2 \lceil \sqrt{n} \rceil - 1) - (2 \lceil \sqrt{n} \rceil - 3) = (\lceil \sqrt{n} \rceil - 2)^2$ vertices.

Hence, after the vertex removal the order of the remaining graph G' decreases at least to $n' \leq (\lceil \sqrt{n} \rceil - 1)^2$ and the claim follows recursively. $\square$

It can easily be seen that the alternative proof yields an algorithm to burn a caterpillar in $\lceil \sqrt{n} \rceil$ steps, though may not necessarily be optimum.

3 The $\mathcal{NP}$-Completeness of the Burning Number Problem for Caterpillars

The $\mathcal{NP}$-completeness of determining the burning number for caterpillars indicates the unstructured nature of the problem as the difficulty or complexity is already hidden in such a simple graph class. Our proof is structured similar to the proof for trees of maximum degree three in [1] and uses a reduction from DISTINCT 3-PARTITION.

Problem: DISTINCT 3-PARTITION
Instance: A set $X = \{a_1, \ldots, a_{3n}\}$ of $3n$ distinct positive integers and a positive integer S, fulfilling $\sum_{i=1}^{3n} a_i = n \cdot S$ with $\frac{S}{4} < a_i < \frac{S}{2}$ for all $1 \leq i \leq 3n$.
Question: Can X be partitioned into n triples each of whose elements sum up to S?

DISTINCT 3-PARTITION is $\mathcal{NP}$-complete in the strong sense as shown in [5], which means the problem remains $\mathcal{NP}$-complete even if S is bounded from above by a polynomial in n.

Theorem 2 BURNING NUMBER *is* $\mathcal{NP}$-*complete for caterpillars of maximum degree three.*

Proof BURNING NUMBER is in $\mathcal{NP}$ as a burning sequence for a graph can be verified in polynomial time by checking whether the whole vertex set is covered by the union of the corresponding burning circles.

To prove the $\mathcal{NP}$-completeness, we reduce DISTINCT 3-PARTITION in polynomial time to BURNING NUMBER. Given an instance for DISTINCT 3-PARTITION as stated above, we denote $m := \max\{a_i \mid a_i \in X\}$, $\underline{m} := \{1, \ldots, m\}$ and $Y := \underline{m} \setminus X$. Transferred to the universe of BURNING NUMBER, we get $X' := \{2a_i - 1 \mid a_i \in X\}$, $S' := 2S - 3$, $\mathcal{O}_m := \{2i - 1 \mid i \in \underline{m}\}$ and $Y' := \mathcal{O}_m \setminus X'$.

Now we construct a caterpillar G of maximum degree three as follows: For each triple whose unknown elements should add up to S we build a path $Q_i^{X'}$ (for all $1 \leq i \leq n$) of order S' and for all numbers in Y (which are not available for the triples) a separate path $Q_i^{Y'}$ (for all $1 \leq i \leq m - 3n$) of order Y'. The resulting path forest

$$\bigcup_{i=1}^{n} Q_i^{X'} \cup \bigcup_{i=1}^{m-3n} Q_i^{Y'}$$

corresponds to $\bigcup_{i=1}^{m} P_{2i-1}$ and can thus be burnt in m steps. Next, we need to connect the graph by using caterpillars to keep the individual paths separated from each other. In order to do so, we need at most $m + 1$ caterpillars $G_1, \ldots, G_{m+1}$ whereby G_i has a spine of length $2(2m + 1 - i) + 1$ with exactly one leg attached to each spine vertex (except the two terminal vertices). The caterpillars and the paths are arranged alternately until only caterpillars are left, which are then placed at the end. The subgraphs are connected through an edge between their end vertices. We denote the longest path in G by P_ℓ and get

$$\ell = \left| \bigcup_{i=1}^{n} V\left(Q_i^{X'}\right) \cup \left| \bigcup_{i=1}^{m-3n} V\left(Q_i^{Y'}\right) \cup \left| \bigcup_{i=1}^{m+1} V\left(P_{2(2m+1-i)+1}\right) \right|$$

$$= \sum_{i=1}^{m}(2i-1) + \sum_{i=1}^{m+1}(2(2m+1-i)+1)$$

$$= \sum_{i=1}^{m}(2i-1) + \sum_{i=m+1}^{2m+1}(2i-1)$$

$$= (2m+1)^2.$$

The inequality in the conjecture is tight for paths; thus $b(G) \geq b(P_\ell) = \left\lceil \sqrt{\ell} \right\rceil = 2m+1$. Due to the strong $\mathcal{NP}$-completeness of DISTINCT 3-PARTITION, we can assume S to be in $\mathcal{O}(n^{\mathcal{O}(1)})$ and as m is bounded by S, the caterpillar G is computed in polynomial time with regard to the input length. Further, we constructed the caterpillar G in such a way that if X can be partitioned into n triples, each of whose elements add up to S (and equivalently $Q_1^{X'}, \ldots, Q_n^{X'}$ can be partitioned into paths $\{P_i \mid i \in X'\}$), lighting the central spine vertex of caterpillar G_i in step i (for $1 \leq i \leq m+1$) and lighting the central vertex of path $P_{2(2m+1-i)+1}$ in step i (for $m+2 \leq i \leq 2m+1$) burns the whole graph in $2m+1$ steps. Consequently, $b(G) \leq 2m+1$ holds and altogether, $b(G) = 2m+1$.

To prove the opposite direction, we assume $b(G) = 2m+1$ and let $(x_1, \ldots, x_{2m+1})$ be an optimal burning sequence for the caterpillar G. First, we can observe that x_i is a spine vertex for all $1 \leq i \leq 2m+1$ and the burning circles have to be pairwise disjoint as ℓ is a square number and $b(P_\ell) = \left\lceil \sqrt{\ell} \right\rceil$. Next, the largest burning circle has to cover G_1 with spine $P_{2(2m+1)-1}$. Otherwise, at least two burning circles are needed which would have to intersect at two spine vertices to cover all legs as pictured in Fig. 6. Inductively, G_i has to be covered with the i-th largest burning circle; thus the central spine vertex of G_i has to be lit in the i-th step for all $1 \leq i \leq m+1$.

Therefore, $\bigcup_{i=1}^{m+1} G_i$ will be burning after $2m+1$ steps induced by $x_1, \ldots, x_{m+1}$ and in the last m time steps $x_{m+2}, \ldots, x_{2m+1}$ have to ignite $\bigcup_{i=1}^{n} Q_i^{X'} \cup \bigcup_{i=1}^{m-3n} Q_i^{Y'} = \bigcup_{i=1}^{m} P_{2i-1}$, i.e., the remaining subpaths need to be covered by

$$\bigcup_{i=m+2}^{2m+1} N_{2m+1-i}[x_i].$$

As seen before the burning circles have to be disjoint; thus $N_{2m+1-i}[x_i]$ has to cover a path of length $2(2m+1-i)-1$ for $m+2 \leq i \leq 2m+1$. Hence, each $Q_i^{Y'}$ for

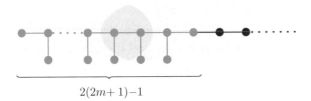

$$2(2m+1)-1$$

Fig. 6 If we do not cover G_1 with the largest burning circles, at least two spine vertices are covered twice

$1 \leq i \leq m - 3n$ is covered by itself and each $Q_i^{X'}$ for $1 \leq i \leq n$ is partitioned in paths of lengths X'. Since $\frac{S}{4} < a_i < \frac{S}{2}$ by assumption, each partition consists of three elements in X', which add up to $2S - 3$. By retranslating this 3-partition of X' to X we obtain the sought-for partition into n triples each of whose elements sum up to S.   □

As caterpillars are exactly the trees of pathwidth one the above theorem provides a statement about the complexity of graphs whose spanning trees are caterpillars.

Corollary 1 BURNING NUMBER *is* $\mathcal{NP}$-*complete for graphs of pathwidth one.*

4 The Burning Number Conjecture for p-Caterpillars

In this section, we turn the study to the more general case of p-caterpillars.

Definition 1 (p-Caterpillar) A p-caterpillar G is a tree in which all vertices are within a distance p of a central spine $P_\ell = \{v_1, \ldots, v_\ell\}$, which is the longest path in G.

Further, r-legs of a given p-caterpillar are defined as disjoint subtrees of $G - P_\ell$ with depth $r - 1$, for $r \leq p$, whose roots are in distance one of the spine. We denote the maximum length of all legs attached to spine vertex v_i by $p_{max}(v_i)$ and the number of all vertices which are connected to the spine via v_i by $p_\Sigma(v_i)$.

Thus, the parameter p indicates the maximum length of the legs and for every tree T there is a p such that T can be regarded as a p-caterpillar. Obviously, a 1-caterpillar denotes a 'common' caterpillar.

Proposition 2 *For a p-caterpillar G it follows that $b(G) \leq \left\lceil \sqrt{\ell} \right\rceil + p$. Thus, for* $\left\lceil \sqrt{n} \right\rceil \geq \left\lceil \sqrt{\ell} \right\rceil + p$ *the conjecture is proven to be true.*

Using a similar idea as in the alternative proof of Theorem 1, we can prove the Burning Number Conjecture for 2-caterpillars.

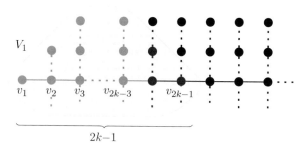

Fig. 7 We remove the grey vertices of the largest burning circle V_1 in a 2-caterpillar

Theorem 3 (Burning Number Conjecture for 2-Caterpillars) *The burning number of a 2-caterpillar G satisfies $b(G) \leq \lceil \sqrt{n} \rceil$.*

Proof As in the alternative proof of Theorem 1 we remove recursively the largest burning circles and thereby intend to reduce the number of vertices to fall below the next smaller square number. If $p_{\max}(v_{2k-2}) \leq 1$ and $p_{\max}(v_{2k-1}) = 0$, we delete the vertices $v_1, \ldots, v_{2k-1}$ together with all adjacent legs and obtain a graph whose vertex number is at most $\lfloor \sqrt{n} \rfloor^2$. In the case $p_{\max}(v_{2k-2}) = 2$ or $p_{\max}(v_{2k-1}) \geq 1$ but $\sum_{i=1}^{2k-3} p_\Sigma(v_i) \geq 2$, removing the vertices $v_1, \ldots, v_{2k-3}$ with their adjacent legs as depicted in Fig. 7 suffices to undercut $\lfloor \sqrt{n} \rfloor^2$ vertices in the remaining graph. Analogously, for $\sum_{i=1}^{2k-2} p_\Sigma(v_i) = 1$ and $p_{\max}(v_{2k-2}) \leq 1$ but $p_{\max}(v_{2k-1}) \geq 1$, we remove $v_1, \ldots, v_{2k-2}$ with all adjacent legs. Hence, it remains to consider the cases

(a) $\displaystyle\sum_{i=1}^{2k-3} p_\Sigma(v_i) = 1$ with $p_{\max}(v_{2k-2}) = 2$ and

(b) $\displaystyle\sum_{i=1}^{2k-3} p_\Sigma(v_i) = 0$ with $p_{\max}(v_{2k-2}) = 2$ or $p_{\max}(v_{2k-1}) \geq 1$.

If in case (a) we additionally have

$$\sum_{i=2k-1}^{(2k-1)+(2k-3)-4} p_\Sigma(v_i) \geq 1 \quad \text{or} \quad p_{\max}\left(v_{(2k-1)+(2k-3)-3}\right) \leq 1,$$

we arrange the two largest burning circles V_1 and V_2 with an overlap of two vertices as outlined in Fig. 8. We delete the vertices $v_1, \ldots, v_{(2k-1)+(2k-3)-4}$ and, if

$$p_{\max}\left(v_{(2k-1)+(2k-3)-3}\right) \leq 1,$$

we also remove vertex $v_{(2k-1)+(2k-3)-3}$ with all its adjacent legs. Hence, at most $n - (2k - 1) - (2k - 3)$ vertices are left.

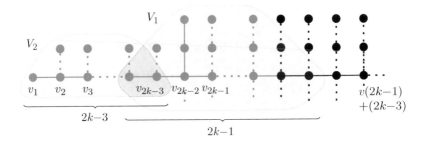

Fig. 8 We arrange the two largest burning circles V_1 and V_2 with an overlap of two vertices

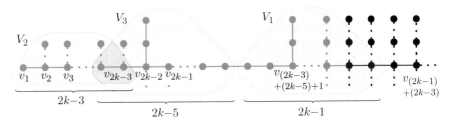

Fig. 9 We delete the grey vertices of the three largest burning circles V_1, V_2 and V_3

If, however, in case (a) we additionally have

$$\sum_{i=2k-1}^{(2k-1)+(2k-3)-4} p_\Sigma(v_i) = 0 \quad \text{and} \quad p_{\max}\left(v_{(2k-1)+(2k-3)-3}\right) = 2,$$

we consider the three largest burning circles and position them as shown in Fig. 9. The removal of $v_1, \ldots, v_{(2k-1)+(2k-3)-4}$ with all adjacent legs yields a graph with at most $n - (2k-1) - (2k-3) - (2k-5) - 1$ vertices.

Lastly, in case (b) we can assume without loss of generality that $\sum_{i=1}^{2k-3} p_\Sigma(v_i) = 0$ with $p_{\max}(v_{2k-2}) = 2$ or $p_{\max}(v_{2k-1}) \geq 1$ holds for both ends of the spine (otherwise we can apply one of the cases above on the other end), i.e., additionally, we have $\sum_{i=1}^{2k-3} p_\Sigma(v_{\ell-i+1}) = 0$. Considering the three largest burning circles again, we place V_3 and V_1 at the beginning of the spine if

$$\sum_{i=2k-2}^{(2k-5)+(2k-1)-2} p_\Sigma(v_i) \geq 2$$

and at the end if

$$\sum_{i=\ell-(2k-2)+1}^{\ell-((2k-5)+(2k-1)-2)+1} p_\Sigma(v_i) \geq 2.$$

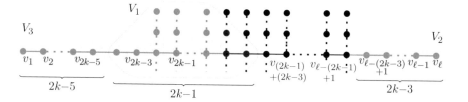

Fig. 10 We delete the grey vertices of the three largest burning circles V_1, V_2 and V_3

As outlined in Fig. 10, we place V_2 at the respective other side of the spine and remove the vertices $v_1, \ldots, v_{(2k-5)+(2k-1)-2}$ as well as $v_\ell, \ldots, v_{\ell-(2k-3)+1}$, and $v_\ell, \ldots, v_{\ell-((2k-5)+(2k-1)-2)+1}$ as well as $v_1, \ldots, v_{2k-3}$, respectively.

In the remaining case, both sums equal one, $p_\Sigma(v_{2k-1}) = p_\Sigma\left(v_{\ell-(2k-1)+1}\right) = 1$ and $p_\Sigma(v_i) = 0$ for all other $(2k-5)+(2k-1)-2$ spine vertices at both ends. Thus, we incorporate V_4, placing it next to V_2 without overlap, and additionally remove $2k - 7$ spine vertices, one of which has an adjacent leg.

This completes the proof of the Burning Number Conjecture for 2-caterpillars. □

Next, we prove the Burning Number Conjecture for 3-caterpillars with at least $2\lceil\sqrt{n}\rceil - 1$ vertices of degree one.

Theorem 4 *The burning number of a 3-caterpillar G with at least $2\lceil\sqrt{n}\rceil - 1$ vertices of degree one satisfies $b(G) \leq \lceil\sqrt{n}\rceil$.*

Proof Assume $G = (V, E)$ to be a minimum counterexample regarding the vertex number n. Hence, $b(G) > \lceil\sqrt{n}\rceil =: k$ and $|L| \geq 2k-1$ with the notation $L := \{v \in V \mid \deg(v) = 1\}$. Deleting all leaves, the remaining graph $G - L$ is a 2-caterpillar, for which the conjecture is proven to be true. Thus

$$b(G - L) \leq \left\lceil\sqrt{n - |L|}\right\rceil \leq \left\lceil\sqrt{n - 2k + 1}\right\rceil \leq \left\lceil\sqrt{k^2 - 2k + 1}\right\rceil \leq k - 1.$$

However, if $G - L$ burns after $\lceil\sqrt{n}\rceil - 1$ steps, using the same burning strategy, G can be burnt in $\lceil\sqrt{n}\rceil$ steps. This contradicts the assumption; so no counterexample exists. □

Finally, we can also prove the conjectured upper bound more general for p-caterpillars with at least $2\lceil\sqrt{n}\rceil - 1$ disjoint legs of length p.

Theorem 5 *For any p-caterpillar G with at least $2\lceil\sqrt{n}\rceil - 1$ disjoint legs of length p, we have $b(G) \leq \lceil\sqrt{n}\rceil$.*

Proof Suppose $G = (V, E)$ is a minimum counterexample regarding p and among these minimal regarding its order n. Now, let L_p be the set of all leaves at the end of p-legs. Then again, $b(G) > \lceil\sqrt{n}\rceil =: k$ and $|L_p| \geq 2k - 1$. Deleting L_p, the remaining graph $G - L_p$ is a $(p - 1)$-caterpillar with at least $2k - 1$ disjoint legs of length $p - 1$, and thus

$$b(G - L_p) \leq \left\lceil \sqrt{n - |L_p|} \right\rceil \leq \left\lceil \sqrt{n - 2k + 1} \right\rceil \leq \left\lceil \sqrt{k^2 - 2k + 1} \right\rceil \leq k - 1.$$

Now, if $G - L_p$ burns after $\left\lceil \sqrt{n} \right\rceil - 1$ steps, G can be burnt in $\left\lceil \sqrt{n} \right\rceil$ steps, a contradiction. □

5   Concluding Remarks

By the results of this paper, it remains to prove the conjecture for p-caterpillars, $p \geq 3$, with less than $2 \left\lceil \sqrt{n} \right\rceil - 1$ disjoint p-legs to complete the proof of the conjectured bound for all connected graphs. Minimum counterexamples for these remaining graph classes can be characterised in great detail. We plan to investigate these characterisations to prove the conjecture in future work.

References

1. Bessy, S., Bonato, A., Janssen, J., Rautenbach, D., Roshanbin, E.: Burning a graph is hard. Discrete Appl. Math. **232**, 73–87 (2017)
2. Bonato, A., Janssen, J., Roshanbin, E.: How to burn a graph. Internet Math. **12**(1–2), 85–100 (2016)
3. Bonato, A., Lidbetter, T.: Bounds on the burning numbers of spiders and path-forests. Theor. Comput. Sci. **794**, 12–19 (2019)
4. Harary, F., Schwenk, A.J.: The number of caterpillars. Discrete Math. **6**(4), 359–365 (1973)
5. Hulett, H., Will, T.G., Woeginger, G.J.: Multigraph realizations of degree sequences: maximization is easy, minimization is hard. Oper. Res. Lett. **36**(5), 594–596 (2008)

An Approximation Algorithm for Network Flow Interdiction with Unit Costs and Two Capacities

Jan Boeckmann and Clemens Thielen

Abstract In the network flow interdiction problem, an interdictor aims to remove arcs of total cost at most a given budget B from a graph with given arc costs and capacities such that the value of a maximum flow from a source s to a sink t is minimized. Although the problem has high applicability in real world problems and is known to be strongly NP-hard, only few polynomial-time approximation algorithms are known. In this paper, we present a $(B + 1)$-approximation algorithm for the special case where arcs have unit costs and may only have a small or a large capacity. Thereby, we develop the first approximation algorithm for a variant of NFI whose approximation ratio only depends on the budget available to the interdictor, but not on the size of the graph.

Keywords Network flow interdiction · Two capacities · Approximation algorithm

1 Introduction

In the network flow interdiction problem (NFI), an interdictor aims to remove arcs of total cost at most a given budget B from a graph with given arc costs and capacities such that the value of a maximum flow from a source s to a sink t is minimized. The problem has first been stated in 1964 [8] and has been widely studied since then due to its numerous applications ranging from supply line disruption to critical infrastructure analysis [6] and drug interdiction [9]. The problem formulation used in most of the literature today has been introduced by Phillips [7], who shows multiple hardness results on different classes of graphs. Furthermore, they present a pseudopolynomial-time algorithm and an FPTAS for the problem on planar

J. Boeckmann (✉) · C. Thielen
Technical University of Munich, TUM Campus Straubing for Biotechnology and Sustainability, Straubing, Germany
e-mail: jan.boeckmann@tum.de; clemens.thielen@tum.de

graphs. Burch et al. [3] and Chestnut and Zenklusen [4] both present algorithms for the general version of NFI that, for any $\epsilon > 0$, either returns a $(1 + \frac{1}{\epsilon})$-approximate solution or a (super-) optimal solution violating the budget by a factor of at most $(1 + \epsilon)$. The algorithm by Chestnut and Zenklusen even works for a generalized class of interdiction problems. So far, the best known polynomial-time approximation algorithm for NFI presented by Chestnut and Zenklusen [5] achieves an approximation ratio of $2(n - 1)$, where n is the number of nodes in the graph. They also present a hardness of approximation result using a reduction from the densest k subgraph problem, which is known to be hard to approximate [2] under certain assumptions.

1.1 Our Contribution

We present a polynomial-time $(B + 1)$-approximation algorithm for a special case of NFI. In this problem variant that we call u-NFI, the arcs have unit costs and arc capacities are restricted to two possible values 1 and $u > 1$. To the best of our knowledge, this is the first algorithm for a variant of NFI to achieve an approximation ratio that only depends on the interdiction budget B, but not on the size of the graph. Moreover, we show that our analysis of the algorithm is essentially tight. The problem u-NFI is strongly **NP**-hard and the best approximation algorithm known for it is the one presented by Chestnut and Zenklusen [5], which achieves an approximation ratio of $(n - 1)$ even for the more general version of NFI where arcs have unit costs but may have arbitrary capacities. Since s can always be separated from t on simple graphs for any budget $B \geq n - 1$ by simply removing the at most $n - 1$ arcs starting in s, the approximation ratio we obtain thus dominates the previously best known approximation ratio for u-NFI on simple graphs.

2 Problem Definition and Structural Results

Let $G = (V, E)$ be a directed graph with nonnegative arc costs and capacities, let $s \neq t$ be two nodes in G, and let $B > 0$ be an interdiction budget. The network flow interdiction problem (NFI) asks for a subset $R \subseteq E$ of arcs of total cost at most B such that the value of a maximum s-t-flow in the network $G_R := (V, E \setminus R)$ is minimized. In this paper, we consider the special case where arcs have unit costs and the arc capacities are restricted to $\{1, u\}$ for some $u > 1$. Here, B can clearly be assumed to be integral and the problem can be formally defined as follows:

Definition 1 (u-NFI)

INSTANCE: A directed graph $G = (V, E)$, two nodes $s \neq t$ in G, a budget $B \in \mathbb{N}_{>0}$, and arc capacities $c : E \to \{1, u\}$, where $u \in \mathbb{Q}_{>1}$.

TASK: Find a subset $R \subseteq E$ of arcs with $|R| \leq B$ such that the value $\mathrm{val}(R)$ of a maximum s-t-flow in the graph $G_R = (V, E \setminus R)$ is minimized.

A proof of strong NP-hardness of the general version of NFI, which also shows strong NP-hardness of u-NFI, can be found in [9]. Arcs having capacity u are called *large* arcs and arcs having capacity 1 are called *small* arcs, and we denote the capacity of an arc $r \in E$ by $c_r := c(r)$. We assume throughout the paper that the arcs are ordered by some arbitrary but fixed ordering, which is used for the purpose of tie breaking. The terms *cut* and *s-t-cut* are used as synonyms and refer to a partition $V = S \dot\cup T$ of the nodes of G such that $s \in S$ and $t \in T$. For any subset $R \subseteq E$ of removed arcs, we use $\delta^+_{G_R}(S)$ to denote the set of arcs starting in $S \subseteq V$ and ending in $V \setminus S$ in the resulting graph $G_R = (V, E \setminus R)$, and say that these arcs are *in the cut* $(S, V \setminus S)$ in G_R. The *capacity* of a cut (S, T) in the graph G_R is defined as the sum of the capacities of the arcs in $\delta^+_{G_R}(S)$, and a *minimum cut* in G_R is a cut of minimum capacity in G_R.

An instance of u-NFI is called *trivial* if its optimum objective value equals zero. By the well-known max-flow min-cut theorem (cf. [1]), it is easy to check in polynomial time whether a given instance is trivial by testing whether a minimum cut in the original graph G with all arc capacities set to one has capacity at most B. In the following, we assume that all instances are non-trivial. The following observation also follows directly from the max-flow min-cut theorem:

Observation 2 *For any solution* $R \subseteq E$, *its objective value* $\mathrm{val}(R)$ *equals the capacity of a minimum s-t-cut in the graph* G_R.

The computation of minimum cuts plays an important role in our algorithm and its analysis. Throughout the paper, we use an arbitrary but fixed (deterministic) algorithm to compute a minimum s-t-cut in a given graph in polynomial time.[1] For a solution R, we denote the minimum cut in the resulting graph G_R computed by this minimum cut algorithm by $C_R = (S_R, T_R)$.

Lemma 3 *For a solution* $R \subseteq E$, *either* $R \subseteq \delta^+_G(S_R)$, *or* $\mathrm{val}(R)$ *can be reduced by removing an arc in* $R \setminus \delta^+_G(S_R)$ *from* R *and adding an arc from* $\delta^+_G(S_R) \setminus R$ *to* R.

Proof Assume that there exists an arc $\hat{r} \in R \setminus \delta^+_G(S_R)$. Since the instance is non-trivial, there must also exist an arc $r' \in \delta^+_G(S_R) \setminus R$. Now let $R' := (R \setminus \{\hat{r}\}) \cup \{r'\}$. By Observation 2, removing $\hat{r}$ from R does not change $\mathrm{val}(R)$, but adding r' to R decreases $\mathrm{val}(R)$ by $c_{r'} > 0$. Therefore, $\mathrm{val}(R') = \mathrm{val}(R) - c_{r'} < \mathrm{val}(R)$. □

Corollary 4 *For any optimal solution* R^{OPT}, *there exists a cut* (S, T) *such that* $R^{OPT} \subseteq \delta^+_G(S)$. □

While any solution R gives rise to a cut C_R, a given cut also gives rise to a solution by using the interdiction budget B to reduce the capacity of the cut as far as possible, which can be easily achieved by first removing as many large arcs

---

[1]For an overview of state-of-the-art minimum cut algorithms, we refer to [1].

from C as possible and then using any remaining budget to remove small arcs. This motivates the following definition:

Definition 5 For an s-t-cut C in G, we define $R_C \subseteq E$ as the solution containing the B arcs of largest capacity from the cut C, where ties are broken by the fixed ordering of the arcs. Furthermore, we define the *value of the cut C* as $\text{val}(C) := \text{val}(R_C)$ and call the cut C *optimal* if R_C is an optimal solution.

Clearly, if an (approximately) optimal cut C is known, the corresponding (approximately) optimal solution R_C can easily be computed in polynomial time. Therefore, the challenge in u-NFI lies in the computation of a cut C whose value $\text{val}(C)$ is as low as possible. Since $\text{val}(C) = \text{val}(R_C)$ depends on the number of large arcs and the number of small arcs in C, this motivates the following definition:

Definition 6 For an s-t-cut C, we denote the number of large arcs in C in the original graph G by $q(C)$, and the number of small arcs by $p(C)$. For $q \in \mathbb{N}$, we define $\mathscr{C}_q$ as the set of cuts with exactly q large arcs in G. A cut of minimum capacity in G amongst all cuts in $\mathscr{C}_q$ is called a *q-min-cut*.

Note that $\mathscr{C}_q$ might be empty for some $q \in \mathbb{N}$, in which case no q-min-cut exists. If $\mathscr{C}_q$ is nonempty, however, the fact that every cut must contain at least $B + 1$ arcs due to the non-triviality of the instance implies that the original capacity of any cut in $\mathscr{C}_q$ can be reduced by the same amount by removing at most B arcs. This yields:

Lemma 7 *Any optimal cut C_{OPT} must be a q-min-cut for $q = q(C_{OPT})$ (and any q-min-cut for $q = q(C_{OPT})$ is then also optimal).* □

In order to find a good solution for u-NFI, our algorithm iteratively computes q-min-cuts for different values of q. The numbers of large arcs in the following two cuts provide bounds on the values of q to consider (as we show in Theorem 10):

Definition 8 We define C_l as the cut with the least number of arcs in G that is found by applying the minimum cut algorithm to G with all arc capacities set to one. Similarly, we let C_m denote the minimum cut in G found by the minimum cut algorithm applied to G with the original arc capacities.

The following observation follows by the same argument as Lemma 7:

Observation 9 *The cut C_l is a $q(C_l)$-min-cut and the cut C_m is a $q(C_m)$-min-cut. Moreover, C_l is optimal if it contains at most B large arcs, and C_m is optimal if it contains at least B large arcs.* □

Due to Observation 9, we assume in the following that $q(C_l) > B$ and $q(C_m) < B$.

Theorem 10

(i) *If there exists a q-min-cut $\hat{C}$ for some $q > q(C_l)$, then $\text{val}(C_l) < \text{val}(\hat{C})$.*
(ii) *If there exists a q-min-cut $\hat{C}$ for some $q < q(C_m)$, then $\text{val}(C_m) < \text{val}(\hat{C})$.*

Thus, there exists an optimal cut that is a q-min-cut for some $q(C_m) \leq q \leq q(C_l)$.

Proof We only show (i). The proof of (ii) is analogous and the existence of an optimal q-min-cut for some $q(C_m) \leq q \leq q(C_l)$ follows directly from (i), (ii), and Lemma 7. So let $\hat{C}$ be a q-min-cut for some $q > q(C_l)$. Then, $q + p(\hat{C}) \geq q(C_l) + p(C_l)$ as C_l is a cut with the least number of arcs. Moreover, $q > q(C_l) > B$ and $u > 1$, so we obtain that

$$\text{val}(C_l) = (q(C_l) - B) \cdot u + p(C_l)$$

$$= (q - B) \cdot u + p(\hat{C}) - \Big((q - q(C_l)) \cdot u + p(\hat{C}) - p(C_l)\Big)$$

$$< \text{val}(\hat{C}) - (q - q(C_l) + p(\hat{C}) - p(C_l)) \leq \text{val}(\hat{C}). \qquad \square$$

3 Capacity-γ-Min-Cuts

Lemma 7 and Theorem 10 clearly motivate to take a closer look at the problem of computing a q-min-cut (or returning that there exists none) for any given q. This problem, however, is strongly NP-hard in general since, by Theorem 10, u-NFI can be solved by computing q-min-cuts for polynomially many values of q.

Corollary 11 *Computing a q-min-cut (or returning that none exists) for some given $q \in \mathbb{N}$ is strongly NP-hard.* $\qquad \square$

The basic idea of our algorithm is to compute q-min-cuts for some values of q by computing minimum cuts in the original graph G when varying the capacity of the large arcs between 1 and u. This motivates the following definition.

Definition 12 For $\gamma \geq 1$, the graph G^γ is defined to be the original graph G with large arcs having capacity γ instead of u. For a cut C, we denote its capacity in the original graph G (with large arcs having capacity u) by $\text{cap}(C)$ and its capacity in G^γ by $\text{cap}^\gamma(C)$. A minimum cut in G^γ is called a *capacity-γ-min-cut*.

Clearly, the cut C_l is a capacity-1-min-cut and the cut C_m is a capacity-u-min-cut.

Lemma 13 *If C^γ is a capacity-γ-min-cut for some $\gamma \geq 1$, then it is a q-min-cut for $q = q(C^\gamma)$.*

Proof Let C^γ be a capacity-γ-min-cut, and let $q = q(C^\gamma)$, so that $C^\gamma \in \mathscr{C}_q$ and $\text{cap}(C^\gamma) = q \cdot u + p(C^\gamma)$. Suppose for the sake of a contradiction that there exists a cut $\hat{C} \in \mathscr{C}_q$ with $\text{cap}(\hat{C}) < \text{cap}(C^\gamma)$. Then it holds that $q \cdot u + p(\hat{C}) = \text{cap}(\hat{C}) < \text{cap}(C^\gamma) = q \cdot u + p(C^\gamma)$ and, therefore, $p(\hat{C}) < p(C^\gamma)$. But this means that $\text{cap}^\gamma(\hat{C}) = q \cdot \gamma + p(\hat{C}) < q \cdot \gamma + p(C^\gamma) = \text{cap}^\gamma(C^\gamma)$, which is a contradiction to C^γ being a capacity-γ-min-cut. $\qquad \square$

The next lemma states that the number of large arcs in a capacity-γ-min-cut decreases monotonically as γ increases.

Lemma 14 *If C^{γ_1} is a capacity-γ_1-min-cut and C^{γ_2} is a capacity-γ_2-min-cut for $1 \leq \gamma_1 < \gamma_2 \leq u$, then $q(C^{\gamma_1}) \geq q(C^{\gamma_2})$.*

Proof Let $q_1 := q(C^{\gamma_1})$, $p_1 := p(C^{\gamma_1})$ and $q_2 := q(C^{\gamma_2})$, $p_2 := p(C^{\gamma_2})$. By definition of the two cuts, we have $\mathrm{cap}^{\gamma_1}(C^{\gamma_1}) \leq \mathrm{cap}^{\gamma_1}(C^{\gamma_2})$ and $\mathrm{cap}^{\gamma_2}(C^{\gamma_2}) \leq \mathrm{cap}^{\gamma_2}(C^{\gamma_1})$, which yields

$$q_1 \cdot \gamma_1 + p_1 = \mathrm{cap}^{\gamma_1}(C^{\gamma_1}) \leq \mathrm{cap}^{\gamma_1}(C^{\gamma_2}) = q_2 \cdot \gamma_1 + p_2, \text{ and} \tag{1}$$

$$q_2 \cdot \gamma_2 + p_2 = \mathrm{cap}^{\gamma_2}(C^{\gamma_2}) \leq \mathrm{cap}^{\gamma_2}(C^{\gamma_1}) = q_1 \cdot \gamma_2 + p_1. \tag{2}$$

Adding (1) and (2) yields $q_1 \cdot \gamma_1 + p_1 + q_2 \cdot \gamma_2 + p_2 \leq q_2 \cdot \gamma_1 + p_2 + q_1 \cdot \gamma_2 + p_1$, which means that $q_1 \geq q_2$ since $\gamma_1 < \gamma_2$. □

We are now ready to state our algorithm in Algorithm 1. It uses the recursive bisection procedure stated in Algorithm 2 in order to compute capacity-γ-min-cuts. Whenever two cuts C_1 and C_2 have been found for γ_1 and γ_2, respectively, the next candidate value $\hat{\gamma}$ is chosen as the value for which the capacities of C_1 and C_2 in $G^{\hat{\gamma}}$ are equal, and a capacity-$\hat{\gamma}$-min-cut $\hat{C}$ is computed. If $\hat{C}$ has a lower capacity than C_1 (and also C_2) in $G^{\hat{\gamma}}$, the bisection method is called recursively for C_1 and $\hat{C}$ and for C_2 and $\hat{C}$. Otherwise, the recursion ends and the cuts C_1 and C_2 are returned.

Lemma 15 *Let $C_1 \neq C_2$ be two cuts for which* `bisection`(C_1, C_2) *is called during the execution of Algorithm 1, and let γ_1, γ_2 be the values of γ for which C_1*

Algorithm 1: Bisection-cut

```
1  Algorithm bisec-cut()
2  |   Compute C_l and C_m
3  |   if q(C_l) ≤ B then
4  |   |   return C_l
5  |   else if q(C_m) ≥ B then
6  |   |   return C_m
7  |   else
8  |   |   return R_C for C being a cut of minimum value val(C) in bisection(C_l, C_m)
9  |   end
```

Algorithm 2: Bisection-procedure

```
1  Procedure bisection(C_1, C_2)
2  |   Let γ̂ := (p(C_2)−p(C_1))/(q(C_1)−q(C_2)) and compute a capacity-γ̂-min-cut Ĉ
3  |   if cap^γ̂(Ĉ) ≤ cap^γ̂(C_1) and q(Ĉ) ∉ {q(C_1), q(C_2)} then
4  |   |   return bisection(C_1, Ĉ) ∪ bisection(Ĉ, C_2)
5  |   else
6  |   |   return {C_1, C_2}
7  |   end
```

and C_2, respectively, have been computed as capacity-γ-min-cuts during the algorithm. Then $1 \leq \gamma_1 \leq \hat{\gamma} \leq \gamma_2 \leq u$.

Proof We show by induction over the recursion tree produced by the recursive calls of the bisection procedure that, whenever $\texttt{bisection}(C_1, C_2)$ is called for two cuts C_1, C_2 as in the claim, then either $1 \leq \gamma_1 < \hat{\gamma} < \gamma_2 \leq u$, or $\hat{\gamma} \in \{\gamma_1, \gamma_2\}$ and no further recursive calls of the bisection procedure are made within $\texttt{bisection}(C_1, C_2)$. Since Algorithm 1 first calls $\texttt{bisection}(C_l, C_m)$ and the cuts C_l, C_m obtained for $\gamma = 1$ and $\gamma = u$, respectively, satisfy the assumptions of the lemma, this will show the claim.

So let $\texttt{bisection}(C_1, C_2)$ be called during the algorithm and assume that the statement holds for all predecessors in the recursion tree. In particular, it holds for the parent in the recursion tree. Without loss of generality, let $\texttt{bisection}(C', C_2)$ be the parent recursion step, where C' is a cut that has been previously computed as a capacity-γ'-min-cut during the algorithm. Then the cut $\hat{C}$ computed in the parent recursion step equals C_1 and is computed as a capacity-γ_1-min-cut. Hence, applying the induction hypothesis for the parent recursion step implies that $1 \leq \gamma' < \gamma_1 < \gamma_2 \leq u$. Now consider the call to $\texttt{bisection}(C_1, C_2)$, which computes a new capacity-$\hat{\gamma}$-min-cut. Since $1 \leq \gamma_1 < \gamma_2 \leq u$, Lemma 14 shows that $q(C_1) \geq q(C_2)$. As $\texttt{bisection}(C_1, C_2)$ has been called, we must also have $q(C_1) \neq q(C_2)$ by line 3 in the bisection procedure, which yields $q(C_1) > q(C_2)$. For the sake of a contradiction, suppose that $\hat{\gamma} < \gamma_1$. By the choice of $\hat{\gamma}$, it holds that $\hat{\gamma} \cdot q(C_1) + p(C_1) = \hat{\gamma} \cdot q(C_2) + p(C_2)$. But by using that $q(C_1) > q(C_2)$, this yields

$$\gamma_1 \cdot q(C_1) + p(C_1) = \hat{\gamma} \cdot q(C_1) + p(C_1) + (\gamma_1 - \hat{\gamma}) \cdot q(C_1)$$
$$> \hat{\gamma} \cdot q(C_2) + p(C_2) + (\gamma_1 - \hat{\gamma}) \cdot q(C_2) = \gamma_1 \cdot q(C_2) + p(C_2),$$

which means that C_1 is not a capacity-γ_1-min-cut. This is a contradiction to the choice of C_1, so we obtain that $\hat{\gamma} \geq \gamma_1$. Along the same lines, one can prove that $\hat{\gamma} \leq \gamma_2$. Consequently, we obtain that $1 \leq \gamma_1 \leq \hat{\gamma} \leq \gamma_2 \leq u$.

It remains to show that, if $\hat{\gamma} \in \{\gamma_1, \gamma_2\}$, no further recursive calls of the bisection procedure are made within $\texttt{bisection}(C_1, C_2)$. Assume without loss of generality that $\hat{\gamma} = \gamma_1$. Since the algorithm used for computing a capacity-$\hat{\gamma}$-min-cut (i.e., a minimum cut in $G^{\hat{\gamma}}$) in line 3 of the bisection procedure is deterministic, it then follows that $\hat{C} = C_1$ and, in particular, $q(\hat{C}) = q(C_1)$. This implies that no further recursion steps are made within $\texttt{bisection}(C_1, C_2)$, which completes the proof. $\qquad\Box$

Using Lemmas 14 and 15, it is now easy to see that Algorithm 1 runs in polynomial time: A single execution of the bisection procedure can be performed in polynomial time since a capacity-$\hat{\gamma}$-min-cut can be obtained by computing a minimum cut in $G^{\hat{\gamma}}$. Moreover, whenever $\texttt{bisection}(C_1, C_2)$ is called in Algorithm 1, Lemma 15 shows that $1 \leq \gamma_1 \leq \hat{\gamma} \leq \gamma_2 \leq u$ for γ_1, γ_2 as in the lemma. Thus, either $\hat{\gamma} \in \{\gamma_1, \gamma_2\}$, in which case the computed capacity $\hat{\gamma}$-min-cut $\hat{C}$ equals C_1 or C_2 since the algorithm used for computing a capacity-$\hat{\gamma}$-min-cut in line 3 of the bisection procedure is deterministic, or $1 \leq \gamma_1 < \hat{\gamma} < \gamma_2 \leq u$, so Lemma 14 yields $q(C_2) \leq q(\hat{C}) \leq q(C_1)$. Thus, since no recursive call to the bisection procedure is made if $q(\hat{C}) \in \{q(C_1), q(C_2)\}$, at most three q-min-cuts can be computed within Algorithm 1 for each $0 \leq q \leq m$ (where m is the number of arcs in G), i.e., there are at most $3(m + 1)$ calls of the bisection procedure.

We now show that the algorithm is a $(B+1)$-approximation algorithm for u-NFI. To this end, first recall that we assume $q(C_l) > B$ and $q(C_m) < B$, so the algorithm does actually call the bisection procedure.

Proposition 16 *Let C_{OPT} be an optimal cut. If Algorithm 1 finds two cuts C, C' such that $B \leq q(C') \leq q(C_{OPT}) \leq q(C)$ or such that $q(C') \leq q(C_{OPT}) \leq q(C) \leq B$, then the algorithm returns an optimal solution.*

Proof First note that, if $q(C_{OPT}) = q(C)$ or $q(C_{OPT}) = q(C')$, then C or C', respectively, is a $q(C_{OPT})$-min-cut by Lemma 13, which is optimal by Lemma 7. Hence, we may assume in the following that $q(C') < q(C_{OPT}) < q(C)$.

Moreover, we may assume without loss of generality that C is the cut in the set bisection(C_l, C_m) with minimum value $q(C)$ such that $q(C) > q(C_{OPT})$, and that C' is the cut in bisection(C_l, C_m) with maximum value $q(C')$ such that $q(C') < q(C_{OPT})$. This means that bisection(C, C') was called at some point in time during the algorithm and returned $\{C, C'\}$ without any further recursive calls.

Case 1: $B \leq q(C') < q(C_{OPT}) < q(C)$.

Let $\hat{\gamma} = \frac{p(C')-p(C)}{q(C)-q(C')}$ be the value for which a capacity-$\hat{\gamma}$-min-cut is computed in bisection(C, C'). Since $q(C_{OPT}) \notin \{q(C), q(C')\}$ and bisection(C, C') returned $\{C, C'\}$ without any further recursive calls, we must have $\text{cap}^{\hat{\gamma}}(C_{OPT}) \geq \text{cap}^{\hat{\gamma}}(C) = \text{cap}^{\hat{\gamma}}(C')$. As $q(C_{OPT}) > q(C')$ and $\hat{\gamma} \leq u$, this yields that also $\text{cap}^u(C_{OPT}) \geq \text{cap}^u(C')$. Therefore, since $q(C_{OPT}) > q(C') \geq B$,

$$\text{val}(C_{OPT}) = \text{cap}^u(C_{OPT}) - B \cdot u \geq \text{cap}^u(C') - B \cdot u = \text{val}(C'),$$

which means that C' is an optimal cut in bisection(C_l, C_m).

Case 2: $q(C') < q(C_{OPT}) < q(C) \leq B$

Then $\text{val}(C) = q(C) + p(C) - B$ and $\text{val}(C_{OPT}) = q(C_{OPT}) + p(C_{OPT}) - B$. With $\hat{\gamma} = \frac{p(C')-p(C)}{q(C)-q(C')}$ as in Case 1, we again have $\text{cap}^{\hat{\gamma}}(C_{OPT}) \geq \text{cap}^{\hat{\gamma}}(C)$. As $q(C) > q(C_{OPT})$ and $\hat{\gamma} \geq 1$, this yields that also $\text{cap}^1(C_{OPT}) \geq \text{cap}^1(C)$. Therefore,

$$\text{val}(C_{OPT}) = q(C_{OPT}) + p(C_{OPT}) - B = \text{cap}^1(C_{OPT}) - B \geq \text{cap}^1(C) - B = \text{val}(C),$$

which means that C is an optimal cut in bisection(C_l, C_m). □

In the following, let C_1 denote a cut in bisection(C_l, C_m) with minimum value $q(C_1)$ such that $q(C_1) > B$, and let C_2 denote a cut in bisection(C_l, C_m)

with maximum value $q(C_2)$ such that $B > q(C_2)$ (both of these cuts exist due to the assumption that $q(C_l) > B$ and $q(C_m) < B$). Note that, by Lemma 16, Algorithm 1 can only return a suboptimal solution if every optimal cut C_{OPT} as in Theorem 10 satisfies $q(C_2) < q(C_{OPT}) < q(C_1)$. Hence, in the following, we focus on this case and fix an optimal cut C_{OPT} with $q(C_2) < q(C_{OPT}) < q(C_1)$.

Lemma 17 *For the numbers of small arcs, it holds that* $p(C_1) < p(C_{OPT}) < p(C_2)$.

Proof If we had $p(C_{OPT}) \leq p(C_1)$, then $q(C_1) > q(C_{OPT})$ would yield $cap^\gamma(C_1) = \gamma \cdot q(C_1) + p(C_1) > \gamma \cdot q(C_{OPT}) + p(C_{OPT}) = cap^\gamma(C_{OPT})$ for any $\gamma \geq 1$, which means that C_1 would not be a capacity-γ-min-cut for any γ and could, thus, not be in bisection(C_l, C_m).

If we had $p(C_{OPT}) \geq p(C_2)$, then, since $q(C_{OPT}) > q(C_2)$, the cut C_2 would contain strictly less large arcs and no more small arcs than C_{OPT}, which means that C_{OPT} could not be an optimal cut.   □

We now prove a central result in our analysis of Algorithm 1, which shows the approximation guarantee of the algorithm conditioned on the existence of an affine function with certain properties. Such an affine function will afterwards be derived from the optimal cut C_{OPT}. To state the result, we need the following definition.

Definition 18 Let $g(\gamma) = q_g \cdot \gamma + p_g$ with constants $q_g, p_g \in \mathbb{Q}$ be an affine function of γ. Then its *interdicted value* is defined as val$(g) := q_g + p_g - B$.

Proposition 19 *Let* $\hat{\gamma} := \frac{p(C_2) - p(C_1)}{q(C_1) - q(C_2)}$. *If there exists an affine function* $g(\gamma) = q_g \cdot \gamma + p_g$ *with*

(a) $q_g \leq B$,
(b) $g(\hat{\gamma}) \geq cap^{\hat{\gamma}}(C_1) = cap^{\hat{\gamma}}(C_2)$,
(c) $q(C_1) - q_g \geq 1$,
(d) $val(g) = val(C_{OPT})$,

then Algorithm 1 is a $(B+1)$*-approximation for u-NFI.*

Proof First note that, by definition of C_1 and C_2, bisection(C_1, C_2) must have been called at some point during the execution of Algorithm 1, so Lemma 15 yields that $1 \leq \hat{\gamma} \leq u$.

The rest of the proof is divided into three steps. First, we show that the algorithm is an r-approximation for some $r > 1$ if val(C_2) is large enough. Afterwards, we show that the algorithm is an r-approximation if val(C_2) is small enough. Finally, we combine the results of the previous steps and prove that, for $r := B + 1$, one of the two conditions on val(C_2) guaranteeing that the algorithm is an r-approximation must always hold.

For the first step, we start by defining a function f_2 by $f_2(\gamma) := \mathrm{cap}^\gamma(C_2) - g(\gamma) = \gamma \cdot (q(C_2) - q_g) + p(C_2) - p_g$. As $g(\hat{\gamma}) \geq \mathrm{cap}^{\hat{\gamma}}(C_1) = \mathrm{cap}^{\hat{\gamma}}(C_2)$ by (b), it holds that

$$0 \geq f_2(\hat{\gamma}) = \hat{\gamma} \cdot (q(C_2) - q_g) + p(C_2) - p_g$$
$$= (q(C_2) - q_g) + p(C_2) - p_g + (\hat{\gamma} - 1) \cdot (q(C_2) - q_g)$$
$$= f_2(1) + (\hat{\gamma} - 1) \cdot (q(C_2) - q_g).$$

This yields that

$$- f_2(1) \geq (\hat{\gamma} - 1) \cdot (q(C_2) - q_g)$$
$$\Rightarrow \quad g(1) - \mathrm{cap}^1(C_2) \geq (\hat{\gamma} - 1) \cdot (q(C_2) - q_g)$$
$$\Rightarrow \quad g(1) \geq (\hat{\gamma} - 1) \cdot (q(C_2) - q_g) + \mathrm{cap}^1(C_2)$$
$$\Rightarrow \quad q_g + p_g - B \geq (\hat{\gamma} - 1) \cdot (q(C_2) - q_g) + q(C_2) + p(C_2) - B$$
$$\Rightarrow \quad \mathrm{val}(g) \geq (\hat{\gamma} - 1) \cdot (q(C_2) - q_g) + \mathrm{val}(C_2)$$
$$\Rightarrow \quad \mathrm{val}(C_{\mathrm{OPT}}) \geq (\hat{\gamma} - 1) \cdot (q(C_2) - B) + \mathrm{val}(C_2) \qquad \text{(by (a) and (d))}$$

Now let $r > 1$. Since the cut C_2 is found by the algorithm, the algorithm is an r-approximation if $\mathrm{val}(C_2) \leq r \cdot \mathrm{val}(C_{\mathrm{OPT}})$ or, equivalently, if $\mathrm{val}(C_{\mathrm{OPT}})/\mathrm{val}(C_2) \geq 1/r$. By the inequality above, this holds if

$$\frac{(\hat{\gamma} - 1) \cdot (q(C_2) - B) + \mathrm{val}(C_2)}{\mathrm{val}(C_2)} \geq \frac{1}{r}$$
$$\Leftrightarrow \quad \frac{(\hat{\gamma} - 1) \cdot (q(C_2) - B)}{\mathrm{val}(C_2)} \geq \frac{1 - r}{r}$$
$$\Leftrightarrow \quad \frac{r}{r - 1} \cdot (\hat{\gamma} - 1) \cdot (B - q(C_2)) \leq \mathrm{val}(C_2) \qquad (3)$$

Similar to the first step, for the second step, define a function f_1 by $f_1(\gamma) := \mathrm{cap}^\gamma(C_1) - g(\gamma) = \gamma \cdot (q(C_1) - q_g) + p(C_1) - p_g$. As $g(\hat{\gamma}) \geq \mathrm{cap}^{\hat{\gamma}}(C_1)$ by (b), it holds that

$$f_1(\hat{\gamma}) \leq 0$$
$$\Rightarrow \quad \hat{\gamma} \cdot (q(C_1) - q_g) + p(C_1) - p_g \leq 0$$
$$\Rightarrow \quad (q(C_1) - q_g) + p(C_1) - p_g + (\hat{\gamma} - 1) \cdot (q(C_1) - q_g) \leq 0$$
$$\Rightarrow \quad f_1(1) + (\hat{\gamma} - 1) \cdot (q(C_1) - q_g) \leq 0$$
$$\Rightarrow \quad f_1(1) \leq (\hat{\gamma} - 1) \cdot (q_g - q(C_1))$$

Rewriting $\text{val}(g) = q(C_1) + p(C_1) + f_1(1) - B$ and using that $\text{val}(g) = \text{val}(C_{\text{OPT}})$ by (d), we, thus, obtain

$$\text{val}(C_{\text{OPT}}) = \text{val}(g) \geq q(C_1) + p(C_1) + (\hat{\gamma} - 1) \cdot (q(C_1) - q_g) - B.$$

Now again let $r > 1$. Then the algorithm is an r-approximation if $\text{val}(C_2) \leq r \cdot \text{val}(C_{\text{OPT}})$, which, by the above, holds if

$$\text{val}(C_2) \leq r \cdot \big(q(C_1) + p(C_1) + (\hat{\gamma} - 1) \cdot (q(C_1) - q_g) - B\big).$$

Using that $q(C_1) - q_g \geq 1$ by (c), the algorithm is, thus, an r-approximation if

$$\text{val}(C_2) \leq r \cdot \big(q(C_1) + p(C_1) + (\hat{\gamma} - 1) - B\big). \tag{4}$$

For the third and last step, we use that, if the algorithm is *not* an r-approximation, then Inequalities (3) and (4) must both be violated, i.e., we must have

$$\frac{r}{r-1} \cdot (\hat{\gamma} - 1) \cdot (B - q(C_2)) > \text{val}(C_2) > r \cdot \big(q(C_1) + p(C_1) + (\hat{\gamma} - 1) - B\big). \tag{5}$$

For the sake of a contradiction, suppose that the algorithm is *not* a $(B + 1)$-approximation. Then, by setting $r := B + 1$ in (5), it holds that

$$\frac{1}{B} \cdot (\hat{\gamma} - 1) \cdot (B - q(C_2)) > q(C_1) + p(C_1) + (\hat{\gamma} - 1) - B \tag{6}$$

If $\hat{\gamma} = 1$, plugging it into (6) yields $0 > q(C_1) + p(C_1) - B$, which contradicts the assumption that the instance is non-trivial. If $\hat{\gamma} > 1$, dividing (6) by $\hat{\gamma} - 1 > 0$ yields

$$\underbrace{\frac{1}{B} \cdot (B - q(C_2))}_{\leq 1} > \underbrace{\frac{q(C_1) + p(C_1) - B}{(\hat{\gamma} - 1)} + 1}_{>0},$$

which is a contradiction (the fraction on the right hand side is strictly positive due to the assumption that the instance is non-trivial). $\qquad\square$

Lemma 20 *If there exists an optimal cut C_{OPT} with $q(C_{\text{OPT}}) \leq B$, then the algorithm is a $(B + 1)$-approximation.*

Proof Define an affine function g by $g(\gamma) := q(C_{\text{OPT}}) \cdot \gamma + p(C_{\text{OPT}})$. It is easy to check that this function fulfills the conditions of Proposition 19. $\qquad\square$

Lemma 21 *If there exists an optimal cut C_{OPT} with $q(C_{\text{OPT}}) > B$, then the algorithm is a $(B + 1)$-approximation.*

Proof We show that the affine function g defined by $g(\gamma) := B \cdot \gamma + p(C_{OPT}) + (q(C_{OPT}) - B) \cdot u$ fulfills the conditions of Proposition 19. Conditions (a) and (c) follow immediately from the definition of g. Condition (d) is fulfilled because $\text{val}(g) = q_g + p_g - B = p(C_{OPT}) + (q(C_{OPT}) - B) \cdot u = \text{val}(C_{OPT})$.

To show Condition (b), we use that $g(u) = \text{val}(g) + B \cdot u = \text{val}(C_{OPT}) + B \cdot u = \text{cap}^u(C_{OPT})$. As $q_g = B < q(C_{OPT})$, this also yields $g(\hat{\gamma}) \geq \text{cap}^{\hat{\gamma}}(C_{OPT})$. Moreover, $\text{cap}^{\hat{\gamma}}(C_{OPT}) \geq \text{cap}^{\hat{\gamma}}(C_1) = \text{cap}^{\hat{\gamma}}(C_2)$: If not, the capacity-$\hat{\gamma}$-min-cut $\hat{C}$ computed during the call to $\texttt{bisection}(C_1, C_2)$ would satisfy $\text{cap}^{\hat{\gamma}}(\hat{C}) \leq \text{cap}^{\hat{\gamma}}(C_{OPT}) < \text{cap}^{\hat{\gamma}}(C_1) = \text{cap}^{\hat{\gamma}}(C_2)$ and, thus, also $q(\hat{C}) \notin \{q(C_1), q(C_2)\}$ since $q(\hat{C}) = q(C_1)$ or $q(\hat{C}) = q(C_2)$ together with $\text{cap}^{\hat{\gamma}}(\hat{C}) < \text{cap}^{\hat{\gamma}}(C_1) = \text{cap}^{\hat{\gamma}}(C_2)$ would imply that C_1 or C_2, respectively, could not be a capacity-γ-min-cut for any γ. □

Theorem 22 *Algorithm 1 is a $(B + 1)$-approximation algorithm for u-NFI.* □

Finally, we provide an example where the approximation ratio is almost tight. The graph in this example is series-parallel, which means that our analysis is almost tight even for series-parallel graphs.

Example 23 Let $G = (V, E)$ be given by $V = \{s, v_1, v_2, t\}$ and $E = E_1 \cup E_2 \cup E_3$, where E_1 consists of $(B+1)^2$ small parallel arcs from s to v_1, E_2 consists of B large and $B + 1$ small parallel arcs from v_1 to v_2, and E_3 consists of $B + 1$ large parallel arcs from v_2 to t.

This graph contains only three cuts. When choosing u large enough compared to B, the minimum cut C_m in G will be the cut $(\{s\}, \{v_1, v_2, t\})$, while the unique optimal cut will be $C_{OPT} = (\{s, v_1\}, \{v_2, t\})$. The least cut C_l is given by the third cut $(\{s, v1, v2\}, \{t\})$. We start by showing that C_{OPT} might not be found by Algorithm 1 when calling bisection(C_l, C_m). To this end, note that we obtain $\hat{\gamma} = B + 1$ in the call to bisection(C_l, C_m), and all three cuts have the same capacity of $(B + 1)^2$ in $G^{\hat{\gamma}}$. Thus, the deterministic minimum cut algorithm might return C_l or C_m as the minimum cut in $G^{\hat{\gamma}}$, in which case C_{OPT} is not found by the algorithm—which we assume in the following.

For the values of the cuts, we have $\text{val}(C_l) = u$, $\text{val}(C_{OPT}) = B + 1$ and $\text{val}(C_m) = B^2 + B + 1$. Thus, for $u > B^2 + B + 1$, the cut C_m will be the cut returned by the algorithm and $\text{val}(C_m) = B^2 + B + 1 > B^2 + B = B \cdot \text{val}(C_{OPT})$, which shows that the algorithm is not a B-approximation algorithm for u-NFI.

4   Conclusion

To the best of our knowledge, the approximation algorithm for u-NFI presented in this paper is the first approximation algorithm for any variant of NFI whose approximation ratio only depends on the budget available to the interdictor, but not on the size of the graph. A noteworthy property of the algorithm is that the budget B is only used for the final choice of which cut to return from the set of cuts computed during the algorithm. This means that a single execution of the algorithm returns a set of cuts that contains a $(B + 1)$-approximate cut for any possible interdiction budget B.

An interesting question for future research is whether the techniques developed here for u-NFI can be generalized in order to construct approximation algorithms for the case of three or more different arc capacities.

References

1. Ahuja, R.K., Magnanti, T.L., Orlin, J.B.: Network Flows. Prentice Hall, Upper Saddle River (1993)
2. Bhaskara, A., Charikar, M., Venkatesan, G., Vijayaraghavan, A., Zhou, Y.: Polynomial integrality gaps for strong SDP relaxations of densest k-subgraph. In: Proceedings of the 23rd ACM-SIAM Symposium on Discrete Algorithms (SODA), pp. 388–405 (2012)
3. Burch, C., Carr, R., Krumke, S.O., Marathe, M., Phillips, C., Sundberg, E.: A decomposition-based pseudoapproximation algorithm for network flow inhibition, chap. 1, pp. 51–68. Kluwer Academic Press, Dordrecht (2003)
4. Chestnut, S.R., Zenklusen, R.: Interdicting structured combinatorial optimization problems with {0, 1}-objectives. Math. Oper. Res. **42**(1), 144–166 (2016)
5. Chestnut, S.R., Zenklusen, R.: Hardness and approximation for network flow interdiction. Networks **69**(4), 378–387 (2017)
6. Murray, A.T., Matisziw, T.C., Grubesic, T.H.: Critical network infrastructure analysis: interdiction and system flow. J. Geograph. Syst. **9**, 103–117 (2007)
7. Phillips, C.: The network inhibition problem. In: Proceedings of the 25th ACM Symposium on the Theory of Computing (STOC), pp. 776–785 (1993)
8. Wollmer, R.: Removing arcs from a network. Oper. Res. **12**(6), 934–940 (1964)
9. Wood, R.K.: Deterministic network interdiction. Math. Comput. Model. **17**(2), 1–18 (1993)

On the Benchmark Instances for the Bin Packing Problem with Conflicts

Tiziano Bacci and Sara Nicoloso

Abstract Many authors, mainly in the context of the Bin Packing Problem with Conflicts, used the random graph generator proposed in "Heuristics and lower bounds for the bin packing problem with conflicts" [M. Gendreau, G. Laporte, and F. Semet, *Computers & Operations Research*, 31:347–358, 2004]. In this paper we show that the graphs generated in this way are not arbitrary but threshold ones. Computational results show that instances of the Bin Packing Problem with Conflicts on threshold graphs are easier to solve w.r.t. instances on arbitrary graphs.

Keywords Bin packing with conflicts · Threshold graphs · Random graph generator

1 Introduction

The Bin Packing Problem with Conflicts ($BPPC$), first introduced in a scheduling context in [14], is defined as follows: given a graph $G = (V, E)$, a nonnegative integer weight w_i for each vertex $i \in V$, and a nonnegative integer B, find a partition of V into k subsets $V_1, \ldots, V_k$, such that the sum of the weights of the vertices assigned to same subset is less than or equal to B, two vertices connected by an edge do not belong to the same subset, and k is minimum.

The minimum value of k will be denoted k_{BPPC}. The graph $G = (V, E)$ is called *conflict graph* and two vertices connected by an edge are said to be *in conflict*.

$BPPC$ generalizes two well known combinatorial optimization problems, the Bin Packing Problem and the Vertex Coloring Problem. In fact, $BPPC$ reduces to Bin Packing when the edge set E of the graph G is empty, and it reduces to Vertex Coloring when $B \geq \sum_{i \in V} w_i$ or when G is complete. Observe that Vertex Coloring

T. Bacci (✉) · S. Nicoloso
Istituto di Analisi dei Sistemi ed Informatica "A. Ruberti", C.N.R., Rome, Italy
e-mail: tiziano.bacci@iasi.cnr.it; sara.nicoloso@iasi.cnr.it

© The Author(s), under exclusive license to Springer Nature Switzerland AG 2021
C. Gentile et al. (eds.), *Graphs and Combinatorial Optimization:
from Theory to Applications*, AIRO Springer Series 5,
https://doi.org/10.1007/978-3-030-63072-0_14

is solvable in linear time on threshold graphs, nevertheless $BPPC$ with a threshold conflict graph is NP-hard because Bin Packing is [10].

In this paper we show that a popular random graph generator [11], widely used in the context of the Bin Packing Problem with Conflicts, generates threshold graphs. Threshold graphs are very special interval graphs, contradicting the authors [11] who claim that "No assumptions are made on the adjacency structure of the graph", and strengthening [28] where the authors recognise the graphs as being arbitrary interval graphs [28].

In Sect. 2 we recall the definition of threshold graphs and discuss some of their peculiar properties, in Sect. 3 we present the generator defined in [11] showing that it produces threshold graphs, in Sect. 4 we analyse the effects of using this generator on instances of Bin Packing Problem with Conflicts. Concluding remarks are discussed in Sect. 5.

2   Threshold Graphs

A graph $G = (V, E)$ is a threshold graph if there exist a real number d (the threshold) and a weight p_x for every vertex $x \in V$ such that (i, j) is an edge iff $(p_i + p_j)/2 \leq d$ (see [12]). W.l.o.g. from now on we assume that $p_x \in [0, 1]$ $\forall x$ (as a consequence it makes sense to choose $d \in [0, 1]$).

According to this definition it follows that a vertex i is connected to all the vertices j such that $p_j \leq 2d - p_i$. Let $N(x)$ denote the set of vertices adjacent to x and let $\deg(x) = |N(x)|$. Then $N(h) \supseteq N(k)$ and $\deg(h) \geq \deg(k)$ if and only if $p_h \leq p_k$.

A threshold graph has many peculiar properties as it is at the same time an interval graph, a co-interval graph, a cograph, a split graph, and a permutation graph. In addition, its complement, where (i, j) is an edge iff $(p_i + p_j)/2 > d$, is a threshold graph too.

W.l.o.g. from now on we assume that the vertices of a threshold graph G are numbered in such a way that $i < j$ if and only if $\deg(i) \geq \deg(j)$. Then the $n \times n$ symmetric adjacency matrix $M = [m_{i,j}]$ of G always appears as in Fig. 1, where an entry 0 is coloured in white and an entry 1 is highlighted in grey, and $m_{i,i} = 0$ for $i = 1, \ldots, n$.

By what above, we observe what follows.

1. For each row i, let $last\_col(i) = \max\{j : m_{i,j} = 1, j = 1, \ldots, n\}$ if $m_{i,1} = 1$, and $last\_col(i) = 0$ if $m_{i,1} = 0$ (see Fig. 1b); hence $last\_col(i) \geq last\_col(i + 1)$.

2. Let $t = \min\{j : m_{j,j+1} = 0, j = 1, \ldots, n\}$. Then the set of vertices $\{1, \ldots, t\}$ induces a maximum clique of size $\omega(G) = t$ (see Fig. 1a). In fact, by definition, $m_{t-1,t} = 1$, thus $last\_col(t - 1) \geq t$ and, by Point 1, $m_{i,j} = 1$ for $i = 1, \ldots, t$ and $j = 1, \ldots, t, i \neq j$.

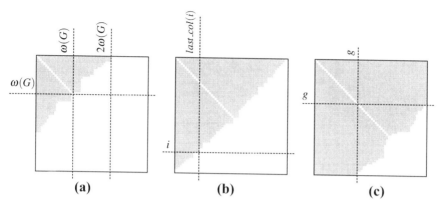

Fig. 1 Examples of adjacency matrices of threshold graphs with $n = 60$ nodes and threshold (**a**) $d = 0.2$, (**b**) $d = 0.5$, and (**c**) $d = 0.7$

3. The set of vertices $\{t, \ldots, n\}$ induces a maximum independent set of size $n - t + 1$. In fact, by definition, $m_{t,t+1} = 0$ and $m_{t,t-1} = 1$ (as $m_{t-1,t} = 1$) thus $last\_col(t) = t - 1$ and $m_{i,j} = 0$ for $i = t, \ldots, n$ and $j = t, \ldots, n$ (see Point 1.).

4. Let $g = last\_col(n)$ (see Fig. 1c). If $g \geq 1$, vertex i, for $i = 1, \ldots, g$, is connected to any other vertex.

5. Recalling that a threshold graph G is a particular interval graph, it is always possible to derive a family of (open) intervals whose intersection graph is G, namely: to each vertex $j = t, \ldots, n$, associate the interval $I_j = (l_j, r_j) = (j - t, j - t + 1)$; to each vertex $j = 1, \ldots, t - 1$, associate the interval $I_j = (l_j, r_j) = (0, r_{last\_col(j)}) = (0, last\_col(j) - t + 1)$ (we remark that $r_j \geq 1$ as $last\_col(j) \geq t$). See an example in Fig. 2.

6. The edge density $\delta = 2|E|/(n(n - 1))$ of G is not equal to the threshold d, generally speaking.

For $n \to \infty$ and $p_1, \ldots, p_n$ uniformly distributed in $[0, 1]$, one has:

7. $\omega(G) = t = nd$.

8. The edge density $\delta = 2|E|/(n(n - 1))$ of G depends on d. Precisely

$$\delta = f(d) = \begin{cases} \frac{2(nd)^2 - nd}{n(n-1)} & \text{for } d \leq 0.5 \\[2mm] \frac{n(n-1) - 2n^2(1-d)^2 - n(1-d)}{n(n-1)} & \text{for } d \geq 0.5 \end{cases}$$

In fact, for $d \leq 0.5$ the $2|E|$ 1's are in the area $A \cup B \cup C$ (see Fig. 3a). In a similar way one can compute the number of 1's in the matrix when $d \geq 0.5$.

9. $g = 0$ when $d \leq 0.5$, and $g = n(2d - 1)$ when $d \geq 0.5$ (see Fig. 3b).

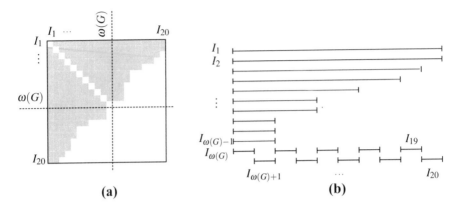

Fig. 2 (a) The adjacency matrix of threshold graph with $n = 20$ nodes and (b) the corresponding opened interval model

According to Points 2 and 3 above, when $d \geq 0.5$, in any optimal solution of $BPPC$ one has $V_i = \{i\}$ for $i = 1, \ldots, g$. The remaining sets V_i for $i \geq g + 1$ can be determined by solving a smaller instance Q defined on the last $n - g$ vertices (observe that the problem becomes simpler and simpler as d increases). The conflict graph of Q is a threshold graph with expected edge density 0.5, thus it contains a maximum clique of expected size $(n - g)/2$. According to Point 7, when $n = 120$ and $d = 0.9$, a lower bound for k_{BPPC} is $nd = 108$ (indeed this value appears in Table 2, column LBO, Size 120, $d = 90$ in [9]).

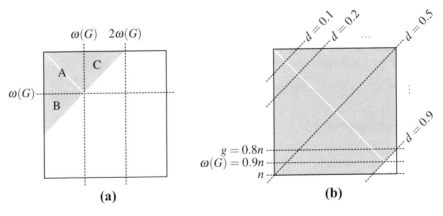

Fig. 3 The expected adjacency matrix when $n \to \infty$ and with threshold (a) $d \leq 0.5$ and (b) $d \geq 0.5$

3 On the Random Graph Generator Proposed in [11]

In [11] the following random graph generator is described: "A value p_i was first assigned to each vertex $i \in V$ according to a continuous uniform distribution on [0, 1]. Each edge (i, j) of G was created whenever $(p_i + p_j)/2 \leq d$, where d is the expected density of G.".

This generator clearly produces a threshold graph whose expected edge density is not d as claimed but it is the one discussed in Points 6 and 8 of Sect. 2.

To get a threshold graph with expected edge density δ one has to set

$$d = \begin{cases} \frac{1+\sqrt{1+8n(n-1)\delta}}{4n} & \text{for } \delta \leq 0.5 \\ 1 + \frac{1-\sqrt{1+8n(n-1)(1-\delta)}}{4n} & \text{for } \delta \geq 0.5 \end{cases}$$

Already for $n \geq 100$ these values can be approximated to $d = \sqrt{\delta/2}$ and $d = 1 - \sqrt{(1-\delta)/2}$, respectively.

The generator in [11] has been improperly used to generate arbitrary graphs [3–9, 13, 15–28, 30, 31]. In particular, the authors in [26] made publicly available "benchmark" instances generated in this way (see http://or.dei.unibo.it/library/bin-packing-problem-conflicts) and used by many authors [4–9, 13, 15–17, 19–21, 25, 27, 28, 30, 31].

Most of the authors using the generator in [11] claim that they group the graphs of their test bed by edge densities, but actually they group the graphs by threshold values. Our analysis of the instances introduced in [26] shows that the relation between the threshold d and the corresponding edge density δ is the following.

d	0.00	0.10	0.20	0.30	0.40	0.50	0.60	0.70	0.80	0.90
δ	0.00	0.02	0.08	0.18	0.32	0.50	0.68	0.82	0.92	0.98

We remark that the values of δ coincide with those which can be computed by the formula of Point 8 in Sect. 2.

4 Computational Results on Different Graph Classes

Since threshold graphs are a subclass of interval graphs, which are in their turn a subclass of arbitrary graphs, we expect that $BPPC$ on threshold graphs is the easiest to solve. To prove our claim we conducted some computational experiments.

By $X(n, \delta)$ we denote a set of ten instances with n vertices, bound $B = 150$, and conflict graph X with expected edge density $\delta \in \{0.02, 0.08, 0.18, 0.32, 0.50, 0.68, 0.82, 0.92, 0.98\}$ (the same densities of the instances used in [26]). In particular, we choose $n \in \{250, 1000\}$ and $X \in \{T, I, A\}$, where $X = T$ (I, A, respectively) when the conflict graph is a threshold (interval, arbitrary, respectively) graph.

The $T(250, \delta)$ and $T(1000, \delta)$ instances are exactly those in the classes 2 and 4 described in [26], respectively. Precisely, given n, the weight of the i-th vertex of the k-th instance of $T(n, \delta)$ is the same for all δ.

As for the $I(n, \delta)$, the weight of the i-th vertex of the k-th instance coincides with the weight of the i-th vertex of the k-th instance of $T(n, \delta)$, and the arbitrary interval conflict graphs have been generated according to the interval graph generator in [2].[1]

As for the $A(n, \delta)$, the weight of the i-th vertex of the k-th instance coincides with the weight of the i-th vertex of the k-th instance of $T(n, \delta)$, and the arbitrary conflict graphs have been generated as in [28]: "*We began with the empty graph. We iteratively selected an item pair* (i, j) *at random (with uniform distribution); then edge* (i, j) *was added to the graph if it was not already defined. The procedure was interrupted as soon as the desired graph density was reached.*".

We solved to optimality the $T(n, \delta)$, $I(n, \delta)$, and $A(n, \delta)$ instances for all n and δ by means of the Vector Packing Solver 3.1.2 (VPS for short) defined in [4], available at http://vpsolver.dcc.fc.up.pt/. This method is based on an arc-flow formulation with side constraints and builds very strong integer programming models that can be given in input to any state-of-the-art mixed integer programming solver. Actually, the arc-flow formulation is derived from a suitable graph which is preliminarily generated and whose size increases rapidly with B. We remark that the algorithm is applied to many classical combinatorial problems: in particular, it is one of the best behaving exact approaches for the instances introduced in [26], which are all solved to optimality within 50 min and with an average runtime of 2 min. For our analysis we solved the integer programming model with Cplex 12.6 on an Intel Core i7-3632QM 2.20 GHz with 16 GB RAM under a Linux operating system, setting a time limit of 600 s for each instance. The instances used in this section and many others can be downloaded at [1].

The computational results are summarized in Table 1, where rows are indexed by δ, and columns by the type of the conflict graph. In the "Opt" columns we report the number of instances, out of ten, solved to optimality within the time limit, and in the "Time" columns the time in seconds required to solve one instance, averaged over the solved instances, only.

The results in the table show that threshold instances T are easier w.r.t. instances with interval conflict graphs, and these latter are easier than those with arbitrary conflict graphs, confirming our claim.

We remark that, as far as we know, no tests on instances of $BPPC$ with arbitrary interval conflict graphs were performed in the literature. The authors in [28] observe that the conflict graphs of the benchmark instances in [26] are interval graphs and not arbitrary graphs (actually they are not arbitrary interval ones). Nevertheless, to

---

[1]The generator in [2] is not able to produce interval graphs with $n = 1000$ and edge density $\delta = 0.98$; in the corresponding cell of Table 1 of the present paper the average edge density of the ten instances is 0.96.

Table 1 Computational results on instances with threshold (T), interval $(I)^1$, and arbitrary (A) conflict graphs

		$n = 250$						$n = 1000$					
		T		I		A		T		I		A	
		Opt	Time	Opt	Time	Opt	Time	Opt	Time	Opt	Time	Opt	Time
δ	0.02	10	1.28	1	138.28	1	206.62	10	76.37	0	–	0	–
	0.08	10	2.75	0	–	0	–	10	292.26	0	–	0	–
	0.18	10	3.37	1	522.39	0	–	10	359.16	0	–	0	–
	0.32	10	3.81	10	201.56	4	340.2	3	444.77	0	–	0	–
	0.50	10	1.00	10	15.31	10	75.16	10	390.94	0	–	0	–
	0.68	10	0.53	10	3.24	10	12.43	10	294.12	0	–	0	–
	0.82	10	0.29	10	2.02	10	5.15	10	222.57	5	543.92	0	–
	0.92	10	0.11	10	1.39	10	2.89	10	197.60	10	453.98	0	–
	0.98	10	0.04	10	1.03	10	1.92	10	199.36	10	366.07	3	561.02

our knowledge, the authors in [28] are the only ones who test their algorithm on instances with arbitrary conflict graphs.

5 Concluding Remarks

In this paper we show that graphs of the BPPC instances considered in [3–9, 13, 15–28, 30, 31] and generated according to [11] are threshold graphs (and not arbitrary ones), and their edge density is not the declared one.

We also show that BPPC instances with threshold conflict graphs are computationally easier to solve than instances with interval or arbitrary conflict graphs (the instances used and many others are available at [1]). This behaviour confirms the behaviour of the computational complexity of many classical combinatorial problems on the three graph classes considered.

We believe that the reduced difficulty of these instances is mainly due to the structure of the conflict graph and not to "*the presence of capacity constraints on the cardinality of the color classes*" as suggested in [8]. This could be ascertained by solving BPPC instances with arbitrary interval conflict graphs and with arbitrary conflict graphs.

We also remark that the authors in [11] claim to use "the procedure described in" [29], but this is not true. In fact, the procedure in [29] generates "*edge* (i, j) *with probability*" $(p_i + p_j)/2$: by doing so it generalizes the uniform random graph generator and outputs arbitrary graphs.

References

1. Bacci, T.: http://www.iasi.cnr.it/~tbacci/ (2017)
2. Bacci, T., Nicoloso, S.: A heuristic algorithm for the bin packing problem with conflicts on interval graphs (2017). arXiv:1707.00496 [math.CO]
3. Basnet, C., Wilson, J.: Heuristics for determining the number of warehouses for storing non-compatible products. Int. Trans. Oper. Res. **12**, 527–538 (2005)
4. Brandão, F., Pedroso, J.P.: Bin packing and related problems: general arc-flow formulation with graph compression. Comput. Oper. Res. **69**, 56–67 (2016)
5. Capua, R., Frota, Y., Vidal, T., Ochi, L.S.: Um algoritmo heurìstico para o problema de bin packing com conflitos (2015). Manuscript http://www.din.uem.br/~ademir/sbpo/sbpo2015/pdf/146111.pdf
6. Capua, R., Frota, Y., Ochi, L.S., Vidal, T.: A study on exponential-size neighborhoods for the bin packing problem with conflicts. J. Heuristics **24**, 667–695 (2018)
7. Clautiaux, F., Khanafer, A., Hanafi, S., Talbi, E.: Le problème bi-objectif de bin-packing avec conflits (2011). Manuscript https://ori-nuxeo.univ-lille1.fr/nuxeo/site/esupversions/49ddd808-5557-41e2-bdfe-73b69e53264d
8. Cornaz, D., Furini, F., Malaguti, E.: Solving vertex coloring problems as maximum weight stable set problems. Discrete Appl. Math. **217**, 151–162 (2017)
9. Elhedhli, S., Li, L., Gzara, M., Naoum-Sawaya, J.: A branch-and-price algorithm for the bin packing problem with conflicts. INFORMS J. Comput. **23**, 404–415 (2011)
10. Garey, M.R., Johnson, D.S.: Computers and Intractability: A Guide to the Theory of NP-Completeness. W.H. Freeman, New York (1979)
11. Gendreau, M., Laporte, G., Semet, F.: Heuristics and lower bounds for the bin packing problem with conflicts. Comput. Oper. Res. **31**, 347–358 (2004)
12. Golumbic, M.C.: Algorithmic Graph Theory and Perfect Graphs. Academic Press, Cambridge (1980)
13. Gschwind, T., Irnich, S.: Dual inequalities for stabilized column generation revisited. INFORMS J. Comput. **28**(1), 175–194 (2016)
14. Jansen, K., Oehring, S.: Approximation algorithms for time constrained scheduling. Inf. Comput. **132**(2), 85–108 (1997)
15. Joncour, C.: Problèmes de placement 2d et application à l'ordonnancement: modélisation par la théorie des graphes et approches de programmation mathématique. Ph.D. Thesis, Université de Bordeaux I (2010). Number 4173
16. Joncour, C., Michel, S., Sadykov, R., Sverdlov, D., Vanderbeck, F.: Column generation based primal heuristics. Electron. Notes Discrete Math. **36**, 695–702 (2010)
17. Jouida, S.B., Krichen, S.: A DSS based on optimizer tools and MTS meta-heuristic for the warehousing problem with conflicts. Inf. Process. Lett. **135**, 14–21 (2018)
18. Jouida, S.B., Ouni, A., Krichen, S.: A multi-start tabu search based algorithm for solving the warehousing problem with conflict. In: Thi, H.A.L., Dinh, T.P., Nguyen, N.T. (eds.) Modelling, Computation and Optimization in Information Systems and Management Sciences: Proceedings of the 3rd International Conference on Modelling, Computation and Optimization in Information Systems and Management Sciences - MCO 2015 - Part II, pp. 117–128. Springer, Berlin (2015)
19. Khanafer, A.: Algorithmes pour des problèmes de bin packing mono- et multi-objectif. Ph.D. Thesis, Université Lille1 (2010). Number 40363
20. Khanafer, A., Clautiaux, F., Hanafi, S., Talbi, E.: The min-conflict packing problem. Comput. Oper. Res. **39**, 2122–2132 (2012)
21. Khanafer, A., Clautiaux, F., Talbi, E.: New lower bounds for bin packing problems with conflicts. Eur. J. Oper. Res. **206**(2), 281–288 (2010)
22. Khanafer, A., Clautiaux, F., Talbi, E.: Tree-decomposition based heuristics for the two-dimensional bin packing problem with conflicts. Comput. Oper. Res. **39**, 54–63 (2012)

23. Maiza, M., Guéret, C.: A new lower bound for bin packing problem with general conflicts graph. In: 1st Doctoriales STIC'09. Université de M'sila, Algérie (2009)
24. Maiza, M., Radjef, M.S.: Heuristics for solving the bin-packing problem with conflicts. Appl. Math. Sci. 5, 1739–1752 (2011)
25. Muritiba, F.A.E.: Algorithms and models for combinatorial optimization problems. Ph.D. Thesis, Alma Mater Studiorum Università di Bologna (2010)
26. Muritiba, F.A.E., Iori, M., Malaguti, E., Toth, P.: Algorithms for the bin packing problem with conflicts. INFORMS J. Comput. 22(3), 401–415 (2010)
27. Sadykov, R.: Modern branch-cut-and-price. Tech. Rep., Operations Research [cs.RO]. Universit de Bor-deaux (2019)
28. Sadykov, R., Vanderbeck, F.: Bin packing with conflicts: a generic Branch-and-Price algorithm. INFORMS J. Comput. 25(2), 244–255 (2013)
29. Soriano, P., Gendreau, M.: Tabu search algorithms for the maximum clique problem. Dimacs Ser. Discrete Math. Theor. Comput. Sci. 26, 221–236 (1996)
30. Wei, L., Luo, Z., Baldacci, R., Lim, A.: A new branch-and-price-and-cut algorithm for one-dimensional bin-packing problems. INFORMS J. Comput. 32(2), 428–443 (2020)
31. Yuan, Y., Li, Y., Wang, Y.: An improved aco algorithm for the bin packing problem with conflicts based on graph coloring model. In: 21th International Conference on Management Science & Engineering, Helsinki (2014)

Directed Zagreb Indices

Barbara M. Anthony and Alison M. Marr

Abstract Zagreb indices for undirected graphs were introduced nearly 50 years ago. Their original development was related to uses in chemistry, but over time mathematicians have also found them to be an interesting topic of study. We define and introduce Zagreb indices for directed graphs, give results that parallel many of the conjectures and theorems that exist for the original Zagreb indices, and produce results specific to the directed graph case.

Keywords Directed graphs · First Zagreb index · Second Zagreb index

1 Introduction

The Zagreb indices were first introduced [4] nearly 50 years ago. Since that time dozens of papers have been written comparing these two indices, finding bounds on their values, and generalizing these indices. The popularity of these indices stems from their applications to chemistry. For a general overview of the history of these indices and their applications to chemistry, see [8]. Additionally, the survey [7] by Liu and You summarizes some of the existing mathematical work in the field.

Let $G = (V, E)$ be a graph. Let $d(v)$ denote the degree of vertex v in the graph G. The classical definitions of the first and second Zagreb indices, developed by Gutman and Trinajstić [4], are as follows:

Definition 1 ([4]) The first Zagreb index on a graph G is defined as

$$M_1(G) = \sum_{v \in V(G)} d(v)^2.$$

B. M. Anthony (✉) · A. M. Marr
Southwestern University, Department of Mathematics and Computer Science, Georgetown, TX, USA
e-mail: anthonyb@southwestern.edu; marra@southwestern.edu

© The Author(s), under exclusive license to Springer Nature Switzerland AG 2021
C. Gentile et al. (eds.), *Graphs and Combinatorial Optimization: from Theory to Applications*, AIRO Springer Series 5,
https://doi.org/10.1007/978-3-030-63072-0_15

The second Zagreb index on a graph G is defined as

$$M_2(G) = \sum_{e=(u,v)\in E(G)} d(u)d(v).$$

In this paper, we define and examine a new generalization of Zagreb indices by defining them on directed graphs. While work has been done on zeroth-order general Randić indices (also motivated by chemistry, giving the sum of bond contributions) on digraphs [9], this is the first time that directed Zagreb indices are being defined and studied. Throughout this work, we use *nodes* and *arcs* when speaking about directed graphs, and *vertices* and *edges* for undirected graphs. The number of nodes or vertices is denoted by n, while the number of arcs or edges in a graph or digraph is denoted by m. We use *digraph* and *directed graph* interchangeably. The *in-degree* of a node u is denoted by $d^-(u)$, and the *out-degree* by $d^+(u)$. A *source* is a node with in-degree zero, and a *sink* is a node with out-degree zero. Let $D = (N, A)$ be a directed graph. The in and out neighborhoods of a node u are defined, respectively, as $N^-(u) = \{v \in N(D)|(v, u) \in A(D)\}$ and $N^+(u) = \{v \in N(D)|(u, v) \in A(D)\}$.

Definition 2 The first Zagreb index on a directed graph D is defined as

$$\vec{M}_1(D) = \sum_{v\in N(D)} d^+(v)d^-(v).$$

The second Zagreb index on a directed graph D is defined as

$$\vec{M}_2(D) = \sum_{e=(u,v)\in A(D)} d^+(u)d^-(v).$$

We allow graphs to be connected or disconnected. We do not allow multiple arcs or loops. We do not allow isolated nodes: just as their inclusion does not alter the undirected Zagreb indices, nor do they change the directed Zagreb indices. When the digraph under consideration is obvious from the context, we may omit it, simply writing $\vec{M}_1$ instead of $\vec{M}_1(D)$ for either of the directed Zagreb indices, and similarly for the graph in undirected Zagreb indices.

An *oriented graph* is a digraph with no bidirected arcs, that is, if (u, v) is an arc in the digraph, (v, u) cannot be an arc in the digraph. A *cycle* is a graph (or digraph with no bidirected arcs) where $n = m \geq 3$. If n is odd, it is an *odd cycle*; otherwise, it is an *even cycle*. When the arcs in a cycle are all oriented in the same direction, it is a *directed cycle*. Given a graph G, we define G^* to be the digraph with bidirected arcs in G^* for every edge in G. Thus, for example, K_2^* is a pair of bidirected arcs. Given a digraph D, let $\overleftarrow{D}$ denote the digraph where the orientation of every arc in D is flipped. We define a *directed path* to be a path in which all arcs are oriented so that the destination of an arc in the path is the origin of the subsequent path arc.

2 Results

This section begins with some fundamental properties of these newly defined directed Zagreb indices. Next, we compare the two directed Zagreb indices and relate these results to previously know relationships about the (undirected) Zagreb indices. We examine the values of $\overrightarrow{M}_1(D)$ and $\overrightarrow{M}_2(D)$ for various cycles, stars, and paths, characterize many categories of graphs in terms of whether or not $\overrightarrow{M}_1$ and $\overrightarrow{M}_2$ are equal, and then explore the possible values for $\overrightarrow{M}_2(D) - \overrightarrow{M}_1(D)$.

2.1 Fundamental Properties of Directed Zagreb Indices

Though $\overrightarrow{M_1}$ is defined in terms of nodes, we show that we can in fact write $\overrightarrow{M_1}$ as a sum over arcs in the digraph. We then observe that flipping the orientation of all arcs in a digraph does not change the directed Zagreb indices. We give explicit formulas for the directed Zagreb indices on regular digraphs. Finally, we show (in Property 4) that the directed Zagreb indices of a disconnected graph are simply the sum of the directed Zagreb indices of the components.

Property 1 An alternative way to write $\overrightarrow{M}_1(D)$ is

$$\overrightarrow{M_1}(D) = \frac{1}{2} \sum_{e=(u,v)\in A(D)} (d^-(u) + d^+(v)).$$

Proof Consider a node $x \in D$. In the proposed alternative, the contribution of the node x to $\overrightarrow{M}_1(D)$ comes from every arc that it is a part of. For those arcs in which x is the origin, we count the number of arcs that enter x, that is, $d^-(x)$, for every arc that starts at x, whose number is $d^+(x)$, giving a total of $d^+(x)d^-(x)$. For those arcs in which x is the destination, we count the number of arcs that leave x, that is, $d^+(x)$, and we count that for every entering arc, namely $d^-(x)$, again giving a total of $d^+(x)d^-(x)$. The division by two handles the double-counting. Ultimately, in the alternative representation, we have counted $d^+(x)d^-(x)$ for every node $x \in D$, precisely matching the definition of $\overrightarrow{M}_1(D)$. $\square$

Property 2 $\overrightarrow{M}_1(D) = \overrightarrow{M}_1(\overleftarrow{D})$ and $\overrightarrow{M}_2(D) = \overrightarrow{M}_2(\overleftarrow{D})$

Property 3 Let $D = (N, A)$ be a regular digraph with $d^+(v) = d^-(v) = k \ \forall v \in N$. Then $\overrightarrow{M}_1(D) = nk^2$ and $\overrightarrow{M}_2(D) = mk^2$.

Corollary 1 *The complete digraph K_n^* has $\overrightarrow{M}_1(K_n^*) = n(n-1)^2$ and $\overrightarrow{M}_2(K_n^*) = n(n-1)^3$.*

Property 4 Let a directed graph D consist of two connected components, digraphs R and S. Then $\overrightarrow{M}_1(D) = \overrightarrow{M}_1(R) + \overrightarrow{M}_1(S)$, and $\overrightarrow{M}_2(D) = \overrightarrow{M}_2(R) + \overrightarrow{M}_2(S)$.

Proof Since nodes in D can be partitioned into nodes in R and nodes in S, $\vec{M}_1(D) = \sum_{v \in N(D)} d^+(v) d^-(v) = \sum_{v \in N(R)} d^+(v) d^-(v) + \sum_{v \in N(S)} d^+(v) d^-(v) = \vec{M}_1(R) + \vec{M}_1(S)$. Analogously, the arcs in D can be partitioned into the arcs in R and the arcs in S, yielding the desired result for $\vec{M}_2$.   □

2.2 Comparing Directed Zagreb Indices

One of the most popular avenues of research when studying the first two Zagreb indices is to compare their values. For undirected Zagreb indices, M_1 and M_2 can be equal, $M_1 > M_2$ or $M_1 < M_2$. In [6], Horoldagva, Das, and Selenge show which classes of graphs fall into each of the three categories. We show that for directed Zagreb indices only two of these options are possible.

Theorem 1 For any directed graph D, $\vec{M}_1(D) \leq \vec{M}_2(D)$.

Proof Proof by induction on the number of arcs in D.

Base case: Trivially, if there are no arcs in D, then $\vec{M}_1(D) = 0 = \vec{M}_2(D)$. For illustration, if D contains a single arc, then $\vec{M}_1(D) = 0 \cdot 1 + 1 \cdot 0 = 0$, and $\vec{M}_2(D) = 1 \cdot 1 = 1$, so $\vec{M}_1(D) < \vec{M}_2(D)$.

Inductive hypothesis: We assume that for any digraph D with k arcs, $\vec{M}_1(D) \leq \vec{M}_2(D)$. Let $D^\wedge$ be a digraph with $k + 1$ arcs. We want to show that $\vec{M}_1(D^\wedge) \leq \vec{M}_2(D^\wedge)$.

Pick an arbitrary arc $e = (u, v) \in D^\wedge$. Removing e from $D^\wedge$ yields a digraph D' with exactly k arcs, and thus $\vec{M}_1(D') \leq \vec{M}_2(D')$ by the inductive hypothesis. Thus, by construction, $e \notin D'$.

We now consider how $\vec{M}_1$ differs between D' and $D^\wedge$. The only terms in the sum which are altered are the terms contributed by the nodes u and v. In $D^\wedge$, the in-degree of node u is unchanged, and its out-degree increases by 1. Thus, the contribution of u to $\vec{M}_1$ was previously $d^-(u) \cdot d^+(u)$, and is now $d^-(u) \cdot (d^+(u)+1)$, showing that the change from the contribution of node u is exactly $d^-(u)$. Similarly, the change from the contribution of node v is exactly $d^+(v)$.

Thus $\vec{M}_1(D^\wedge) = \vec{M}_1(D') + d^-(u) + d^+(v)$.

Calculating $\vec{M}_2(D^\wedge)$, since $e \notin D'$, $\vec{M}_2(D^\wedge)$ is precisely $\vec{M}_2(D')$ plus the contribution from arc e, both to the new summand term from e, and potential increases to existing arcs in D'.

The new arc e generates a contribution of $(d^+(u) + 1)(d^-(v) + 1)$, along with additional nonnegative contributions to the terms for arcs leaving u and entering v, namely

$$\sum_{x \in N^+(u)} d^-(x) + \sum_{y \in N^-(v)} d^+(y).$$

Thus

$$\vec{M}_2(D^\wedge) = \vec{M}_2(D') + d^+(u) + d^-(v) + d^+(u)d^-(v) + 1 + \sum_{x \in N^+(u)} d^-(x) + \sum_{y \in N^-(v)} d^+(y)$$

$$\geq \vec{M}_1(D') + d^+(u) + d^-(v) + d^+(u)d^-(v) + 1 + \sum_{x \in N^+(u)} d^-(x) + \sum_{y \in N^-(v)} d^+(y)$$

$$= \vec{M}_1(D^\wedge) + d^+(u)d^-(v) + 1 + \sum_{x \in N^+(u)} d^-(x) + \sum_{y \in N^-(v)} d^+(y)$$

$$\geq \vec{M}_1(D^\wedge) \text{ since all terms are nonnegative.}$$

$\square$

We now establish an explicit connection between the classical Zagreb indices on an undirected graph, and the directed Zagreb indices on the corresponding digraph with bidirected arcs for all edges in the undirected graph. That result, Proposition 1, combined with Theorem 1 lead us to an alternative proof of a known result, that for undirected graphs the first Zagreb index is at most twice the second Zagreb index. While the result was already presented in [2], we highlight it here in Corollary 2 because of how it further illustrates the connection between the directed and undirected Zagreb indices.

Proposition 1 *Let G be an arbitrary undirected graph. Then $M_1(G) = \vec{M}_1(G^*)$ and $2 \cdot M_2(G) = \vec{M}_2(G^*)$.*

Proof Recall that G^* is the directed graph with bidirected arcs in G^* for every edge in G. By construction of G^*, for any node $v \in G^*$ arising from a vertex $x \in G$, $d^+(v) = d^-(v) = d(x)$. The equality of $M_1(G) = \vec{M}_1(G^*)$ follows immediately, and $2 \cdot M_2(G) = \vec{M}_2(G^*)$ because there are two arcs in G^* for every edge in G. $\square$

Corollary 2 *Let G be an arbitrary undirected graph. Then, $M_1(G) \leq 2M_2(G)$.*

As reported in Caporossi et al. [1], experiments with the AutoGraphiX system led to a conjecture that for undirected Zagreb indices $\dfrac{M_1(G)}{n} \leq \dfrac{M_2(G)}{m}$, which Pierre Hansen presented at the second meeting of the International Academy of Mathematical Chemistry in 2006. However, while it was shown the following year by Hansen and Vukicević [5] that the relationship always holds true for chemical graphs, they show that the conjecture is not true for general graphs. One such instance provided in [5] consists of a disconnected graph whose two components were a $K_{1,6}$ and C_3. We show that there is a natural transformation of that graph into the digraph $K_{1,6}^* \cup C_3^*$ that likewise disproves the analogous inequality for digraphs.

Lemma 1 *There exists a digraph such that $\dfrac{\vec{M}_1(D)}{n} > \dfrac{\vec{M}_2(D)}{m}$.*

Fig. 1 A directed Zagreb instance where the first index divided by the number of nodes exceeds the second index divided by the number of arcs

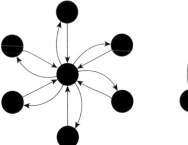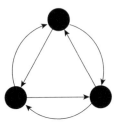

Proof Let $D = K_{1,6}^* \cup C_3^*$, as shown in Fig. 1. Observe that $\vec{M}_1(D) = 36 + 6 \cdot 1 + 3 \cdot 4 = 54$, where the contributions come from the center of the star, the leaves of the star, and the nodes of the cycle, respectively. Observe also that $\vec{M}_2(D) = 6 \cdot 6 + 6 \cdot 6 + 6 \cdot 4 = 96$, with contributions from the six arcs directed out of the center of the star, the six arcs directed into the center of the star, and the six arcs in the $\vec{C}_3^*$. Since the graph consists of 10 nodes and 18 arcs, $\dfrac{\vec{M}_1(D)}{n} = 5.4 > 5.333 = \dfrac{\vec{M}_2(D)}{m}$. $\quad\square$

2.3 Bounds on Directed Zagreb Indices

Considering all orientations on a particular graph, we can create bounds on the possible values for $\vec{M}_1(D)$ and $\vec{M}_2(D)$.

Proposition 2 *For any orientation of a $K_{1,n}$ (with no bidirectional arcs), $0 \leq \vec{M}_1(K_{1,n}) \leq \lfloor \frac{n^2}{4} \rfloor$ and $\lceil \frac{n^2}{2} \rceil \leq \vec{M}_2(K_{1,n}) \leq n^2$.*

Proof For the lower bound for $\vec{M}_1$, consider a star $K_{1,n}$ where all arcs are directed into the center. Then $\vec{M}_1(K_{1,n}) = 0$. Clearly a negative value is not possible.

For the upper bound for $\vec{M}_1$, consider a star $K_{1,n}$ where $\lfloor \frac{n}{2} \rfloor$ of the arcs are directed into the center, and the rest ($\lceil \frac{n}{2} \rceil$ arcs) are directed out of the center. This is the largest that $\vec{M}_1$ can be as the only contribution to $\vec{M}_1$ is at the center, and it is maximized when the in-degree and out-degree are as close as possible. Then $\vec{M}_1(K_{1,n}) = \lfloor \frac{n}{2} \rfloor * \lceil \frac{n}{2} \rceil$. If $n = 2s$ for some positive integer s, then $\lfloor \frac{n}{2} \rfloor * \lceil \frac{n}{2} \rceil = s * s = s^2 = n^2/4 = \lfloor \frac{n^2}{4} \rfloor$. If $n = 2s + 1$ for some positive integer s, then $\lfloor \frac{n}{2} \rfloor * \lceil \frac{n}{2} \rceil = \lfloor s + \frac{1}{2} \rfloor * \lceil s + \frac{1}{2} \rceil = s(s + 1) = s^2 + s = \lfloor \frac{4s^2+4s+1}{4} \rfloor = \lfloor \frac{n^2}{4} \rfloor$.

For $\vec{M}_2$, each arc contributes one times the in-degree (or out-degree) of the center. Arcs directed into the center contribute the in-degree of the center, and arcs directed out of the center contribute the out-degree of the center. Since the in-degree and out-degree of the center sums to n, $\vec{M}_2$ is maximized when either the in-degree or out-degree is maximized; that is, if all arcs are directed into the center of the star, or all are directed out of center of the star, $\vec{M}_2 = n^2$. Similarly, it is minimized when each is smallest, namely one is $\lfloor \frac{n}{2} \rfloor$ and the other is $\lceil \frac{n}{2} \rceil$. If $n = 2s$ for some positive integer s, then $\overrightarrow{M}_2 = \lfloor \frac{n}{2} \rfloor^2 + \lceil \frac{n}{2} \rceil^2 = s^2 + s^2 = 2s^2 = \lceil \frac{n^2}{2} \rceil$. If $n = 2s + 1$, then $\overrightarrow{M}_2 = \lfloor \frac{n}{2} \rfloor^2 + \lceil \frac{n}{2} \rceil^2 = \lfloor s + \frac{1}{2} \rfloor^2 + \lceil s + \frac{1}{2} \rceil^2 = s^2 + (s+1)^2 = 2s^2 + 2s + 1 = \frac{4s^2 + 4s + 1 + 1}{2} = \frac{n^2 + 1}{2} = \lceil \frac{n^2}{2} \rceil$. The result follows in either case. □

Proposition 3 *For any oriented P_n, $0 \le \overrightarrow{M}_1(P_n) \le n - 2$ (where there is some orientation which yields each possible integral value) and $n - 1 \le \overrightarrow{M}_2(P_n) \le 4n - 8$.*

Proof If arcs in P_n alternate directions, then $\overrightarrow{M}_1(P_n) = 0$. Clearly a negative value is not possible. Each endpoint of the path is either a source or a sink and thus does not contribute to $\overrightarrow{M}_1$. Each interior node on the path has either two arcs pointed in, contributing nothing, two arcs pointing out, contributing nothing, or one arc pointing in and one arc pointing out, contributing 1 to $\overrightarrow{M}_1$. Thus their sum, $\overrightarrow{M}_1(P_n)$ is maximized at $n - 2$ when all arcs are oriented in the same direction on the path, and integral values between the bounds can be obtained by the appropriate number of interior nodes with one arc pointing in and one arc pointing out.

For $\overrightarrow{M}_2(P_n)$, each of the $n - 1$ arcs must contribute at least 1, and the lower bound of $n - 1$ is achieved when all arcs are oriented in the same direction on the path. Each arc can contribute at most 4 to $\overrightarrow{M}_2(P_n)$, which happens only if at each node the in-degree and out-degree on the path are both 2. The number of such occurrences is maximized when the arcs in P_n alternate directions. and all but the first and last arcs thus contribute 4, yielding $\overrightarrow{M}_2(P_n) = 4n - 8$. □

We next give results about when $\overrightarrow{M}_1(D) = 0$ and when $\overrightarrow{M}_1(D) \neq 0$.

Lemma 2 $\overrightarrow{M}_1(D) = 0$ *if and only if every node in D is either a source or a sink.*

Proof $\overrightarrow{M}_1(D) = 0$ means that each node contributes 0 to the sum which means either $d^+(v) = 0$ or $d^-(v) = 0$ for every node in D. Hence, each node is either a source or a sink. And if each node is a source or sink that implies that either $d^+(v) = 0$ or $d^-(v) = 0$ for every node v in D and hence $\overrightarrow{M}_1(D) = 0$. □

Proposition 4 *If a graph G has an odd cycle, then $\overrightarrow{M}_1(D) \neq 0$.*

Proof Consider an odd cycle C in G. There is no possible orientation of the arcs in C such that every node in C will be a sink or a source. That is, by a simple parity argument, some node must have an arc entering it and an arc leaving it. Thus, the directed graph D does not consist only of sources and sinks, and by Lemma 2, $\overrightarrow{M}_1(D) \neq 0$. □

Proposition 5 *If D contains no odd cycles, then there is an orientation of the arcs in D so that $\overrightarrow{M}_1(D) = 0$.*

Proof Since the digraph has no odd cycles, either it has no cycles, or its only cycles are even. We consider those cases separately.

Suppose the digraph has no cycles. Take a longest path in the tree and orient adjoining arcs in opposite directions. When all arcs on that path have been oriented, return to any node on that path that is incident to unoriented arcs, and orient any adjacent arcs in the same direction as all the others at that node (either all into or all out of the node) and then continue down each of those paths and orient adjoining arcs in opposite directions. Repeat until all arcs are oriented. Then, by construction every node in D is either a source or a sink, and by Lemma 2, $\overrightarrow{M}_1(D) = 0$.

Now suppose the digraph has at least one even cycle. Pick a largest even cycle, and orient adjoining arcs in that cycle in opposite directions. Then, continue this process on all remaining even cycles and/or paths until the graph clearly has only sources and sinks. If there are any arcs within this previously oriented cycle, they must only be connecting nodes that are an odd distance apart on the original cycle. Hence, the arcs inside the cycle can be oriented to keep all nodes being sources and sinks. Any paths that connect to a node of the original cycle and are not inside the original cycle can be oriented as described above starting with the same direction as the node where the path begins. Any additional cycles that might be adjoining the original cycle can also be oriented to keep all nodes sources and sinks as they are also even.   □

2.4   Equality of Directed Zagreb Indices

We seek to fully characterize instances where $\overrightarrow{M}_1 = \overrightarrow{M}_2 \neq 0$. First we show that we need only focus on connected digraphs. Then we show that directed cycles and K_2^* have this property. However, we then show that digraphs where this equality holds are quite limited. We conjecture that directed cycles, K_2^*, and digraphs that are a disjoint union of these digraphs are in fact the only digraphs for which equality of $\overrightarrow{M}_1 = \overrightarrow{M}_2 \neq 0$ holds. Proving this conjecture remains an open question, but we make progress in that direction by showing that no digraph with a source and a sink will have $\overrightarrow{M}_1 = \overrightarrow{M}_2$, nor will oriented trees, nor a directed cycle plus an additional arc, nor cycles that are oriented but not directed.

Lemma 3 *If a disconnected graph has $\overrightarrow{M}_1 = \overrightarrow{M}_2 \neq 0$, then each of its connected components must also have $\overrightarrow{M}_1 = \overrightarrow{M}_2 \neq 0$.*

Proof By Property 4, the directed Zagreb indices of each component sum to the directed Zagreb index of the overall graph. Since Theorem 1 ensures that $\overrightarrow{M}_1 \leq \overrightarrow{M}_2$ for every digraph, the only way that the overall digraph can have $\overrightarrow{M}_1 = \overrightarrow{M}_2$ is thus if for each component equality holds.   □

Lemma 4 *The directed cycle C_n has $\overrightarrow{M}_1(C_n) = \overrightarrow{M}_2(C_n) \neq 0$.*

Proof By Property 3, since the in-degree and out-degree of every node is $k = 1$, $\overrightarrow{M}_1(D) = nk^2 = n$ and $\overrightarrow{M}_2(D) = mk^2 = m$. Since $m = n$ in C_n with $n \geq 3$, the result is immediate.   □

Lemma 5 *K_2^* has $\overrightarrow{M}_1(K_2^*) = \overrightarrow{M}_2(K_2^*) \neq 0$.*

Conjecture 1 The directed cycle and K_2^* (or graphs consisting solely of directed cycles and K_2^*) are the only graphs in which $\vec{M}_1 = \vec{M}_2 \neq 0$.

The following results lend support to the conjecture.

Property 5 It is NOT true that inserting an arc will always increase $\vec{M}_2$ by more than it increases $\vec{M}_1$.

Proof Consider a unidirectional path, that starts at node v_0 and ends at node v_{n-1}. Suppose we then insert a directed arc from v_{n-1} to v_0. The increase in $\vec{M}_1$ is 2, with one each contributed at v_0 and v_{n-1}. The increase in $\vec{M}_2$ is 1, the contribution from the new arc, as the sum from the other arcs does not change. □

Theorem 2 *Any digraph with a source and a sink cannot have* $\vec{M}_1 = \vec{M}_2$.

Proof Proof by contradiction. Let D be a digraph with a node u that is a source, and a node v that is a sink. Suppose that $\vec{M}_1 = \vec{M}_2$. Insert an arc from v to u. The increase in $\vec{M}_2$ is exactly 1, since the arc (v, u) contributes 1, but v as a sink had no other arcs out of it, and u as a source had no other arcs into it. But the increase in $\vec{M}_1$ is more than 1, since the increase is precisely the number of arcs into v (which is at least 1 as a sink) plus the number of arcs out of u (again, at least 1 as a source). Since the increase in $\vec{M}_1$ is more than the increase in $\vec{M}_2$, the values $\vec{M}_1$ and $\vec{M}_2$ could not have been equal, contradicting the original assumption. □

Since every oriented tree must contain both a source and a sink, we have the following corollary. We include the proof that every oriented tree must contain both a source and a sink for completeness.

Corollary 3 *Any oriented tree T with $n \geq 2$ has* $\vec{M}_1(T) < \vec{M}_2(T)$.

Proof Suppose our tree T has no sinks. Pick an arbitrary node, and follow an oriented edge (in the appropriate direction) out of that node. Repeat. Either we arrive at a node that has out-degree 0, which is thus a sink, or we return to a node we have already visited, which would mean there is a cycle, which is not possible in a tree.

Suppose instead our tree T has no source nodes. Reverse the orientation of all edges. Then our reversed graph would be a tree with no sinks. However, by the above argument, that is again impossible.

Thus, since every oriented tree has a source and a sink, we cannot have $\vec{M}_1 = \vec{M}_2$. □

Theorem 3 *Any digraph D which consists of solely a directed cycle and one additional arc has* $\vec{M}_1(D) < \vec{M}_2(D)$.

Proof First, recall that Lemma 4 ensures that $\vec{M}_1 = \vec{M}_2$ for any directed cycle. A graph that consists of a directed cycle and one additional arc can be constructed by the addition of an arc in one of the following ways:

1. as a disconnected arc,
2. as a chord in the cycle,

3. as an arc directed inward (or outward) into one node of the cycle and the other node would be a new node, or
4. an arc going in the opposite direction of one of the current arcs in the cycle.

Let e be the new arc in each case below.

Case 1   [Disconnected arc]: $\vec{M}_1(D+e) = \vec{M}_1(D)$ and $\vec{M}_1(D+e) = \vec{M}_2(D)+1$, by Property 4 hence $\vec{M}_1(D+e) < \vec{M}_2(D+e)$.

Case 2   [Chord in the cycle]: $\vec{M}_1(D+e) = \vec{M}_1(D)+1+1$ as two nodes will now be contributing 2 instead of 1. Similarly, $\vec{M}_2(D+e) = \vec{M}_2(D)+4+1+1$ where the 4 is from the new arc and the two 1s are from the additional in/out degree at the endpoints. Again, since $\vec{M}_1(D) = \vec{M}_2(D)$, $\vec{M}_1(D+e) < \vec{M}_2(D+e)$.

Case 3   [New node]: $\vec{M}_1(D+e) = \vec{M}_1(D)+1$ as only one node within the cycle will have a changed in- (or out-) degree. Similarly, $\vec{M}_2(D+e) = \vec{M}_2(D)+2+1$ where the 2 comes from the new arc and the 1 is how much the one arc in the cycle will change by.

Case 4   [Opposite direction]: $\vec{M}_1(D+e) = M_1(D)+2$ and $\vec{M}_2(D+e) = M_1(D)+4+1+1$ where the 4 is the new arc's contribution and each 1 is the amount two different arcs in the cycle will change.

In all cases, we see $\vec{M}_1(D+e) > \vec{M}_2(D+e)$.   □

Theorem 4 *For any cycle C that is oriented but not directed, $\vec{M}_1(C) < \vec{M}_2(C)$.*

Proof Consider a cycle C of length n that is oriented but not directed. Let s be the number of maximal directed paths of length 1 in the cycle. Let t be the number of maximal directed paths of length greater than 1, but less than n in the cycle. Recall that since C is not directed, no directed path in the cycle can have length more than $n - 1$.

To calculate $\vec{M}_1(C)$, first note that any node will either contribute 1 or 0. The node contributes 1 if the node is part of a unidirectional path (with in-degree and out-degree both one) and contributes 0 if it is a place where the direction of arcs in the cycle changes (that is, the in-degree and out-degree are not equal). The direction will change at $s + t$ places (where the paths change direction) and thus $\vec{M}_1(C) = n - (s+t) = n - s - t$. Note: $s+t$ must be even as it counts the number of direction changes and you cannot change direction an odd number of times and have a cycle. Furthermore, $s + t > 0$ as the cycle C is not unidirectional.

For $\vec{M}_2(C)$, any edge that is a path of length 1 will contribute 4 to the sum. Any path of length greater than 1 and less than n will have two arcs that each contribute 2 (the arcs at the start/end of the path). And, any arcs remaining will each contribute one to the sum. This gives: $\vec{M}_2(C) = 4s + 4t + n - s - 2t = n + 3s + 2t$.

Since s and t are nonzero, $n + 3s + 2t > n - s - t$ which ensures that $\vec{M}_1(C) < \vec{M}_2(C)$.   □

2.5 Differences of Zagreb Indices

For undirected graphs, differences between the first two Zagreb indices were first studied in-depth by Furtula, Gutman, and Ediz in [3], who introduced the idea of the reduced second Zagreb index and studied this difference mainly on trees. In 2016, Das, Horoldagva, and Selenge [6] completely characterized which (undirected) graphs have $M_1 - M_2 = 0$, $M_1 - M_2 > 0$ and $M_1 - M_2 < 0$. In the directed case, we already know $\overrightarrow{M}_2 - \overrightarrow{M}_1 \geq 0$, and can in fact show that $\overrightarrow{M}_2 - \overrightarrow{M}_1$ can equal any nonnegative integer.

We know from Lemma 4 that $\overrightarrow{M}_2 - \overrightarrow{M}_1$ can equal zero. In addition we can also get $\overrightarrow{M}_2 - \overrightarrow{M}_1 = 1$ as if $D = P_n$ with all the arcs oriented in the same direction, $\overrightarrow{M}_2 - \overrightarrow{M}_1 = n - 1 - (n - 2) = 1$. We also know we can make $\overrightarrow{M}_2 - \overrightarrow{M}_1$ arbitrarily large by noting that $\overrightarrow{M}_2 - \overrightarrow{M}_1 = 4n$ for any $n \geq 3$ by using $D = C_n^*$, the cycle with all bidirectional edges present or we can get $\overrightarrow{M}_2 - \overrightarrow{M}_1 = n^2$ for $K_{1,n}$ with all the arcs directed out of the center. While these examples provide motivation that all nonnegative integer values are possible for this difference, the following theorem gives a construction technique for producing a digraph with any desired difference.

Theorem 5 *For all $s \in \mathbb{N}$, there exists a directed graph with $\overrightarrow{M}_2 - \overrightarrow{M}_1 = s$.*

Proof Let D be the digraph $K_{1,n}$ with x edges directed into the center and k edges directed out of the center where $x + k = n$ and $k \leq x$. Consider the collection of $k + 1$ digraphs $\{D = D_0, D_1, D_2, \ldots, D_k\}$ where D_i is the digraph formed from D by connecting i of the arcs directed out of the center to i different arcs directed into the center. An example of this construction can be seen in Fig. 2.

Fig. 2 A construction technique for digraphs with all possible values for the difference between the two Zagreb indices. The inclusion of any subset of the dotted edges leads to one of the digraphs in the collection

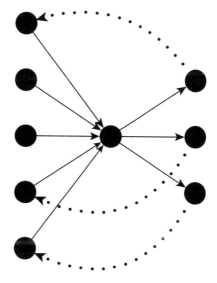

In general, $\vec{M}_2(D_i) - \vec{M}_1(D_i) = (x^2 + k^2 + i) - (xk + 2i) = x^2 + k^2 - xk - i$. Considering the case where $x = k$, $\vec{M}_2(D_i) - \vec{M}_1(D_i) = (2k^2 + i) - (k^2 + 2i) = k^2 - i$ and i ranges from 0 to k. Hence, in this case we get all the integers in the interval $[k^2 - k, k^2]$. Now, consider the case where $x = k + 1$. Similar calculations give $\vec{M}_2(D_i) - \vec{M}_1(D_i) = (k + 1)^2 + k^2 - k(k + 1) - i = k^2 + k + 1 - i$ and i ranges from 0 to k, so we get all the integers in the interval $[k^2 + 1, k^2 + k + 1]$. And if we move up to the next value for k, (so now $x = k + 1$ and k becomes $k + 1$), we get $\vec{M}_2(D_i) - \vec{M}_1(D_i) = (k + 1)^2 - i$ with i ranging from 0 to $k + 1$. So, we get the next interval to be $[k^2 + k, k^2 + 2k + 1]$. And, thus the overlap of intervals continues and we continue to increase the upper bound. If we plug in $k = 1$, we see that we start the interval at $[0, 1]$ and hence can get any nonnegative integer values since these intervals line up and/or overlap and increase without bound.   □

3   Conclusions and Open Questions

In this paper, we introduce the definition of first and second Zagreb indices on directed graphs. Initial propositions are given, relationships between the two indices are explored, and several classes of digraphs are studied in depth. We showed that the difference between $\vec{M}_2$ and $\vec{M}_1$ can take on any nonnegative integer value and state a conjecture on when this difference is zero. In particular, we believe that in all cases other than a directed cycle or K_2^* or disconnected combinations thereof, the difference between $\vec{M}_1$ and $\vec{M}_2$ is non-zero and inserting additional arcs or nodes will not result in equality of $\vec{M}_1$ and $\vec{M}_2$.

Another avenue of future research is motivated by Sect. 2.2. There we discuss $\vec{M}_1/n \leq \vec{M}_2/m$. While this was shown to not be true for all digraphs, could it be true for all connected digraphs? Or even possibly for all digraphs where not all arcs are bidirectional?

Finally, since directed Zagreb indices do not have the same chemistry motivations of undirected Zagreb indices, they could be defined in many other ways or other indices described on undirected graphs could be generalized for digraphs. New definitions would prompt new results, propositions, and relationships, leading to additional areas for mathematical exploration of indices on digraphs.

References

1. Caporossi, G., Hansen, P., Vukicević, D.: Comparing Zagreb indices of cyclic graphs. Commun. Math. Comput. Chem. **63**, 441–451 (2010)
2. Fath-Tabar, G.H.: Old and new Zagreb indices of graphs. Commun. Math. Comput. Chem. **65**, 79–84 (2011)
3. Furtula, B., Gutman, I., Ediz, S.: On difference of Zagreb indices. Discrete Appl. Math. **178**, 83–88 (2014)

4. Gutman, I., Trinajstić, N.: Graph theory and molecular orbitals. Total φ-electron energy of alternant hydrocarbons. Chem. Phys. Lett. **17**, 535–538 (1972)
5. Hansen, P., Vukicević, D.: Comparing the Zagreb indices. Croat. Chem. Acta **80**, 165–168 (2007)
6. Horoldagva, B., Das, K.C., Selenge, T.A.: Complete characterization of graphs for direct comparing Zagreb indices. Discrete Appl. Math. **215**, 146–154 (2016)
7. Liu, B., You, Z.: A survey on comparing Zagreb indices. Commun. Math. Comput. Chem. **65**, 581–593 (2011)
8. Nikolić, S., Kovačević, G., Miličević, A., Trinajstić, N.: The Zagreb indices 30 years after. Croatica Chem. Acta **76**, 113–124 (2003)
9. Volkmann, L.: Sufficient conditions on the zeroth-order general Randić index for maximally edge-connected digraphs. Commun. Combin. Optim. **1**, 1–13 (2016)

Edge Tree Spanners

Fernanda Couto, Luís Cunha, and Daniel Posner

Abstract A tree t-spanner of a graph G is a spanning tree T of G in which any two adjacent vertices of G have distance at most t in T. The line graph $L(G)$ of a graph G is the intersection graph of the edges of G. We define the edge tree t-spanner of a graph G as a spanning tree T of $L(G)$ in which any two edges that share an endpoint in G have distance at most t in T. Although determining if G has a tree 3-spanner is an open problem for more than 20 years, we settle that deciding if a graph G has an edge tree 3-spanner is polynomial-time solvable. As a consequence, we present polynomial time algorithms for the edge tree t-spanner problem for several graph classes such as trees, join of graphs, split graphs, P_4-tidy, and $(1, 2)$-graphs. Moreover, we establish that deciding whether a graph G has an edge tree 8-spanner is NP-complete, even if G is bipartite.

Keywords Tree t-spanner · Edge tree t-spanner · Polynomial time algorithms · NP-completeness · Line graphs · Graph classes

1 Introduction

The problem of looking for a spanning tree with constraints on the vertices' or edges' distances is a combinatorial challenge with many applications and approaches [1, 11]. A *tree t-spanner* of a graph G is a spanning tree T of G in which any two adjacent vertices of G have distance at most t in T. A graph G having a tree t-spanner is called a *t-admissible* graph. The smallest t for which a graph G is t-admissible is the *stretch index of G* and is denoted by $\sigma_T(G)$ (or

F. Couto (✉) · D. Posner
Universidade Federal Rural do Rio de Janeiro, Nova Iguaçu, Brazil
e-mail: fernandavdc@ufrrj.br; posner@cos.ufrj.br

L. Cunha
Universidade Federal Fluminense, Niterói, Brazil
e-mail: lfignacio@ic.uff.br

© The Author(s), under exclusive license to Springer Nature Switzerland AG 2021
C. Gentile et al. (eds.), *Graphs and Combinatorial Optimization:
from Theory to Applications*, AIRO Springer Series 5,
https://doi.org/10.1007/978-3-030-63072-0_16

simply $\sigma(G)$). The t-*admissibility* problem aims to decide whether a given graph G has $\sigma(G) \leq t$. The problem of determining the tree stretch index, i.e. *the minimum stretch spanning tree problem* (MSST) has been studied by establishing bounds on $\sigma(G)$ or developing the computational complexity of the decision version of MSST for several graph classes [2–4]. Cai and Corneil [2] proved that t-admissibility is NP-complete, for $t \geq 4$, whereas 2-admissible graphs can be recognized in polynomial-time. However, the characterization of 3-admissible graphs is still an open problem.

The characterization for 2-admissible graphs [2], stated in Theorem 1, deals with triconnected components of a connected graph, defined as any maximal subgraph that does not contain two vertices whose removal disconnects the graph (the authors also consider K_2 and K_3 as triconnected components). A *nonseparable* graph is a graph without a *cut vertex*, i.e., a vertex whose removal disconnects the graph. A *star* with $n + 1$ vertices is the complete bipartite graph $K_{1,n}$. A v-*centered star* is a star centered on v, that is a universal vertex. Similarly, a bi-star is a graph such that there is an edge uv and every edge of E shares an endpoint with uv. Hence, uv is a *universal edge* of the bi-star. A uv-*centered bi-star* is a bi-star centered on a universal edge uv.

Theorem 1 ([2]) *A nonseparable graph G is 2-admissible if and only if G contains a spanning tree T such that for each triconnected component H of G, $T \cap H$ is a spanning star of H.*

Given a graph G, its *line graph* $L(G)$ is obtained as follows: $V(L(G)) = E(G)$; $E(L(G)) = \{\{uv, uw\}|uv, uw \in E(G)\}$. I.e., each edge of G is a vertex of $L(G)$ and if two edges share an endpoint, then their corresponding vertices are adjacent in $L(G)$. The *distance between two edges* e_1 and e_2 of G, for $e_1, e_2 \in E(G)$ is the distance between their corresponding vertices in $L(G)$.

We define the *edge tree t-spanner* of a graph G as a spanning tree T of $L(G)$ such that, for any two adjacent edges of G, their distance is at most t in T. Therefore, an edge tree t-spanner of G is a tree t-spanner of $L(G)$.

A graph G that has an edge tree t-spanner is called *edge t-admissible*. The smallest t for which G is an edge t-admissible graph is the *edge stretch index of G*, and is denoted by $\sigma'_T(G)$ (or simply $\sigma'(G)$). The *edge t-admissibility* problem aims to decide whether a given graph G has $\sigma'(G) \leq t$. Figure 1 depicts the relation between the edge tree spanner of a graph and the tree spanner of its line graph.

An immediate consequence of MSST is that the property of being t-admissible graph is not hereditary, i.e., if G is t-admissible then there may exist a subgraph H of G that is not t-admissible. Indeed, the addition of a universal vertex u to any t-admissible graph results in a 2-admissible graph by a u-centered star.

On the other hand, regarding the edge tree t-spanner, in Sect. 3 we prove that being an edge 3-admissible graph is a hereditary property, and based on that, we are able to decide whether G is edge 3-admissible in polynomial time. Moreover, in Sect. 4 we determine polynomial time algorithms to obtain the edge stretch index for some edge 4-admissible and edge 5-admissible classes, such as split graphs, join graphs, P_4-tidy graphs and $(1, 2)$-graphs. In Sect. 5, we prove that

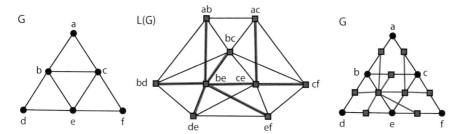

Fig. 1 A graph G, a tree 3-spanner of $L(G)$ in red, and G with the related edge 3-spanner in red

edge 8-admissibility is NP-complete for $(2, 0)$-graphs, i.e. bipartite graphs. In Sect. 6, we present concluding remarks. Next (Sect. 2), we relate admissibility and edge admissibility problems, presenting immediate consequences and preliminary results.

2 Admissibility Versus Edge Admissibility for Graph Classes

Since induced cycles in a graph G correspond to cycles of the same length in $L(G)$, we have that $\sigma'(C_n) = \sigma(C_n) = n - 1$. Although cycle graphs satisfy $\sigma' = \sigma$, for several other classes the stretch index is different of the edge stretch index.

For instance, trees are 1-admissible and the unique edge 1-admissible graphs are the ones such that their line graphs are trees. Since line graphs are claw-free, then path graphs are the unique edge 1-admissible graphs. In Proposition 1 we determine the edge stretch index of trees.

Proposition 1 *Let G be a tree. If G is a path graph then $\sigma'(G) = 1$, otherwise $\sigma'(G) = 2$.*

Proof Note that if G is a path, then $L(G)$ is a path and $\sigma'(G) = 1$. For any other tree there is a vertex of degree at least 3, implying a complete subgraph of length at least 3 in $L(G)$. Each internal node u of G correspond to a maximal complete subgraph of $L(G)$ of size $d_G(u)$ and two of such maximal complete subgraphs share at most a vertex in $L(G)$. Hence, any triconnected component of $L(G)$ is a complete subgraph and satisfies Theorem 1.   □

Since the study of edge tree spanners is equivalent to the study of tree spanners of line graphs, and deciding whether a graph is 2-admissible is polynomial-time solvable, Theorem 1 implies Corollary 1.

Corollary 1 *Edge 2-admissibility is polynomial-time solvable.*

The edge stretch index of cycle graphs and complete graphs are useful to characterize edge 3-admissible graphs, as discussed in Sect. 3.

Complete graphs are 2-admissible, however their line graphs are not. In order to prove that $\sigma'(K_n) = 4$, from Lemma 1 we have that $\sigma'(K_5) \leq 4$, and it is possible to prove that K_5 is not edge 3-admissible, as highlighted below.

To prove that K_5 is not edge 3-admissible, one can verify by a case analysis that it is not possible obtain a spanning tree T such that $T \cap L(K_5)$ has at least 3 internal nodes. Clearly, $T \cap L(K_5)$ cannot have more than 3 internal nodes, because otherwise the edge factor of such a tree would be at least 4. Moreover, it is not possible obtain a spanning tree T such that $T \cap L(K_5)$ is a bi-star or it is a tree with three internal nodes whose leaves at distance 4 in T are not adjacent in $L(K_5)$.

In Sect. 3 we prove that being edge 3-admissible is a hereditary property for induced subgraphs (Lemma 2), then Corollary 3 states that $\sigma'(K_n) = 4$, for $n \geq 5$.

A graph G has a *distance two dominating edge* uv if every edge of $E(G)$ has a vertex in $N[u] \cup N[v]$ as one of its endpoints, where $N[x]$ is the *closed neighborhood of* x, i.e. $N[x] = N(x) \cup \{x\}$. Moreover, G has two adjacent distance two dominating edges uv and vw if every edge of $E(G)$ has a vertex in $N[u] \cup N[v] \cup N[w]$ as one of its endpoints.

Lemma 1 *A graph G with a distance two dominating edge uv has $\sigma'(G) \leq 4$.*

Proof Since G has a distance two dominating edge uv, there is a spanning tree with diameter at most four of $L(G)$ with the vertex uv as its root, the vertices $\{ux \mid ux \in E(G)\} \cup \{vy \mid vy \in E(G)\}$ adjacent to uv, and the remaining vertices of $L(G)$ adjacent to some vertex in $\{ux \mid ux \in E(G)\} \cup \{vy \mid vy \in E(G)\}$. □

Figure 2 depicts graphs with distance two dominating edges and their edge tree 4-spanners, as the proof of Lemma 1. A graph is *split* if its vertex set can be partitioned into a stable set and a clique. The *join* between two graphs G_1 and G_2 results in the graph G such that $V(G) = V(G_1) \cup V(G_2)$ and $E(G) = E(G_1) \cup E(G_2) \cup \{uv \mid u \in V(G_1)$ and $v \in V(G_2)\}$.

Several graph classes can be constructed by join and complement of join operations, i.e. *union* operations. Cographs are the P_4-free graphs, i.e. graphs without a P_4 as an induced subgraph, and G is a cograph iff it has the following recursive definition: (i) G is a K_1; (ii) G is a join of cographs; (iii) G is a union of cographs. A generalization of cographs are the graphs with few P_4's, such as P_4-sparse and P_4-tidy [7].

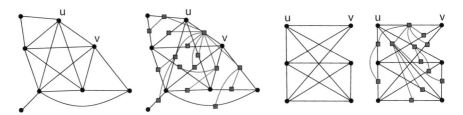

Fig. 2 A split graph and a join graph with their edge tree 4-spanners

A graph is P_4-*sparse* if for each set of 5 vertices, there is at most one induced P_4. A graph is P_4-*tidy* if for each induced P_4 of G, say P, there is at most one vertex $v \in V(G) \setminus V(P)$ such that $V(P) \cup \{v\}$ induces at most two P_4's in G. P_4-tidy generalizes P_4-sparse graphs, and G is a P_4-tidy graph iff it has the following recursive definition: (i) G is P_5, C_5, $\overline{P_5}$, or K_1; (ii) G is a join of P_4-tidy graphs; (iii) G is a union of P_4-tidy graphs; (iv) G is a spider; (v) G is an almost spider. A graph is a *spider* graph if its vertex set can be partitioned into $\mathcal{S}$, $\mathcal{K}$ and $\mathcal{R}$ such that (i) $\mathcal{K}$ is a clique ($\mathcal{K}$ is called *body*), $\mathcal{S}$ is a stable set and $|\mathcal{S}| = |\mathcal{K}| \geq 2$; (ii) each vertex of $\mathcal{R}$ ($\mathcal{R}$ is called *head*) is adjacent to all vertices of $\mathcal{K}$ and is non-adjacent to any vertex of $\mathcal{S}$; (iii) There is a bijection $f : \mathcal{S} \mapsto \mathcal{K}$ such that, for all $x \in \mathcal{S}$, either $N(x) = \{f(x)\}$, or $N(x) = \mathcal{K} - \{f(x)\}$. A graph is an *almost-spider* graph if it can be constructed from a spider graph $G = (\mathcal{S}, \mathcal{K}, \mathcal{R})$ by adding a vertex v' which is either a false twin of v or a true twin of v, such that $v \in \mathcal{S} \cup \mathcal{K}$ [10].

Split graphs, join graphs and P_4-tidy graphs are 3-admissible [3, 4]. Corollary 2 follows from Lemma 1 and: for split graphs, any clique's edge is distance two dominating; for join graphs between G_1 and G_2, any uv such that $u \in V(G_1)$ and $v \in V(G_2)$ is distance two dominating; for P_4-tidy graphs, any edge between the head and the body is distance two dominating.

Corollary 2 *Split graphs, join graphs and P_4-tidy graphs are edge 4-admissible.*

Since 3-admissibility is still open and t-admissibility is NP-complete, for $t \geq 4$, we are interested to establish the computational complexity of determining the edge stretch index. In Sect. 3, we prove that edge 3-admissibility is polynomial-time solvable, and as an immediate consequence, we are able to determine in polynomial time the edge stretch index for any edge 4-admissible graph, such as split graphs, join graphs and P_4-tidy graphs (Corollary 6).

3   Edge 3-Admissibility Is Polynomial-Time Solvable

Lemma 2 *Edge 3-admissibility is a hereditary property for induced subgraphs.*

Proof Assume that there is an edge 3-admissible graph G with an induced subgraph H such that H is not edge 3-admissible. W.l.o.g. let G' be an induced subgraph of G such that: $|V(G')| = |V(H)| + 1$, $u \in V(G') \cap V(H)$; G' is edge 3-admissible; H is edge k-admissible for $k \geq 4$; T' is an edge tree 3-spanner of G'; and T is an edge k-tree spanner of H with $k \geq 4$. In any edge tree k-spanner T of H there is a path P with $k + 1$ vertices using edges of T and an edge of G' not in T between the two endpoints of this path (see Fig. 3a that considers $k = 5$). Since G' is edge 3-admissible, the addition of the vertex u must remove a part of that path P from T. For the sake of contradiction, assume T'' is a tree that contains at least three internal nodes among the edges incident to u. Since these edges have u as endpoint, then the leaves that are at distance 4 in T'' correspond to adjacent edges in G', a contradiction. Therefore, the edges incident to u must be a bi-star in T' (see Fig. 3b).

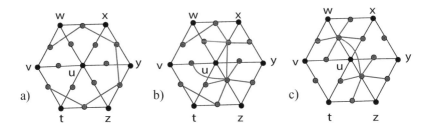

Fig. 3 (**a**) $V(H) = \{v, w, x, y, z, t\}$ and a path P in red. (**b**) In red a bi-star satisfying Case 1. (**c**) In red a bi-star satisfying Case 2

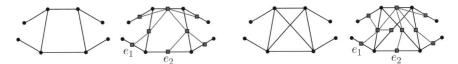

Fig. 4 C_4 and K_4 whose vertices have degree at least 3 and 4 in G, resp. Note that $d_T(e_1, e_2) = 4$

W.l.o.g. assume that u is adjacent to all vertices of G related to the path P of T. The edges of the bi-stars cover at most four vertices of P. We have two cases: Case 1: the bi-star connects consecutive vertices of P. In this case it does not reduce the distance between the vertices of P in T' (e.g. see Fig. 3b, the distance between vw and vt is 5 in T') and T' is not an edge tree 3-spanner, a contradiction; Case 2: the bi-star connects non-consecutive vertices of P. In this case it does reduce the distance between vertices of P, however, the vertex xy between this non consecutive vertex of P is connected to leaves of the two centers of the bi-star in $L(G)$, which implies that T' is not edge 3-admissible, a contradiction (Fig. 3c).   □

Corollary 3 *Any complete graph K_n has $\sigma'(K_n) = 4$, for $n \geq 5$.*

Proof Since $\sigma'(K_5) = 4$ (Sect. 2) and for $n \geq 5$, K_n has a K_5 as an induced subgraph, then, by Lemma 2, we have that K_n are not edge 3-admissible, for $n \geq 5$. Furthermore, complete graphs have a distance two dominating edge, hence by Lemma 1, $\sigma'(K_n) \leq 4$, for $n \geq 5$, and the result follows.   □

Line graphs of K_n are complement of *Kneser graph* $KG_{n,2}$ [8], then $\sigma(\overline{KG_{n,2}}) = 4$.

Note that C_k and K_k, for $k \geq 5$ are not subgraphs of edge 3-admissible graphs. See Fig. 4 for examples of C_4 and K_4 where all vertices have degree at least 3 and 4 in G, resp. Suppose H is an induced C_4 (or K_4) in G. In $L(G[H])$ there must be a path through all $L(C_4)$'s vertices (or through four $L(K_4)$'s vertices) and one more vertex corresponding to an edge that does not belong to the C_4 (to the K_4) in H. Hence, it implies that $\sigma'(H) \geq 4$, and Corollary 4 follows.

Corollary 4 *Let G be an edge 3-admissible graph. If $X \in \{C_4, K_4\}$ is an induced subgraph of G, then there is a vertex $v \in V(X)$ such that $N_G(v) \subseteq V(X)$.*

By Corollary 4, any edge 3-admissible graph has vertices of degree 2 and 3 in each induced C_4's and K_4, resp. Hence, Construction 2 presents a way to break C_4's and K_4's into P_5's and K_3's, resp., in order to present a stronger necessary condition in Lemma 4.

Construction 2 *Let G be a graph that satisfies: G does not have induced C_k nor K_k, for $k \geq 5$, as induced subgraphs; for each induced C_4 there is a vertex of degree two in G; and for each induced K_4 there is a vertex of degree three in G. We construct a graph H from G as follows:*

1. *each induced $C_4 = a, b, c, d, a$, for $d_G(a) = 2$, is transformed into a $P_5 = a, b, c, d, a'$ by adding a new vertex a' and the edge da', and removing the edge da;*
2. *each induced $K_4 = \{a, b, c, d\}$, for $d_G(a) = 3$, is transformed into three complete graphs K_3 by adding a new vertex a' and: removing edge ba; adding edges ba' and ca'.*

Lemma 3 *A graph G is edge 3-admissible if and only if the graph H from Construction 2 is edge 3-admissible.*

Proof If G is edge 3-admissible, then all edges of an edge tree 3-spanner of G are used to obtain a spanning tree of H and we do not increase the edge stretch index from G to H, because, by construction, we are not increasing a maximum path between any two adjacent vertices of G in H. If H is edge 3-admissible, then all edges of an edge tree 3-spanner of H are used for a spanning tree of G and, since we are identifying vertices that belong only to C_4's or K_4's in G, such identification does not affect cycles that give the edge tree 3-spanner of H and does not increase such index of G by the used edges of H.   □

A *k-tree* is a graph obtained from a K_{k+1} by repeatedly adding vertices in such a way that each added vertex v has exactly k neighbors defining a clique of size $k+1$. A *partial k-tree* is a subgraph of a k-tree [9].

Lemma 4 *Let G be an edge 3-admissible graph. If H is the graph obtained from G in Construction 2, then H is a chordal partial 2-tree graph.*

Proof If G is edge 3-admissible with $X \in \{C_4, K_4\}$ as an induced subgraph, then, by Corollary 4, X must have at least one vertex a such that $N(a) \subseteq X$. Based on that, in Construction 2 we obtain a graph without C_4's nor K_4's. Since, by Lemma 3, the transformed graph H from an edge 3-admissible graph G is also edge 3-admissible, we have that the length of any clique is at most 3 and it does not have C_k, for $k \geq 4$. Since chordal graphs with maximum clique of length 3 are partial 2-tree [9], we have that H is a chordal partial 2-tree graph.   □

By Lemma 4, edge tree 3-spanner graphs are formed by 2-trees where either an edge or a vertex connects two 2-trees. Hence, for the former case such edge is a bridge and for the later case it is a cut vertex of the graph. Lemmas 5 and 6 present conditions that force spanning trees correspond to edge 3-admissible graphs.

Lemma 5 *Given an edge 3-admissible graph G and two 2-trees A_1 and A_2 connected by a bridge uv, such that $|V(A_i)| > 3$ for $i \in \{1, 2\}$, then for any edge 3-spanner T, uv is a pendant vertex in $T[A_1 \cup \{u, v\}]$, i.e. $d_{T[A_1 \cup \{u,v\}]}(uv) = 1$.*

Proof Assume $u \in A_1$, u, x, y is a triangle and $v \in A_2$. Suppose $d_{T[A_1 \cup \{u,v\}]}(uv) \geq 2$, hence xy must be adjacent to either ux or to uy in T. W.l.o.g., let xy be adjacent to uy, then, there is an edge wx in A_1 which implies the distance between wx and xy to be equal to 4 by a path through uv, a contradiction. $\square$

Each bridge forces a unique way to obtain an edge tree 3-spanner of G. Hence, by Lemma 5, assume G is 2-*edge connected*, i.e. there is not a bridge in G. Otherwise, we consider each connected component separately after the bridges removal of G.

Now, consider the case that G has a cut vertex. Let a *windmill graph* $Wd(3, n)$ be the graph constructed for $n \geq 2$ by identifying n copies of K_3 at a universal vertex. Since an edge 3-admissible graph is partial 2-tree, we have that if there is a cut vertex u in G, then $G[N_G[u]]$ contains a windmill graph $Wd(3, d)$, for $2 \leq d \leq \frac{d_G(u)}{2}$. Let a *diamond graph* be a K_4 minus an edge. Each K_3 of a windmill centered in u has two vertices of degree 2, or it has a cut vertex of G distinct of u, or it belongs to a diamond graph of G.

Lemma 6 *Let G be 2-edge connected graph with a cut vertex u and edge 3-admissible. If the associated windmill graph $Wd(3, n)$ centered in u satisfies $n \geq 3$, then u belongs to at most 2 diamonds in G.*

Proof Assume that u is center of the windmill graph $Wd(3, 3)$ and it belongs to 3 diamonds D_1, D_2 and D_3 in G. We prove that G is not edge 3-admissible, and then it implies that if G is edge 3-admissible, then u does not belong to more than 3 diamonds for every $n \geq 3$, either, because the hereditary property proved in Lemma 2.

Note that $L(H)$, for $H = Wd(3, 3) \cup D_1 \cup D_2 \cup D_3$, is composed by a K_6 and the addition of three other subgraphs, named B_1, B_2 and B_3, constructed by a join between a vertex and a C_4. Moreover, each edge of a perfect matching of the K_6, $\{e_1, e_2, e_3\}$, is identified to an edge of B_1, B_2 and B_3 that belongs to the C_4s, resp. Suppose that $L(H)$ is 3-admissible, hence for any tree 3-spanner T of $L(H)$ we have that $T \cap L(H)$ is a fl-centered bi-star, for f and l being any two K_6's vertices. Since any vertex of the K_6 belongs to exactly one of the other three subgraphs added to it, i.e. each K_6's vertex belongs to either B_1, B_2 or B_3, then at least two adjacent vertices of $L(H)$ are adjacent to leaves of the fl-centered bi-star, implying $\sigma'(H) = 4$. $\square$

If there is a vertex u that belongs to $Wd(3, 2)$ then there are two solutions in $T \cap Wd(3, 2)$, less than isomorphism. Consider a $Wd(3, 2)$ such that $V(Wd(3, 2)) = \{u, v, w, v', w'\}$ such that u, v, w and u, v', w' induce K_3's. Note that an edge tree 3-spanner $T \cap Wd(3, 2)$ can be formed as follows: Case 1: $\{uv, uw\}$, $\{uv, vw\}$, $\{uv, uv'\}$, $\{uv', uw'\}$, $\{uv', v'w'\}$; Case 2: $\{uv, uw\}$, $\{uv, vw\}$, $\{uv, uv'\}$, $\{uv, uw'\}$, $\{uv', v'w'\}$. Any other edge tree spanner of $Wd(3, 2)$ is not edge tree 3-spanner.

Although a $Wd(3, 2)$ graph centered in u may have two spanning trees, if each triangle also belongs to a diamond, let D_1 and D_2 be such diamonds with vertices $V(D_1) = \{u, v, w, x\}$ and $V(D_1) = \{u, v', w', x'\}$, then the previous Case 1 is the unique edge tree 3-spanner for $T \cap Wd(3, 2)$, less than isomorphism.

Furthermore, let $H = Wd(3, 2) \cup D_1$ be formed by a $Wd(3, 2)$ centered in u with vertices $V(Wd(3, 2)) = \{u, v, w, v', w'\}$ such that vw belongs to the diamond D_1 with vertices $V(D_1) = \{v, w, s, t\}$, then we have that H is not edge 3-admissible, which can be verified by conditions above and a simple case analyses.

Hence, we have presented necessary conditions of a 2-edge connected graph G satisfying Construction 2 to be edge 3-admissible when it has a cut vertex.

Now, consider G a biconnected graph. Theorem 2 characterizes such graphs. The *diameter* of a graph G is the greatest distance between any pair of vertices, and is denoted by $D(G)$.

Theorem 2 *Given G a biconnected graph with $D(G) \leq 3$. We have that $\sigma'(G) \leq 3$ if and only if either there is distance two dominating edge $e_1 = uv$ or for any edges $e_1 = uv$, $e_2 = uw$, and $e_3 \notin N(u) \cup N(v) \cup N(w)$, e_3 is adjacent to edges only of $N(v)$ (or equivalently, only of $N(w)$).*

Proof If G has a dominating edge, for $D(G) \leq 3$, then $\sigma'(G) \leq 3$ by a uv centered bi-star. Or, if any edge is not dominated by e_1 but it is adjacent to edges only of $N(v)$, then in the solution spanning tree such vertex is adjacent to a leaf of v and it does not turn $\sigma'(G) \geq 4$ because it is not adjacent to leaves of u. Assume that G is edge 3-admissible, there is not a distance two dominating edge and there is an edge e_3, such that $e_3 \notin N(u) \cup N(v) \cup N(w)$ that is adjacent to edges of $N(v)$ and $N(w)$. In this case e_3 is connected to leaves of the two centers of the bi-star in $L(G)$, which implies that T' is not edge 3-admissible, a contradiction. $\square$

Note that Theorem 2 gives another argument on the lower bound of Corollary 3, since a K_n does not satisfy conditions of Theorem 2.

Corollary 5 *Edge 3-admissibility is polynomial-time solvable.*

4   Edge Stretch Index for Split and Generalized Split Graphs

Since $\sigma'(G) \leq 4$ for graphs with a distance two dominating edge (Theorem 1), the polynomial time algorithm for edge 3-admissible of Corollary 5 also works for these graphs and their subclasses, such as split graphs, join graphs and P_4-tidy graphs. I.e., we know whether these graphs have $\sigma'(G) = 2$, $\sigma'(G) = 3$ or $\sigma'(G) = 4$.

Corollary 6 *Edge t-admissibility is polynomial-time solvable for split graphs, join graphs and P_4-tidy graphs.*

As presented in Corollary 6, we are able to determine the edge stretch index for split graphs. Split graphs can be generalized as the (k, ℓ)-graphs, which are the

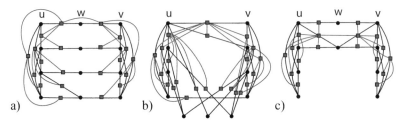

Fig. 5 Cases of $(1, 2)$-graphs and the corresponding edge tree spanners. (**a**) an edge 5-admissible graph. (**b**) and (**c**) are edge 4-admissible graphs

graphs that the vertex set can be partitioned into k stable sets and ℓ cliques. The (k, ℓ)-graphs are also denoted as the generalized split graphs [5].

In [4], the dichotomy P *versus* NP-complete on deciding the stretch index for (k, ℓ)-graphs was partially classified. One of the open problems regarding MSST is to establish the computational complexity for $(1, 2)$-graphs. Next, we prove that the edge stretch index for $(1, 2)$-graphs can be determined in polynomial time.

We denote a $(1, 2)$-graph as a graph $G = (V, E)$ where V is partitioned into $V = \mathcal{K}_1 \cup \mathcal{K}_2 \cup S$, such that each $\mathcal{K}_i$ induces a clique and S is a stable set.

Lemma 7 *If G is a $(1, 2)$-graph, then G is edge 5-admissible.*

Proof Since G is connected, there is a path between a vertex $u \in \mathcal{K}_1$ and $v \in \mathcal{K}_2$ by an edge uv or by a $P_3 = u, w, v$. Figure 5 depicts the cases of $(1, 2)$-graphs and their edge 5-tree spanners. In Fig. 5a there is an induced C_6 by two vertices of each clique and two vertices of S, implying a non-edge in any tree, hence $\sigma'(G) \leq 5$. □

Theorem 3 *A $(1, 2)$-graph $G = (\mathcal{K}_1 \cup \mathcal{K}_2 \cup S, E)$ has $\sigma'(G) \leq 4$ if and only if G has a distance two dominating edge or two adjacent distance two dominating edges that are adjacent to at least one edge of each pair of edges incident to a vertex of S such that one endpoint of an edge of this pair is in $\mathcal{K}_1$ and another one in $\mathcal{K}_2$.*

Proof From Lemma 1, if G has a distance two dominating edge, then G is edge 4-admissible. Moreover, if G has two distance two dominating edges e_1 and e_2 adjacent to at least one edge of each pair of edges incident to a vertex of S such that one endpoint of an edge of this pair is in $\mathcal{K}_1$ and an endpoint of the other edge is in $\mathcal{K}_2$, one obtain an edge tree 4-spanner T of G by selecting any spanning tree of $L(G)$ that maximizes the degrees of these two distance two dominating edges in T.

Conversely, for the sake of contradiction assume that G does not have such distance two dominating edges and T is an edge tree 4-spanner of G. Since G is connected, there is a vertex of S adjacent to both $\mathcal{K}_1$ and $\mathcal{K}_2$ and we can select these two edges of S to be two distance two dominating edges of G. Therefore, for all distance two dominating edges e_1 and e_2 of G we have two edges e_i and e_f incident to a vertex of S such that these edges are both not adjacent to e_1 and e_2. Therefore, in the best case scenario these two edges are adjacent to edges e'_1 and e'_2

adjacent to e_1 and e_2. However, we have a path in T e_i e'_1 e_1 e_2 e'_2 e_f with these two edges e_i and e_f sharing an endpoint, which implies that T is not an edge 4-tree spanner of G.   □

Corollary 7 *Edge t-admissibility is polynomial-time solvable for $(1, 2)$-graphs.*

5   Edge 8-Admissibility Is **NP**-Complete for Bipartite Graphs

Next, we present a polynomial time transformation from 3-SAT [6] to edge 8-admissibility for $(2, 0)$-graphs, i.e. bipartite graphs.

Construction 3 *Given an instance $I = (U, C)$ of 3-SAT we construct a graph G as follows. We add a P_2 with labels x and x' to G. For each variable $u \in U$ we add a C_8 to G with three consecutive vertices labeled as u, m_u, and $\overline{u}$ and the other five consecutive vertices labeled as u_1 to u_5. For each u_i, $i = 1, \ldots, 5$, u and $\overline{u}$ we add a pendant vertex. For each variable $u \in U$ we add the edge xm_u to G. For each clause $c_1 = (u, v, w) \in C$, we add two vertices vertex c_1 and c'_1 to G and the edges $c_1 c'_1$, $c_1 u$, $c_1 v$, and $c_1 w$. For each variable $u \in U$ we add a P_4 to G with endpoints labeled p_{u1} and p_{u4} and the edges $p_{u1}x$ and $p_{u4}m_u$.*

Figure 6 depicts an example of a graph obtained from a 3-SAT instance.

The key idea of the proof of Theorem 4 is that, for each variable $u \in U$, we have exactly one edge in the edge tree 8-spanner T which is near to x and u or $\overline{u}$. We relate this proximity to a true assignment of that literal. Next, we require that at least one edge incident to each clause to be connected to a true literal. Otherwise, if they are all false literals, we end up with two of the edges incident to that clause being vertices of $L(G)$ with distance at least 9 in T.

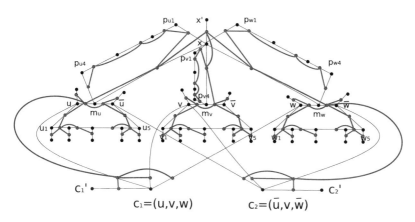

Fig. 6 Graph obtained from Construction 3 on the instance $I = (\{u, v, w\}, \{(u, v, w), (\overline{u}, v, \overline{w})\})$ and an edge tree 8-spanner of it in red

Theorem 4 *Edge 8-admissibility is NP-complete for bipartite graphs.*

Proof By construction, G is bipartite. Moreover, not only the problem is in NP, but also the size of the graph G, obtained from Construction 3 on an instance $I = (U, C)$ of 3-SAT, is polynomially bounded by the size of I. We prove that G is edge 8-admissible if and only if there is a truth assignment to I. Consider a truth assignment of $I = (U, C)$. We obtain an edge tree 8-spanner T of G as follows (see Fig. 6).

Add to T the edges: $\{x'x, xm_u \mid u \in U\}$; $\{xm_u, m_u u \mid u \in U$ and u is true$\}$ or $\{xm_u, m_u \bar{u} \mid u \in U$ and $\bar{u}$ is true$\}$; $\{um_u, \bar{u}m_u \mid u \in U\}$; For each clause select a true literal and add to T: $\{c'c, uc \mid c$ is a clause with the selected true literal $u\}$;

$\{uc, um_u \mid c$ is a clause with the selected true literal $u\}$;

$\{\bar{u}c, \bar{u}m_u \mid c$ is a clause with the selected true literal $\bar{u}\}$;

$\{uc, vc \mid c$ is a clause with the selected trueliteral u and v is other literal of $c\}$;

For each variable $u \in U$ add to T the edges: $\{m_u p_{u_4}, p_{u_4} p_{u_3}\}$; $\{p_{u_4} p_{u_3}, p_{u_3} p_{u_2}\}$; $\{p_{u_3} p_{u_2}, p_{u_2} p_{u_1}\}$; $\{p_{u_2} p_{u_1}, p_{u_1} x\}$; $\{p_{u_1} x, xm_u\}$; $\{um_u, uu_1\}$; $\{\bar{u}m_u, \bar{u}u_5\}$; $\{uu_1, u_1 u_2\}$; $\{u_3 u_4, u_4 u_5\}$; $\{u_4 u_5, \bar{u}u_5\}$; and each pendant G is added to a solution tree as Fig. 6.

Consider an edge tree 8-spanner T of G (resp. tree 8-spanner of $L(G)$), we present a truth assignment of $I = (U, C)$. First we claim that for each variable $u \in U$, there is exactly one of these two edges in T: $\{xm_u, um_u\}$ and $\{xm_u, \bar{u}m_u\}$. Assume that both edges are in T. There are in $L(G)$ two adjacent vertices $u_i u_{i+1}$ and $u_{i+1} u_{i+2}$ of the cycle C_9 of variable u with distance 9 in T, a contradiction. Now, assume that both edges are not in T. We consider two cases. If there are no edges $p_{u_4} m_u, um_u$ or $p_{u_4} m_u, \bar{u}m_u$, then there are in $L(G)$ two adjacent vertices $p_{u_4} m_u$ and um_u (or $\bar{u}m_u$) with distance at least 9 in T, since it is necessary to make a path passing through xx', a contradiction. Otherwise, there is an edge $p_{u_4} m_u, um_u$ or $p_{u_4} m_u, \bar{u}m_u$. In both cases, let $c_1 = (u, v, w)$ be a clause that contains u, there are in $L(G)$ two adjacent vertices $c_1 v, vv_1$ that have distance at least 9 in T, a contradiction.

Hence, relate the edge $\{xm_u, um_u\}$ or $\{xm_u, m_u \bar{u}\}$ in T for each variable $u \in U$ to a true assignment to the literal u or $\bar{u}$. Assume that there is a clause with three false literals $c_3 = (x, y, z)$. No matter how we connect the vertices $c'_3 c_3, c_3 x, c_3 y$ and $c_3 z$ in T, two of them have distance at least 9 in T, a contradiction. Therefore, each clause has at least one true literal, and this is a truth assignment of I. □

Construction 3 can be adapted in order to prove that edge $2k$-admissibility is NP-complete, for $k \geq 5$. It can be obtained by subdividing the edge $m_u x$ and the cycles corresponding to each variable u.

6   Concluding Remarks

We have obtained the edge stretch index of some graph classes, or equivalently, the stretch index of line graphs, such as gridline graphs (line graphs of bipartite graphs); complement of Kneser graphs $KG_{n,2}$ (line graphs of complete graphs); and line graphs of (k, ℓ)-graphs. Although deciding the 3-admissibility is open for more than 20 years, we characterize the edge 3-admissible graphs in polynomial time, and we also prove that edge 8-admissibility is NP-complete, even for bipartite graphs. Hence, some open questions arise, such as determine the computational complexity of edge t-admissibility for $4 \leq t \leq 7$, and $t = 2k + 1, k \geq 4$.

Acknowledgments This study was partially supported by Coordenação de Aperfeiçoamento de Pessoal de Nível Superior—Brasil (CAPES)—Finance Code 001.

References

1. Bhatt, S., Chung, F., Leighton, T., Rosenberg, A.: Optimal simulations of tree machines. In: 27th Annual Symposium on Foundations of Computer Science, pp. 274–282. IEEE, Piscataway (1986)
2. Cai, L., Corneil, D.G.: Tree spanners. SIAM J. Discrete Math. **8**(3), 359–387 (1995)
3. Couto, F., Cunha, L.F.I.: Tree t-spanners of a graph: minimizing maximum distances efficiently. In: 12th COCOA, Lecture Notes in Computer Science, vol. 11346, pp. 46–61 (2018)
4. Couto, F., Cunha, L.F.I.: Hardness and efficiency on minimizing maximum distances for graphs with few P4's and (k, ℓ)-graphs. Electron. Notes Theor. Comput. Sci. **346**, 355–367 (2019)
5. Couto, F., Faria, L., Gravier, S., Klein, S.: Chordal-(2, 1) graph sandwich problem with boundary conditions. Electron. Notes Discrete Math. **69**, 277–284 (2018)
6. Garey, M.R., Johnson, D.S.: Computers and Intractability: A Guide to the Theory of NP-Completeness. W. H. Freeman Co., New York (1979)
7. Giakoumakis, V., Roussel, F., Thuillier, H.: On P4-tidy graphs. Discr. Math. Theoretical Comput. Sci. **1**, 17–41 (1997)
8. Godsil, C., Royle, G.: Kneser graphs. In: Algebraic Graph Theory, pp. 135–161. Springer, New York (2001)
9. Heggernes, P.: Treewidth, partial k-trees, and chordal graphs. INF334-Advanced algorithmical techniques, Department of Informatics, University of Bergen (2005)
10. Jamison, B., Olariu, S.: P-components and the homogeneous decomposition of graphs. SIAM J. Discrete Math. **8**(3), 448–463 (1995)
11. Peleg, D., Ullman, J.D.: An optimal synchronizer for the hypercube. SIAM J. Comput. **18**(4), 740–747 (1989)

Sequence Graphs: Characterization and Counting of Admissible Elements

Sammy Khalife

Abstract We present a family of graphs implicitly involved in sequential models, which are obtained by adding edges between elements of a discrete sequence appearing simultaneously in a window of size w, and study their combinatorial properties. First, we study the conditions for a graph to be a sequence graph. Second, we provide, when possible, the number of sequences it represents. For $w = 2$, unweighted 2-sequence graphs are simply connected graphs, whereas unweighted 2-sequence digraphs form a less trivial family. The decision and counting for weighted 2-sequence graphs can be transformed by reduction into Eulerian graph problems. Finally, we present a polynomial time algorithm to decide if an undirected and unweighted graph has the said property for $w \geq 3$. The question of NP-hardness is left opened for other cases.

Keywords Graphs · Combinatorics · Representations

1 Introduction

The graphs we are interested in this paper, referred to as sequence graphs, represent the co-occurrences (potentially oriented) of the elements in a sequence appearing simultaneously in a window of constant size w. These structures encode information of several sequential models, in particular for natural language [4, 7, 9], supplementing the information of bag-of-words representations, which are invariant to any permutation. They also have been used for biological sequences, namely for protein visualization or protein-protein interaction prediction [2, 8]. In this work, we are interested in two main questions; first the question of recognition of such graphs, and second, the counting of corresponding sequences.

S. Khalife (✉)
LIX CNRS Ecole Polytechnique, Institut Polytechnique de Paris, Palaiseau, France
e-mail: khalife@lix.polytechnique.fr

© The Author(s), under exclusive license to Springer Nature Switzerland AG 2021   209
C. Gentile et al. (eds.), *Graphs and Combinatorial Optimization:*
from Theory to Applications, AIRO Springer Series 5,
https://doi.org/10.1007/978-3-030-63072-0_17

1.1 Definitions and Problem Statement

In the following, let $x = x_1, x_2, \ldots, x_p$ be a finite sequence of discrete elements among a finite vocabulary X. Without loss of generality, we can suppose that $X = \{1, \ldots, n\}$, let $I_p = \{1, \ldots, p\}$ and let $\mathbb{N}^*$ be the set of strictly positive integers.

Definition 1 $G = (V, E)$ is the graph of the sequence x with window size $w \in \mathbb{N}^*$ if and only if $V = \{x_i \mid i \in I_p\}$, and

$$(i, j) \in E \iff \exists (k, k') \in I_p^2, \ |k - k'| \le w - 1, \ x_k = i \text{ and } x_{k'} = j \tag{1}$$

For digraphs, Eq. (1) is replaced by

$$(i, j) \in E \iff \exists (k, k') \in I_p^2, \ k \le k' \le k + w - 1, \ x_k = i \text{ and } x_{k'} = j \tag{2}$$

Finally, a weighted sequence digraph G is endowed with the matrix $\Pi(G) = (\pi_{ij})$ such that:

$$\pi_{ij} = \mathsf{Card} \, \{(k, k') \in I_p^2 \mid k \le k' \le k + w - 1, \ x_k = i \text{ and } x_{k'} = j\} \tag{3}$$

By convention, a weighted (undirected) sequence graph is endowed with $\Pi = (\pi_{ij})$, $\pi_{ij} = \pi'_{ij} + \pi'_{ji}$ if $i \ne j$ and π'_{ij} otherwise, where π' verifies Eq. (3).

We say that x is a w-admissible sequence for G if G is the graph of the sequence x. G is referred to as the w-sequence graph of x with window size w.

π_{ij} represents the number of co-occurrences of i and j in a window of size w. Hence, the graph of a sequence x is unique for a given w. In the following, we use $G_w(x)$ as a shorthand for the w-sequence graph of x. In the weighted and directed case, it can be obtained with Algorithm 1.

Algorithm 1: Construction of a weighted sequence digraph

Data: Sequence x of length p, window size w, $p \ge w \ge 2$
Result: $(G_w(x), \Pi)$
1 $V \leftarrow \emptyset$;
2 $d \leftarrow$ number of distinct elements of x;
3 Initialize $\Pi = (\pi_{i,j})$ to $d \times d$ matrix of zeros;
4 **for** $i = 1 \rightarrow p - 1$ **do**
5 $\quad$ $V \leftarrow V \cup \{x_i, x_{i+1}\}$;
6 $\quad$ **for** $j = i + 1 \rightarrow \min(i + w - 1, p)$ **do**
7 $\quad\quad$ $\mid$ $\pi_{x_i, x_j} \leftarrow \pi_{x_i, x_j} + 1$;
8 $\quad$ **end**
9 **end**
10 Return V, Π

If G is not oriented, one should replace line 7 of Algorithm 1 by the "symmetrized" update:

$$\text{if } \pi_i \neq \pi_j: \quad \alpha \leftarrow \pi_{x_i,x_j}, \qquad \pi_{x_i,x_j} \leftarrow \alpha + 1, \qquad \pi_{x_j,x_i} \leftarrow \alpha + 1$$
$$\text{else}: \quad \pi_{x_i,x_i} \leftarrow \pi_{x_i,x_i} + 1 \tag{4}$$

The procedure in Algorithm 1 defines a correspondence between the sequence set S_X into the graph set $\mathscr{G}: \phi_w: S_X \to \mathscr{G}, x \mapsto G_w(x). G \in \text{Im}\,\phi_w$ exactly means that G is a w-sequence graph. For a given w, the two problems we address in this paper are the characterization (or recognition) of w-sequences graph, and the counting of the number of their w-admissible sequences.

1.2 Related Work

Despite their relations with co-occurrences based models for language [1, 7, 9], no such combinatorial questions were investigated in computational linguistics which we believe to be of interest, namely to understand the degree of ambiguity of these models. Besides, such structures have been partially studied in the Distance Geometry (DG) literature before, mostly to do with proteins, where an "atom window" can be defined by using the protein backbone [6]. However, the type of graph studied in Distance geometry does not refer directly to the results we are investigating in this paper. Indeed, the necessary and sufficient conditions for which such study would apply are:

- each element of the sequence x is associated with a unique vertex (which is not the case we investigate here, since a symbol can be repeated several times but only one vertex is created)
- the absence of loops

As a consequence, the results mentioned in the DG survey [6] do not apply to the present case.

1.3 Notations

In the following, we use $\mathscr{M}_d(\mathbb{N})$ as a shorthand for the square $d \times d$ matrices over the set of natural integers, $\text{Tr}(M)$ for the trace of a matrix M, and $\text{Sp}(M)$ for its set of eigenvalues.

2   2-Sequence Graphs

In this section, we consider $w = 2$. Algorithm 1 encodes each adjacency in the sequence x as an edge in $G_w(x)$. Obviously, the simplest case concerns undirected graphs as stated in the:

Proposition 1 *Let $G = (V, E)$ be an unweighted and undirected graph with $|V| > 1$. Then, the following assertions are equivalent:*

(i)   G is connected
(ii) G has a 2-admissible sequence
(iii) G admits an infinite number of 2-admissible sequences

Proof If G is connected, a sequence is obtained by visiting all edges, for instance using a list of arbitrary sequences and shortest paths. The other implications are immediate.   □

For digraphs, the previous characterization is wrong, even with strong connectivity. A counter example is given in Fig. 1. However, strong connectivity remains a sufficient condition:

Proposition 2 *Let $G = (V, E)$ be an unweighted digraph. If G is strongly connected then $G \in \operatorname{Im} \phi_2$. Moreover, a 2-admissible sequence can start or end at any given vertex of G.*

Proof Straightforward, similarly to (i) $\Longrightarrow$ (ii) for Proposition 1.   □

Proposition 3 *Let $G = (V, E)$ be an unweighted digraph. If G is Eulerian or semi-Eulerian, then $G \in \operatorname{Im} \phi_2$.*

Proof If G is Eulerian or semi-Eulerian, there exists a walk going through all edges, this walk defines a 2-admissible sequence.   □

Again the converse of Proposition 3 does not hold as depicted in Fig. 2. First, it is natural to consider the case of directed acyclic graphs (DAGs):

Fig. 1 G has 1 2 3 as a 2-admissible sequence but is not strongly connected

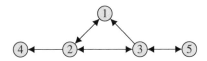

Fig. 2 G has 3 5 3 1 2 1 2 3 2 4 as a 2-admissible sequence but is not Eulerian nor semi-Eulerian

Proposition 4 *Let $G = (V, E)$ be a DAG. G is a 2-sequence graph if and only if it is a directed path, i.e. G is a directed tree where each node has at most one child and at most one parent. In this case, G has a unique 2-admissible sequence.*

Proof If G is a directed path, since G is finite, it admits a source node. Therefore a 2-admissible sequence is obtained by simply going through all vertices from the source node. This is obviously the only one.

Conversely, let us suppose G is a DAG and a 2-sequence graph. If G is not a directed path, there are two cases: either there exists a vertex having two children, or two parents. Let s be a vertex having 2 distinct children c_1 and c_2. This is not possible since there cannot be a walk going through (s, c_1) and (s, c_2): G would have a cycle otherwise. Finally a vertex v cannot have two parents p_1 and p_2: if a 2-admissible sequence existed, it would have to go through (p_1, v) and (p_2, v), creating a cycle, hence the contradiction.   □

Every directed graph G is a DAG of its strongly connected components. In the following, let $R(G)$ be the DAG obtained by contracting the strongly connected components of G.

Proposition 5 *Let $G = (V, E)$ be a digraph. If G is a 2-sequence graph then $R(G)$ is a 2-sequence graph.*

Proof Let G be a 2-sequence graph, and let us suppose that $R(G)$ is not a 2-sequence graph. Since $R(G)$ is a (weakly) connected DAG, then using Proposition 4, it cannot be a directed path, so $R(G)$ has either a node having two children or two parents. Let S be a node of $R(G)$ having at least 2 distinct children C_1 and C_2. This means that there exist three distinct corresponding nodes in V, s, v_1 and v_2 such that $(s, v_1) \in E$ and $(s, v_2) \in E$. Since G is a 2-sequence graph, there exists a walk covering (s, v_1) and (s, v_2), such walk would make S, C_1 and C_2 the same node in $H(G)$, hence the contradiction. The case for which a vertex has two parents is dealt with similarly.   □

The converse of Proposition 5 does not hold as depicted in Fig. 3, which motivates the following definition.

Definition 2 Let G be a digraph, and $R^+(G)$ be the weighted DAG obtained from $R(G)$, such that the weight of an edge is the number of distinct arcs from two strongly connected components in G.

Theorem 1 *Let $G = (V, E)$ be an unweighted digraph.*

Fig. 3 G is not a 2-sequence graph while $R(G)$ is. (**a**) G. (**b**) $R(G)$

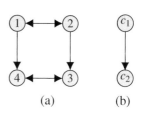

(a)   (b)

G is a 2-sequence graph if and only if $R^+(G)$ is a directed path and its weights are all equal to 1.

Proof If G is a 2-sequence graph, $R(G)$ is a 2-sequence graph using Proposition 5. Also Proposition 4 implies that $R(G)$ and $R^+(G)$ are directed paths. Moreover, if $R^+(G)$ had a weight strictly greater than 1, then there would be strictly more than one edge between two strongly connected components C_1 and C_2. All these edges go in the same direction otherwise $C_1 \cup C_2$ would be part of a larger strongly connected component. This is a contradiction since any 2-admissible sequence would have to go from C_1 to C_2 and then come back to C_1 (or conversely) and $C_1 \cup C_2$ would again be part of a larger strongly connected component.

Conversely, let us suppose $R^+(G)$ is a a directed path and its weights are equal to one. First, there exists a walk $x_1, \ldots, x_p$ covering all edges of $R^+(G)$ verifying: (i) $\forall i, x_i \in V$ or x_i represents a strongly connected component of G, (ii) there is only one edge in G between from x_i to x_{i+1} and (iii) x has no repetition, i.e. there is no common vertex in G between x_i and x_{i+1}. We construct a 2-admissible sequence y for G by means of the following procedure.

Initialisation: If $x_1 \in V$, we simply set $y \leftarrow x_1$. Otherwise, x_1 corresponds to a strongly connected component C_1 of G and we add to y any 2-admissible sequence of C_1.

For $i \in \{1, .., p - 1\}$:

- If $(x_i, x_{i+1}) \in E$: we add x_{i+1} to the sequence y.
- If $x_i \in V$ and x_{i+1} is a strongly connected component C_i of G: By assumption, there exists only one edge of G from x_i to a vertex of C_i, say c_0^i. Since C_i is strongly connected, using Proposition 2, C_i has a walk going through all of its edges and starting in c_0^i, say $c_0^i, \ldots, c_p^i$. We add $c_0^i, \ldots, c_p^i$ to y.
- If x_i corresponds to a strongly connected component C_i and $x_{i+1} \in V$: we perform similar operations by stopping on the single node of C_i that has a edge to x_{i+1} (this is possible thanks to Proposition 2).
- x_i and x_{i+1} both correspond to strongly connected components C_i and C_{i+1}, there exists only one edge between in E between C_i and C_{i+1}, say $e_i = (v_i, v_{i+1})$. We can complete y by a walk from the last vertex visited which belong to C_i and v_i, and then by a 2-admissible sequence through C_{i+1} starting in v_i and ending in v_{i+1}.

The process stops when $i = p - 1$, and all edges are covered by the sequence y. ☐

Therefore, an algorithm to decide if a digraph is a 2-sequence graph is obtained by extracting its strongly connected components (there exist linear time algorithms e.g. [10]), and to count the number of distinct edges between these.

Corollary 1 *Let G be an unweighted digraph. The possible numbers of 2-admissible sequences for G is exactly $\{0, 1, +\infty\}$. Moreover, G admits a unique 2-admissible sequence if and only if G is a directed path.*

Proof Let G a be 2-sequence graph. G verifies the characterization of Theorem 1. If $R(G)$ has a vertex C representing a strongly connected component of G (or

Fig. 4 G is strongly connected but is not a 2-sequence graph

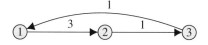

a vertex with a loop), then by adding an arbitrary number of cycles in C to the admissible sequence y (cf. Proof 2), the new sequence is still admissible. Otherwise, if every vertex of $R(G)$ is in V without self-loops in E, then G is a DAG. Using Proposition 4, y is the unique 2-admissible sequence. $\qquad\square$

2.1 Weighted 2-Sequence Graphs

The weighted case cannot be treated similarly due to the constraint 3. A counterexample is depicted in Fig. 4. Moreover, a weighted graph has a finite number of admissible sequences. This property can be seen using Proposition 6 below.

Proposition 6 *If a graph is a weighted w-sequence graph, all of its admissible sequences have the same length.*

Proof Let x be a w-admissible sequence for G of length p. If G is a digraph, Algorithm 1 is incrementing $(p - w + 1)(w - 1) + \frac{(w-1)(w-2)}{2}$ times the total weight, therefore:

$$\sum_{i,j} \pi_{ij} = (p - w + 1)(w - 1) + \frac{(w - 1)(w - 2)}{2} \qquad (5)$$

If $w \geq 2$, this yields: $p = w - 1 - \frac{w-2}{2} + \frac{1}{(w-1)} \sum_{i,j} \pi_{ij}$

Otherwise, if G is undirected, the weights matrix obtained with Algorithm 1 does not yield Eq. (5), due to the update of Eq. (4). The weights on the diagonal remain the same, but the others are multiplied by 2, hence the formula:

$$\sum_{i,j} \pi_{ij} + \mathrm{Tr}(\Pi) = 2(p - w + 1)(w - 1) + (w - 1)(w - 2) \qquad (6)$$

leading to $p = \frac{1}{2(w-1)}[\sum_{i,j} \pi_{ij} + \mathrm{Tr}(\Pi)]$. $\qquad\square$

Corollary 2 *Let G be a weighted w-sequence digraph, and Π its weights matrix. If w even, then $(w - 1) \mid \sum_{i,j} \pi_{ij}$.*

Corollary 3 *Let G be a w-sequence (undirected) graph and Π its weights matrix. Then $2(w - 1) \mid \sum_{i,j} \pi_{ij} + \mathrm{Tr}(\Pi)$.*

Definition 3 Let $\psi(G)$ be the auxiliary multigraph with the same vertices as $G = (V, E)$ and with π_{ij} edges between $(i, j) \in V^2$.

Due to the previous study, the characterization of weighted 2-sequence graphs using $\psi(G)$ is immediate. A semi-Eulerian graph is a graph that admits a Eulerian walk (instead of cycle for Eulerian graphs).

Theorem 2 *If G is a weighted graph (directed or not), with $\Pi(G) \in \mathcal{M}_d(\mathbb{N})$, then:*
$G \in \operatorname{Im} \phi_2 \iff \psi(G)$ *is connected and semi-Eulerian.*

Proof $G \in \operatorname{Im} \phi_2$ means that there is a trail going through each edge $(i, j) \in E$ exactly π_{ij} times. This trail corresponds to a semi-Eulerian path in $\psi(G)$. $\qquad \square$

2.2 Counting 2-Admissible Sequences for Weighted Graphs

Proposition 7 sums up the results for the counting problem of a weighted graph:

Proposition 7 *Counting the number of 2-sequences for a weighted graph is #P-complete. However, if G is a weighted digraph with $\Pi(G) \in \mathcal{M}_d(\mathbb{N})$, then the number p_2 of 2-admissible sequences is given by:*

$$p_2 = \frac{t(\psi(G))}{\prod_{e \in E} \pi_e!} \prod_{v \in V} \left(\deg_{\psi(G)}(\psi(v)) - 1 \right)! \tag{7}$$

where $t(G)$ is the number of spanning trees of a graph G. If L is the Laplacian matrix of G, then $t(G)$ is given by $t(G) = \prod_{\substack{\lambda_i \in \operatorname{Sp}(L) \\ \lambda_i \neq 0}} \lambda_i$.

Proof Given a 2-admissible sequence of G, the choice of a corresponding Eulerian path in $\psi(G)$ is the choice of $\sigma = (\tau_1, \ldots, \tau_{|E|})$ of $|E|$ permutations of $\{1, \ldots, \pi_e\}$ representing the visit order in $\psi(G)$. $G \mapsto \psi(G)$ being bijective, counting Eulerian paths in an undirected graph is #P-complete [3], hence so is the problem of counting the 2-sequences of a weighted graph. BEST [11] and Matrix tree [5] theorems allow to derive formula (7) which guarantees in that the problem on digraphs is in P. $\qquad \square$

To use formula (7), $\deg_{\psi(G)}(\psi(v))$ can be obtained using the following formula:
$deg_{\psi(G)}(\psi(v)) = \sum_{n \in V} \pi_{nv} + \sum_{n \in V} \pi_{vn}$.
The results are summed up in Table 1.

Table 1 Results for various instances of our problems ($w = 2$)

Problem	Undirected		Directed	
	Unweighted	Weighted	Unweighted	Weighted
Nb. sequences	(P) $\{0, +\infty\}$	#P-hard	(P) $\{0, 1, +\infty\}$	(P) BEST Theorem
$G \in \operatorname{Im} \phi_2$?	G connected	$\psi(G)$ Eulerian or semi Eulerian	Theorem 1	$\psi(G)$ Eulerian or semi Eulerian

3 What Happens If $w > 2$?

The characterization of 3-graphs is not the same as for 2-graphs, as the counter-example in Fig. 5a shows: the depicted graph has no loop so there must at least one clique of size 3, which is not the case. Similarly, Fig. 5b depicts a counter example for directed graphs: G does not have loop, so if it had a 3-admissible sequence, such sequence must be of the form $\{1\,2\,3\,1\ldots, 1\,3\,2\,1\ldots, 2\,3\,1\,2\ldots, 3\,2\,1\,3\ldots, 2\,1\,3\,2\ldots\}$ but then $(2, 1)$ would form an edge.

Similarly to the procedure in Sect. 2.1, we will use an auxiliary graph built on G. Let $H(G) = (E, E_H)$ be the new graph obtained with the following procedure. Two edges $e = (v_1, v_2)$, $f = (v_3, v_4)$ of E are connected in $H(G)$ if and only if (An illustration is given Fig. 6):

$$v_2 = v_3 \text{ and } (v_1, v_4) \in E \tag{8}$$

Therefore, by definition, a walk P in $H(G)$ is always of the form:

$$P = (t_1, t_2), \ldots, (t_{p-1}, t_p) \text{ s.t } \forall i \in \{1, \ldots, p-1\}, \ (t_i, t_{i+1}) \in E \tag{9}$$

It is clear that if $H(G)$ is a 2-graph, then G is a 3-graph since there is a walk going through all edges of $H(G)$. However, the converse is not true as depicted in Fig. 7. In order to determine if $G = (V, E)$ has an admissible sequence for any w, a procedure is to recursively merge pairs of vertices, maintaining constraints defined below. These constraints are similar to Eq. (8). We adopt the following notations, $u_{i,j} = (u_i, u_j)$ and $u_{1:k} = (u_1, \ldots, u_k)$. The iterative procedure (for $w \geq 3$) is summed up in 10.

(a)   (b)

Fig. 5 Counter-examples for $w = 3$. (**a**) G is connected but does not have any 3-admissible sequence. (**b**) G is strongly connected but does not have any 3-admissible sequence

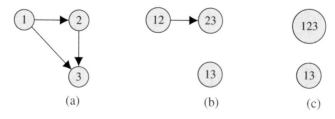

(a)   (b)   (c)

Fig. 6 Reduction on a simple example ($w = 3$). (**a**) Original graph G. (**b**) Graph H. (**c**) DAG $R(H)$

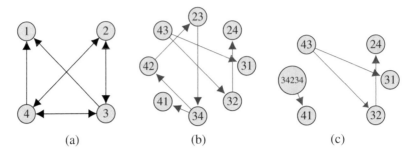

Fig. 7 Procedure to find a 3-admissible sequence. 34234, 41: is 3-admissible, with authentic sequence 3 4 2 3 4 1. (**a**) Original graph G. (**b**) Graph H is not a 2-sequence graph. (**c**) DAG $R(H^{(1)})$

Namely, $\forall k \in \{2, \ldots, w - 2\}$, one has

$$E^{(k)} = \{u_{1:k+1} \in V^{k+1} \mid u_{1:k} \in E^{(k-1)}, u_{2:k+1} \in E^{(k-1)} \land (u_1, u_{k+1}) \in E\} \tag{10}$$

Let $H^{(k)} = (E^{(k)}, E^{(k+1)})$, it can be defined recursively through:

$$H^{(0)} = G \qquad\qquad \forall k \in \mathbb{N}^*, \quad H^{(k)} = f(H^{(k-1)}) \tag{11}$$

where f transforms edges into vertices and creates edges between new vertices that verify Eq. (10). It should be noted that $H(G)$ is directed if and only if G is.

Definition 4 Let u be a vertex of $H^{(k)}$ for $k \in \mathbb{N}$, $u = (u_1, \ldots, u_k, u_{k+1})$, where $u_j \in V$ for each j. The sequence $u_1, \ldots, u_{k+1}$ is the authentic sequence of u. We also call an authentic sequence of a walk on $H^{(k)}$: $P = (x_1, \ldots, x_{k+1}), (x_2, \ldots, x_{k+2}), \ldots, (x_v, \ldots, x_{v+k})$ the sequence $x_1, x_2, \ldots, x_{v+k}$.

In order to obtain admissible sequences of length p, the computation of $H^{(p)}$ requires p iterations, and the number of vertices and edges of $H^{(k)}$ can increase during iterations (the complete graph is an example for which theses numbers increase quadratically).

Proposition 8 *Let $x = x_1, \ldots, x_p$ be a w-admissible sequence of a graph (or digraph) $G = (V, E)$. If $w \leq p$, then x is an authentic sequence of a walk of length $p - w + 1$ on $H^{(w-2)}$.*

Proof Let $x = x_1, \ldots, x_p$ be a w-admissible sequence of G. Let P be a walk on $H^{(w-2)}$, and $P[i]$ be the i-th element of P, $P[i] \in H^{(w-2)}$: $P[i] = (P[i]_1, \ldots, P[i]_{w-1})$.

Let us suppose that $w \leq p$ (which we can always do), and let us show the following property by induction on k:

$$\forall k \in \{w - 1, \ldots, p\}, \ \exists \text{ walk } P \text{ on } H^{(w-2)},$$

$$x_{1:k} = P[1]_1, P[2]_1, \ldots, P[k - (w - 1)]_1, P[k + 1 - (w - 1)]_{1:(w-1)} \tag{12}$$

- Initialisation: $k = w - 1$. By construction of $H^{(w-2)}$, $x_{1:w-1}$ is the authentic sequence of "static walk": $P = P[1] = x_{1:w-1} \in H^{(w-2)}$.
- Induction: let us suppose the property is verified for $k \in \{w-1, \ldots, p-1\}$, i.e. there exists a walk P on $H^{(w-2)}$ such that:

$$x_{1:k} = P[1]_1, P[2]_1, \ldots, P[k-(w-1)]_1, P[k+1-(w-1)]_{1:(w-1)}$$

Since x is w-admissible, then by definition:

$$\forall i \in \{k+1-(w-1), \ldots, k\}, \forall j \in \{i+1, \ldots, \min\{k+1, i+w-1\}\} : (x_i, x_j) \in E$$

Therefore, by definition of $H^{(w-2)}$, $\xi^{k+1} = x_{k+1-(w-1)}, \ldots, x_{k+1} \in H^{(w-2)}$.

Let $P[k+2-(w-1)] \stackrel{\wedge}{=} \xi^{k+1}$, then $P[k+2-(w-1)]_{1:(w-1)} = x_{k+1-(w-1)}, \ldots, x_{k+1}$. Besides, from the induction assumption: $\forall i \in \{1, \ldots, k-(w-1)\}$, $P[i]_1 = x_i$. This ensures that: $x_{1:(k+1)} = P[1]_1, P[2]_1, \ldots, P[k+1-(w-1)]_1, P[k+2-(w-1)]_{1:(w-1)}$ which ends the induction and the proof. $\square$

Theorem 3 *Let G be a graph and $w \in \mathbb{N}^* - \{1, 2\}$. If G is undirected and unweighted then deciding if G is a w-sequence graph is in P.*

Proof It is possible to compute the connected components of $H^{(w-2)}$, say $C_1, \ldots, C_m$, in polynomial time. For each $i \in \{1, \ldots, m\}$, it is possible to construct walks covering all edges in polynomial time (for instance iteratively using shortest paths). Let $W_1, \ldots, W_m$ be such walks and $X_1, \ldots, X_m$ their respective authentic sequences. Using Proposition 8, G is a w-sequence graph if and only if there exists a walk $\tilde{W}_{i_0}$ on some C_{i_0} creating exactly the edges of G. However, W_{i_0} creates more edges than any walk on C_{i_0} by construction.

In conclusion, the assertion: $\exists i \in \{1, \ldots, m\}$, $\phi_w(X_i) = G$ is a characterization of G being a w-sequence graph. This assertion is decidable in polynomial time since for all i, computing $\phi_w(X_i)$ requires a polynomial number of operations. $\square$

For digraphs, the analogue of the aforementioned procedure would consist in enumerating all paths in the DAG $R(H^{(w-2)})$. However, the number of paths can be exponential, even for a sequence graph. For the sake of completeness, we will prove that the reduction by strongly connected components preserves admissibility.

Lemma 1 *Let x be a walk on $H^{(w-2)}$ whose authentic sequence is w-admissible for its corresponding unweighted graph G. If x goes through a strongly component C of $H^{(w-2)}$, adding any supplementary path of C to x lets x w-admissible. Any graph generated by a walk on $H^{(w-2)}$ can be generated by a walk on $R(H^{(w-2)})$.*

Proof Let $P = P[1], \ldots, P[r]$ be a walk on $H^{(w-2)}$ going through a strongly connected component C, with an arbitrary ordering of its vertices, i.e. $C = \{c_1, \ldots, c_m\}$. This means $\exists (m_0, i_0) \in \{1, \ldots, m\} \times \{1, \ldots, r-1\}$ s.t $P[i_0] = c_{m_0}$ and $(c_{m_0}, P[i_0+1]) \in E$. Let $\mathscr{C} = c_{m_0}, c_{j_1}, \ldots, c_{j_v}$ be a path in C with $(c_{j_v}, P[i_0+1]) \in E$. Let Q be the new path: $Q = P[1], \ldots, P[i_0], c_{j_1}, \ldots, c_{j_v}, P[i_0 +$

Algorithm 2: A recognition algorithm for unweighted digraphs

Data: Graph G, window width w
Result: (Boolean, empty set or w-admissible sequence)
1 Build $H^{(w-2)}$ recursively (e.g. with 11);
2 Construct $R_H^w = R(H^{(w-2)})$;
3 **for** *source-sink path of* R_H^w **do**
4   **if** *authentic sequence of path is w-admissible for G* **then**
5 | return (True, sequence)
6   **end**
7 **end**
8 return (False, $\emptyset$);

$1], \ldots, P[r]$. By construction of $H^{(w-2)}$, the edges created by any walk on $H^{(w-2)}$ are in E, so Q is still admissible.

Let us label every node of $R(H^{(w-2)})$ representing a strongly connected component of $H^{(w-2)}$ by any $2-$admissible sequence (one exists thanks to Proposition 2). A walk on $H^{(w-2)}$: $x_1, \ldots, x_p$ can be met by a walk on $R(H^{(w-2)})$ using the following procedure:

For $i \in \{1, \ldots, p-1\}$:

- if $x_i, x_{i+1} \in E$, we keep x_i and x_{i+1}
- if $x_i \in V$ and x_{i+1} is in a strongly connected component of $H^{(w-2)}$ (but a node of $R(H^{(w-2)})$), represented by $c_1, \ldots, c_{C_i}$, then a path from x_{i+1} to c_1 exists since the component is strongly connected: $x_{i+1}, p_1, \ldots, p_m, c_1$. We keep x_i, x_{i+1}, $p_1, \ldots, p_m, c_1, \ldots, c_{C_i}$. Using the aforementioned result, this does not perturb admissibility.
- if $x_{i+1} \in V$ and x_i is in a strongly connected component of H^{w-2}, we proceed similarly (x_i and x_{i+1} are swapped).
- if both x_{i+1} and x_i are strongly connected components of H^{w-2}, we add intermediary nodes to connected both components similarly.

□

4 Conclusion

In this preliminary study, we considered two main combinatorial problems: the recognition problem of sequences graphs, and the counting of their realizations. Solving the second problem totally solves the first one, but in the trivial case $w = 2$, the first one is "simpler": the recognition problem of sequence graphs is P for $w = 2$ for any data instance, but the counting problem is #P-hard for weighted graphs. This justifies the distinction of these problems from a computational point of view.

Furthermore, for $w > 2$, the recognition problem is in P for one configuration (unweighted graphs), but the complexity classes of the other instances are left opened, and so are the counting problems for $w > 3$. A possible lead to answer these questions would be to investigate forbidden patterns in a sequence graph. Finally, it should be noted that the abstraction of sequences graphs exactly coincides with the graphs implicitly involved in co-occurrence models or point wise-mutual information models [1, 7, 9], used as input of algorithms to construct word representations. In these models, representations are ambiguous if the given weighted graph has several realizations. Therefore, other extensions of this work would be to propose scalable algorithms (or at least, for reasonable values of w and length of the sequences) to count and explicit realizations, in order to obtain more information about the degree of ambiguity in these models.

References

1. Arora, S., Li, Y., Liang, Y., Ma, T., Risteski, A.: A latent variable model approach to PMI-based word embeddings. Trans. Assoc. Comput. Linguist. **4**, 385–399 (2016)
2. Asgari, E., Mofrad, M.R.: Continuous distributed representation of biological sequences for deep proteomics and genomics. PLOS One **10**(11) (2015)
3. Brightwell, G., Winkler, P.: Counting Eulerian circuits is #p-complete. In: Proceedings of the Second Workshop on Analytic Algorithmics and Combinatorics (2005)
4. Broder, A.Z., Glassman, S.C., Manasse, M.S., Zweig, G.: Syntactic clustering of the web. Comput. Netw. ISDN Syst. **29**(8–13), 1157–1166 (1997)
5. Chaiken, S.: A combinatorial proof of the all minors matrix tree theorem. SIAM J. Algebraic Discrete Methods **3**(3), 319–329 (1982)
6. Liberti, L., Lavor, C., Maculan, N., Mucherino, A.: Euclidean distance geometry and applications. SIAM Rev. **56**(1), 3–69 (2014)
7. Mikolov, T., Yih, W.T., Zweig, G.: Linguistic regularities in continuous space word representations. In: Proceedings of the 2013 Conference of the North American Chapter of the Association for Computational Linguistics: Human Language Technologies, pp. 746–751 (2013)
8. Ng, P.: dna2vec: Consistent vector representations of variable-length k-mers. arXiv preprint arXiv:1701.06279 (2017)
9. Pennington, J., Socher, R., Manning, C.D.: Glove: Global vectors for word representation. In: Proceedings of the 2014 Conference on Empirical Methods in Natural Language Processing (EMNLP), pp. 1532–1543 (2014)
10. Sharir, M.: A strong-connectivity algorithm and its applications in data flow analysis. Comput. Math. Appl. **7**(1), 67–72 (1981)
11. van Aardenne-Ehrenfest, T., de Bruijn, N.: Circuits and trees in oriented linear graphs. In: Classic Papers in Combinatorics, pp. 149–163. Springer, Berlin (2009)

On Solving the Time Window Assignment Vehicle Routing Problem via Iterated Local Search

Lucas Burahem Martins, Manuel Iori, Mayron César O. Moreira, and Giorgio Zucchi

Abstract In this paper, we propose a combined algorithm based on an Iterated Local Search (ILS) and a mathematical model to solve the Time Window Assignment Vehicle Routing Problem (TWAVRP). The TWAVRP appears when the volume of customer demands is uncertain and time windows should be allocated to customers so as to minimize expected travel costs. Our goal is to find a heuristic strategy that can efficiently improve the current TWAVRP solution methods in the literature. For this purpose, we first use an ILS algorithm to generate feasible sets of routes. Then, we invoke a Mixed Integer Linear Programming formulation that assigns time windows to customers and selects the subset of routes of minimum expected cost. Computational results performed on benchmark instances show that our algorithm is competitive with respect to the literature, especially for instances with more than 45 customers.

Keywords Vehicle routing problem · Time window assignment · Iterated Local Search · Mathematical model

L. B. Martins (✉) · M. C. O. Moreira
Federal University of Lavras, Department of Computer Science, Lavras, Minas Gerais, Brazil
e-mail: lucas.martins@estudante.ufla.br; mayron.moreira@ufla.br

M. Iori
University of Modena and Reggio Emilia, Department of Sciences and Methods for Engineering, Reggio Emilia, Italy
e-mail: manuel.iori@unimore.it

G. Zucchi
University of Modena and Reggio Emilia, Marco Biagi Foundation, Modena, Italy

R&D Department, Coopservice S.coop.p.a, Reggio Emilia, Italy
e-mail: giorgio.zucchi@unimore.it; giorgio.zucchi@coopservice.it

© The Author(s), under exclusive license to Springer Nature Switzerland AG 2021
C. Gentile et al. (eds.), *Graphs and Combinatorial Optimization: from Theory to Applications*, AIRO Springer Series 5,
https://doi.org/10.1007/978-3-030-63072-0_18

223

1 Introduction

Vehicle routing is a class of problems that appears in several combinatorial optimization studies due to their practical relevance, mainly in the areas of retail and transport [22]. The classical *Vehicle Routing Problem* (VRP) calls for shipping freight to customers located along a distribution network by means of a fleet of capacitated vehicles, with the aim of minimizing the delivery costs. Since its introduction in the 1950s, several variations of the VRP have become prominent in the literature. Among these, we mention the *Vehicle Routing Problem with Time Windows* (VRPTW), where a given interval of time in which deliveries should occur is associated with each customer.

Inspired by retail distribution networks, [20] introduced the Time Window Assignment Vehicle Routing Problem (TWAVRP). The TWAVRP appears when the volume of customer demands is uncertain and time windows should be allocated to the customers located along a distribution network, so as to minimize the expected travel costs. In the TWAVRP, each endogenous time window, that has a fixed-width, must be associated within the exogenous time window of the customer. The exogenous time windows is represented by the arrival and departure limits of a customer. According to [16], the TWAVRP can be defined as a *two-stage stochastic optimization problem*. Given a set of customers to be visited within a regular period, the first stage decisions are to assign a set of time windows to customers, before demand is known. In the second stage, after requests are revealed for each day, delivery schedules respecting the assigned time windows must be designed.

The TWAVRP faced in this work is part of a research whose focus is to give an efficient and accurate solution for a routing problem faced by an Italian company providing logistics services in several distribution fields. One of the characteristics presented in the particular routing problem faced by the company is the presence of a time window assignment decision phase. Our purpose is to help the company to minimize the actual delivery time and the total cost of the service they offer. We decided to start our research by first looking at the combinatorial aspect of the TWAVRP, with the aim of focusing later on its application to the company case study. In particular, the main contribution of this paper is to provide an answer to the following question: "Is there a heuristic strategy that can efficiently solve the TWAVRP as defined by Dalmeijer and Spliet [7], Spliet and Gabor [20]?". For this purpose, we propose an algorithm that generates a set of routes by invoking an *Iterated Local Search* (ILS) metaheuristic, and then selects the most appropriate routes through an auxiliary mathematical formulation.

The remainder of the paper is structured as follows. Section 2 presents a brief literature review concerning the VRPTW and the TWAVRP. Section 3 formally describes the TWAVRP and presents a mathematical model. Section 4 reports the methodology that we developed to solve the TWAVRP. Section 5 shows the computational experiments that we performed. Finally, Sect. 6 presents some conclusions and some future research perspectives.

2 Brief Literature Review

Due to the academic interest in VRP variants and to the need of solving difficult real-world problems, researchers have been focusing more and more on realistic VRP, studying the class of so-called *Rich Vehicle Routing Problems* (RVRP). The RVRP class deals with realistic objective functions, uncertainty, dynamism, and a variety of practical constraints related to time, distance, heterogeneous fleet, inventory and scheduling problems, to mention a few. The reader is referred to [4, 23, 24] for more details about RVRP variants.

In this work, we deal with the TWAVRP, a problem that has characteristics resembling the VRPTW. The VRPTW is a generalization of the VRP involving appropriate time intervals for performing services, called time windows. In this class of problems, customer service can only be started within the time window defined by the customer [8]. Recently, several interesting VRPTW applications have been addressed in the literature. Among these, we mention the delivery of food [1], the electric vehicle recharging problem [11], and the use of anticipated deliveries in pharmaceutical distribution [12].

In literature, we found some examples of classical approaches of VRPTW that consider Branch-and-Price [3, 8] and Tabu Search algorithms [5, 6], and also more recent studies with applications at delivery food [1] and electric vehicles recharging problem [11].

The TWAVRP has characteristics that make it even harder to solve in practice than the VRPTW [18]. Several methods have been developed for its solution in recent years. The problem was formally introduced by [20] that considered a finite number of scenarios with certain probabilities of realization. The authors proposed a mathematical model involving a large number of variables and solved it using a Branch-Price-and-Cut algorithm. Computational experiments on 40 instances involving 10, 15, 20, and 25 customers proved the efficiency of the proposed algorithm.

Moreover, [19] tackled a variant of the TWAVRP where, for each customer, a time window is selected from a set of possibilities. To solve the problem, they implemented a Branch-Price-and-Cut and a Tabu Search. The results they obtained on a new set of instances showed that an approach considering five scenarios led to an average cost reduction of about 3.6% compared to a single-scenario approach.

In their paper, [7] addressed the TWAVRP through a Branch-and-Cut algorithm. They considered branching strategies based on a set of precedence inequalities. The effectiveness of the algorithm was demonstrated through numerical experiments and comparisons with the literature.

Inspired by a large European food retailer, [16] applied the TWAVRP to a real food distribution case study, involving around 200 customers, with time windows defined according to the product segments. This problem considers both the traveled distance and fleet requirements costs. Solution method uses three phases: Route Generation; Initial Solution Construction; and Improvement by a Matheuristic.

Finally, [18] proposed a mathematical formulation for a TWAVRP variant that includes time-dependent travel times. To deal with this new problem, they applied a Branch-Price-and-Cut algorithm. Computational tests were run on artificial instances having up to 25 customers. The best solution value they found was only 0.55% higher, on average, than the optimal solution value.

3 Formal Problem Definition

We deal with the same problem described by Dalmeijer and Spliet [7]. Consider a set of customers denoted by $H = \{1, 2, \ldots, n\}$. A graph $G = (N, A)$ models the network of this problem, where $N = H \cup \{0, n + 1\}$ is the overall set of nodes and 0 and $n + 1$ represent, respectively, the departure and arrival depot nodes of all routes. A set A_H of arcs indicates the connections between any pair of customer nodes $i, j \in H$. Similar to set N, we denote $A = A_H \cup \{(0, j) \cup (j, n + 1)$ for all $j \in H\}$, as the overall set of arcs connecting customers and depot nodes. Each arc $(i, j) \in A$ has an associated travel time t_{ij} and a travel cost c_{ij}. The travel times are non-negative and respect the triangular inequality ($t_{ij} \leq t_{ik} + t_{kj}$ for all i, j, and k), and the same applies to travel costs.

An unlimited set of homogeneous vehicles with capacity Q is available at the departure depot. We consider a set Ω of demand scenarios, each having probability of occurrence p_ω, for $\omega \in \Omega$, in such a way that $\sum_{\omega \in \Omega} p_\omega = 1$. Each customer $j \in H$ has a demand in scenario $\omega \in \Omega$ given by $0 \leq q_j^\omega \leq Q$.

Each customer $j \in H$ has to be assigned to an endogenous time window of width w_j, which must be selected in a fixed exogenous time window $[e_j, l_j]$ provided in input, where $l_j - e_j \geq w_j$. A time window $[e_0, l_0]$ represents the opening hours of the departure depot. Similarly, a time window $[e_{n+1}, l_{n+1}]$ represents the opening hours of the arrival depot. The objective function consists in minimizing the expected traveled cost over all scenarios, which is given by $\min \sum_{\omega \in \Omega} p_\omega \sum_{(i,j) \in A} c_{ij} x_{ij}^\omega$, where x_{ij}^ω is a binary variable that takes the value 1 if arc $(i, j) \in A$ is traveled in scenario ω, 0 otherwise. For each scenario $\omega \in \Omega$, a route is feasible if the exogenous time window constraints are satisfied, the capacity constraints are satisfied and the customer j must be visited after the service time at customer i added to the travel time t_{ij} in the case that the customer j is visited after i.

4 An Iterated Local Search-Based Algorithm

The heuristic method proposed in this paper is outlined in Algorithm 1. It is based on two successive phases, the first used to generate routes and the second used to select a subset of routes having minimum cost. A similar idea was adopted by Moreira and Costa [15], who efficiently solved a quite different combinatorial optimization problem involving job rotation schedules in assembly lines with heterogeneous workers. Our method is composed of two parts. First, we generate a pool of feasible routes, minimizing the total cost of each scenario (Lines 4–6), subject to vehicle capacity constraints and exogenous time windows. Then, we call an auxiliary *Mixed Integer Linear Programming* (MILP) formulation to select the most appropriate routes of the set, so as to optimize the total cost over all scenarios (Line 7) by respecting the generated endogenous time windows. The reference framework of Phase 1 is the ILS introduced by Lourenço et al. [14]. Such ILS has four components: (i) initial solution generator; (ii) local search procedure; (iii) perturbation; and (iv) acceptance criterion. The choice of this metaheuristic derives from the fact that it has been successfully applied in several combinatorial optimization problems [2, 9, 17], including a number of VRP variants [10, 21]. Moreover, it contains fewer parameters to be fine-tuned with respect to other metaheuristics. In Line 5, we represent the ILS by function $ILS(I_\omega, \alpha, n_{iter})$, which returns the set of routes obtained by the execution of the metaheuristic after receiving in input data I_ω, that is all the data from scenario ω. Note that parameter α corresponds to the perturbation factor, whereas n_{iter} gives the number of iterations without improvements. Next, we explain each component of the ILS and of the subsequent mathematical formulation used to select the final set of routes.

Algorithm 1: Main algorithm

---

1 **Input:** I (instance)
2 **Output:** $(s, f(s))$ (solution, and its objective function)
3 $P \leftarrow \emptyset$;   ▷ Empty pool of routes
4 **foreach** $\omega \in \Omega$ **do**
5 | $P \leftarrow P \cup ILS(I_\omega, \alpha, n_{iter})$;   ▷ Generating the set of routes for each scenario
6 $s \leftarrow RSM(P, I)$;   ▷ Route Selector Model (RSM)
7 **return** $(s, f(s))$;

4.1 Iterated Local Search (ILS)

Algorithm 2 gives the heuristic invoked to create the initial solution for the proposed ILS. The algorithm is inspired by the greedy strategy presented by Zhigalov [25]. Let $\tilde{H}_\omega$ be the set of all customers in scenario $\omega \in \Omega$, that is, all customers demands in that scenario. First, $\tilde{H}_\omega$ is sorted according to the earliest start time of the exogenous time window (i.e., e_i, for $i \in \tilde{H}_\omega$) of the customers (Line 4). The main loop consists of Lines 5–17, and terminates when all customers have been assigned. In each iteration, an empty route is opened (Line 6), and the highest priority customers (according to the sorting in Line 4) are appended to the route, one at a time, if such assignment respects vehicle capacity and time window constraints (Lines 7–10). Feasibility is checked by invoking the *infeasible($\mathscr{R}$)* procedure. If the current route is feasible, customer j is included in the route ($\mathscr{R}$) under construction and then removed from $\tilde{H}_\omega$ (Line 13).

Algorithm 2: Constructive Heuristic (CH)

---

1 **Input:** I (data set), H_ω (set of all available customers for a data set I on scenario ω)
2 **Output:** s (feasible solution)
3 $s \leftarrow \emptyset$;
4 $\tilde{\mathscr{H}} \leftarrow sort(H_\omega)$;   ▷ sort customers in non-descending order of earliest exogenous time window
5 **while** $\tilde{\mathscr{H}} \neq \emptyset$ **do**
6     $\mathscr{R} \leftarrow \emptyset$;
7   **foreach** $j \in \tilde{\mathscr{H}}$ **do**
8        $\mathscr{R} \leftarrow \mathscr{R} \cup \{j\}$;
9   **if** $infeasible(\mathscr{R}) = true$ **then**
10          $\mathscr{R} \leftarrow \mathscr{R}\backslash\{j\}$;
11   **else**
12          $\tilde{\mathscr{H}} \leftarrow \tilde{\mathscr{H}}\backslash\{j\}$;
13    $s \leftarrow s \cup \mathscr{R}$;
14 **return** s

The Local Search (LS) method is composed of six elementary neighborhoods:

N1 *Relocate intra-route*: change position of a customer in a route;
N2 *Swap intra-route*: swap two customer positions in a route;
N3 *2-opt*: invert a sequence of customers allocated to the same route;
N4 *Relocate inter-route*: relocate a customer to a different route in the same scenario;
N5 *Swap inter-route*: exchange two customers allocated in different routes, in the same scenario;
N6 *Cross inter-route*: split two routes at given points and exchange their remaining parts.

The LS method invokes the neighborhoods according to the procedure shown in Algorithm 3. Given a solution s, a list $NL(s)$ of neighborhoods is initialized according to the inter-route neighborhoods (N4, N5, and N6). If s' is feasible and the distance performed, represented by function $f(s')$, decreases compared to the current solution (Line 7), an intra-route search procedure (N1, N2, and N3) is performed over s' (Lines 9–13). If the intra-route procedure improves s', the current solution $\tilde{s}$ is used to replace s^* (Line 13). The process terminates when no inter-neighborhood can return an improvement.

Starting from a solution s^*, the Perturbation method invokes a list $NL(s^*)$ of possible neighborhood moves according to the inter-route neighborhoods (N4, N5, and N6). A percentage α of neighborhoods in $NI(s^*)$ is randomly chosen and applied to s^*. Regarding the Acceptance criterion, we accept only solutions that are better than the current one. Algorithm 4 summarizes the ILS that is applied to each scenario of Phase 1.

Algorithm 3: Local Search method (LS)

---

1 Input: s (feasible solution)
2 Output: s^* (best feasible solution found)
3 $s^* \leftarrow s$;
4 foreach $N \in NL(s^*)$ ▷ $NL(s^*)$: list of inter-neighborhoods of solution s^*
5 do
6   **foreach** $s' \in N$ **do**
7   **if** $f(s') < f(s^*)$ **and** feasible$(s') = $ **true then**
8 $s^* \leftarrow s'$;
9   **foreach** $N \in NI(s^*)$ ▷ $NI(s^*)$: list of intra-neighborhoods of solution s'
10   **do**
11   **foreach** $\tilde{s} \in N$ **do**
12   **if** $f(\tilde{s}) < f(s^*)$ **and** feasible$(\tilde{s}) = $ **true then**
13 $s^* \leftarrow \tilde{s}$;
14 return s^*

Algorithm 4: Iterated Local Search (ILS)

---

1 Input: H_ω (data set), α (perturbation factor), n_{iter} (number of iterations)
2 Output: $\mathscr{P}$ (set of feasible solutions found)
3 $s^* \leftarrow \emptyset$;   ▷ Best solution found so far (take $f(s^*) = +\infty$)
4 $s \leftarrow CH(H, H_\omega)$;   ▷ H_ω: set of available customers of data set H
5 $s_{ls} \leftarrow LS(s)$;
6 $\mathscr{P} \leftarrow s_{ls} \cup s$;   ▷ Initializing the set of feasible solutions
7 $s^* \leftarrow s_{ls}$;
8 $count \leftarrow 0$
9 while $count \neq n_{iter}$ **do**
10 $s' \leftarrow Perturbation(s^*, \alpha)$;
11 $s_{ls} \leftarrow LS(s')$;
12 $\mathscr{P} \leftarrow \mathscr{P} \cup s' \cup s_{ls}$;
13   **if** $f(s') < f(s^*)$ **then**
14 $s^* \leftarrow s'$;
15 $count \leftarrow 0$;
16   **else**
17 $count \leftarrow count + 1$;
18 return $\mathscr{P}$;

4.2 Route Selector Model

The ILS algorithm generates a set P of viable routes for each scenario $\omega \in \Omega$ (see Algorithm 1). Note that all routes in P respect the capacity and time-windows constraints. We built a MILP formulation, called *Route Selector Model* (RSM), whose aim is to choose the most appropriate subset of routes from P, assigning an endogenous time window to each customer, over all scenarios.

To present the RSM, we take from P: (i) f_{jr}^ω as the starting time of service on customer j on the route r in scenario ω; (ii) c_r^ω as the cost to choose a route $r \in P$ in scenario ω; and (iii) x_{jr}^ω as a binary parameter equal to one if customer j belongs to route $r \in R_\omega$ in scenario ω, 0 otherwise. Consider u_r^ω as a binary variable equal to one if route $r \in P$ is selected, 0 otherwise, and y_i as a continuous variable that measures the starting time of the endogenous time window of customer $i \in \widetilde{H}_\omega$. Recall that, as indicated above, w_i gives the time window width of customer i. The RSM is as follows:

$$\min \sum_{\omega \in \Omega} p_\omega c_r^\omega u_r^\omega \tag{1}$$

subject to

$$\sum_{r \in R_\omega} x_{jr}^\omega u_r^\omega = 1 \qquad \forall j \in H, \omega \in \Omega \tag{2}$$

$$\sum_{r \in R_\omega} f_{jr}^\omega x_{jr}^\omega u_r^\omega \geq y_j \qquad \forall j \in H, \omega \in \Omega \tag{3}$$

$$\sum_{r \in R_\omega} f_{jr}^\omega x_{jr}^\omega u_r^\omega \leq y_j + w_i \qquad \forall j \in H, \omega \in \Omega \tag{4}$$

$$y_j \in [e_j, l_j - w_j] \qquad \forall j \in H, \omega \in \Omega \tag{5}$$

$$u_r^\omega \in \{0, 1\} \qquad \forall \omega \in \Omega, r \in R_\omega. \tag{6}$$

The model optimizes the total cost of the selected routes. Constraints (2) indicate that each customer has to be served in all scenarios by a single route. Constraints (3)–(4) establish the endogenous time windows. Domain variables are presented by Constraints (5)–(6).

5   Computational Experiments

We performed a set of computational experiments aimed at assessing the performance of the ILS-based algorithm that we developed for the TWAVRP. The algorithms were implemented in *Python 3.7.4*, using the MILP solver *Gurobi 8.1.1* for the development of the RSM (described in Sect. 4.2), running a single thread for a time limit of 3600 s on each instance. All experiments were performed on a PC Intel i7, 3.5 GHz with 16 GB RAM, which is similar to the computer used by Dalmeijer and Spliet [7].

To generate the pool of routes, Algorithm 4 was executed five times on each instance. This number was tuned through preliminary tests in which we obtained a good trade-off between quality and computational effort. Furthermore, this value allowed the algorithm to make good use of its stochastic components. The number of iterations without improvements (n_{iter}) and the perturbation percentage (α) were fine-tuned through the *Irace* package [13]. For that purpose, we generated 200 training instances by using the instance generator proposed by Dalmeijer and Spliet [7]. The values returned by the *Irace* package at the end of this test were $n_{iter} = 100$ and $\alpha = 0.35$.

5.1 Instances

We use the set of TWAVRP instances proposed by Spliet et al. [18]. Each instance considers a different combination of number of customers, vehicle capacity, demand for each scenario, probability of each scenario, size of exogenous and endogenous time windows, travel costs, and travel times. In this way, the data set comprises ninety instances divided into two classes: small instances and large ones. Small instances contain four sets of ten instances each, having 10, 15, 20, and 25 customers, respectively. Large instances contain five sets of ten instances each, with 30, 35, 40, 45, and 50 customers, respectively. The customer's coordinates were generated as uniformly distributed over a square with sides of length five. The depot is located in the center of the square. Each instance includes demands for each customer in three scenarios with equal probability of occurrence. Exogenous time windows are distributed as follows: a time window [10, 16] is given to 10% of the customers; [7, 21] to 30% of the customers; and [8, 18] to the remaining 60%. The width of the endogenous time window is set to $w_i = 2$ for all customers. The costs and the travel times between the nodes were obtained by calculating the Euclidean distances between their coordinates.

5.2 Results

The experiments compare our ILS based-algorithm with the *Branch-and-Cut* (B&C) proposed by Dalmeijer and Spliet [7], which can be considered the state-of-art method for the solution of the TWAVRP. The results that we obtained are summarized in Table 1. They are aggregated for instances having the same quantity of customers (first column). The remaining columns contain the average and the standard deviation of each measure, with the exception of the B&C time as it was executed just once. Columns B&C, ILS, and ILS + RSM, under the group CPU time (seconds), give the computational times spent by, respectively, the algorithm by Dalmeijer and Spliet [7], the ILS for the construction of the pool of routes, and the overall Algorithm 1 including ILS and RSM. The column called Gap (%) indicates the gap of the solution value found overall repetitions concerning the best solution value. Column Gap *(%) indicate the standard deviation of the gap for the solution value found overall repetitions concerning the best-known values obtained by the B&C method.

Regarding instances that have between 10 to 35 customers, we can observe that our method found relative average deviations from 1.5% to 0.16% compared with

Table 1 Average results aggregated by number of customers (10 instances per line, 5 ILS executions per instance)

Instance	CPU time (seconds)			Gaps	
N. customers	B&C	ILS	ILS + RSM	Gap*(%)	Gap(%)
10	0.1	4.50 ± 0.29	6.61 ± 0.56	0.34 ± 1.00	0.41 ± 1.02
15	4.5	16.50 ± 1.17	26.25 ± 1.86	0.00 ± 0.18	0.11 ± 0.25
20	2.2	39.06 ± 2.01	80.30 ± 7.49	0.02 ± 0.05	0.06 ± 0.10
25	12.4	68.48 ± 2.03	153.29 ± 18.56	0.06 ± 0.14	0.27 ± 0.78
30	544.0	107.27 ± 3.40	284.38 ± 12.62	0.04 ± 0.10	0.28 ± 0.39
35	1531.7	161.59 ± 9.48	501.77 ± 97.94	0.02 ± 0.13	0.29 ± 0.42
40	3252.0	224.33 ± 6.11	749.92 ± 41.11	0.10 ± 0.52	0.72 ± 0.73
45	3600.0	289.34 ± 28.78	990.15 ± 172.79	−0.69 ± 0.83	−0.18 ± 1.61
50	3600.0	372.98 ± 24.41	1743.16 ± 261.71	−1.89 ± 0.12	−1.62 ± 1.31

the B&C solutions in the worst case, and that the total time of the five executions of ILS + RSM is higher than the B&C execution time. In the group of larger instances (45–50 customers), the ILS + RSM outperforms the results found in literature concerning both best-found and average solution values of the five performed tests.

Table 2 highlights the behavior of our method on the 20 larger instances having 45 and 50 customers. We report the lower bound and upper bound obtained in [7] (columns LB and UB, respectively), and the best (column Best) and average (column Avg) solution values found by our ILS + RSM method. In the problems with 45 customers, both methods were competitive, each finding the best results for about half of the cases. Our method improved the solution cost obtained by the B&C for all instances with 50 customers, both considering columns Best and Avg. We estimate that the diversity of routes caused by different local search operators was beneficial for the performance of the ILS + RSM algorithm for these most difficult instances. Overall, we can conclude that the ILS + RSM is a good heuristic method for moderate and large size instances of the TWAVRP. Our research will now focus on adapting it to the real-world case study that motivated our study, so as to embed possible complicating constraints and solve even larger instances.

6 Conclusions and Future Research Avenues

We studied the *Time Windows Assignment Vehicle Routing Problem* (TWAVRP), a VRP variant that appears when the volume of customer demands is uncertain and visits over multiple days should be planned. The objective is to create routes that minimize expected travel costs, assigning a time window over all scenarios to each customer, and respecting the vehicle capacity. Our interest in this problem derives from a real-world case study. We decided to begin our research with the development of a good and flexible metaheuristic, and to test it on the benchmark TWAVRP instances, so as to check if good-quality solutions can be found within reasonable computational efforts.

Table 2 Results for instances with 45–50 customers (best UB values appear in bold)

| Instance | | B&C by Dalmeijer and Spliet [7] | | ILS + RSM | |
#	N. customers	LB	UB	Best UB	Avg. UB
71	45	49.52	51.78	**51.22**	**51.41**
72	45	50.73	**52.13**	51.86	52.94
73	45	41.50	**41.70**	41.95	42.24
74	45	47.25	**47.84**	47.96	48.16
75	45	48.77	49.86	**49.47**	50.02
76	45	48.38	52.09	**49.90**	**50.03**
77	45	50.09	**51.18**	**51.18**	51.25
78	45	52.02	53.95	**53.35**	**53.74**
79	45	47.45	**48.21**	48.27	48.69
80	45	49.57	**50.57**	50.61	50.78
81	50	56.81	58.85	**58.16**	**58.29**
82	50	51.50	53.20	**52.98**	**53.03**
83	50	57.45	60.67	**58.77**	**58.89**
84	50	52.31	56.38	**54.09**	**54.23**
85	50	53.74	56.07	**55.06**	**55.26**
86	50	51.68	54.76	**53.02**	**53.16**
87	50	52.47	54.14	**53.81**	**53.87**
88	50	54.82	56.91	**56.27**	**56.36**
89	50	59.23	61.51	**60.32**	**60.62**
90	50	57.68	59.55	**58.95**	**59.23**

To this aim, we proposed an *Iterated Local Search* (ILS) algorithm that generates a pool of feasible routes for each scenario, and a mathematical model, called *Route Selector Model* (RSM), that chooses the most appropriate routes, among those created, in order to minimize total costs and indicate the time windows for the customers. We compared the results of our algorithm (ILS + RSM) with the Branch-and-Cut proposed by Dalmeijer and Spliet [7]. The ILS + RSM presented competitive results, concerning both solution quality and computational effort, in particular for the larger size instances involving 45 and 50 customers.

Interesting avenues of further research concern: (i) incorporating new complicating constraints deriving from the real-world case study in the metaheuristic; (ii) testing other neighborhood-based metaheuristics as generators of routes; (iii) testing multiple calls to the RSM with different pools of routes. This last avenue is motivated by the fact that in our tests the RSM converged quickly to the incumbent solution, so there is hope to find good solution values by invoking it multiple times.

Acknowledgments The authors would like to thank the support of Coopservice S.Coop.p.A (Italy), of UNIMORE (Italy) under grant FAR 2018, and of the National Council for Scientific and Technological Development (CNPq, Brazil).

References

1. Amorim, P., Parragh, S.N., Sperandio, F., Almada-Lobo, B.: A rich vehicle routing problem dealing with perishable food: a case study. TOP **22**, 489–508 (2014)
2. Avci, M., Topaloglu, S.: A multi-start iterated local search algorithm for the generalized quadratic multiple knapsack problem. Comput. Operat. Res. **83**, 54–65 (2017)
3. Azi, N., Gendreau, M., Potvin, J.: An exact algorithm for a vehicle routing problem with time windows and multiple use of vehicles. European J. Operat. Res. **202**(3), 756–763 (2010)
4. Caceres-Cruz, J., Arias, P., Guimarans, D., Riera, D., Juan, A.A.: Rich vehicle routing problem: survey. ACM Comput. Surv. **47**(2), 32:1–32:28 (2014)
5. Ceschia, S., Gaspero, L.D., Schaerf, A.: Tabu search techniques for the heterogeneous vehicle routing problem with time windows and carrier-dependent costs. J. Sched. **14**(6), 601–615 (2011)
6. Cordeau, J.F., Laporte, G., Mercier, A.: A unified tabu search heuristic for vehicle routing problems with time windows. J. Oper. Res. Soc. **52**(8), 928–936 (2001)
7. Dalmeijer, K., Spliet, R.: A branch-and-cut algorithm for the time window assignment vehicle routing problem. Comput. Oper. Res. **89**, 140–152 (2018)
8. Desrochers, M., Desrosiers, J., Solomon, M.: A new optimization algorithm for the vehicle routing problem with time windows. Operat. Res. **40**(2), 342–354 (1992)
9. Gunawan, A., Lau, H.C., Lu, K.: An iterated local search algorithm for solving the orienteering problem with time windows. In: G. Ochoa, F. Chicano (eds.) Evolutionary Computation in Combinatorial Optimization, pp. 61–73. Springer International Publishing, Cham (2015)
10. Haddadene, S.R.A., Labadie, N., Prodhon, C.: A GRASP X ILS for the vehicle routing problem with time windows, synchronization and precedence constraints. Expert Syst. Appl. **66**, 274–294 (2016)
11. Keskin, M., Çatay, B.: A matheuristic method for the electric vehicle routing problem with time windows and fast chargers. Comput. Oper. Res. **100**, 172–188 (2018)
12. Kramer, R., Cordeau, J-F., Iori, M.: Rich vehicle routing with auxiliary depots and anticipated deliveries: An application to pharmaceutical distribution. Transport. Res. Part E **129**, 162–174 (2019)
13. López-Ibáñez, M., Dubois-Lacoste, J., Cáceres, L.P., Birattari, M., Stützle, T.: The irace package: iterated racing for automatic algorithm configuration. Oper. Res. Perspect. **3**, 43–58 (2016)
14. Lourenço, H.R., Martin, O.C., Stützle, T.: Iterated local search. In: G.K. F. Glover (ed.) Handbook of Metaheuristics, pp. 320–353. Springer, Berlin (2003)
15. Moreira, M.C.O., Costa, A.M.: Hybrid heuristics for planning job rotation schedules in assembly lines with heterogeneous workers. Int. J. Prod. Eco. **141**(2), 552–560 (2013)
16. Neves-Moreira, F., da Silva, D.P., Guimares, L., Amorim, P., Almada-Lobo, B.: The time window assignment vehicle routing problem with product dependent deliveries. Transport. Res. E-Log. **116**, 163–183 (2018)
17. Nogueira, B., Pinheiro, R.G.S., Subramanian, A.: A hybrid iterated local search heuristic for the maximum weight independent set problem. Optim. Lett. **12**(3), 567–583 (2018)
18. Spliet, R., Dabia, S., Woensel, T.V.: The time window assignment vehicle routing problem with time-dependent travel times. Transp. Sci. **52**(2), 261–276 (2018)
19. Spliet, R., Desaulniers, G.: The discrete time window assignment vehicle routing problem. Europ. J. Oper. Res. **244**(2), 379–391 (2015)
20. Spliet, R., Gabor, A.F.: The time window assignment vehicle routing problem. Transp. Sci. **49**(4), 721–731 (2014)
21. Subramanian, A., dos Anjos Formiga Cabral, L.: An ILS based heuristic for the vehicle routing problem with simultaneous pickup and delivery and time limit. In: J. van Hemert, C. Cotta (eds.) Evolutionary Computation in Combinatorial Optimization, pp. 135–146. Springer, Berlin (2008)

22. Toth, P., Vigo, D.: Vehicle Routing: Problems, Methods, and Applications, 2nd edn. SIAM, Philadelphia (2014)
23. Vidal, T., Crainic, T.G., Gendreau, M., Prins, C.: Heuristics for multi-attribute vehicle routing problems: a survey and synthesis. European J. Oper. Res. **231**(1), 1–21 (2013)
24. Vidal, T., Crainic, T.G., Gendreau, M., Prins, C.: A unified solution framework for multi-attribute vehicle routing problems. European J. Oper. Res. **234**(3), 658–673 (2014)
25. Zhigalov, G.: VRPTW. GitHub repository: https://github.com/donfaq/VRPTW, last commit: 0261f086d1610c2da3152540d0e535a4eaeef76c (2018)

Synchronized Pickup and Delivery Problems with Connecting FIFO Stack

Michele Barbato, Alberto Ceselli, and Nicolas Facchinetti

Abstract In this paper we introduce a class of routing problems where pickups and deliveries need to be performed in two distinct regions, and must be synchronized by considering the presence of a first-in-first-out channel linking them. Our research is motivated by applications in the context of automated warehouses management. We formalize our problem, defining eight variants which depend on the characteristics of both the pickup and delivery vehicles, and the first-in-first-out linking channel. We show that all variants are in general **NP**-hard. We focus on two of these variants, proving that relevant sub-problems can be solved in polynomial-time. Our proofs are constructive, consisting of resolution algorithms. We show the applicability of our results by computational experiments on instances from the literature.

Keywords Optimization in manufacturing · Routing · FIFO loading · Dynamic programming

1 Introduction

In the field of logistics the analysis of routing problems plays a fundamental role: when applied to real-world situations, it may lead to considerable cost savings for transportation companies operating over large regions, see [8].

Routing problems are often encountered on a smaller scale too. As a relevant case, in this paper we introduce the family of *synchronized pickup and delivery problems with connecting first-in-first-out stack* (SPDP-FS). It is inspired by the production system of an industrial partner on smart cosmetics manufacturing [1]. Overall, problems in such a setting concern the transportation of items between a pickup area and a delivery area inside the same factory. The two areas communicate

M. Barbato (✉) · A. Ceselli · N. Facchinetti
Università degli Studi di Milano, Dipartimento di Informatica, OptLab, Milan, Italy
e-mail: michele.barbato@unimi.it; alberto.ceselli@unimi.it; nicolas.facchinetti@studenti.unimi.it

© The Author(s), under exclusive license to Springer Nature Switzerland AG 2021
C. Gentile et al. (eds.), *Graphs and Combinatorial Optimization:*
from Theory to Applications, AIRO Springer Series 5,
https://doi.org/10.1007/978-3-030-63072-0_19

through a first-in-first-out (FIFO) conveyor. As a consequence the pickup and the delivery routes cannot be independently optimized. Instead they must be coordinated so as not to violate the FIFO policy, and potentially synchronized. In particular, in the application inspiring this work, the pickup vehicle is an automated crane which collects items from the warehouse; at the end of each trip, that is a sequence of pickup operations in which the crane starts empty, moves reaching and incrementally loading items finally bringing them to an unloading spot, the items are put on the conveyor in the same order they have been collected. The delivery vehicle is an automated shuttle which must wait for loading until a batch of items is put on the conveyor, but then might be free to follow any order to deliver them. Since the vehicle capacities are finite and typically much lower than the overall number of items to pickup and deliver (e.g., at most 2 or 3 items can be transported by a single trip) the resulting problem of optimizing the pickup and delivery routes while satisfying the constraints arising from the FIFO rule does not reduce in general to classical routing problem as, e.g., the travelling salesman problem [3].

In this paper we provide a modeling of the above situation by introducing SPDP-FS variants, each depending on the degree of freedom the vehicles have when loading items to and from the conveyor with respect to the order of visits of the pickup and delivery locations.

Contributions and Outline In Sect. 2 we formally define eight SPDP-FS variants and show that they are all in general **NP**-hard. We consider the case in which the capacities of the pickup and delivery vehicles are part of the input and study cases in which they are fixed parameters. In Sect. 3 we focus on the variants in which the order of the items on the conveyor must coincide with both their pickup and delivery orders. We study how to complete partial solutions. To this aim we first present an algorithm to construct a conveyor order which is consistent with two given sets of pickup and delivery trips; our algorithm thus determines the feasibility of the given sets of trips. Next, we present a dynamic programming algorithm to determine the optimal pickup and delivery trips which are consistent with a given conveyor order. In order to evaluate the suitability of the latter algorithm as a sub-routine for more sophisticated SPDP-FS heuristics, in Sect. 4 we test it on instances from the literature [7].

2 Modelling and Complexity

Throughout the paper, let $n \in \mathbb{Z}_{>0}$ and let $G = (V, A)$ be a complete digraph on the vertex set $V = \{0, 1, 2, \ldots, n\}$. The *pickup network* is a weighted digraph $D^1 = (G, c^1)$ with arc cost function $c^1 \colon A \to \mathbb{R}_+$ and whose vertices in $V \setminus \{0\}$ represent the positions of n items identical in terms of size. Intuitively, D^1 represents a storage area (a warehouse in the real application) for raw materials demanded by the production units of a factory. The *delivery network* is represented by another weighted digraph $D^2 = (G, c^2)$ with arc cost function $c^2 \colon A \to \mathbb{R}_+$. Intuitively, D^2

represents a production area. We assume that the production unit located at vertex $i \in V \setminus \{0\}$ of D^2 must receive the item located at vertex $i \in V \setminus \{0\}$ of D^1; in other words, items and production units are in one-to-one correspondence.

The items are collected by a vehicle of finite capacity k_1 by visiting the corresponding vertices of D^1. The capacity is finite, and typically lower than $|V \setminus \{0\}|$: in order to pick up all items, the vehicle performs several trips, where a *trip* is a sequence of vertices forming a simple directed cycle starting and ending at vertex 0. Once items have been collected, they are delivered in an analogous manner in D^2 by a second vehicle of capacity k_2.

Overall, a solution (P, D) consists of an ordered sequence of pickup trips $P = (p_1, p_2, \ldots, p_\ell)$ and of an ordered sequence of delivery trips $D = (d_1, d_2, \ldots, d_m)$ that must be synchronized: indeed the items are passed from the pickup area to the delivery area by means of a FIFO stack (a horizontal conveyor in the real application) whose input (resp. output) extremity is located at vertex 0 of D^1 (resp. D^2). Thus items collected by a pickup trip must be put on the stack before those collected by subsequent pickup trips. By the FIFO policy a delivery trip d_i preceding another delivery trip d_j must deliver items that, on the stack, precede those delivered by d_j.

For a trip t we indicate as $a \in t$ any arc between vertices which are consecutive in t. For every trip t and for $R = 1, 2$ let $c^R(t) = \sum_{a \in t} c^R(a)$ and for every sequence of pickup trips P let $c^1(P) = \sum_{i=1}^{\ell} c^1(p_i)$; symmetrically, $c^2(D) = \sum_{i=1}^{m} c^2(d_i)$. The objective of a problem in the SPDP-FS family is to find a solution (P, D) that minimizes $c^1(P) + c^2(D)$.

Example 1 In Fig. 1 we illustrate a small instance of a problem belonging to the SPDP-FS family. A solution is depicted in boldface: the pickup vehicle (left) performs first trip $(1, 5, 4)$ then trip $(2, 3)$. Hence items in $\{1, 4, 5\}$ must be delivered before items in $\{2, 3\}$. In the example they are in fact arranged in the order of visit. This has an impact on the delivery solution. In the most constrained setting, the delivery vehicle is forced to the inconvenient trip $(1, 5, 4)$ to respect such an order, as shown in Fig. 1. In fact, the order of visits is not the only degree of freedom of the vehicles, which may or may not be allowed to permute items on the stack after pickup or before delivery. This is what we call the *permutation* variant. For example,

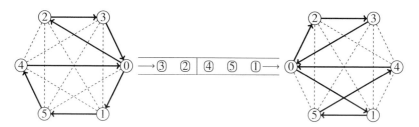

Fig. 1 Schematic illustration of the automated warehouse. The pickup area is on the left, delivery area on the right. Both vehicles are assumed to have capacity 3. No permutations and no overlaps are allowed

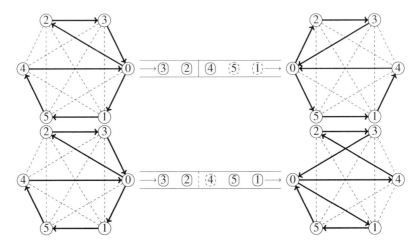

Fig. 2 Schematic illustration of the automated warehouse, when permutations are allowed (top) and when overlaps are allowed (bottom)

in Fig. 2 (top) the delivery vehicle is allowed to retrieve the items $\{1, 4, 5\}$ from the stack, changing their order from $(1, 5, 4)$ to $(5, 1, 4)$ thereby removing the need of a detour. Such an option may be symmetrically given to the pickup vehicle.

Besides permuting, delivery vehicles may or may not be allowed to mix items from subsequent pickup batches. This is what we call the *overlap* variant. For example, in Fig. 2 (bottom) the delivery vehicle is allowed to deliver $(1, 5)$, wait for the next pickup batch to arrive, and finally deliver $(4, 2, 3)$, possibly improving delivery cost. Permutation and overlap may or may not be allowed simultaneously, giving more optimization potential at the expense of higher complexity in the real world process. For instance, the trip $(4, 2, 3)$ in the last example may be replaced by $(4, 3, 2)$, thus removing a detour. Pickup permutation and delivery permutation are not equivalent. For instance, a delivery trip like $(5, 4, 3)$ is feasible only due to pickup permutations, while a delivery trip like $(5, 2, 4)$ only due to delivery permutations.

The SPDP-FS Variants As previously mentioned, the synchronization between two sequences of pickup and delivery trips in a SPDP-FS solution may depend not only on the order in which items are collected, but also on how items are arranged on the stack. We therefore identify eight variants mixing the possibility of permuting items by each vehicle (or not) and the presence of constraints allowing (or disallowing) delivery trips to be overlapping different pickup trips. When permutation is allowed we are essentially modelling situations where the items can be loaded or unloaded without observing any particular order; when overlapping is disallowed, we are modelling situations where the stack must be empty before a new pickup trip can take place.

Formally, for every trip t let $V(t)$ be the set of vertices other than 0 visited by t. Let $P = (p_1, p_2, \ldots, p_\ell)$ be an ordered sequence of pickup trips such that the $V(p_i)$ form a partitioning of $V \setminus \{0\}$. P induces a partial order: if $v, w \in V \setminus \{0\}$, we write $v \prec_P w$ if $v \in p_i$ and $w \in p_j$ for some $1 \leq i < j \leq \ell$ and we write $v \not\prec_P w$ otherwise. A P-sequence, instead, is the (totally) ordered sequence of vertices in $V \setminus \{0\}$, in the order they are visited in P. Identical definitions hold for a partial order $\prec_D$ and a D-sequence w.r.t. a fixed ordered sequence of delivery trips $D = (d_1, d_2, \ldots, d_m)$. (P, D) forms a feasible SPDP-FS solution if it satisfies:

$$|V(p_i)| \leq k_1 \qquad\qquad \forall i = 1, 2, \ldots, \ell \qquad (1)$$

$$|V(d_j)| \leq k_2 \qquad\qquad \forall j = 1, 2, \ldots, m \qquad (2)$$

$$V(p_1), V(p_2), \ldots, V(p_\ell) \text{ partition } V \setminus \{0\} \qquad (3)$$

$$V(d_1), V(d_2), \ldots, V(d_m) \text{ partition } V \setminus \{0\} \qquad (4)$$

Intuitively, SPDP-FS OVERLAP variants are those in which the FIFO stack is *not* required to be empty before a new pickup trip can take place. As a result, delivery trips can mix items from different pickup batches, as long as they are adjacent in the stack. Formally, the OVERLAP variants are defined as follows:

NO-PERMUTATION,OVERLAP.   The pair (P, D) is a feasible solution if and only if it satisfies (1)–(4) and the P-sequence is identical to the D-sequence.

PERMUTATION,OVERLAP.   The pair (P, D) is a feasible solution if and only if it satisfies (1)–(4) and for every $v, w \in V \setminus \{0\}$ such that $v \prec_P w$ it also holds $w \not\prec_D v$.

DELIVERY PERMUTATION,OVERLAP.   The pair (P, D) is a feasible solution if and only if it satisfies (1)–(4) and:

- for every $j = 1, 2, \ldots, m$, $V(d_j)$ is a set of elements which are consecutive in the P-sequence;
- for every $v, w \in V \setminus \{0\}$, if $v \prec_D w$ then v precedes w in the P-sequence.

PICKUP PERMUTATION,OVERLAP.   The pair (P, D) is a feasible solution if and only if it satisfies (1)–(4) and for every $v, w \in V \setminus \{0\}$ such that $v \prec_P w$ we also have that v precedes w in the D-sequence.

Symmetrically, in NO-OVERLAP variants the FIFO stack is required to be empty before new pickup trips, and therefore delivery trips do not mix items from different pickup trips. Their models are obtained from the four above by further adding the condition: $\forall j = 1, 2, \ldots, m$, $\exists$ a unique $i \in \{1, 2, \ldots, \ell\}$ s.t. $V(d_j) \subseteq V(p_i)$.

When the capacity values k_1 and k_2 are part of the input all SPDP-FS variants above are **NP**-hard. We reduce from the *Euclidean travelling salesman problem* (Euclidean-TSP) of which an instance is given by a complete graph $H = (W, E)$ with each vertex $w \in W$ corresponding to a point $\pi(w)$ in the Euclidean plane. Letting $c(v, w)$ be the Euclidean distance between $\pi(v)$ and $\pi(w)$, the Euclidean-TSP asks to find a minimum weight Hamiltonian tour on the weighted graph (H, c).

Note that function c is *metric*, that is, it satisfies the *triangle inequality* $c(u, v) + c(v, w) \geq c(u, w)$. Nonetheless, the Euclidean-TSP is known to be **NP**-hard [6].

Proposition 1 *If k_1 and k_2 are part of the input, the SPDP-FS is **NP**-hard.*

Proof Let (H, c) be a weighted complete graph on n vertices defining an Euclidean-TSP instance. We define a SPDP-FS instance on $D^1 = (H, c^1)$ and $D^2 = (H, c^2)$ with $k_1 = n$, $k_2 = 1$ and $c^1(u, v) = c(u, v)$ and $c^2(u, v) = 0$ for all u, v vertices of H. Since c^1 is metric and $k_1 = n$ a minimum cost Hamiltonian tour P of D^1 is also a minimum cost pickup sequence satisfying (1). Let the P-sequence be $(v_1, v_2, \ldots, v_n)$. The delivery sequence $D = ((v_1), (v_2), \ldots, (v_n))$ has cost 0; moreover (P, D) is feasible for all SPDP-FS variants. Hence the optimal value of the given SPDP-FS instance is the optimal value of the starting Euclidean-TSP instance.

Some variants of the SPDP-FS become solvable in polynomial time when the capacities k_1 and k_2 are very specific fixed values. The most immediate case is when $k_1 = k_2 = 1$ (independently on the variant): here all trips of a sequence visit exactly one vertex $v \in V \setminus \{0\}$. Then, all pairs of sets $\mathscr{P} = \{p_1, p_2, \ldots, p_\ell\}$ of pickup trips and $\mathscr{D} = \{d_1, d_2, \ldots, d_m\}$ of delivery trips, when an arbitrary (but identical) ordering is applied to both, yield a feasible pair (P, D) of pickup and delivery sequences for every variant. Moreover, the cost of such a solution does not depend on the chosen order.

The cases with larger fixed values of k_1 and k_2 are more involved. In fact, we are able to provide a polynomial-time algorithm only for the NO-OVERLAP variants with $k_1, k_2 \in \{1, 2\}$.

Proposition 2 *If $k_1, k_2 \in \{1, 2\}$ there exists a polynomial-time algorithm solving the NO-OVERLAP variants of the SPDP-FS.*

Proof We start by calculating for all $v \in V \setminus \{0\}$ the total cost $c[v]$ of visiting only v in a pickup trip and a delivery trip, i.e., $c[v] = c^1((v)) + c^2((v))$. We also compute, for every $v, w \in V \setminus \{0\}$, the minimum total cost $c[v, w]$ of visiting v and w in the same pickup trip and of delivering them in any manner by satisfying the NO-OVERLAP variant under consideration. Let $t[v, w]$ denote the sequences of pickup and delivery trips attaining value $c[v, w]$. The whole pre-processing phase takes $O(n^2)$ time.

Now, for every $v \in V \setminus \{0\}$ we let v' and v'' be two copies of v. We consider H' and H'' two complete graphs on vertex sets $W' = \{v' : v \in V \setminus \{0\}\}$ and $W'' = \{v'' : v \in V \setminus \{0\}\}$ respectively. From the union of H' and H'' we create a new graph H obtained by further linking vertices v' and v'' for every $v \in V \setminus \{0\}$. Finally, we associate weights s to the edges of H with $s_e = c[v, w]$ if $e = \{v', w'\}$ for some $v, w \in V \setminus \{0\}$, $s_e = c[v]$ if $e = \{v', v''\}$ for some $v \in V \setminus \{0\}$ and $s_e = 0$ otherwise. A perfect matching of minimum weight in (H, s) corresponds to an optimal solution for the considered NO-OVERLAP variant: just perform the trips in $t[v, w]$ and trip (u) for every $u, v, w \in V \setminus \{0\}$ such that $\{v', w'\}$ and $\{u', u''\}$ respectively belong to such a perfect matching. The proposition holds since a perfect matching of minimum weight can be computed in polynomial time on every graph [4].

The algorithm provided in the proof of Proposition 2 easily generalizes to all SPDP-FS variants if $k_1, k_2 \in \{1, 2\}$ and at least one capacity value is 1. However, when $k_1 = k_2 = 2$, the same algorithm does not work outside NO-OVERLAP variants.

The last observation suggests that NO-OVERLAP variants may have some structural feature which is not in common with the others. Supporting this intuition, it is not hard to prove that (P, D) is a feasible solution to a PERMUTATION,NO-OVERLAP problem if and only if it is a feasible solution to the corresponding PICKUP PERMUTATION,NO-OVERLAP problem. This immediately implies that if (P, D) is a feasible solution to the DELIVERY PERMUTATION,NO-OVERLAP problem then it is a solution to the corresponding PICKUP PERMUTATION,NO-OVERLAP problem. Such a phenomenon does not occur on OVERLAP variants.

In fact, we have analysed the inclusion relations between the sets of solutions of the eight variants building a full set-inclusion hierarchy. To keep the focus of the paper we omit these results. However, our main conclusion is that NO-PERMUTATION variants take a relevant place in such a hierarchy, and are therefore good candidates to start a structural investigation on the whole SPDP-FS family.

3 Algorithms for Sub-problems of NO-PERMUTATION Variants

From now on we restrict to NO-PERMUTATION variants; in particular, we consider relevant subproblems arising when only part of a solution is given, and either feasibility must be checked or optimal completion must be found.

NO-PERMUTATION Feasibility Let us assume to have a set $\mathscr{P} = \{p_1, p_2, \ldots, p_\ell\}$ of pickup trips and a set $\mathscr{D} = \{d_1, d_2, \ldots, d_m\}$ of delivery trips such that (1) and (2) hold. The *feasibility problem* is to independently find an ordering of the elements of $\mathscr{P}$ and $\mathscr{D}$ so that the resulting sequences P and D represent a feasible solution, or to prove that no such ordering exists.

We start with the NO-PERMUTATION,OVERLAP case. Let us denote as *starting* each vertex which is the first vertex of two trips in $\mathscr{P} \cup \mathscr{D}$. A trip t' is said to be *contained* in t if it is a subsequence of t. Two trips $t = (u_1, u_2, \ldots, u_k, v_1, v_2, \ldots, v_\ell)$ and $t' = (v_1, v_2, \ldots, v_\ell, w_1, w_2, \ldots, w_m)$ are said *overlapping*: they respectively end and start by a same non-empty subsequence; note that there can be at most one trip overlapping with a given trip t. We can solve the feasibility problem for $\mathscr{P}$ and $\mathscr{D}$ in polynomial time. We omit the details of the algorithm, but its general idea is to consider a set $T = \mathscr{P} \cup \mathscr{D}$, and then to pick a starting vertex v and one of the trips t containing it, removing both t and all trips t' contained in t from T, to either update t with a possible trip overlapping with t or pick another starting vertex and to iterate, until no update is possible. The problem is feasible if and only if $T = \emptyset$ at the end of this procedure.

In the NO-PERMUTATION,NO-OVERLAP case we first check that every pair of trips $p \in \mathscr{P}$ and $d \in \mathscr{D}$ such that $V(p) \cap V(d) \neq \emptyset$ also satisfies $V(d) \subseteq V(p)$. If at least one pair violates the condition then $\mathscr{P}$ and $\mathscr{D}$ are infeasible. Otherwise we run the iterative algorithm for the NO-PERMUTATION,OVERLAP case.

NO-PERMUTATION Splitting Now we turn our attention to the following subproblem: a stack configuration is given as input (together with data), and the task is to find pickup and delivery sequences of minimum cost which are *consistent with the given configuration*. In other words, we assume that the pickup and delivery sequences are fixed, while the depot return operations forming trips remain to be optimized. Let $S = (v_1, \ldots, v_n)$ be a sequence of elements in $V \setminus \{0\}$. It represents the order of the corresponding items on the stack. A *splitting* of S is a partition of S into subsequences that respect its order. For example, if $S = (1, 2, 3, 4, 5)$ one of its splitting is $((1, 2), (3, 4, 5))$, while $((3, 4, 5), (1, 2))$ is not because it does not respect the order of S. Here we consider both NO-PERMUTATION variants, thus the trips of P and D in a feasible SPDP-FS solution (P, D) are splittings of $(v_1, v_2, \ldots, v_n)$. The *optimal sequence splitting problem* is to find two splittings P and D of $(v_1, v_2, \ldots, v_n)$, minimizing $c^1(P) + c^2(D)$ and such that (P, D) is feasible for the considered NO-PERMUTATION variant. Up to renaming the vertices, we give polynomial-time algorithms to solve this problem for $S = (1, 2, \ldots, n)$.

For what concerns NO-PERMUTATION,NO-OVERLAP, we use an acyclic network $\mathscr{N}$ whose vertices are labelled (i, j) with $0 \leq j \leq i \leq n$. For every $i = 0, 1, \ldots, n$ there are arcs from vertex (i, i) to all vertices (k, i) with $i < k \leq \min\{i + k_1, n\}$. For every $i = 1, 2, \ldots, n$ and $j = 1, 2, \ldots, i - 1$ there are arcs from vertex (i, j) to all vertices (i, k) with $j < k \leq \min\{j + k_2, i\}$. Vertex (i, j) is interpreted as a state in the construction of the optimal pickup and delivery splittings P and D, namely it indicates that all vertices from 1 to i have been visited by some pickup trip and all vertices from 1 to j have been visited by some delivery trip. Thus, the arc from (i, i) to $(i+k, i)$ for some $1 \leq k \leq k_1$ corresponds to extending the current state (i, i) into state $(i + k, i)$ by performing $p = (i + 1, i + 2 \ldots, i + k)$ as next pickup trip. Similarly, the arc from (i, j) to $(i, j + k)$ corresponds to performing $d = (j + 1, j + 2, \ldots, j + k)$ as next trip in the optimal delivery sequence D. By construction, $|V(p)| \leq k_1$, $|V(d)| \leq k_2$ for all pickup and delivery trips p and d as above. Hence a path from $(0, 0)$ to (n, n) corresponds to a pair (P, D) of pickup and delivery splittings of S; moreover, the P-sequence and D-sequence of these splittings coincide. Finally, representing $\mathscr{N}$ as in Fig. 3 (left) the arcs traversing $\mathscr{N}$ "vertically" only leave from vertices on the "diagonal". This property guarantees that each trip in the delivery splitting is contained in a trip of the pickup splitting. To see this, let us fix a path in $\mathscr{N}$. This gives a splitting of S into pickup and delivery trips. Let $d = (j_1 + 1, \ldots, j_2)$ be one such a delivery trip. It corresponds to the arc of the path linking (i, j_1) to (i, j_2) for some $i \geq j_2$. Letting j_0 be the smallest index such that (i, j_0) is traversed by the chosen path, the vertex preceding (i, j_0) in the path is necessarily (j_0, j_0). It follows that the pickup splitting of S contains

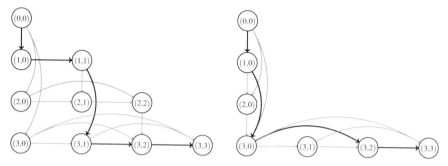

Fig. 3 Networks $\mathcal{N}$ for the optimal sequence splitting problem with $n = k_1 = k_2 = 3$ in the cases NO-PERMUTATION,NO-OVERLAP (left) and NO-PERMUTATION,OVERLAP (right). The boldface path in the left network corresponds to the splittings $P = ((1), (2, 3))$ and $D = ((1), (2), (3))$. The boldface path in the right network corresponds to the splitting $P = ((1), (2, 3))$ and $D = ((1, 2), (3))$

the trip $(j_0 + 1, \ldots, i)$. Since $j_0 \leq j_1$ and $i \geq j_2$ we get $V(d) \subseteq V(p)$. It follows that every path in $\mathcal{N}$ from $(0, 0)$ to (n, n) corresponds to a splitting of S which is also a NO-PERMUTATION,NO-OVERLAP solution. In fact the latter is a one-to-one correspondence as it can be easily verified by an analogous reasoning. An example is given in Fig. 3 (left).

In a pre-processing step we calculate, in polynomial time, the cost $c^1[i_1, i_2]$ of the trip corresponding to each arc between (i_1, j) and (i_2, j) as well as the cost $c^2[j_1, j_2]$ of the trip corresponding to each arc between (i, j_1) and (i, j_2). Associating these costs to the corresponding arcs in $\mathcal{N}$ we obtain an acyclic digraph with nonnegative weights. A path of minimum weight from $(0, 0)$ to (n, n) gives an optimal splitting of S by the arc-trip correspondence explained above. Being $\mathcal{N}$ acyclic, very efficient algorithms can be used to compute such a path [5, Sect. 24.2].

For what concerns NO-PERMUTATION,OVERLAP, instead, we use a network $\mathcal{N}$ whose vertices are labelled $(i, 0)$ and (n, j) for $0 \leq i, j \leq n$. For every $i = 0, 1, \ldots, n$ there is an arc from vertex $(i, 0)$ to all vertices $(k, 0)$ with $i < k \leq \min\{i + k_1, n\}$; similarly, for every $j = 0, 1, \ldots, n$ there is an arc from (n, j) to all vertices (n, k) with $j < k \leq \min\{j + k_2, n\}$. An example is given in Fig. 3 (right). The arc-trip correspondence is the same as in the NO-PERMUTATION,NO-OVERLAP case. Associating costs to the arcs as we did for that variant, we compute an optimal pair of pickup and delivery splittings by finding a minimum cost shortest path in $\mathcal{N}$.

4 Computational Results

Finally, we carried out experiments to assess the practical applicability of the optimal splitting algorithms presented in Sect. 3 for both NO-PERMUTATION variants.

Algorithms We designed dynamic programming algorithms which are able to compute an optimal sequence splitting when a stack configuration is given.

Their structure is simple: once the appropriate network $\mathcal{N}$ is built, we consider the vertices in the topological order induced by their labels, that is, row-by-row and, for each row, column-by-column. Then, we assign *partial splitting costs* to each vertex, with a procedure similar to [5, Sect. 24.2]. The partial splitting cost of the starting vertex $(0, 0)$ is initialized to 0. When a vertex is considered, its partial splitting cost is set by looking at all its incoming arcs, and computing the minimum among the cost of each arc plus the partial splitting cost of the corresponding starting vertex. At the end of this procedure, the partial splitting cost of the vertex (n, n) corresponds to an optimal (complete) one. A corresponding solution can be found by keeping track of the arcs defining the minimum.

For testing, we took as starting stack configuration the sequence of items corresponding to an optimal Hamiltonian tour of a complete graph having one vertex for each item, and one arc (i, j) between each pair of items i and j, whose cost is the sum of pickup and delivery distances between i and j. Clearly, an optimal splitting of such a Hamiltonian tour is not guaranteed to produce and optimal SPDP-FS solution. Therefore we have also experimented with a simple 2-opt local search mechanism: we generate the full 2-opt neighborhood of the current solution and score it by solving an optimal splitting problem for each of its elements. We take the best solution in the neighborhood and we iterate, until a local minimum is reached.

As a trivial bound, we computed the optimal Hamiltonian tours of pickup and delivery item graphs independently, and summed up their values: this is clearly a relaxation of the problem, neglecting both the vehicle capacities and the effect of the linking stack. Our algorithms have been implemented in C++ and compiled with gcc version 7.2 with full optimization options. Optimal Hamiltonian tours have been computed by means of the Concorde library [2]. Tests were run on a linux PC equipped with an i7-3630QM CPU running at 2.40 GHz, single thread.

Datasets We considered a dataset of pickup and delivery instances originally designed for the double travelling salesman problem with multiple stacks [7]. Our dataset includes three classes of instances, respectively containing 33, 66 and 132 *pairs* of pickup and delivery vertices. Ten instances are given for each class. Distances were set as the euclidean ones computed from the input coordinates, rounded to the nearest integer and finally corrected with an all-pair shortest path procedure to ensure triangle inequalities to be respected. For each instance having 33 (resp. 66) pairs of vertices, we considered pickup vehicle capacities from 3 to 33 (resp. 66) in steps of 3, and delivery vehicle capacities from 3 to the pickup capacity value in steps of 3. For each instance having 132 pairs of vertices we considered instead capacities from 6 to 132 in steps of 6, and delivery vehicle capacities from 6 to the pickup capacity value in steps of 6. This procedure yielded a dataset of 11,440 instances for each variant.

Computing Time The computing time for solving a single optimal splitting problem was always negligible (below $0.01s$ for both variants, independently on

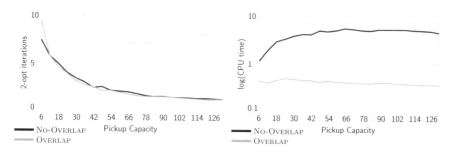

Fig. 4 Performance of heuristics

the instance size and capacity). The computing time for the full 2-opt local search process was also negligible on instances with either 33 or 66 pairs of vertices, independently from the pickup and delivery capacity values. Therefore, in Fig. 4 (left) we report the computing time (y axis) for the full 2-opt process only on instances with 132 pairs of vertices (using logarithmic scale, on both variants), averaged by pickup capacity value (x axis); in Fig. 4 (right) we report instead the number of 2-opt moves required to reach a local minimum (y axis), on the same instances and performing the same aggregation. Values related to the OVERLAP variant are depicted in grey, those related to NO-OVERLAP in black.

As expected, higher capacity values yield higher computing times: the optimal splitting graphs are more dense, and more checks are needed in the dynamic programming algorithms. The number of 2-opt moves required to reach a local minimum decreases as the capacity increases. We argue that higher capacity values yield less overhead costs for the vehicle to return to the stack; as a consequence, being the routing part of the cost dominant, the Hamiltonian tour is already similar to a local minimum. The increase of overall CPU time is not monotone: for very high pickup capacity values the reduced number of 2-opt moves balances the higher CPU time needed during a single iteration.

Solutions Quality We denote as IS the value of the initial solution, as DA the sum of costs of arcs leaving and returning to the depot for intermediate stops in such a solution, as OS the solution value after 2-opt, and TB the value of the trivial bound.

Table 1 Quality of heuristics and bounds

Variant	Size	Avg. overhead	Avg. impr.	Avg. opt.
NO-OVERLAP	33	10.05%	0.84%	47.14%
	66	8.17%	0.61%	54.27%
	132	5.10%	0.41%	59.65%
OVERLAP	33	6.50%	0.76%	45.09%
	66	5.56%	0.52%	53.00%
	132	3.66%	0.33%	59.07%

In Table 1 we report in turn, for each class of instances, for both OVERLAP and NO-OVERLAP variants, the average fraction of the initial solution cost for the vehicle to return to the stack with intermediate stops (overhead, DA/IS), the average overall improvement yielded by 2-opt (impr., (TO-IS)/TO) and the average gap between the value of the solution found by 2-opt and the trivial bound (opt., (TO-TB)/TO).

The results show that the overhead decreases as the number of items increases; this is most probably a statistical side effect: when the set of items is large, and the depot is in the baricenter of the item locations (as in our instances) the probability of passing near the depot in a random connection is higher. The average improvement yielded by 2-opt is always very low, and decreasing as the number of items increases. At the same time, the gap between local minima solution values and the trivial bound is very large, and increases as the number of items increases. The values in these three columns make us conjecture that by computing an optimal Hamiltonian tour, and then optimally split it, good heuristic solutions can be achieved. However, we suspect the trivial bound to be very poor, thereby asking for better lower bounding procedures.

5 Conclusions

From a modelling point of view, we restricted the complexity of SPDP-FS from industry by considering two peculiar features: the possibility of changing the order of items during loading or unloading operations to and from the conveyor, and the possibility of starting pickup trips even if the conveyor is not empty.

We have shown that all variants arising from the combination of these features are **NP**-hard, although some of them admit polynomial-time resolution algorithms when the capacities of the vehicles are fixed to small values i.e., 1 or 2.

By focusing on the NO-PERMUTATION variants, we could find polynomial-time algorithms for two relevant sub-problems: checking feasibility when the set of pickup and delivery trips are given (but the order of items in the conveyor is unknown), and optimizing the pickup and delivery trips when the order of items in the conveyor is given.

Our algorithms proved to be effective also from an experimental point of view: tests on instances of the double travelling salesman problem with multiple stacks showed that a single sub-problem resolution can be carried out in fractions of a second; their embedding in a simple local search algorithm provided promising results.

Since, on the contrary, trivial lower bounds appear very weak, our current research is focused on the design of good quality ones.

Acknowledgments This research was partially funded by Regione Lombardia, grant agreement n. E97F17000000009, project AD-COM, and Università degli Studi di Milano, Dipartimento di Informatica, Piano Sostegno alla Ricerca 2016–2020.

References

1. AD-COM: ADvanced Cosmetic Manufacturing (2020). https://ad-com.net/
2. Applegate, D.L., Bixby, R.E., Chvatal, V., Cook, W.J.: Concorde (2003). http://www.math. uwaterloo.ca/tsp/concorde.html
3. Applegate, D.L., Bixby, R.E., Chvatal, V., Cook, W.J.: The Traveling Salesman Problem: A Computational Study. Princeton University Press, Princeton (2006)
4. Edmonds, J.: Maximum matching and a polyhedron with 0, 1-vertices. J. Res. Natl. Bur. Stan. B **69**(125–130), 55–56 (1965)
5. Leiserson, C.E., Rivest, R.L., Cormen, T.H., Stein, C.: Introduction to Algorithms, vol. 6. MIT Press, Cambridge (2001)
6. Papadimitriou, C.H.: The Euclidean travelling salesman problem is NP-complete. Theor. Comput. Sci. **4**(3), 237–244 (1977)
7. Petersen, H.L., Madsen, O.B.: The double travelling salesman problem with multiple stacks– formulation and heuristic solution approaches. Eur. J. Oper. Res. **198**(1), 139–147 (2009)
8. Toth, P., Vigo, D.: The Vehicle Routing Problem, 2nd edn. SIAM, Philadelphia (2014)

A Comparison Between Simultaneous and Hierarchical Approaches to Solve a Multi-Objective Location-Routing Problem

Aydin Teymourifar, Ana Maria Rodrigues, and José Soeiro Ferreira

Abstract This paper deals with a multi-objective location-routing problem (MO-LRP) and follows the idea of sectorization to simplify the solution approaches. The MO-LRP consists of sectorization, sub-sectorization, and routing sub-problems. In the sectorization sub-problem, a subset of potential distribution centres (DCs) is opened and a subset of customers is assigned to each of them. Each DC and the customers assigned to it form a sector. Afterward, in the sub-sectorization stage customers of each DC are divided into different sub-sector. Then, in the routing sub-problem, a route is determined and a vehicle is assigned to meet demands. To solve the problem, we design two approaches, which adapt the sectorization, sub-sectorization and routing sub-problems with the non-dominated sorting genetic algorithm (NSGA-II) in two different manners. In the first approach, NSGA-II is used to find non-dominated solutions for all sub-problems, simultaneously. The second one is similar to the first one but it has a hierarchical structure, such that the routing sub-problem is solved with a solver for binary integer programming in MATLAB optimization toolbox after solving sectorization and sub-sectorization sub-problem with NSGA-II. Four benchmarks are used and based on a comparison

A. Teymourifar (✉)
INESC TEC – Institute for Systems and Computer Engineering, Technology and Science, Porto, Portugal
e-mail: aydin.teymourifar@inesctec.pt

A. M. Rodrigues
INESC TEC – Institute for Systems and Computer Engineering, Technology and Science,
CEOS.PP – Center for Organizational and Social Studies of Porto Polytechnic, Porto, Portugal
e-mail: ana.m.rodrigues@inesctec.pt

J. S. Ferreira
INESC TEC – Institute for Systems and Computer Engineering, Technology and Science, FEUP
– Faculty of Engineering, University of Porto, Porto, Portugal
e-mail: jsf@inesctec.pt

© The Author(s), under exclusive license to Springer Nature Switzerland AG 2021
C. Gentile et al. (eds.), *Graphs and Combinatorial Optimization:*
from Theory to Applications, AIRO Springer Series 5,
https://doi.org/10.1007/978-3-030-63072-0_20

251

between the obtained results it is shown that the first approach finds more non-dominated solutions. Therefore, it is concluded that the simultaneous approach is more effective than the hierarchical approach for the defined problem in terms of finding more non-dominated solutions.

Keywords Location-routing problems · Sectorization · Routing · Evolutionary algorithm · Non-dominated sorting genetic algorithm · Binary integer programming

1 Introduction

Sectorization is generally related to geographical aspects and has many applications in dividing a large political territory or districts of sales, airspace, municipality, healthcare, electric power, emergency service, internet networking, police patrol, public transportation network, social facilities, collection and transportation of solid waste in municipalities, etc., into smaller regions [2, 4, 5, 11, 12]. The equilibrium of load, distance, client, contiguity, and compactness are the criteria which are generally considered in sectorization [14]. The concept of sectorization, is similar to clustering though have significant differences. Clustering strives for inner homogeneity of data while sectorization aims at the outer similarity. Therefore models for solving both problems are in general not compatible [6].

One of the problems that is related to territorial design is the location-routing problem (LRP). In the literature, it is stated that LRP consists of two difficult problems as the facility location problem (FLP) and also the vehicle routing problem (VRP) [7]. Many methods have been proposed to solve different types of LRP, which is an NP-hard problem [13]. In this work, we deal with a multi-objective location-routing problem (MO-LRP), where there are a set of potential DCs and also a set of customers in different geographical locations with known demands. Unlike previous studies, we model it as a problem consisting of sectorization, sub-sectorization, and routing sub-problems. In the sectorization stage, a subset of potential DCs is selected to be opened and customers are divided among them. Each DC and its assigned customers form a sector. In the sub-sectorization stage the customers of each DC are divided into sub-sectors. To meet their demands a fleet is considered. Starting from and returning to a DC, a route is defined for each sub-sector [8, 14–16].

Previously MO-LRP has been solved based on sectorization concept; for example, Barreto et al. [1] integrate some hierarchical and non-hierarchical clustering techniques into a sequential heuristic algorithm to solve this problem. Martinho et al. [8] propose a method consisting of pre-sectorization, sectorization, routing, and multi-criteria evaluation phases to deal with multi-criteria and large dimensions of the capacitated location-routing problem (CLRP). Different from these studies, we design two new approaches adapting the non-dominated sorting genetic algorithm (NSGA-II) with sectorization, sub-sectorization and routing sub-problems and solve MO-LRP by them. In the first one, all sub-problems are solved simultaneously with NSGA-II. The second approach is similar to the first one however, it has a hierarchical structure such that after creation of sectors and sub-sectors with NSGA-II, routes are defined using a binary integer programming solver for the obtained Pareto solutions. It should be noted that the operators of NSGA-II used in both approaches are the same as in the algorithm proposed by Deb et al. [3]. Also, in this study, new chromosome representation, crossover and mutation operators are designed and used in NSGA-II to solve MO-LRP, which comprise sectorization, sub-sectorization and routing stages. The operators can be used in similar evolutionary algorithms to solve the problem based on the sectorization concept.

A comparison is made over four benchmarks, based on the number of non-dominated solutions acquired by the approaches. The results show that the first approach is able to find more non-dominated solutions.

The other sections of the study are such that the problem and proposed approaches are described in Sects. 2 and 3. The experimental results, conclusion and future works are the last two sections of the work.

2 Problem Description

In this section, we describe the problem, which is solved by both approaches. In the problem, there are some potential DCs and also some customers. A subset of DCs is opened and a subset of customers is assigned to each of them. Then routes are defined to meet the demands of customers. Each route starts from a DC and returns to the same DC by visiting a subset of the assigned customers. There is a cost to open each DC and it is desired to minimize the total cost of opening DCs. As described in Sect. 1, we name each opened DC and the customers assigned to it a sector. The resulting sectors are desired to be balanced both in terms of customers' demands and distance on routes, which are defined as the standard deviations of demands and distances in sub-sectors. In addition, in the formed sub-sectors, customers are desired to be quite close to the center, which is defined as the compactness of sub-sectors.

In Fig. 1, an illustrative example is presented, where DCs and customers are shown with squares and circles, respectively. The squares shown in green and blue are the opened ones. Each DC and the assigned customers form a sector, which are shown with the same color. In this example, each sector is divided into two sub-sectors, and a route is defined for each one. For instance, the routes denoted by dark and light green, are determined for the sub-sectors with green DC.

Some of the terminology and notations used in the paper are summarized in Table 1.

Fig. 1 An illustrative example

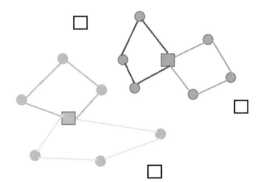

Table 1 Used notations

f_1	Total cost to open DCs
f_2	Standard deviation of demands in sub-sectors
f_3	Standard deviation of compactness of sub-sectors
f_4	Standard deviation of distances in sub-sectors
$i, j \in \bar{I} = \{1, \dots, I\}$	Index of all customers
$k \in \bar{K} = \{1, \dots, K\}$	Index of DCs
$m \in \bar{M} = \{1, \dots, M\}$	Index of sub-sectors
CO_k	Opening cost of DC k
DE_i	Demand of customer i
DE^m	Total demand of customers in sub-sector m
DS_{ij}	Distance of path from customer i to customer j
DS^m	Total distance along the route in sub-sector m
CE^m	Total distance between the centroid and customers in sub-sector m
CE^m_{max}	Distance between the centroid and farthest customer in sub-sector m
CP^m	Compactness of sub-sector m
VC	Capacity of each vehicle
x_k	Decision variable about if DC k opened or not
y_i^m	Decision variable about if customer i belongs to sub-sector m or not
z_{ij}^m	Decision variable about if there is a path from customer i to customer j on the defined route for sub-sector m

As defined in Eq. 1, f_1 is the total cost of opening DCs.

$$f_1 = \sum_{l=1}^{L} CO_k \times x_k \qquad (1)$$

where

$$x_k = \begin{cases} 1, & \text{if } DC \; k \; is \; opened \\ 0, & \text{otherwise.} \end{cases}$$

For each sector, the sub-problems of sub-sectorization and routing are solved. Sectors and sub-sectors are expected to be balanced in terms of demand and distance. So, the objective functions of sub-sectorization and routing sub-problems are the standard deviations of demands, distances and compactness in sub-sectors, defined as in Eqs. 2, 4 and 3.

$$f_2 = \sqrt{\frac{1}{M-1} \sum_{m=1}^{M} (DE^m - \bar{DE})^2} \tag{2}$$

where $DE^m = \sum_{i=1}^{I} DE_i \times y_i^m$ and $\bar{DE} = \frac{\sum_{m=1}^{M} DE^m}{M}$ and

$$y_i^m = \begin{cases} 1, & \text{if } customer\ i\ belongs\ to\ sub-sector\ m \\ 0, & \text{otherwise.} \end{cases}$$

$$f_3 = \sqrt{\frac{1}{M-1} \sum_{m=1}^{M} (CP^m - \bar{CP})^2} \tag{3}$$

where $CP^m = \frac{CE^m}{CE_{max}^m}$ and $\bar{CP} = \frac{\sum_{m=1}^{M} CP^m}{M}$.

To calculate CE_{max}^m the coordinates of the centre point of each sub-sector are considered as the average of the coordinates of the customers in the sub-sector.

$$f_4 = \sqrt{\frac{1}{M-1} \sum_{m=1}^{M} (DS^m - \bar{DS})^2} \tag{4}$$

$$DS^m = \sum_{m=1}^{M} \sum_{i=1}^{I} \sum_{j}^{I} DS_{ij} \times z_{ij}^m \text{ and } \bar{DS} = \frac{\sum_{m=1}^{M} DS^m}{M}.$$

$$z_{ij}^m = \begin{cases} 1, & \text{if } the\ path\ from\ customer\ i\ to\ customer\ j\ is\ on\ a\ defined\ route\ in\ sub-sector\ m \\ 0, & \text{otherwise.} \end{cases}$$

It is assumed that all customers are connected with each other.

There are also some constraints; each customer must be only assigned to one sector, which is imposed by Constraint 5.

$$\sum_{m=1}^{M} y_i^m = 1, \ \forall i \in \bar{I} \tag{5}$$

One vehicle is allocated to a sub-sector. It is assumed that the eets are homogeneous, i.e. the capacity of the vehicles is the same. Therefore, there is no need for a decision variable for assigning the vehicles. However, the total demand of customers in each sector must be less than or equal to the capacity of each vehicle, which is imposed by Constraint 6.

$$DE^m \leq VC, \ \forall m \in \bar{M} \tag{6}$$

In addition, the number of sub-sectors must be less than or equal to the number of vehicles.

3 Solution Approaches and Algorithms

In this section, the solution approaches are described, which are also summarized in Fig. 2. In both approaches, sectorization, sub-sectorization and routing sub-problems are solved sequentially, and in this sense, both have a hierarchical structure. But in the first approach, Pareto optimality is evaluated inside of NSGA-II considering all objective functions, simultaneously. However, in the second approach, sectorization and sub-sectorization sub-problems are solved with NSGA-II and Pareto optimality is evaluated considering the total cost to open DCs, the standard deviations of demands and compactness in sub-sectors. For the non-dominated solutions obtained in this way, the routing sub-problem is solved with a solver to minimize the standard deviation of distance in sub-sectors. In the second approach, since the problem is solved in two different stages, it is called as a hierarchical approach.

At first, using the weighted sum method the problem is transformed to a single-objective one and it is solved with a single-objective genetic algorithm (SOGA). The purpose is to make comparison with the multi-objective one and also to create the initial populations of NSGA-II in the two approaches. Using weights w_i, the objective function of the single-objective problem is defined as in Eq. 7.

$$Min \ f = w_1 \times f_1 + w_2 \times f_2 + w_3 \times f_3 + w_4 \times f_4 \tag{7}$$

<table>
<tr><td>

First approach (simultaneous)
Do sectorization, sub-sectorization
and routing with NSGA-II

Second approach (hierarchical)
Do sectorization and sub-sectorization with NSGA-II
Define routes for the non-dominated solutions
obtained with NSGA-II

</td><td>

NSGA-II:
Generate initial population
Sort population with non-dominated sorting
and crowding distance calculation

While stopping criteria not reached {
    Generate new population with selection,
        crossover and mutation operators
    Sort population
    Evaluate offspring
}
Return non-dominated solutions

</td></tr>
</table>

Fig. 2 General steps of the approaches

To generate initial population of the SOGA, randomly a subset of potential DCs is opened and the corresponding objective function is calculated. Each customer is assigned to the nearest open DC and in this way sectors are formed. If no customer is assigned to a DC, it is removed from the opened DCs. The sectors are divided into sub-sectors, taking the constraints of capacity and number of vehicles into account. It is supposed that a vehicle is allocated to each sub-sector and the fleets are homogeneous.

During iterations, new individuals of the SOGA are derived using crossover and mutation operators. The objective functions of the new individuals formed in this process may change, in which case they must be recalculated. The used chromosome structure is seen in Fig. 3, where each column represents a DC and the subjacent rows, shows the related sectors and sub-sectors. Figure 3a and b show an example chromosome before and after sub-sectorization. In Fig. 3, '{}' is used to show that the corresponding DC is not open and therefore no customer is assigned to it. Using the nearest neighbor heuristic, a traveling salesman problem (TSP) is solved for each sub-sectors. A route is defined starting from a DC, visiting the nearest customer at each stage, repeating this process until all customers are visited, and returning to the DC. After solving TSP for each sub-sector, the customers in the second line are written sequentially.

In the designed crossover operator, which is seen in Fig. 4, a subset of customers is selected and is replaced in sectors of children according to the parents. For example, as shown in red in Fig. 4, customers 1, 3 and 9 are selected. In parent 2, customer 3 is in the sector related to DC 1 and customers 1 and 9 are in the sector related to DC 4. Therefore, in child 1, these customers are placed in the sectors related to DCs 1 and 4. A similar process is done for child 2 considering parent 1. In children, customers are placed in random positions. The order of customers in the representation of sectors in chromosomes affects the formation of sub-sectors.

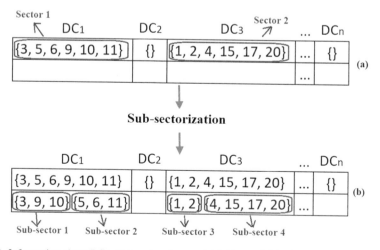

Fig. 3 Information about DCs, assigned customers (**a**) before and (**b**) after sub-sectorization

Parent 1:

DC₁	DC₂	DC₃	DC₄
{9,3,5}	{}	{4,1,2,10,7}	{8,6}
...	...	...	...

Parent 2:

DC₁	DC₂	DC₃	DC₄
{2,7,3}	{4,5}	{10,8,6}	{9,1}
...	...	...	...

Child 1:

DC₁	DC₂	DC₃	DC₄
{3,5}	{}	{4,2,10,7}	{8,9,6,1}
...	...	...	...

Child 2:

DC₁	DC₂	DC₃	DC₄
{2,7,9,3}	{4,5}	{10,6,8,1}	{}
...	...	...	...

Fig. 4 The used crossover

As seen in Fig. 5, three types of mutations are used. The operator seen in Fig. 5a is similar to the single-point crossover but it is a mutation because it is applied to a single chromosome. In this operator, two sectors of a chromosome are selected and a process similar to single-point crossover is performed. Using the mutation operator shown in Fig. 5b, an open DC is randomly selected to close and its customers are assigned to another DC, which is randomly selected to open. If all DCs are open, one of them is closed randomly and its customers are added to another DC, which is also chosen randomly. As mentioned before, the order of customers in the representation of sectors in chromosomes is important and affects the formation of sub-sectors'. In the mutation operation shown in Fig. 5c a sector is selected randomly, and the order of customers in the representation of sectors in the chromosome is changed randomly. To apply the mutation, one of these three operations is randomly selected and performed.

Both crossover and mutation operators are applied to the part of chromosomes that represents sectors. Sub-sectors are created from sectors, taking into account the constraints of the number and capacity of vehicles.

The final population obtained after finishing the SOGA iterations is used as the initial population of NSGA-II in both approaches. NSGA-II applies the same crossover and mutation operators with SOGA. In NSGA-II, during iterations, the parent and offspring populations are selected and merged. Using the non-dominated sorting and crowding distance calculation, Pareto fronts are formed and then the new population is chosen using the selection operator. The general steps of the implemented NSGA-II can also be found in [17].

As seen in Fig. 2, the two approaches are similar, and in both the sectorization and sub-sectorization steps are done with NSGA-II. In the first approach, routing is also done in NSGA-II. For this aim, using the nearest neighbor heuristic, a TSP is solved for each sub-sectors. But in the second approach the routing problem is solved outside of NSGA-II. It is performed when the iterations of the algorithm are finished. This process is done once and only for non-dominated solutions obtained by NSGA-II. For this aim, a TSP is solved for each formed sub-sectors using the *intlinprog* function, which is a binary integer linear programming solver in MATLAB optimization toolbox.

We suppose that each sub-sector is a complete graph, whose vertices are the DC and customers. A directional link between vertices is named a trip. Tours consist of a combination of trips. If there is a trip between two vertices on a route, the

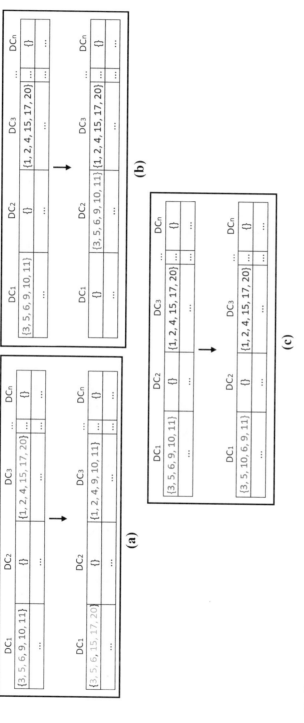

Fig. 5 The used mutations

corresponding binary variable equals 1 and in otherwise it is 0. In this way, the binary decision variables of the model are defined as all possible trips. The distance of each trip is calculated and to be minimized, the objective function of the routing problem is the total distance of the resulting trips, which is also defined as in Eq. 4.

The *intlinprog* function can deal with both equality and inequality linear constraints. In the applied routing problem there are two types of linear equality constraints. The first one ensures that in each sub-sector the total number of trips is equal to the number of vertices, while the second one makes certain that each vertex has connected to two trips [10].

During the routing, sub-tours may occur, which are disconnected loops instead of a continuous path. In the applied method, iteratively, sub-tours are detected for each obtained solution and inequality constraints are used to prevent them. This process can be summarized as: suppose that s vertices create a sub-tour. In this case, there are s links that connect them to each other. The corresponding constraint provides that the number of links between the vertices is less than or equal to $s - 1$. A more detailed description of the used routing method at this stage can be found in [10].

4 Experimental Results

We implement the approaches described in Sect. 3 in MATLAB R2019b environment on an Intel Core i7 processor, 1.8 GHz with 16 GB of RAM. The parameters used in both NSGA-II and also SOGA are: *population size = 200, number of iterations = 300, crossover rate = 0.6* and *mutation rate = 0.1*. The weights defined in Eq. 7 are: $w_1 = w_2 = 1$, $w_3 = 20$, to generate initial populations of NSGA-II in the first and second approached $w_4 = 1$ and $w_4 = o$, respectively. The reason to choose $f_3 = 20$ is that in some trial runs, values were generally as 20 times less than the others. Equal weights are given for the other objective-functions.

Indicating as *Number of customers × Number of vehicles × Number of possible DCs*, four benchmarks are generated as $15 \times 5 \times 3$, $30 \times 10 \times 6$, $60 \times 20 \times 12$ and $120 \times 40 \times 24$. We use the discrete uniform distributions as $U(10, 100)$ to create customers' demands. After defining the demands of customers, the total demand is calculated and then the capacity of each vehicle is defined as *1.3 × round (total demand/number of vehicles)*. Furthermore, the coordinates of both customers and DCs in two dimensions are generated according to the normal distribution as $N(50, 10)$. The opening cost of each DC is created according to $U(10, 15)$.

For each benchmark, when the solutions achieved by two approaches are compared with each other, some of them dominate others. So, for each benchmark, we again do a domination check between solutions that two approaches get. To do a comparison, we divide the number of non-dominated solutions (*NC*) obtained by each of the approaches into the number of all non-dominated solutions. The acquired value is shown by e. Related results are presented in Table 2.

Table 2 Comparison between the approaches

Benchmark	NC_{all}	First approach		Second approach	
		NC	e	NC	e
$15 \times 5 \times 3$	31	25	81%	6	19%
$30 \times 10 \times 6$	37	33	89%	4	11%
$60 \times 20 \times 12$	49	42	86%	7	14%
$120 \times 40 \times 24$	62	53	85%	9	15%

As shown in Table 2, the first approach in all of the benchmarks achieves significantly better results than the second one according to the value of e. Therefore, it can be considered as an efficient approach to solving all stages of MO-LRP simultaneously with an effective algorithm as NSGA-II.

Excluding benchmark 1, in the other ones, the best solution found by SOGA is among the non-dominated solutions obtained NSGA-II. Similar results are obtained when the initial solution is not taken from SOGA and is derived in NSGA-II. But, in this case, more non-dominated solutions are found. For example, in this way, 81 more non-dominated solutions are found applying the first approach for benchmark 1, however, the variance of the value of the objective functions increases. Even in this case, the best solution found with SOGA is among the non-dominated solutions.

All details about the benchmark and the obtained non-dominated solutions are accessible via the corresponding author's email address.

5   Conclusion and Future Work

In this paper, we designed two approaches to solve MO-LRP. The problem consists of four objective functions, which are the total cost of opening DCs, the standard deviations of demands, distances and compactness in sub-sectors. Unlike previous studies, to solve the problem, we adapted sub-problems of sectorization, sub-sectorization and routing into two approaches. In the first approach, sectorization, sub-sectorization and routing sub-problems were solved simultaneously with NSGA-II. But in the second approach, there was a hierarchical structure such that the routing problem was solved for non-dominated solutions obtained after performing sectorization and sub-sectorization with NSGA-II. For this aim, a TSP solved using a function in the MATLAB optimization toolbox for each formed sub-sector.

The approaches are applied for four benchmarks and the achieved results are compared based on the number of non-dominated solutions. According to the acquired results, the most important outcome of this study can be summarized like this: the simultaneous approach for this problem is more effective in terms of finding more non-dominated solutions.

The sectorization and sub-sectorization stages in both approaches are mixed-integer quadratic programming optimization problems. As in many other softwares, it is also possible to solve such models in MATLAB. For example, *qpprob* function is one of the options that can be used for this aim but in case of using this function the nonlinear part of the objective function must be added as a constraint. The reason is that MATLAB, currently, does not have a solver for non-linear objective functions [9]. In future studies, it is planned to propose new methods by using linearization techniques as well as using the non-linear part of the objective functions as the constraints.

In this study, new chromosome representation, crossover and mutation operators designed to use in NSGA-II, which can be applied in similar algorithms. In future studies, more effective operators will be proposed.

The weights used to transform the multi-objective problem into a single-objective one affect the results. In further studies, more detailed works will be carried out on this matter.

Sectorization is more appropriate for solving large-scale problems. In future studies, larger benchmarks will be derived and solved.

Acknowledgments This work is financed by the ERDF—European Regional Development Fund through the Operational Programme for Competitiveness and Internationalisation—COMPETE 2020 Programme and by National Funds through the Portuguese funding agency, FCT—Fundao para a Ciłncia e a Tecnologia within project POCI-01-0145-FEDER-031671.

The authors would like to thank the editor and the anonymous referees for their valuable comments which helped to significantly improve the manuscript.

References

1. Barreto, S., Ferreira, C., Paixão, J., Santos, B.S.: Using clustering analysis in a capacitated location-routing problem. Eur. J. Operat. Res. **179**(3), 968–977 (2007)
2. Camacho-Collados, M., Liberatore, F., Angulo, J.M.: A multi-criteria police districting problem for the efficient and effective design of patrol sector. Eur. J. Oper. Res. **246**(2), 674–684 (2015)
3. Deb, K., Pratap, A., Agarwal, S., Meyarivan, T.: A fast and elitist multiobjective genetic algorithm: Nsga-ii. IEEE Trans. Evol. Comput. **6**(2), 182–197 (2002)
4. Filipiak, K.A., Abdel-Malek, L., Hsieh, H.N., Meegoda, J.N.: Optimization of municipal solid waste collection system: case study. Pract. Period. Hazard. Toxic Radioact. Waste Manage. **13**(3), 210–216 (2009)
5. Ghiani, G., Lagana, D., Manni, E., Musmanno, R., Vigo, D.: Operations research in solid waste management: A survey of strategic and tactical issues. Comput. Oper. Res. **44**, 22–32 (2014)
6. Kalcsics, J., Nickel, S., Schröder, M.: Towards a unified territorial design approach applications, algorithms and GIS integration. TOP **13**(1), 1–56 (2005)
7. Litvinchev, I., Cedillo, G., Velarde, M.: Integrating territory design and routing problems. J. Comput. Syst. Sci. Int. **56**(6), 969–974 (2017)
8. Martinho, A., Alves, E., Rodrigues, A.M., Ferreira, J.S.: Multicriteria location-routing problems with sectorization. In: Congress of APDIO, the Portuguese Operational Research Society, pp. 215–234. Springer, Berlin (2017)
9. MathWorks, M.I.Q.P.P.O.P.B.: https://www.mathworks.com/help/optim/examples/mixed-integer-quadratic-programming-portfolio-optimization.html. Accessed 21 April 2020
10. MathWorks, T.S.P.P.B.: https://www.mathworks.com/help/optim/examples/travelling-salesman-problem.html. Accessed 21 April 2020

11. McLeod, F., Cherrett, T.: Quantifying the transport impacts of domestic waste collection strategies. Waste Manag. **28**(11), 2271–2278 (2008)
12. Mourão, M.C., Nunes, A.C., Prins, C.: Heuristic methods for the sectoring Arc routing problem. Eur. J. Oper. Res. **196**(3), 856–868 (2009)
13. Oudouar, F., Lazaar, M., El Miloud, Z.: A novel approach based on heuristics and a neural network to solve a capacitated location routing problem. Simul. Model. Pract. Theory **100**, 102064 (2020)
14. Rodrigues, A.M., Ferreira, J.S.: Measures in sectorization problems. In: Operations Research and Big Data, pp. 203–211. Springer, Berlin (2015)
15. Rodrigues, A.M., Ferreira, J.S.: Sectors and routes in solid waste collection. In: Operational Research, pp. 353–375. Springer, Berlin (2015)
16. Rodrigues, A.M., Soeiro Ferreira, J.: Waste collection routing limited multiple landfills and heterogeneous fleet. Networks **65**(2), 155–165 (2015)
17. Teymourifar, A., Ozturk, G., Bahadir, O.: A comparison between two modified nsga-ii algorithms for solving the multi-objective flexible job shop scheduling problem. Univers. J. Appl. Math. **6**(3), 79–93 (2018)

Piecewise Linear Valued Constraint Satisfaction Problems with Fixed Number of Variables

Manuel Bodirsky, Marcello Mamino, and Caterina Viola

Abstract Many combinatorial optimisation problems can be modelled as *valued constraint satisfaction problems*. In this paper, we present a polynomial-time algorithm solving the valued constraint satisfaction problem for a fixed number of variables and for piecewise linear cost functions. Our algorithm finds the infimum of a piecewise linear function and decides whether it is a proper minimum.

Keywords Valued constraint satisfaction · Linear programming · Fixed dimension · Infinite-domain optimisation · Piecewise linear cost functions

1 Introduction

The input of a *valued constraint satisfaction problem*, or *VCSP* for short, is a finite set of cost functions depending on a given finite set of variables, and the computational task is to find an assignment of values for the variables that minimises the sum of the cost functions. Many computational optimisation problems arising in industry, business, manufacturing, and science, can be modelled as a VCSP.

VCSPs have been extensively studied in the case of cost functions defined on a fixed finite set (the *domain*). The computational complexity of solving VCSPs depends on the set of allowed cost functions and has recently been classified if the domain is finite [3, 12–14, 21]: every finite domain VCSP is either polynomial-time tractable or it is NP-complete.

M. Bodirsky (✉) · C. Viola
Technische Universität Dresden, Dresden, Germany
e-mail: manuel.bodirsky@tu-dresden.de; caterina.viola@tu-dresden.de

M. Mamino
Università di Pisa, Pisa, Italy
e-mail: marcello.mamino@dm.unipi.it

© The Author(s), under exclusive license to Springer Nature Switzerland AG 2021
C. Gentile et al. (eds.), *Graphs and Combinatorial Optimization:*
from Theory to Applications, AIRO Springer Series 5,
https://doi.org/10.1007/978-3-030-63072-0_21

However, many outstanding combinatorial optimisation problems can be formulated as VCSPs only by allowing cost functions defined on infinite domains, e.g., the set $\mathbb{Q}$ of rational numbers.

Despite the interest in concrete VCSPs over the set of rational numbers and over other infinite numeric domains (e.g., the integers, the reals, or the complex numbers), VCSPs over infinite domains have not yet been investigated systematically. The class of VCSPs for all sets of cost functions defined on arbitrary infinite domains is too large to allow general complexity results. Indeed, every computational problem is polynomial-time Turing-equivalent to the VCSP for a suitable set of cost functions over an infinite domain [1]. Therefore, we need to restrict the class of cost functions that we focus on. One restriction that captures a great variety of theoretically and practical interesting optimisation problems is the class of all *piecewise linear (PL)* cost functions over $\mathbb{Q}$, i.e., $\mathbb{Q}$-valued[1] partial functions, whose graph is the union of linear half spaces. In general, the VCSP for PL cost functions is NP-complete. Indeed, the containment in NP follows from the fact that the VCSP for the class of *all* PL cost functions is equivalent to the existential theory of $(\mathbb{Q}; \leq, +, 1)$, which is in NP (see [2]). The NP-hardness follows from the fact that there exist NP-complete problems, e.g., the MINIMUM CORRELATION CLUSTERING, and the MINIMUM FEEDBACK ARC SET problem (see [6, 9]), which can be formulated as instances of the VCSP for PL cost functions.

In this paper, we prove that the restriction of the VCSP for *all* PL cost functions to instances with a fixed number of variables is polynomial-time tractable. The restriction to a fixed number of variables has been studied for several problems in computational optimisation, and usually this kind of restriction has led to an improvement in the computational complexity: two remarkable examples of this situation are the combinatorial polynomial-time algorithm of Megiddo [16] to solve the restriction to a fixed number of variables of LINEAR PROGRAMMING (in its full generality, Linear Programming can be solved in polynomial-time (see, e.g., [8, 10, 20]), but all the known algorithms rely on approximation procedures); and the algorithm of Lenstra [15] to solve the restriction of Integer Programming Feasibility (which is an NP-complete problem in its full generality) to a fixed number of variables.

2 Preliminaries

We adopt the following notation: $\mathbb{Q}$ denotes the set of rational numbers, and x_i denotes the i-th component of a tuple x. We start with some preliminaries on the cost functions that we want to take into account.

---

[1]In the PL setting the domains $\mathbb{Q}$ and $\mathbb{R}$ are interchangeable; we only require the coefficients of the cost functions to be rational as we need to manipulate them computationally.

Definition 1 A *cost function over* $\mathbb{Q}$ is a function $f : \mathbb{Q}^n \rightarrow \mathbb{Q} \cup \{+\infty\}$, for a positive integer n. Here, $+\infty$ is an extra element with the expected properties that for all $c \in \mathbb{Q} \cup \{+\infty\}$

$$(+\infty) + c = c + (+\infty) = +\infty$$

$$\text{and } c < +\infty \text{ iff } c \in \mathbb{Q}.$$

A cost function $f : \mathbb{Q}^n \rightarrow \mathbb{Q} \cup \{+\infty\}$ can also be seen as a partial function such that f is not defined on $x \in \mathbb{Q}^n$ if, and only if, $f(x) = +\infty$.

Definition 2 A set $C \subseteq \mathbb{Q}^d$ is a *polyhedral set* if it is the intersection on finitely many (open or closed) half spaces, i.e., it can be specified by a conjunction of finitely many linear constraints, i.e., for some $r \in \mathbb{N}$ there exist linear functions $f_i : \mathbb{Q}^d \rightarrow \mathbb{Q}$, for $1 \leq i \leq r$, such that

$$C = \left\{ x \in \mathbb{Q}^d \mid \bigwedge_{i=1}^{p}(f_i(x) \leq 0) \wedge \bigwedge_{i=p+1}^{q} (f_i(x) < 0) \wedge \bigwedge_{i=q+1}^{r} (f_i(x) = 0) \right\}.$$

Observe that non-empty polyhedral sets in $\mathbb{Q}^d$ are, in particular, convex sets.

A polyhedral set $C \subseteq \mathbb{Q}^d$ is *open* if it is the intersection of finitely many open half spaces, i.e., for some $p \in \mathbb{N}$ there exist $f_i : \mathbb{Q}^d \rightarrow \mathbb{Q}$ linear functions, $1 \leq i \leq p$, such that

$$C = \{x \in \mathbb{Q}^d \mid \bigwedge_{i=1}^{p}(f_i(x) < 0)\}.$$

Similarly, a polyhedral set $C \subseteq \mathbb{Q}^d$ is *closed* if it is the intersection of finitely many closed half spaces, i.e., for some $q \in \mathbb{N}$ there exist $f_i : \mathbb{Q}^d \rightarrow \mathbb{Q}$ linear functions, $1 \leq i \leq p$, such that

$$C = \left\{ x \in \mathbb{Q}^d \mid \bigwedge_{i=1}^{p}(f_i(x) \leq 0) \wedge \bigwedge_{i=p+1}^{q} (f_i(x) = 0) \right\}.$$

A polyhedral set $C \subseteq \mathbb{Q}^n$ is *bounded* if it is bounded as a subset of $\mathbb{Q}^n$. We remark that the infimum of a linear function in a closed and bounded polyhedral set is a proper minimum; while the infimum of a linear function in an open or unbounded polyhedral set is attained only if the linear function is constant.

Definition 3 ([18], Definition 2.47) A function $f : \mathbb{Q}^d \rightarrow \mathbb{Q} \cup \{+\infty\}$ is a *piecewise linear (PL)* if its domain, $\text{dom}(f)$, can be represented as the union of finitely many polyhedral sets, relative to each of which $f(x)$ is given by a linear expression, i.e., there exist finitely many mutually disjoint $C_1, \ldots, C_m$ polyhedral sets such that

$\bigcup_{i=1}^{m} C_i = \text{dom}(f) \subseteq \mathbb{Q}^d$, and

$$f(x_1, \ldots, x_d) = \begin{cases} a_0^i + a_1^i x_1 + \cdots + a_d^i x_d & \text{if } (x_1, \ldots, x_d) \in C_i \\ +\infty & \text{if } (x_1, \ldots, x_d) \in \mathbb{Q}^d \setminus \text{dom}(f) \end{cases}$$

where $a^i = (a_0^i, a_1^i, \ldots, a_d^i) \in \mathbb{Q}^{d+1}$, for $1 \leq i \leq m$. Piecewise linear functions are sometimes called *semilinear* functions.

We are now ready to formally define the computational problem that we want to focus on.

Definition 4 Let d be a positive integer, and let $V := \{x_1, \ldots, x_d\}$ be a set of variables. An *instance I* of the *valued constraint satisfaction problem (VCSP) for PL cost functions with variables in V* consists of an expression ϕ of the form

$$\sum_{i=1}^{m} f_i(x_1^i, \ldots, x_{\text{ar}(f_i)}^i)$$

where $f_1, \ldots, f_m$ are finitely many PL cost functions and all the x_j^i are variables from V. The task is to find the *infimum cost* of ϕ, defined as

$$\inf_{\alpha : V \to \mathbb{Q}} \sum_{i=1}^{m} f_i(\alpha(x_1^i), \ldots, \alpha(x_{\text{ar}(f_i)}^i)),$$

and to decide whether it is attained, i.e., whether it is a proper minimum or not.

The computational complexity of the VCSP for PL cost functions with variables from a fixed set V depends on how the cost functions are represented in the input instances. We fix a representation of cost functions which is strictly related both to the mathematical properties of piecewise linear functions and to the algorithmic procedures and mathematical tools that we want to use to deal with them.

Definition 5 (Representation of PL Cost Functions) We assume that a PL cost function is given by a list of linear constraints, specifying the polyhedral sets, and a list of linear polynomials and $+\infty$ symbols, defining the value of the function relatively to each polyhedral set. The linear constraints and the linear polynomials are encoded by the list of their rational coefficients, and $+\infty$ is represented by a special symbol. The constants for (numerators and denominators of) rational coefficients for linear constraints and linear polynomials are represented in binary.

LINEAR PROGRAMMING is an example of a problem which can be formulated in our setting; it is also a tool that plays an important role later in the paper.

Definition 6 LINEAR PROGRAMMING (*LP*) is an optimisation problem with a linear objective function and a set of linear constraints imposed upon a given set of underlying variables. A linear program has the form

$$\text{minimise} \quad \sum_{j=1}^{n} c_j x_j$$

$$\text{subject to} \quad \sum_{j=1}^{n} a_j^i x_j \le b^i, \quad \text{for } i \in \{1, \ldots, m\}.$$

This problem has n variables x_j (ranging over the rationals or over the real numbers) and m linear inequalities constraints. The coefficients c_j, a_j^i, and b^i are rational numbers, for all $j \in \{1, \ldots, n\}$, and all $i \in \{1, \ldots, m\}$. The linear constraints, $\sum_{j=1}^{n} a_j^i x_j \le b^i$, specify a polyhedral set, namely the *feasibility polytope* over which the objective function has to be optimised.

An algorithm solving LP either finds a point in the feasibility polytope where the objective function has the smallest value if such a point exists, or it reports that the instance is infeasible (in this case we assume that the output of the algorithm is $+\infty$), or it reports that the infimum of the objective function is $-\infty$ (in this case we assume that the output of the algorithm is $-\infty$).

The LINEAR PROGRAM FEASIBILITY *problem*, *LPF*, is a decision problem having the form of a standard linear program but without any objective function to minimise. The output of an algorithm solving LPF is "no" or "yes", respectively, depending on whether the polyhedral set defined by the linear constraints is empty or not. Both LP and LPF can be solved in polynomial time (see, e.g., [8, 10, 20]).

In the remainder of the paper, we use the fact that LINEAR PROGRAM FEASIBILITY for a set of linear constraints containing also strict inequalities can be solved in polynomial time. Given a set of linear constraints l and a linear expression obj, we denote by $LPF(l)$ the LPF instance defined by the linear constraints in l, and we denote by $LP(l, \text{obj})$ the LP instance defined by the linear constraints in l and by the objective function obj.

We remark that in LP and LPF the feasibility polytope is defined by weak linear inequalities, i.e., by linear constraints of the form $\sum_{j=1}^{n} a_j x_j \le b$. The feasibility of a set of linear constraints containing also strict linear inequalities (i.e., of the form $\sum_{j=1}^{n} a_j x_j < b$) can be solved by solving a linear number of linear programs, as shown in [7], where the authors give a polynomial-time algorithm deciding the feasibility of a set of *Horn disjunctive linear constraints*. However, the feasibility of a set of linear constraints containing strict and weak linear inequalities can be decided by solving only one LP instance.

Lemma 1 (Motzkin Transposition Theorem [17, 19]) *Let $A \in \mathbb{Q}^{k_1 \times d}$, and $B \in \mathbb{Q}^{k_2 \times d}$ be matrices such that $\max(k_1, d) \ge 1$. The system*

$$\begin{cases} Ax < 0 \\ Bx \le 0 \end{cases}$$

has a solution $x \in \mathbb{Q}^d$ if, and only, if the system

$$\begin{cases} A^T y + B^T z = 0 \\ y \geq 0, \ z \geq 0 \end{cases}$$

does not admit a solution $(y, z) \in \mathbb{Q}^{k_1 + k_2}$ such that $y \neq (0, \ldots, 0)$.

Proposition 1 *The* LINEAR PROGRAM FEASIBILITY *problem (LPF) for a finite set of strict or weak linear inequalities is polynomial-time many-one reducible to LP and, therefore, it can be solved in polynomial time.*

Proof Let us assume that the linear constraints in the input consist of k_1 strict inequalities, and k_2 weak inequalities, i.e., we have to check the satisfiability of the following system

$$\begin{cases} \sum_{i=1}^{d} a_{j,i} x_i + a_{j,d+1} < 0 & \text{for } 1 \leq j \leq k_1 \\ \sum_{i=1}^{d} b_{j,i} x_i + b_{j,d+1} \leq 0 & \text{for } 1 \leq j \leq k_2, \end{cases} \tag{1}$$

Let us first observe that the system (1) is equivalent to the following one

$$\begin{cases} \sum_{i=1}^{d+1} a_{j,i} t_i < 0 & \text{for } 1 \leq j \leq k_1 \\ -t_{d+1} < 0 \\ \sum_{i=1}^{d+1} b_{j,i} t_i \leq 0 & \text{for } 1 \leq j \leq k_2. \end{cases} \tag{2}$$

Indeed, if $(t_1, \ldots, t_d, t_{d+1})$ is a solution for (2), then $(x_1, \ldots, x_d)$ with $x_i := \frac{t_i}{t_{d+1}}$ is a solution for (1); vice versa if $(x_1, \ldots, x_d)$ is a solution for (1), then $(x_1, \ldots, x_d, 1)$ is a solution for (2). Let us consider the following linear program

$$\text{minimise} \qquad \sum_{j=1}^{k_1+1} (-y_j)$$

$$\text{subject to} \qquad A^T y + B^T z$$

$$-y \leq 0 \tag{3}$$

$$-z \leq 0,$$

with variables $y_1, \ldots, y_{k_1+1}, z_1, \ldots, z_{k_2}$, where $A \in \mathbb{Q}^{(k_1+1) \times (d+1)}$ is the matrix such that $(A)_{ji} = a_{ji}$ for $1 \leq j \leq k_1$ and $1 \leq i \leq d+1$, and such that the $(k_1 + 1)$-th row of A is $(0, \ldots, 0, -1)$; and the matrix $B \in \mathbb{Q}^{(k_2) \times (d+1)}$ is such that $(B)_{ji} = b_{ji}$.

Observe that the linear program (3) can be computed in polynomial time (in the size of the input). By Lemma 1, the system (2) is satisfiable if, and only if, the feasibility polytope determined by the linear constraints in (3) does not admit a solution $(y, z) \in \mathbb{Q}^{(k_1+1)+k_2}$ such that $y \neq (0, \ldots, 0)$. If the output of the algorithm for LP on instance (3) is $+\infty$ or a tuple having 0 in the first $k_1 + 1$ coordinates, then the system (1) is satisfiable, and therefore we accept. Otherwise, if the output is $-\infty$ or a tuple $(y, z) \in \mathbb{Q}^{(k_1+1)+k_2}$ such that $y \neq (0, \ldots, 0)$, then the system (1) is not satisfiable and we reject. $\qquad \square$

3 PL VCSPs with Fixed Number of Variables

We exhibit a polynomial-time algorithm that solves the VCSP for PL cost functions having variables from a fixed finite set. We assume that the input VCSP instance is given as a sum of PL cost functions represented as in Definition 5. Our algorithm computes the infimum of the objective function, and specifies whether it is attained, i.e., whether it is a proper minimum.

The following theorem uses an idea that appeared in [2], Observation 17.

Theorem 1 *Let V be a finite set of variables. Then there is a polynomial-time algorithm that solves the* VCSP *for PL cost functions having variables in V.*

Proof We prove that Algorithm 1 correctly solves the VCSP for PL cost functions with variables in V in polynomial time. An input of an instance of the VCSP is a representation of an objective function ϕ as the sum of a finite number of given cost functions, $f_1, \ldots, f_n$, applied to some of the variables in $V = \{x_1, \ldots, x_d\}$, that is,

$$\phi(x_1, \ldots, x_d) = \sum_{i=1}^{n} f_i(x^i),$$

where $x^i \in V^{\mathrm{ar}(f_i)}$ for $1 \leq i \leq d$. We can assume that the cost function f_i is defined for every $x \in \mathbb{Q}^d$ by

$$f_i(x) = \begin{cases} \sum_{j=1}^{d} a_j^{i,l} x_j + b^{i,l} & \text{if } C_{i,l}(x), \text{ for some } 1 \leq l \leq m_i \\ +\infty & \text{otherwise.} \end{cases}$$

For every $1 \leq l \leq m_i$ the formulas $C_{i,l}(x)$ have the following form:

$$C_{i,l}(x) = \bigwedge_{j=1}^{p} (h_j^{i,l}(x) \leq 0) \wedge \bigwedge_{j=p+1}^{q} (h_j^{i,l}(x) < 0) \wedge \bigwedge_{j=q+1}^{r} (h_j^{i,l}(x) = 0),$$

for some $p, q, r \in \mathbb{N}$ and for some linear polynomials $h_j^{i,l} : \mathbb{Q}^d \to \mathbb{Q}$, where $1 \leq j \leq r$. We assume that the cost functions f_i are represented as in Definition 5.

Algorithm 1 first extracts the list of linear polynomials $p_1, \ldots, p_k$ that appear in the finite set of linear constraints defining some cost function f_i, i.e.,

$$\{p_1, \ldots, p_k\} := \bigcup_{i=1}^{n} \bigcup_{l=1}^{m_i} \bigcup_{j} \{h_j^{i,l}\}.$$

Observe that the linear polynomials $p_1, \ldots, p_k$ decompose the space $\mathbb{Q}^d$ into σ polyhedral sets, where

$$\sigma \leq \tau_d(k) = \sum_{i=0}^{d} 2^i \binom{k}{i} \tag{4}$$

and that this bound is tight, i.e., $\sigma = \tau_d(k)$, whenever the hyperplanes defined by $p_i(x) = 0$, for $1 \leq i \leq k$, are in general position.

Inequality (4) can be verified by induction on the number of hyperplanes, k. Clearly, for all $d \in \mathbb{N}$, one hyperplane divides $\mathbb{Q}^d$ into $3 = 2^0 + 2^1$ polyhedral sets. Suppose now that $k \geq 3$ and that Inequality (4) is true for every d and for at most $k - 1$ hyperplanes. Suppose that the k hyperplanes are in general position (we get in this way the upper bound $\tau_d(k)$). Observe that, by adding the hyperplanes one-by-one, the k-th hyperplane intersects at most $\tau_{d-1}(k - 1)$ of the polyhedral sets obtained until the previous step. In fact, this number is equal to the number of polyhedral sets in which a hyperplane, that is a subspace of dimension $d - 1$, is divided by $k - 1$ subspaces of dimension $d - 2$.

Suppose that we know how the space is decomposed into polyhedral sets by the hyperplanes $p_1(x) = 0, \ldots, p_{k-1}(x) = 0$. Adding $p_k(x) = 0$ to the list of hyperplanes decomposing the space, each one of the polyhedral sets intersecting it is divided in three polyhedral sets (corresponding to $p_k(x) < 0$, $p_k(x) = 0$, and $p_k(x) > 0$, respectively). Summing up, at every step we add to the "old polyhedral sets" (i.e., polyhedral sets obtained until the previous step) two more polyhedral sets for each of the old ones intersecting $p_k(x) = 0$, then it follows that

$$\tau_d(k) = \tau_d(k - 1) + 2\tau_{d-1}(k - 1).$$

Using this equality and the inductive hypothesis we obtain

$$\tau_d(k) = 2 \sum_{i=0}^{d-1} 2^i \binom{k-1}{i} + \sum_{i=0}^{d} 2^i \binom{k-1}{i} = \sum_{i=1}^{d} 2^i \left(\binom{k-1}{i-1} + \binom{k-1}{i} \right) + 1$$

$$= \sum_{i=1}^{d} 2^i \binom{k}{i} + 1 = \sum_{i=0}^{d} 2^i \binom{k}{i}.$$

In particular, the number σ of polyhedral sets is bounded by a linear polynomial in k, and the Algorithm 1 produces a tree, that a priori has 3^k branches, but that actually has $O(k)$ branches.

The algorithm computes the list of all non-empty polyhedral sets by computing at most $\sum_{i=1}^{k-1} \tau_d(i)$ instances of linear program feasibility, and then it computes the infimum of the objective function in every non-empty polyhedral set by computing at most $3\tau_d(k)$ linear programs. Observe that the only closed and bounded non-empty polyhedral sets computed by Algorithm 1 are 0-dimensional subspaces, i.e., points, and that all the other polyhedral sets computed are open or unbounded. Therefore, in order to check whether the infimum in a polyhedral set C is a minimum, it is enough to check whether the objective function in C is constant, that is whether the infimum in C is equal to the supremum in C. This is done by our algorithm solving at most $3\tau_d(k)$ further linear programs. The linear expression of the objective function in a polyhedral set can be computed by running a number of LINEAR PROGRAM FEASIBILITY instances that is polynomial in the size of the input instance. Globally, the running time of Algorithm 1 is polynomial in the size of the input. □

4 Conclusion and Future Work

We have provided a polynomial-time algorithm solving the VCSP for piecewise linear cost functions having a fixed number of variables. In the future, we would like to continue this line of research by studying the computational complexity of the VCSP with a fixed number of variables and *semialgebraic* cost functions. A function $f : \mathbb{R}^n \rightarrow \mathbb{R} \cup \{+\infty\}$ is called semialgebraic if its domain can be represented as the union of finitely many *basic semialgebraic sets* (see [2]) of the form $\{x \in \mathbb{R}^n \mid \chi(x)\}$ where χ is a conjunction of (weak or strict) polynomial inequalities with integer coefficients, relative to each of which $f(x)$ is given by a polynomial expression with integer coefficients.

The VCSP for all semialgebraic cost function is equivalent to the existential theory of the reals (see [2]), which is in PSPACE (see [4]). The restriction of the feasibility problem associated with a semialgebraic VCSP to a fixed number of variables is polynomial-time tractable by cylindrical decomposition (cf. [5]). However, we do not know whether with this approach can solve our optimisation problem in polynomial time. Another contribution related with our open problem was given in [11] by Khachiyan and Porkolab, who proved that the problem of minimising a convex polynomial objective function with integer coefficients over a fixed number of integer variables, subject to polynomial constraints with integer coefficients that define a convex region, can be solved in polynomial time in the size of the input.

Algorithm 1: Algorithm for PL VCSPs with a fixed number of variables

---

Input: $\phi(x) = f_1(x) + \cdots + f_n(x)$ with $f_i(x) = f_{ij}(x)$ if $x \in C_{ij}$, and the C_{ij}'s each given
      as a finite set of linear conditions, for $1 \leq j \leq n_i$, and $1 \leq i \leq n$.

Output: (val, attr) where val is the value of the infimum of the objective function, and attr is a
      string which specifies whether val is attained (attr $=$ "min") or not (attr $=$ "inf").

$\{p_1, \ldots, p_k\} :=$ the set of all the linear functions appearing in the C_{ij}'s;

$L := \{\{\}\}$ (the set of polyhedral sets in which the p_i's divide the space);

for $i = 1, \ldots, k$ **do**
    for *each l in L* **do**
        $l_{-1} := l \cup \{p_i < 0\}$;
        $l_0 := l \cup \{p_i = 0\}$;
        $l_1 := l \cup \{-p_i < 0\}$;
        for $j = -1, 0, 1$ **do**
            if $LPF(l_j) = yes$ **then**
                $L := (L \setminus \{l\}) \cup \{l_j\}$
            end
        end
    end
end

val $:= +\infty$;

attr $:''$ inf ";

for *each l in L* **do**
    $l_c := \{\}$ (the closure of l);
    for *each $c \in l$* **do**
        if *c is of the form $(p < 0)$* **then**
            $l_c := l_c \cup \{p \leq 0\}$
        else
            $l_c := l_c \cup \{c\}$
        end
    end
    for $i = 1, \ldots, n$ **do**
        $g_i := +\infty$;
        for $j = 1, \ldots, n_i$ **do**
            if $LPF(l \cup C_{ij}) = yes$ **then**
                $g_i(x) := f_{ij}(x)$
            end
        end
    end
    obj $:= \sum_{i=1}^{n} g_i(x)$;
    $m :=$ LP(l_c, obj) (the infimum of ϕ in l);
    $M := -$LP$(l_c, -\text{obj})$ (the supremum of ϕ in l);
    if $m < $ val **then**
        if $m = M$ **then**
            attr $:=$ "min"(the infimum is attained iff ϕ is constant in l)
        else
            attr $:=$ "inf"
        end
    end
end

return (val, attr);

Acknowledgments The authors have received funding from the European Research Council (ERC) under the European Union's Horizon 2020 research and innovation programme (grant agreement No 681988, CSP-Infinity). The third author has been supported by DFG Graduiertenkolleg 1763 (QuantLA).

References

1. Bodirsky, M., Grohe, M.: Non-dichotomies in constraint satisfaction complexity. In: Proceedings of the 35th International Colloquium on Automata, Languages and Programming (ICALP), pp. 184–196. Springer, Berlin (2008). https://doi.org/10.1007/978-3-540-70583-3_16

2. Bodirsky, M., Mamino, M.: Constraint satisfaction problems over numeric domains. In: Krokhin, A., Zivny, S. (eds.) The Constraint Satisfaction Problem: Complexity and Approximability, vol. 7, pp. 79–111. Schloss Dagstuhl–Leibniz-Zentrum fuer Informatik, Dagstuhl (2017). http://drops.dagstuhl.de/opus/volltexte/2017/6958

3. Bulatov, A.A.: A dichotomy theorem for nonuniform CSPs. In: Proceedings of the 58th IEEE Annual Symposium on Foundations of Computer Science (FOCS), Berkeley, pp. 319–330. (2017). https://ieeexplore.ieee.org/document/8104069

4. Canny, J.: Some algebraic and geometric computations in PSPACE. In: Proceedings of the 20th ACM Annual Symposium on Theory of Computing (STOC), pp. 460–467. ACM, New York (1988). https://doi.org/10.1145/62212.62257

5. Collins, G.E.: Quantifier elimination for real closed fields by cylindrical algebraic decomposition: a synopsis. SIGSAM Bull. **10**(1), 10–12 (1976). https://doi.org/10.1145/1093390.1093393

6. Garey, M.R., Johnson, D.S.: Computers and Intractability; A Guide to the Theory of NP-Completeness. W. H. Freeman & Co., New York (1990). https://dl.acm.org/doi/book/10.5555/574848

7. Jonsson, P., Bäckström, C.: A unifying approach to temporal constraint reasoning. Artif. Intell. **102**(1), 143–155 (1998). https://doi.org/10.1016/S0004-3702(98)00031-9

8. Karmarkar, N.: A new polynomial-time algorithm for linear programming. Combinatorica **4**(4), 373–395 (1984). http://dx.doi.org/10.1007/BF02579150

9. Karp, R.M.: Reducibility among combinatorial problems. In: Miller, R.E., Thatcher, J.W., Bohlinger, J.D. (eds.) Complexity of Computer Computations: Proceedings of a Symposium on the Complexity of Computer Computations, pp. 85–103. Springer, Boston (1972). https://doi.org/10.1007/978-1-4684-2001-2_9

10. Khachiyan, L.: A polynomial algorithm in linear programming. Dokl. Akad. Nauk SSSR **244**, 1093–1097 (1979). https://doi.org/10.1016/0041-5553(80)90061-0

11. Khachiyan, L., Porkolab, L.: Integer optimization on convex semialgebraic sets. Discrete & Comput. Geom. **23**, 207–224 (2000). https://doi.org/10.1007/PL00009496

12. Kolmogorov, V., Krokhin, A.A., Rolinek, M.: The complexity of general-valued CSPs. In: Proceedings of the 56th IEEE Annual Symposium on Foundations of Computer Science (FOCS), pp. 1246–1258 (2015). https://doi.org/10.1109/FOCS.2015.80

13. Kolmogorov, V., Thapper, J., Živný, S.: The power of linear programming for general-valued CSPs. SIAM J. Comput. **44**(1), 1–36 (2015). https://doi.org/10.1137/130945648

14. Kozik, M., Ochremiak, J.: Algebraic properties of valued constraint satisfaction problem. In: Proceedings of the 42nd International Colloquium on Automata, Languages, and Programming (ICALP), Part I, pp. 846–858 (2015). https://doi.org/10.1007/978-3-662-47672-7_69

15. Lenstra, H.W.: Integer programming with a fixed number of variables. Math. Oper. Res. **8**(4), 538–548 (1983). http://www.jstor.org/stable/3689168

16. Megiddo, N.: Linear programming in linear time when the dimension is fixed. J. ACM **31**(1), 114–127 (1984). http://doi.acm.org/10.1145/2422.322418

17. Motzkin, T.S.: Beiträge zur theorie der linearen ungleichungen. Ph.D. Thesis, University of Basel, Azriel (1936). https://www.springer.com/gb/book/9780817630874. English Translation: Contributions to the theory of linear inequalities, RAND Corporation Translation 22, The RAND Corporation, Santa Monica, California, 1952. Reprinted in: Theodore S. Motzkin: Selected Papers (D. Cantor, B. Gordon, B. Rothschild, eds.), Birkhäuser, Boston (1983), pp. 1–80

18. Rockafellar, R.T., Wets, R.J.B.: Variational Analysis, vol. 317. Springer, Berlin (1998). https://sites.math.washington.edu/~rtr/papers/rtr169-VarAnalysis-RockWets.pdf

19. Roos, K.: Linear optimization: theorems of the alternative. In: Floudas, C.A., Pardalos, P.M. (eds.) Encyclopedia of Optimization, pp. 1878–1881. Springer, Boston (2009). https://doi.org/10.1007/978-0-387-74759-0_334

20. Wright, M.H.: The interior-point revolution in optimization: history, recent developments, and lasting consequences. Bull. Am. Math. Soc. **42**, 39–56 (2005). https://doi.org/10.1090/S0273-0979-04-01040-7

21. Zhuk, D.: A proof of CSP dichotomy conjecture. In: Proceedings of the 58th IEEE Annual Symposium on Foundations of Computer Science (FOCS), pp. 331–342 (2017). http://ieee-focs.org/FOCS-2017-Papers/3464a331.pdf

A Lagrangian Approach to Chance Constrained Routing with Local Broadcast

Matteo Cacciola, Antonio Frangioni, Laura Galli, and Giovanni Stea

Abstract Mobile cellular networks play a pivotal role in emerging Internet of Things (IoT) applications, such as vehicular collision alerts, malfunctioning alerts in Industry-4.0 manufacturing plants, periodic distribution of coordination information for swarming robots or platooning vehicles, etc. All these applications are characterized by the need of routing messages within a given local area (geographic proximity) with constraints about both timeliness and reliability (i.e., probability of reception). This paper presents a Non-Convex Mixed-Integer Nonlinear Programming model for a routing problem with probabilistic constraints on a wireless network. We propose an exact approach consisting of a branch-and-bound framework based on a novel Lagrangian decomposition to derive lower bounds. Preliminary experimental results indicate that the proposed algorithm is competitive with state-of-the-art general-purpose solvers, and can provide better solutions than existing highly tailored ad-hoc heuristics to this problem.

Keywords Chance-constrained optimization · Mixed-integer nonlinear programming · Internet-of-things · Mobile network routing · Local broadcast · Lagrangian relaxation · Bundle methods · Branch-and-bound

1 Introduction

Long Term Evolution Advanced (LTE-A) technology for cellular networks is the new forefront in the context of transmission networks for location-based broadcast services, such as advertising, smart-city applications, and Internet-of-Things (IoT)

M. Cacciola · A. Frangioni · L. Galli (✉)
Dipartimento di Informatica, Università di Pisa, Pisa, Italy
e-mail: m.cacciola1@studenti.unipi.it; antonio.frangioni@unipi.it; laura.galli@unipi.it

G. Stea
Dipartimento di Ingegneria dell'Informazione, Università di Pisa, Pisa, Italy
e-mail: giovanni.stea@unipi.it

© The Author(s), under exclusive license to Springer Nature Switzerland AG 2021
C. Gentile et al. (eds.), *Graphs and Combinatorial Optimization: from Theory to Applications*, AIRO Springer Series 5,
https://doi.org/10.1007/978-3-030-63072-0_22

277

deployments. Yet, some new IoT services, such as vehicular collision alerts and augmented-reality live games, require *low latency* and *high reliability*, as well as the possibility to target an area defined by the application itself rather than the cell coverage. While traditional LTE-A tools can support these services, they do so at a rather large cost in terms of energy. In fact, on the one hand, LTE's built-in Multicast/Broadcast mechanism was originally devised for broadcasting multimedia, and therefore unsuitable to this task because it is static: the message transmission format, the target area and the period of broadcast transmissions must all be selected statically. On the other hand, having the base station (antenna), called eNodeB (eNB) in the LTE terminology, relay messages to all the User Equipment (UEs) in a target area using unicast downlink (DL) transmissions (one per targeted UE) would require too many DL resources, hence too much energy. For this reason, recently, a new communication framework has been proposed. We consider a network of mobiles (UEs) which are under the control of a single eNB, as shown in Fig. 1.

The eNB can send them information using DL (i.e., vertical) transmissions. Information can also travel through device-to-device (D2D) links (i.e. horizontal broadcast transmissions originated at UEs). Vertical links are reliable but costly, and should be avoided if possible. By contrast, horizontal transmissions are free (from the eNB viewpoint), but not reliable: there is no ARQ mechanism involved, and it is impractical to try and ascertain which UEs, in the neighbourhood of the transmitter, have successfully decoded a message. However, UEs can act as multi-hop relays: horizontal transmissions are scheduled by the eNB, which issues grants to the UEs that may transmit. This allows to model the probability that a certain horizontal transmission is successful with a reasonable accuracy, given the position of the UEs, the transmission power of the transmitter, and the

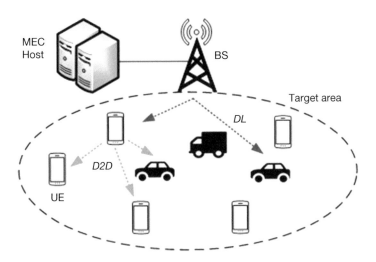

Fig. 1 System model

modulation and coding scheme adopted for transmission. This yields a new *Chance-Constrained Unicast-Multicast Routing Problem* (CCUMRP): select which vertical and horizontal multi-hop transmissions to choose in order to guarantee that all UEs receive the information with a certain level of reliability within a certain time limit, at minimum energy cost. We propose a Non-Convex Mixed-Integer Nonlinear Programming (MINLP) model for CCUMRP, together with an ad-hoc Lagrangian decomposition approach to compute lower bounds that separates the variable of the problem in a somewhat unusual fashion. We use the latter as the basis of a Branch-and-Bound (B&B) approach that we computationally compare both with state-of-the-art, general-purpose exact solvers and with highly tailored ad-hoc heuristics for the problem.

We model the *system* as a graph $G = (N, A)$, where $N = \{0\} \cup N'$ (0 being the eNB and N' representing the UEs) with $n = |N'|$, while the arc set $A = A' \cup A''$ consists of two types of arcs:

- *vertical* arcs A' of the form $(0, i)$ for all $i \in N'$, representing a DL transmission between the eNB the UE i having probability 1 to be decoded successfully at i but high energy cost;
- *horizontal* arcs A'' of the form (i, j) for $i \neq j \in N'$, representing a D2D transmission from i to j having probability $0 < P_{ij} < 1$ to be decoded successfully at j, but low (energy) cost. (Probabilities are independent.)

In the initial stage of the process the eNB transmits the message to a subset of UEs via DL (i.e., vertical transmission), while in the following stages only horizontal transmissions are allowed. A node $i \in N'$ can only issue an horizontal transmission at a given stage if granted permission from the eNB. At most M grants can be assigned in each stage, to ensure that the ensuing transmissions are not mutually interfering. The problem is therefore to transmit the message to the entire floorplan with the following constraints:

1. To ensure timeliness of reception of the message to all UEs, the broadcast process must be over in k stages, with k known a priori. Because the first round is clearly "special" (vertical transmissions from eNB to UEs), it is useful to define the set $K' = \{2, \ldots, k\}$ of "normal" stages (horizontal transmissions between UEs).
2. To ensure reliability of reception, at the end of the broadcast process, each UE must possess the message with at least a given probability α.

The main *objective* is to reduce the number of vertical transmissions. However, besides them, we also need to decide which UEs should transmit when (i.e., in which stage), so we must define the schedule of the grants that the eNB has to issue in order to compose the broadcast schedule. A secondary objective is to issue the least possible numbers of grants. In our model it is actually easy to generalize this by considering node-and-stage weighted grants costs β_i^h ($i \in N'$, $h \in K'$), e.g., depending on the type of node i and/or its remaining battery power.

2   Mathematical Model and Decomposition

We define the following set of *variables*:

- binary x_i for $i \in N'$ indicating whether node i is selected in the initial set of UEs that are reached by the vertical transmission from eNB at stage 1;
- continuous $p_i^h \in [0, 1]$ for $i \in N'$ and $h \in K$ indicating the total probability that node i has been reached at all stages up to h;
- binary g_i^h for $i \in N'$ and $h \in K'$ indicating whether node i is selected to receive a grant for broadcasting at stage h.

The MINLP *model* of CCUMRP is as follows:

$$\min \ \sum_{i \in N'} x_i + \sum_{h \in K'} \sum_{i \in N'} \beta_i^h g_i^h \tag{1}$$

$$p_i^1 = x_i \qquad\qquad\qquad\qquad\qquad\qquad\qquad\qquad i \in N' \tag{2}$$

$$p_i^k \geq \alpha \qquad\qquad\qquad\qquad\qquad\qquad\qquad\qquad i \in N' \tag{3}$$

$$1 - p_i^h \geq (1 - p_i^{h-1}) \prod_{(j,i) \in A''} (1 - g_j^h p_j^{h-1} P_{ji}) \quad i \in N' \ , \ h \in K' \tag{4}$$

$$\sum_{i \in N'} g_i^h \leq M \qquad\qquad\qquad\qquad\qquad\qquad\qquad h \in K' \tag{5}$$

$$x_i \in \{0, 1\} \qquad\qquad\qquad\qquad\qquad\qquad\qquad i \in N' \tag{6}$$

$$0 \leq p_i^h \leq 1 \qquad\qquad\qquad\qquad\qquad\qquad i \in N' \ , \ h \in K \tag{7}$$

$$g_i^h \in \{0, 1\} \qquad\qquad\qquad\qquad\qquad\qquad i \in N' \ , \ h \in K' \tag{8}$$

The objective function (1) minimizes the number of initial vertical transmissions used in the first stage ($h = 1$) and the cost of grants issued during the subsequent stages ($h \in K'$); it is therefore intended that $\beta_i^h \ll 1$. Constraints (2) ensure that all initially targeted nodes are certainly reached. Constraints (3) impose that each UE node $i \in N'$ is ultimately (at stage k) reached with probability at least α; clearly, it would be trivial to generalize the model by allowing node-specific thresholds α_i. Constraints (5) bound the total number of grants available at each stage (again, it would be trivial to let M depend on h). Finally, the constrains characterizing the model are the *nonlinear nonconvex* (4) ones, expressing the probability that node i at stage h has *not* yet received the message.

Clearly, the problem would be almost trivial were it not for (4); therefore, it is those we will concentrate upon. Taking logarithms and noting that $g_j^h = 0 \implies \log(1 - g_j^h p_j^{h-1} P_{ji}) = 0$ they can be reduced to

$$\log(1 - p_i^h) \geq \log(1 - p_i^{h-1}) + \sum_{(j,i) \in A''} g_j^h \log(1 - p_j^{h-1} P_{ji}) \qquad (9)$$

which is at least linear with respect to variables g_j^h. However, the logarithm is ill-defined when $p_i^h = 1$, which certainly happens at least whenever $x_i = 1$. We therefore consider a *restriction* of the problem by selecting a constant $\bar{p} < 1$ "arbitrarily close to 1", replacing (2) and (7), respectively, with

$$p_i^1 = x_i \bar{p} \qquad\qquad i \in N' \qquad (10)$$

$$0 \leq p_i^h \leq \bar{p} \qquad\qquad i \in N' \;,\; h \in K' \qquad (11)$$

Clearly, each feasible solution of the new model is feasible for the original one as well, and by choosing $\bar{p}$ appropriately the practical difference between the two is poised to be minimal. Finally, let us mention for future reference that for the second stage (i.e., $h = 2$) constraints (9) can be written in the form

$$\log(1 - p_i^2) \geq \log(1 - \bar{p})x_i + \sum_{(j,i) \in A''} g_j^2 \log(1 - P_{ji})x_j \qquad i \in N' \;, \qquad (12)$$

whose useful property is that the right-hand side does not contain any continuous variable (the p_i^1 having been substituted with the x_i). Therefore, (4) can be replaced by (9) for $h \in K' \setminus \{2\}$ and by (12) for $h = 2$. Nor that this, by itself, makes the constraints significantly easier to deal with. However, it allows us to propose a decomposition approach to compute globally valid lower bounds. In particular, we present a *Lagrangian decomposition* of the MINLP formulation

$$\min \; \sum_{i \in N'} x_i + \sum_{h \in K'} \sum_{i \in N'} \beta_i^h g_i^h$$

$$(10) \;,\;\; (3) \;,\;\; (12) \;,\;\; (9) \;,\;\; (5) \;,\;\; (6) \;,\;\; (11) \;,\;\; (8)$$

The idea is to relax constraints (12) and (9) with *Lagrangian multipliers* $\lambda_i^h \geq 0$ for $i \in N'$ and $h \in K'$. In so doing, the problem is decomposed into k separate sub-problem; this is clearly due to the fact that (12)/(9) are the only constraints that link the variables of one stage to those of the following one. One may expect that each sub-problem has the variables corresponding to one specific level $h \in K$, but in fact the decomposition is somewhat different, and perhaps somewhat unusual. Indeed, each sub-problem actually has variables "of one kind" for one stage h and variables "of another kind" for the subsequent stage $h + 1$ (if any). This is due to the terms $g_j^h \log(1 - p_j^{h-1} P_{ji})$ in (9) (and, similarly, $g_j^2 \log(1 - P_{ji})x_j$ in (12)) that link together variables g_j^h with variables p_j^{h-1} (x_j). We will now describe the sub-problems. Due to the special nature of the first stage, the corresponding sub-problem clearly has a particular structure. Not having a subsequent stage, the sub-problem corresponding to the last stage also has a peculiar form. All the sub-problems corresponding to intermediate stages rather share the same structure.

The *first sub-problem* ($h = 1$) contains the x_i variables (that substitute for the probability variables p_i^1) of the first stage and the grant variables g_i^2 of the second stage, reading

$$\min \ \sum_{i \in N'} \left[x_i + \beta_i^2 g_i^2 + \lambda_i^2 \left(\log(1 - \bar{p})x_i + \sum_{(j,i) \in A''} g_j^2 \log(1 - P_{ji})x_j \right) \right] \quad (13)$$

$$\sum_{i \in N'} g_i^2 \leq M$$

$$x_i \ , \ g_i^2 \in \{0, 1\} \qquad\qquad\qquad\qquad\qquad\qquad\qquad\qquad i \in N'$$

Collecting like terms of (13), and observing that there is no point in setting $g_i^2 = 1$ if $x_i = 0$, yields:

$$\min \ \sum_{i \in N'} \left[(1 + \lambda_i^2 \log(1 - \bar{p}))x_i + \left(\beta_i^2 + \sum_{(i,j) \in A''} \lambda_j^2 \log(1 - P_{ij}) \right)g_i^2 \right]$$

$$\sum_{i \in N'} g_i^2 \leq M$$

$$g_i^2 \leq x_i \qquad\qquad\qquad\qquad\qquad\qquad\qquad\qquad\qquad i \in N'$$

$$x_i \ , \ g_i^2 \in \{0, 1\} \qquad\qquad\qquad\qquad\qquad\qquad\qquad i \in N'$$

Since all nonlinear operations are applied to constants, the problem is linear. Furthermore, the special structure of the constraints ensures that, despite the variables being integer-valued, it can easily be solved in $\mathcal{O}(n \log n)$.

Next, each of *sub-problems* ($2 < h < k$) contains grant variables g_i^{h+1} of stage $h + 1$ and probability variables p_i^h of stage h, reading

$$\min \ \sum_{i \in N'} \left[(\lambda_i^{h+1} - \lambda_i^h) \log(1 - p_i^h) + \right.$$
$$\left. \left(\beta_i^{h+1} + \sum_{(i,j) \in A''} \lambda_j^{h+1} \log(1 - p_i^h P_{ij}) \right)g_i^{h+1} \right] \quad (14)$$

$$0 \leq p_i^h \leq \bar{p} \qquad\qquad\qquad\qquad\qquad i \in N' \qquad\qquad$$

$$\sum_{i \in N'} g_i^{h+1} \leq M \qquad\qquad\qquad\qquad\qquad\qquad\qquad\qquad (15)$$

$$g_i^{h+1} \in \{0, 1\} \qquad\qquad\qquad\qquad\qquad i \in N' \qquad (16)$$

Clearly, in this problem each variable p_i^h only interacts with the others via the single term in which it is multiplied by the corresponding g_i^{h+1}. The term is highly nonlinear, but still one can consider the corresponding function

$$f_i^h(p, g) = (\lambda_i^{h+1} - \lambda_i^h) \log(1-p) + \left(\beta_i^{h+1} + \sum_{(i,j) \in A''} \lambda_j^{h+1} \log(1 - pP_{ij}) \right)g \ .$$

By computing the two costants $p_i^{h,g} = \text{argmin}\{ f_i^h(p, g) : 0 \le p \le \bar{p} \}$ for $g \in \{0, 1\}$, the sub-problem can be rewritten as

$$\min \left\{ \sum_{i \in N'} f_i^h(p_i^{h,1}, 1)g_i^{h+1} + f_i^h(p_i^{h,0}, 0)(1 - g_i^{h+1}) : (15), (16) \right\}$$

and, therefore, again easily solved in $\mathcal{O}(n \log n)$. The crux of the subproblem therefore lies in the computation of $p_i^{h,1}$ and $p_i^{h,0}$. Computing the latter is trivial, as it reduces to minimizing on $p \in [0, \bar{p}]$ the monotone function $(\lambda_i^{h+1} - \lambda_i^h) \log(1 - p)$; the optimum necessarily lies in one of the two extremes. Finding $p_i^{h,1}$, instead, requires to tackle a more complex *one-dimensional minimization problem* of the form

$$\min \left\{ f(p) = c \log(1 - p) + \sum_{i \in N'} a_i \log(1 - pb_i) : 0 \le p \le \bar{p} \right\} \quad (17)$$

where, $a_i = \lambda_i^{h+1} \ge 0, 0 \le P_{ji} = b_i < 1, c = \lambda_i^{h+1} - \lambda_i^h$ is unrestricted in sign, and whose solution is discussed below.

Finally, the remaining *Lagrangian term for k* is

$$\min \left\{ \sum_{i \in N'} -\lambda_i^k \log(1 - p_i^k) : \alpha \le p_i^k \le \bar{p} \right\}$$

that is separable over i; being the objective convex ($\lambda_i^k \ge 0$), the optimum is in the left endpoint $p_i^k = \alpha$.

The crucial step is clearly the ability to efficiently solve the *one-dimensional problem* (17). Yet, if $c \ge 0$ then the problem is trivial: $f(p)$ is a decreasing function with $\lim_{p \to 1^-} f(p) = -\infty$, so the minimum is attained at $\bar{p}$. We will therefore concentrate on the case where $c < 0$ instead, for which we will prove that there is *at most one* critical point $p_0 \in [0, \bar{p}]$; moreover, the minimum is either attained at 0 or p_0. Indeed, $f(p)$ is the sum of the increasing function $c \log(1 - p)$ ($c < 0$) with vertical asymptote at $p = 1$, and n decreasing functions $a_i \log(1 - pb_i)$ ($a_i \ge 0$) with vertical asymptotes at $p = 1/b_i > 1$ (since $b_i < 1$). Hence, clearly as $p \to 1$ the increasing function dominates: $\lim_{p \to 1^-} f(p) = +\infty$, and $f(p)$ has to be strictly increasing "close to" $\bar{p}$. As p approaches 0, instead, the behaviour depends on the a_i values. In particular, we prove the following two cases: *either* the function is decreasing in 0 and becomes increasing "closer to" $\bar{p}$, which implies that the minimum is attained in the interior, *or* the function is increasing in 0 and remains so in the whole interval, which implies that the minimum is attained at 0.

Lemma 1 *If $c < 0, 0 \le b_i \le 1$ and $a_i \ge 0$ then there exists* at most one *critical point $p_0 \in [0, 1)$ such that $f'(p_0) = 0$, and $f(p)$ is strictly increasing in $p_0 < p \le 1$.*

Proof Consider $f'(p) = -\frac{c}{1-p} - \sum_{i \in N'} \frac{a_i b_i}{1-p b_i}$, we have

$$f'(p) \geq 0 \quad \Longleftrightarrow \quad -\sum_{i \in N'} \frac{(1-p)a_i b_i}{(1-p b_i)c} = h(p) \leq 1 \ .$$

It is now immediate to see that

$$h'(p) = -\sum_{i \in N'} \frac{a_i b_i}{c} \frac{b_i - 1}{(1-p b_i)^2} \leq 0$$

for all $p \in [\,0,\,1\,]$. This means that there can be at most one point $p_0 \in [\,0,\,1\,)$ such that $f'(p_0) = 0$, and therefore $f(p)$ is *strictly increasing* in $(\,p_0,\,1\,)$. □

We now analyse convexity of f, showing that if the function is non-convex then there is exactly one point $\hat{p}$ in which the second derivative changes its sign, and that the function is convex in $[\,\hat{p},\,1\,]$.

Lemma 2 *If $c < 0$, $0 \leq b_i \leq 1$ and $a_i \geq 0$, then there exists* at most one *point $\hat{p} \in [\,0,\,1\,)$ with $f''(\hat{p}) = 0$, and $f(p)$ is* convex *in $\hat{p} \leq p \leq 1$.*

Proof Along the same lines, for $f''(p) = -\frac{c}{(1-p)^2} - \sum_{i \in N'} \frac{a_i b_i^2}{(1-p b_i)^2}$ we have

$$f''(p) \geq 0 \quad \Longleftrightarrow \quad -\sum_{i \in N'} \frac{(1-p)^2 a_i b_i^2}{(1-p b_i)^2 c} = h(p) \leq 1$$

which similarly yields

$$h'(p) = -\sum_{i \in N'} \frac{a_i b_i^2}{c} 2(1-p)(1-p b_i) \frac{b_i - 1}{(1-p b_i)^4} < 0$$

for all $p \in [\,0,\,1\,]$. Again, this implies that if $f''(\hat{p}) = 0$ for some $\hat{p} \in [\,0,\,1\,)$, then $f''(p) \geq 0$ (i.e., f is convex) for all $\hat{p} \leq p \leq 1$. □

To recap, the following *three* cases can happen:

1. $f(p)$ is increasing in $[\,0,\,\bar{p}\,]$, hence the minimum is 0;
2. $f(p)$ is decreasing in 0 but convex in $[\,0,\,\bar{p}\,]$, hence the minimum is in the interior of the interval;
3. $f(p)$ is decreasing in 0 and convex in $[\,\hat{p},\,\bar{p}\,]$ for some $\hat{p} > 0$, hence the minimum lies in the interval $[\,\hat{p},\,\bar{p}\,]$;

that are represented in Figs. 2, 3, and 4, respectively.

From an algorithmic viewpoint, such a function can be efficiently globally minimized using a simple globalization of *Newton's method*. We keep an interval $[\,p_-,\,p_+\,]$ such that $f'(p_-) < 0$ and $f'(p_+) > 0$ (initialized as $[\,0,\,\bar{p}\,]$, unless

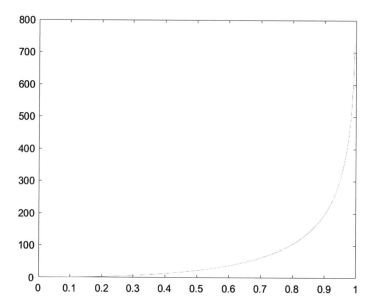

Fig. 2 Increasing

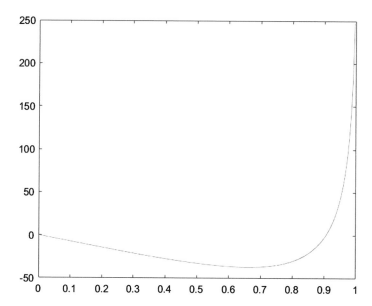

Fig. 3 Convex

$f'(0) \geq 0$ in which case we immediately terminate). If $f''(p_-) < 0$ (f is non convex at p_-) we use a simple bisection rule to find a point $p_- < p' < p_+$, we compute $f'(p')$ and shrink the interval accordingly. Otherwise (f is convex at

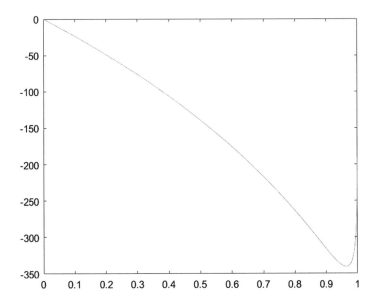

Fig. 4 Concave, then convex

p_-) we compute Newton's step, and we accept it if it belongs to the interval and it shrinks it enough; otherwise we revert to the simple bisection rule. This is clearly convergent, and typically quadratically so in the tail. Note that in our function the minimum is often close to 1, so instead of using a standard bisection we use the point $p' = p_- + 3/4(p_+ - p_-)$, as this typically leads to faster initial convergence.

3 Algorithmic Approaches and Experiments

Due to space restrictions, we briefly discuss the algorithmic approach that we developed using the proposed model and decomposition method.

It is well-known that for each choice of $\lambda \in \mathbb{R}^{2n}_+$, the solution of the corresponding *Lagrangian relaxation* provides a valid global lower bound on the optimal value of the original problem. To find the best possible Lagrangian relaxation, one then has to solve the *Lagrangian Dual* problem, i.e., maximize over all $\lambda \geq 0$ the *Lagrangian function* consisting of the sum of the k terms previously described. The efficiency of the solution process obviously depend on the specific algorithm used to solve the Lagrangian Dual; in our case we use the freely available implementation of the *(generalized) proximal Bundle method* [2] already used with success in other applications (e.g., [4, 5]) provided by the NDOSolver/FiOracle suite of C++ solvers for NonDifferentiable Optimization problems developed by the Department of Computer Science of the University of Pisa [9]. We refer to [2] and [9] for details on Bundle methods.

A fundamental component of any partial enumeration approach are the *heuristics* used to produce good feasible solutions that can be used to prune nodes of the decision tree (and that ultimately provide the returned best solution). We do this, potentially at each iteration, using both the integer, but (typically) not feasible, solution that we obtain by computing the Lagrangian function, and the continuous, but (quickly) "almost feasible", convexified solution that can be obtained as a by-product of solving the master problem in the Bundle method [3]. Actually, exploiting both synergistically has been shown to be useful in some applications [1, 6].

Since both upper and lower bounds obtained with the methods previously discussed are not very tight, we implemented an *implicit enumeration* (Branch-and-Bound) algorithm in order to obtain better gaps.

We tested the model on the realistic scenarios constructed with the help of the SimuLTE simulator developed at the Department of Information Engineering of the University of Pisa [10]. The tool allows to create many instances of the problem tuning the main parameters of interests; in our experiments we mainly concentrated on the number of UEs, on the radius (in meters) of the geographical region of interest, and on the required coverage probability α.

We compared our Lagrangian-based B&B with the state-of-the-art, general-purpose MINLP solver BARON [8] 18.11.12, as well as with a highly-tailored combinatorial heuristic available in SimuLTE and described in [7]. For BARON, we scaled the objective function by a factor of 5 so that all the coefficients are integer, allowing it to also exploit integrality to round up the lower bound. All codes have been compiled with g++ 7.4.0 and ran single-threaded on a machine sporting a 16-core Intel Xeon5120 CPU@2.20 GHz and 64 Gb RAM, running Ubuntu 18.04. The results are reported in the following Tables, with two different time limits: 300 s and 3000 s. The instances are characterized by the number of UEs ("#"), the radius ("r") and the covering probability ("α"). For both exact methods we report the total running time ("time") if they terminated before the time limit, and "–" otherwise, plus the total number of B&B nodes ("nodes"). We also report the *inherent gap* ("gap"), i.e., $(UB - LB)/\max\{1, LB\}$ (in percentage), where UB and LB are the best upper and lower bound on the optimal value produced by the corresponding algorithm at termination. To better represent the relative quality of the upper and lower bounds, we also separately report the *primal gap* ("pgap") $(UB - \underline{UB})/\max\{1, \underline{UB}\}$ and the *dual gap* ("dgap") $(\overline{LB} - LB)/\max\{1, LB\}$ (in percentage), where $\underline{UB}$ and $\overline{LB}$ are, respectively, the best (lowest) known upper bound and best (highest) known lower bound on the optimal value of the instance. Note that since the largest and hardest instances were not solved within 3000 s, a 0 primal or dual gap does necessarily means that the corresponding UB/LB are the optimal value, but only that they are the best ever found in our experiments.

The results in Table 1 clearly show how challenging CCUMRP is. Only 10-UEs instances can be all solved to optimality by our approach within the 5-min time

Table 1 Computational results, time limit 300 s

Instances			BARON					B&B					CH
#	r	α	Time	Nodes	gap	pgap	dgap	Time	Nodes	gap	pgap	dgap	pgap
10	100	0.92	4.86	1	0.00	0.00	0.00	0.59	20	0.00	0.00	0.00	0.00
10	100	0.95	3.07	1	0.00	0.00	0.00	0.58	20	0.00	0.00	0.00	0.00
10	100	0.96	3.43	1	0.00	0.00	0.00	0.67	20	0.00	0.00	0.00	0.00
10	250	0.92	4.92	1	0.00	0.00	0.00	0.44	20	0.00	0.00	0.00	0.00
10	250	0.95	75.39	1	0.00	0.00	0.00	0.73	20	0.00	0.00	0.00	0.00
10	250	0.96	31.32	1	0.00	0.00	0.00	0.46	20	0.00	0.00	0.00	71.4
10	500	0.92	80.67	84	0.00	0.00	0.00	193.8	12, 323	0.00	0.00	0.00	71.4
10	500	0.95	44.45	52	0.00	0.00	0.00	44.00	2717	0.00	0.00	0.00	40.0
10	500	0.96	383.4	1597	0.00	0.00	0.00	229.1	5130	0.00	0.00	0.00	46.7
10	750	0.92	269.2	1778	0.00	0.00	0.00	153.42	2402	0.00	0.00	0.00	29.4
10	750	0.95	–	715	4.00	0.00	4.00	208.5	6880	0.00	0.00	0.00	38.5
10	750	0.96	–	717	13.0	0.00	13.0	29.87	1026	0.00	0.00	0.00	50.0
10	1000	0.92	1.78	1	0.00	0.00	0.00	79.81	2913	0.00	0.00	0.00	26.9
10	1000	0.95	1.42	1	0.00	0.00	0.00	210.0	13, 754	0.00	0.00	0.00	36.4
10	1000	0.96	0.82	1	0.00	0.00	0.00	1.91	120	0.00	0.00	0.00	63.6
25	100	0.92	–	1	3780	3050	120	121.6	164	0.00	0.00	0.00	0.00
25	100	0.95	–	17	100	0.00	100	98.45	130	0.00	0.00	0.00	0.00
25	100	0.96	–	18	80.0	0.00	80	57.56	84	0.00	0.00	0.00	0.00
25	250	0.92	–	1	3780	3050	120	12.83	58	0.00	0.00	0.00	0.00
25	250	0.95	–	1	3780	2600	140	11.30	58	0.00	0.00	0.00	0.00
25	250	0.96	–	1	3780	2600	140	10.04	56	0.00	0.00	0.00	0.00
25	500	0.92	–	1	3780	1354	260	–	1995	40.0	7.69	30.0	30.8
25	500	0.95	–	1	3780	1354	260	–	1983	23.1	23.1	0.00	69.2
25	500	0.96	–	1	3780	1250	260	–	1569	23.1	14.3	0.00	42.9
25	750	0.92	–	2	1160	456	80	–	642	40.0	2.94	8.00	32.4
25	750	0.95	–	5	1081	425	88	–	1263	23.3	2.78	0.00	22.2
25	750	0.96	–	4	1081	425	88	–	1004	23.3	2.78	0.00	22.2
25	1000	0.92	–	12	330	210	25	–	1332	14.6	3.28	0.00	36.2
25	1000	0.95	–	10	311	205	26	–	1185	12.5	1.61	3.57	29.0
25	1000	0.96	–	12	294	200	25	–	951	12.3	1.59	5.26	47.6
50	100	0.92	–	1	6280	5133	40	–	283	100	0.00	20.0	0.00
50	100	0.95	–	1	6280	5133	40	–	283	100	0.00	20.0	0.00
50	100	0.96	–	1	6280	5133	40	–	283	100	0.00	20.0	0.00
50	250	0.92	–	1	780	550	80	–	283	60.0	0.00	20.0	0.00
50	250	0.95	–	1	6280	4386	80	–	283	80.0	0.00	20.0	0.00
50	250	0.96	–	1	6280	4386	80	–	284	80.0	0.00	20.0	0.00
50	500	0.92	–	1	6280	2143	140	–	283	180	0.00	40.0	21.4
50	500	0.95	–	1	6280	1993	160	–	284	114	0.00	14.3	20.0
50	500	0.96	–	1	6280	1993	160	–	283	87.5	0.00	0.00	20.0
50	750	0.92	–	1	6280	913	200	–	292	230	6.45	0.00	3.20

(continued)

Table 1 (continued)

Instances			BARON					B&B					CH
#	r	α	Time	Nodes	gap	pgap	dgap	Time	Nodes	gap	pgap	dgap	pgap
50	750	0.95	–	1	6280	772	260	–	291	192	5.56	0.00	22.2
50	750	0.96	–	1	6280	749	260	–	290	185	0.00	0.00	18.9
50	1000	0.92	–	1	6280	398	560	–	283	220	1.59	40.0	11.1
50	1000	0.95	–	1	6280	376	600	–	280	156	4.55	11.1	28.8
50	1000	0.96	–	1	6280	355	700	–	280	154	2.90	25.0	31.9

limit; it is generally more efficient than BARON (which fails to solve two) except for very large r, where BARON closes at root node. Interestingly, the combinatorial heuristic—which is the state-of-the-art for the problem up until this work—provides solutions that can be in excess of 50% off the optimum, although of course does so in orders-of-magnitude less time. When the size of the instances grows, BARON is basically unable to solve the problem except in a handful of cases, providing both lower and especially upper bounds that are of no practical value. Our approach cannot be exactly deemed to be very successful, with final gaps up to 40% with 25 users and even in excess of 200% with 50 users; however, it still produces the best solutions and lower bounds.

Moving to the time limit of 3000 s, depicted in Table 2, confirms that our approach at least scales much better than BARON; the much faster bound computation allows to enumerate more B&B nodes, which ultimately results in much better upper and lower bounds. In particular, we are able to solve about half of the instances with 25 users to optimality, with the other half ending with "reasonable" gaps (at least, if compared with these of BARON, both lower and upper, and with the upper bounds provided by the combinatorial heuristic). All in all, our approach is only partly successful. In particular, the lower bound is not particularly tight, which limits the size of the instances that can be practically solved. However, it at least provides a way to assess the performances of the heuristics approaches which, due to the extremely tight time limits (a handful of milliseconds) imposed by the application, are probably the only practical way of approaching the problem. Hopefully, the information provided by our approach will allow to better identify the limits of the current heuristics, and develop better ones.

Table 2 Computational results, time limit 3000 s

Instances			BARON					B&B					CH
#	r	α	Time	Nodes	gap	pgap	dgap	Time	Nodes	gap	pgap	dgap	pgap
10	100	0.92	5.13	1	0.00	0.00	0.00	0.59	20	0.00	0.00	0.00	0.00
10	100	0.95	3.12	1	0.00	0.00	0.00	0.58	20	0.00	0.00	0.00	0.00
10	100	0.96	3.48	1	0.00	0.00	0.00	0.67	20	0.00	0.00	0.00	0.00
10	250	0.92	4.69	1	0.00	0.00	0.00	0.44	20	0.00	0.00	0.00	0.00
10	250	0.95	75.22	1	0.00	0.00	0.00	0.73	20	0.00	0.00	0.00	0.00
10	250	0.96	31.12	1	0.00	0.00	0.00	0.46	20	0.00	0.00	0.00	71.4
10	500	0.92	82.39	84	0.00	0.00	0.00	193.7	12, 323	0.00	0.00	0.00	71.4
10	500	0.95	46.08	52	0.00	0.00	0.00	44.0	2717	0.00	0.00	0.00	40.0
10	500	0.96	383.4	1597	0.00	0.00	0.00	229.1	5130	0.00	0.00	0.00	46.7
10	750	0.92	269.2	1778	0.00	0.00	0.00	153.42	2402	0.00	0.00	0.00	29.4
10	750	0.95	439.0	911	0.00	0.00	0.00	208.5	6880	0.00	0.00	0.00	38.5
10	750	0.96	1456	2605	0.00	0.00	0.00	29.9	1026	0.00	0.00	0.00	50.0
10	1000	0.92	1.78	1	0.00	0.00	0.00	79.81	2913	0.00	0.00	0.00	26.9
10	1000	0.95	1.66	1	0.00	0.00	0.00	210.0	13, 754	0.00	0.00	0.00	36.4
10	1000	0.96	0.90	1	0.00	0.00	0.00	1.91	120	0.00	0.00	0.00	63.6
25	100	0.92	–	71	100	0.00	100	121.6	164	0.00	0.00	0.00	0.00
25	100	0.95	–	94	80.0	0.00	80.0	98.5	130	0.00	0.00	0.00	0.00
25	100	0.96	–	109	80.0	0.00	80.0	57.6	84	0.00	0.00	0.00	0.00
25	250	0.92	–	107	80.0	0.00	80.0	12.8	58	0.00	0.00	0.00	0.00
25	250	0.95	–	21	100	0.00	100	11.3	58	0.00	0.00	0.00	0.00
25	250	0.96	–	18	100	0.00	100	10.0	56	0.00	0.00	0.00	0.00
25	500	0.92	–	29	3680	1354	160	–	5300	27.3	7.69	18.2	30.8
25	500	0.95	–	28	3050	1354	117	–	7552	7.69	7.69	0.00	69.2
25	500	0.96	–	39	3050	1250	117	–	6404	23.1	14.3	0.00	42.9
25	750	0.92	–	35	950	456	50.0	–	4295	34.6	2.94	3.85	32.4
25	750	0.95	–	52	845	425	50.0	–	8314	23.3	2.78	0.00	22.2
25	750	0.96	–	49	845	425	50.0	–	4485	26.7	5.56	0.00	22.2
25	1000	0.92	–	82	294	210	14.6	–	12, 406	12.7	1.64	0.00	36.1
25	1000	0.95	–	83	286	205	18.4	–	11, 378	10.5	1.61	1.75	29.0
25	1000	0.96	–	104	49.0	20.6	17.7	–	10, 330	6.67	1.59	0.00	47.6
50	100	0.92	–	11	100	0.00	20	–	2805	80.0	0.00	0.00	0.00
50	100	0.95	–	1	6280	5133	40	–	2804	80.0	0.00	0.00	0.00
50	100	0.96	–	1	6280	5133	40	–	2803	80.0	0.00	0.00	0.00
50	250	0.92	–	1	80.0	0.00	40	–	2795	40.0	0.00	0.00	0.00
50	250	0.95	–	1	100	0.00	40	–	2796	60.0	0.00	0.00	0.00
50	250	0.96	–	1	100	0.00	40	–	2794	60.0	0.00	0.00	0.00
50	500	0.92	–	1	6280	2143	140	–	2773	100	0.00	0.00	21.4
50	500	0.95	–	1	6280	1993	160	–	2779	87.5	0.00	0.00	20.0
50	500	0.96	–	1	6280	1993	160	–	2763	87.5	0.00	0.00	20.0
50	750	0.92	–	8	3040	913	0.00	–	2947	230	6.45	0.00	3.20

(continued)

Table 2 (continued)

Instances			BARON					B&B					CH
#	r	α	Time	Nodes	gap	pgap	dgap	Time	Nodes	gap	pgap	dgap	pgap
50	750	0.95	–	10	3040	773	30.0	–	2942	177	0.00	0.00	22.2
50	750	0.96	–	8	3040	749	30.0	–	2955	185	0.00	0.00	18.9
50	1000	0.92	–	20	1470	398	40.0	–	2825	125	0.00	0.00	11.1
50	1000	0.95	–	13	1395	376	42.9	–	2880	127	3.03	0.00	28.8
50	1000	0.96	–	14	1327	355	59.1	–	2784	97.1	0.00	0.00	31.9

Acknowledgments The authors gratefully acknowledge the partial financial support from the Italian Ministry of Education, University and Research (MIUR), under the project "Nonlinear and combinatorial aspects of complex networks" (grant PRIN 2015B5F27W).

References

1. Borghetti, A., Frangioni, A., Lacalandra, F., Nucci, C.A.: Lagrangian heuristics based on disaggregated bundle methods for hydrothermal unit commitment. IEEE Trans. Power Syst. **18**(1), 313–323 (2003)
2. Frangioni, A.: Generalized Bundle Methods. SIAM J. Optim. **13**(1), 117–156 (2002)
3. Frangioni, A.: About Lagrangian Methods in integer optimization. Ann. Oper. Res. **139**(1), 163–193 (2005)
4. Frangioni, A., Gorgone, E.: Generalized bundle methods for sum-functions with "Easy" components: applications to multicommodity network design. Math. Program. **145**(1), 133–161 (2014)
5. Frangioni, A., Lodi, A., Rinaldi, G.: New approaches for optimizing over the semimetric polytope. Math. Program. **104**(2–3), 375–388 (2005)
6. Frangioni, A., Gentile, C., Lacalandra, F.: Solving unit commitment problems with general ramp constraints. Int. J. Electr. Power Energy Syst. **30**, 316–326 (2008)
7. Nardini, G., Stea, G., Virdis, A.: Scheduling multihop D2D transmissions for guaranteed message dissemination. Working paper (2020, Submitted)
8. The BARON solver. https://minlp.com/baron
9. The `ndosolver/fioracle` project. https://gitlab.com/frangio68/ndosolver_fioracle_project
10. Virdis, A., Stea, G., Nardini, G.: Simulating LTE/LTE-advanced networks with simulte. In: Obaidat, M.S., Ören, T., Kacprzyk, J., Filipe, J. (eds.), Simulation and Modeling Methodologies, Technologies and Applications, vol. 402, pp. 83–105. Advances in Intelligent Systems and Computing. Springer, Berlin (2015)

A Metaheuristic Approach for Biological Sample Transportation in Healthcare

Paolo Detti, Garazi Zabalo Manrique de Lara, and Mario Benini

Abstract In this paper, a real-world transportation problem is addressed, concerning the collection and the transportation of biological sample tubes from draw centers to the main hospital in Bologna, Italy. Blood and other biological samples are collected in different centers during morning hours. Then, they are transported to the main hospital for their analysis by a fleet of vehicles. Each sample has a given lifetime, i.e., a deadline. If a sample cannot arrive to the hospital before the deadline either is discarded or a *stabilization* process must be carried out in on of the dedicated facilities called Spoke Centers. After stabilization, a sample can be delivered to the main hospital by a new deadline. Transfers of samples are allowed at Spoke Centers. If a sample is delivered by a vehicle to a Spoke Center to be processed, it can be picked up from the Spoke Center after the stabilization by a different vehicle for the delivery to the main hospital. An Adaptive Large Neighborhood Search Algorithm is developed and tested. Computational experiments on different sets of instances based on real-life data are presented.

Keywords Vehicle routing problem · Healthcare · Transfers · Adaptive Large Neighborhood Search

1 Introduction

The transportation problem addressed in this paper arises from a real-world healthcare application, concerning the reorganization of the transportation of biological sample tubes from draw centers to the main hospital in Bologna, Italy, hereafter called HUB. The problem basically consists in collecting and routing blood and other biological samples from different locations and in delivering them to the HUB for the analysis.

P. Detti (✉) · G. Z. M. de Lara · M. Benini
University of Siena, Siena, Italy
e-mail: detti@dii.unisi.it; mbenini@diism.unisi.it

© The Author(s), under exclusive license to Springer Nature Switzerland AG 2021
C. Gentile et al. (eds.), *Graphs and Combinatorial Optimization:*
from Theory to Applications, AIRO Springer Series 5,
https://doi.org/10.1007/978-3-030-63072-0_23

293

Blood and other biological samples are drawn out in different centers, during morning hours. Then, each sample must be transported to the HUB for the analysis by a deadline. In fact, samples have a limited lifetime, i.e., a deadline, and have to arrive to the main hospital by the deadline. If a sample cannot arrive to the HUB before the end of its lifetime either is discarded or is *stabilized* to give it an extra lifetime. After stabilization, a sample can be delivered to the main hospital by a new deadline. The stabilization process is performed in dedicated facilities called *Spoke Centers*. The overall objective of the problem is to minimize the number of samples arriving late at the HUB and subsequently minimize the total distance traveled by the vehicles. Furthermore, since a Spoke Center is highly expensive in terms of machines and dedicated personnel, at a strategic level, a further objective is to evaluate whether some of the Spoke Centers can be closed. The problem has been first addressed in [1], where a mixed integer linear programming problem has been proposed and solved on small instances generated from real data. In the literature there are different works that address blood sample collection problems, [4, 5, 10, 11]. Grasas et al. [4] consider the problem of sample collection and transportation from different collection points to a core laboratory for testing in Spain. The problem is modeled as a variant of the capacitated vehicle routing problem with open routes and route length constraints, and a heuristic based on a genetic algorithm is proposed. A similar problem is addressed in [10] where a mobile blood collection system is designed and a routing problem is proposed with the aim of transporting blood samples from blood mobile draw centers to the depot. A mathematical model and a 2-stage IP based heuristic algorithm are proposed to solve the problem. In [5], a vehicle routing problem is addressed for blood transportation between hospitals or donor/client sites. A hybrid meta-heuristic algorithm including genetic algorithms and local search is developed able to reduce the cost and the response time for emergency. In [11], the problem of allocating units of blood from a regional blood transfusion centre to the hospitals of its area is considered. The problem is formulated as a multiobjective transportation problem.

Spoke Centers can be modeled as transfer points, where samples may be "transferred" from one vehicle to another, after the end of the stabilization process. Indeed, the vehicle delivering a sample to a Spoke Center does not have to wait until the end of the stabilization process, but may depart after dropping the samples off. Transportation problems with transfers have been addressed in the literature [3, 6, 8]. In [8], the pickup-and-delivery problem in which transfers are allowed is addressed, and mixed integer-programming formulations are proposed and evaluated. The pickup-and-delivery problem with transfers is also addressed in [6]. The authors propose heuristics capable of efficiently inserting requests through transfer points and embed them into an Adaptive Large Neighborhood Search (ALNS) scheme. The approach is evaluated on real-life instances. Cortes et al. [3] address Dial-a-Ride problems where passengers may be transferred from one vehicle to another at specific locations. A mathematical programming formulation is presented and a solution method based on Benders decomposition is proposed.

In this paper, an Adaptive Large Neighborhood Search based algorithm is proposed able to tackle all the characteristics of the problem. The framework of the algorithm is based on the approach proposed by Ropke and Pisinger in [9] for the pickup and delivery problem with time windows, and specific developments for the problem under study have been introduced. A computational campaign on different sets of instances based on real-life data is presented. A comparison between the solutions obtained with the developed algorithm and the real solutions show the effectiveness of the proposed approach, able to attain solutions with at most 3% of samples delivered after their deadline, without any use of Spoke Centers.

The paper is organized as follows. In Sect. 2 a detailed description of the problem is presented. Section 3 describes the ALNS algorithm used to tackle the problem. The description of the real-life data and the computational results are reported in Sect. 4, and finally conclusions are given in Sect. 5.

2 Problem Description

In the addressed problem, a set of transportation requests, i.e., biological samples, must be carried from draw centers to the HUB. A fleet of vehicles, located in geographically distributed depots is available to perform the transportation requests. Given the small dimension of the samples, vehicles can be considered with unlimited capacity. The problem consists in assigning the transportation requests to the vehicles and in finding the routing of each vehicle in such a way that: (1) the number of samples arriving on time is maximized; (2) the total traveled distance is minimized. Additional constraints must be taken into account, regarding the arrival to the HUB of the samples and the fulfillment of the time windows on the pickup locations. In fact, a sample must be delivered to the hospital before a pre-specified time span from its withdrawal. In other words, each sample has a lifetime, defining a deadline in which the sample has to be delivered to the HUB. The pickup of the samples can be performed during the opening hours of the draw centers, and pickup and delivery operations require given times to load or unload a request. If a request cannot be delivered to the HUB by the deadline, a stabilization process can be performed that will give an extra lifetime to the biological sample. The stabilization process is made in geographically distributed Spoke Centers. Observe that, in a Spoke Center, a vehicle can depart after dropping the samples off, and another vehicle can pass to pick up the sample after the end of the stabilization process. Figure 1 shows a scheme of the two transportation modes of a request: either directly from the draw center to the HUB or first from the draw center to a Spoke Center and then to the HUB. In the remainder of this section, notation is introduced and a formal definition of the problem is given. Let $G = (V, A)$ be a complete directed graph, where $V = \{1, \ldots, n, n + 1, n + 2, \ldots, n + s + 2, n + s + 3, \ldots, n + s + m + 3\}$ is the node set and $A = \{(i, j) : i, j \in V\}$ is the arc set. The nodes in $P = \{1, \ldots, n\}$ are the pickup nodes of the transportation requests, the node $n + 1$, also denoted as H, is the delivery node, i.e., the HUB. The nodes in $SP = \{n + 2, \ldots, n + s + 2\}$ correspond to the Spoke Centers and the nodes in $DEP = \{n + s + 3, \ldots, n + s + m + 3\}$ are the depots of the vehicles. For each

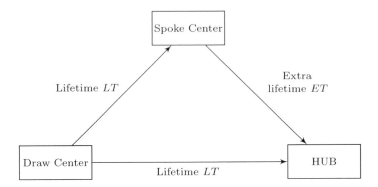

Fig. 1 Transportation modes of a sample

node i in P, LT_i is the lifetime of request i, ET_i is the extra lifetime gained if i is processed in a Spoke Center, and $[e_i, l_i]$ is the related pickup time window. The time window $[e_i, l_i]$ of each node $i \in P$ indicates that request i can only be picked up from its pickup location between time e_i and l_i. A vehicle is allowed to arrive at the location of i before the start of the time window, but it has to wait until e_i to begin the loading operation. Finally, let $K = \{1, \ldots, k\}$ be the set of vehicles.

3 An Adaptive Large Neighborhood Search Algorithm

In this section, the ALNS-based algorithm developed for the addressed problem is presented. The basic framework of the algorithm is based on the approach proposed in [9] for the pickup and delivery problem with time windows, where specific developments for the problem under study have been introduced. The algorithm has two main steps. First an initial solution, say s_0, is generated not necessarily feasible, through a fast insertion heuristic. Afterwards the solution is iteratively destroyed and repaired until a maximum number of iterations is reached. More precisely, at each iteration, the current solution is destroyed through a *destroy heuristic* and repaired by an *insertion heuristic* in order to find a different, possibly better, solution. If the new solution is accepted under given acceptance criteria, the solution is saved as the new current solution, and the algorithm continues with the next iteration.

During our ALNS algorithm, the destroying and repairing operators are only applied to solutions in which Spoke Centers are not used. In fact, the algorithm can be employed in two modes: (1) either without any use of Spoke Centers; (2) or with the use of Spoke Centers. The first mode has been designed to evaluate the possibility of excluding Spoke Centers from the overall process. In the second mode, the algorithm attempts to insert Spoke Centers periodically (i.e., at the end of each segment as described in Sect. 3.4) by an ad hoc procedure, in order to solve infeasibilities due to the violation of lifetime constraints of the current solution, if any. Then, the solution with Spoke Centers is saved if better than those obtained so far, Spoke Centers are removed from the current solution and the research process continues (on the current solution). The performances of the two modes are evaluated by an experimental campaign in Sect. 4.

3.1 Evaluation Function and Route Length Minimization

Although the ALNS method has been originally designed to explore feasible solutions [9], in our algorithm, we allow the exploration of infeasible solutions in order to facilitate the search in the solution space. More precisely, during the algorithm the violation of the two main set of constraints of the problem is allowed, namely, the constraints on the lifetimes (i.e., deadlines) and on the time windows of the requests. The violations of these constraints are weighted in the evaluation function by penalty positive coefficients. More formally, given a solution s of the problem, the evaluation function tackled by ALNS has two main components and is equal to $f(s) = f_1(s) + f_2(s)$. The function $f_1(s)$ corresponds to the total distance of the routes. The term $f_2(s)$ is a penalization component of the form $f_2(s) = \alpha t(s) + \beta w(s)$, where $t(s)$ and $w(s)$ represent the total violation of solution s with respect to lifetimes and time windows constraints, respectively, and α and β are positive penalty coefficients. In more detail, given a solution s of the problem, constraint violations are calculated as in (1) and (2), where, for each pickup request $i \in P$:

- T_i and D_i are the times in s, in which the request is ready to be transported and is delivered to the main hospital, respectively, if i is not stabilized;
- if i is stabilized, T_{ST_i} and D_{ST_i} are the times in s, in which the request is ready to be transported after the stabilization and is delivered to the main hospital, respectively;
- $\delta(i)$ is 1 if i is stabilized in a Spoke Center and 0 otherwise, and A_i is the time in which i is picked up by a vehicle from its pickup location. (Note that, $(a)^+ = a$ if $a > 0$ and 0 otherwise.)

$$t(s) = \sum_{i \in P} \left([(D_i - T_i) - LT_i]^+ (1 - \delta(i)) + [(D_{ST_i} - T_{ST_i}) - ET_i]^+ \delta(i) \right) \quad (1)$$

$$w(s) = \sum_{i \in P} (A_i - l_i)^+. \quad (2)$$

Initially, coefficients α and β are set to given values α_0 and β_0, respectively. At each iteration, they are modified by a factor $1 + \eta$, where $\eta > 0$, as follows. If the lifetime constrains are violated (are not violated) by the current solution s, then $\alpha = \alpha(1 + \eta)$ is set ($\alpha = \alpha/(1 + \eta)$ is set). The same rule is applied to β with respect to time windows constraints.

In order to minimize the length of the routes, a simple procedure is applied to each solution s found during the algorithm. As in [2], the procedure tries to postpone the starting of a route as much as possible while lifetime constraints are not violated and the violation of the time window constraints is not increased. The procedure is only applied to routes not containing any violation of lifetime constraints.

3.2 Destruction and Repairing Operators

In the ALNS algorithm developed for the problem, destruction operations are performed by two removal heuristics. The heuristics are described below.

Worst Removal
The Worst Removal heuristic has been proposed in [9] and is based on calculating the change in the objective function value when one of the selected requests is removed. From the current solution s, n_R requests are deleted in a semi-random way. The probability that a request is removed from the solution is proportional to the improvement that the request produces for the objective function.

Random Removal
This heuristic randomly deletes n_R different requests from the solution.

The repairing operations are performed by the two heuristics described below.

Best Insertion
Best Insertion is a construction heuristic based on greedy criteria. At each iteration, the insertion cost of each request is calculated for each possible position. The request with the minimum insertion cost is then inserted in the position with the smallest cost. The heuristic terminates when all the requests are added to the solution. The heuristic stops when all requests are routed or none can be inserted [9].

Regret-2
This heuristic is based on the notion of regret used, for example, in [7] for the vehicle routing problem with time windows (VRPTW), and also used in [9]. The Regret-2 heuristic is based on the idea of inserting first the requests that will produce a bigger increase of the objective function, when are not inserted immediately. At each iteration, the request to be inserted is the one that maximizes the increase of the objective function, called *regret* value. The procedure is repeated until all the requests are inserted.

3.3 Adaptive Component and Heuristic Selection

At each iteration of the ALNS algorithm, a pair of destroy-repair heuristics (described in Sects. 3.2) is selected with a certain probability. The selection is made using a *roulette wheel selection principle* [9], described in the following. A weight w_h and a score π_h is assigned to each (destroy or repair) heuristic h. During the algorithm, heuristic h is chosen with the following probability $w_h / \sum_j w_j$.

The weights w_h are automatically adjusted during the algorithm using the scores. The scores π_h store the information on the performance of each heuristic. A heuristic producing better results will have a higher score. During the search, the ALNS algorithm is divided into *segments* that correspond to I iterations. (In our experiments we set $I = 100$.) The score of each heuristic is set to 0 at the beginning of each segment, and, at each iteration of the segment, is increased of σ_1, σ_2 or σ_3, where σ_1, σ_2 and σ_3 are algorithm's parameters, as follows. If the solution produced by the applied destroy/repair heuristic pair is new and better than the best known solution, σ_1 is used; if the solution produced by the applied heuristic pair has not been yet accepted and better than the current solution then σ_2 is used; and if the solution produced by the applied heuristic pair, completely new, has not been yet accepted and is worse than the current solution σ_3 is used. The scores of the pair of destroy and repair heuristics are updated equally, as we cannot specify whether the destroy or the repair heuristic was the one with the good performance. At the beginning of the algorithm the same weight is assigned to all the heuristics. At the end of each segment t, the weight of each heuristic h is updated for the next segment $t + 1$ as follows: $w_{h,t+1} = w_{ht}(1 - r) + r \frac{\pi_h}{\theta_h}$, where π_h is the score of the heuristic h at the end of segment t and θ_h is the number of times the heuristic has been used so far. The parameter r controls the sensitivity of the algorithm to the weight changes. Note that, if $r = 0$ the same weight will be used during all the algorithm.

In order not to get stuck in a local minima, a *simulated annealing* approach is used as acceptance criteria. Let s be the current solution, a new generated solution s' will be accepted as the new current solution with probability $e^{(f(s') - f(s))/T}$, where $T > 0$ is the *temperature*. The initial temperature is $T = T_0$ and decreases at each iteration according to the expression $T = T \cdot c$, where $0 < c < 1$ is the *cooling rate*. The initial temperature T_0 depends on the initial solution and is set as follows. Let $f(s_0)$ be the cost of the initial solution of the ALNS algorithm, then the initial temperature is calculated in such a way that if the solution is $w\%$ worse than the current solution, the solution is accepted with a probability of 0.5. More precisely the initial temperature is set as $T_0 = \frac{-w \cdot f(s_0)}{\log(0.5)}$.

3.4 Spoke Insertion Procedure

In this section we describe how the insertion of the Spoke Centers, if needed, is performed during the algorithm. (Recall that, a request stabilized in a Spoke Center gains an extra lifetime.) Given a solution s in which no Spoke Center is used, the procedure attempts to insert Spoke Centers in s. The objective is that of resolving violations of lifetime constrains, if any. During the algorithm, the Spoke Insertion procedure is executed after each *segment*, i.e. after $I = 100$ iterations, if a violation on a lifetime of a request exists. Each Spoke Center is modeled as a pickup node and a delivery node. In the delivery node the samples are left for the stabilization and in the pickup node the samples are picked up by a vehicle (after the stabilization process) to deliver them to the HUB. In the procedure, the requests violating the lifetime constraints and not already stabilized are first detected in the current solution s. Then, Spoke Centers are inserted in the routes containing such requests as follows. If a route only contains one request violating the lifetime constraint, say request i_p, the Spoke Center minimizing the sum of the distance to node i_p and to the subsequent request i_{p+1} in the route is chosen. The delivery node of the Spoke is inserted right after the node i_p. If a route has more than one request violating the lifetime constraints, the insertion of a Spoke Center is performed as described above by considering the last of such requests in the route. If all the lifetime violations are solved with this insertion the procedure ends. If not, the procedure will start again by choosing the last request in the route violating the lifetime constraints. At the end of the Spoke Insertion procedure the new solution is stored if contains fewer lifetime violations of the best solution found so far. Then, the ALNS algorithm continues on the next segment starting by the initial solution s generated at the end of the segment (before the application of the Spoke Insertion procedure).

Algorithm 1

ALNS Algorithm

Generate Initial Solution, s_0
$\quad s_{best} \leftarrow s_0$ (best solution without Spoke Centers)
$\quad s_{cur} \leftarrow s_0$
Repeat
$\quad$ **Selection** *Destroy* and *Repair* heuristics
$\quad\quad s \leftarrow s_{cur}$
$\quad\quad s \leftarrow Destroy(s)$
$\quad\quad s \leftarrow Repair(s)$
$\quad\quad$ **if** $f(s) < f(s_{best})$
$\quad\quad\quad s_{best} \leftarrow s$
$\quad\quad\quad s_{cur} \leftarrow s$
$\quad\quad$ **else**
$\quad\quad\quad$ **if** $Accepted(s, s_{cur})$
$\quad\quad\quad\quad s_{cur} \leftarrow s$
$\quad\quad\quad$ **end if**
$\quad\quad$ **end if**
$\quad\quad$ **if** end of segment **and** Spoke Centers can be used
$\quad\quad\quad s_{sp} \leftarrow InsertSpoke(s_{best})$
$\quad\quad\quad$ **if** $RTviol(s_{sp}) < RTviol(s_{spbest})$ **or** $s_{spbest} = \emptyset$
$\quad\quad\quad\quad s_{spbest} \leftarrow s_{sp}$ (best solution with Spoke Centers)
$\quad\quad\quad$ **end if**
$\quad\quad$ **end if**
Until a maximum number of iterations is reached
if Spoke Centers cannot be used
$\quad$ Apply the Post-Processing Procedure on s_{best} **return** s_{best}
else
$\quad$ Apply the Post-Processing Procedure on s_{spbest} **return** s_{spbest}

end

3.5 Post-Processing Procedure

A post-processing procedure is applied in order to improve the final solution returned by the algorithm. The post-processing procedure is focused on re-assigning the requests belonging to the same draw center that are not transported together (in the same route) with other requests produced in the same draw center. The procedure basically tries to assign to a given vehicle as many samples as possible of the same draw center, as long as this choice allows to fix the violations of deadlines' constraints. The requests are re-assigned only if the overall deadline violation with the re-assignment is less or equal than before.

4 Instance Description and Experimental Results

An experimental campaign has been performed on different sets of real data provided by the Local Healthcare Agency of Bologna, and arising from the metropolitan area of Bologna, Italy. In the area, there are 46 (blood and biological) geographically distributed draw centers. The opening days and hours of each center depend on the center itself and vary from Monday to Saturday, from 7 am to 10 am. According to the real data, in a center, a sample is drew every 3 min, approximately. The number of available vehicles is 26 during weekdays and 16 on Saturday. Vehicles are located in 8 different depots. Each day, 12 Spoke Centers are available for the stabilization process. The lifetime of a sample is of 120 min. Hence, from its withdrawal, a sample i must be delivered to the main hospital or to a Spoke Center within $LT_i = 120$ min. The stabilization process takes 30 min and gives to the stabilized sample $ET_i = 90$ min of (extra) lifetime to arrive to the main hospital (see Fig. 1). Hence, after the end of the stabilization process, a sample must be delivered to the HUB within $ET_i = 90$ min. The service time required by a vehicle to load or unload samples at a draw center or at a Spoke Center is $st_i = 10$ min. The number of draw centers available each weekday ranges from 30 to 36 from Monday to Friday and is equal to 16 on Saturday. From the real data described above, 4 sets of instances have been generated, called Set 1, Set 2, Set 3 and Set 4, by grouping samples in different ways. Each set contains 6 instances (one for each day of the week, from Monday to Saturday). More precisely, recalling that a biological sample is produced every 3 min, we organize biological samples in batches and generate instances in which a batch is created every either 10, 15, 20 or 30 min of activity of a draw center. Hence, when batches are generated every 15 min, each batch will contain 5 samples (possibly except for the last batches generated at each center). The lifetime of a batch, i.e., the deadline, is computed taking into account the time of the production of the first sample included in the batch. Therefore the bigger the batch is the smaller the lifetime of the batch is. The earliest limit of the time window assigned to each batch is set equal to the ending production time of the last sample of the batch and the latest limit of the

Table 1 Number of requests per set each day

Set	Time interval	Monday	Tuesday	Wednesday	Thursday	Friday	Saturday
1	30	104	109	100	118	100	58
2	20	147	155	140	167	141	82
3	15	208	220	200	238	200	116
4	10	312	329	300	356	300	174

time window is always set to 30 min after the closing time of the draw center. Thus, the time window width of a batch varies depending on the samples assigned to it. Table 1 shows the number of requests of each instance of Sets $1 - 4$, corresponding to the number of batches obtained by grouping samples every 30, 20, 15 and 10 min respectively, from Monday to Saturday, according to the procedure described above. The time interval used to group the samples is shown in the second column of the table. Hence, by grouping samples every 30 min we get 6 instances (one for each day from Monday to Saturday) with a number of batches, i.e., requests, ranging from 58 to 118. Note that, such a variation depends on the opening times of the draw centers, varying with the day of the week. By grouping samples every 10 min we get the biggest instances with a number of requests ranging from 174 to 356.

As detailed in Sect. 3, the proposed ALNS algorithm has different parameters. By a trial and error preliminary testing phase, $\alpha_0 = 0.05$ and $\beta_0 = 0.3$ have been in set in the evaluation function. The other parameters of the algorithm (introduced in Sect. 3.3) have been set as in [9]: $r = 0.1$, $\sigma_1 = 33$, $\sigma_2 = 9$, $\sigma_3 = 19$, $c = 0.99975$ and $w = 0.05$. At each iteration of the algorithm, 33% of the requests are removed and then re-inserted. The maximum number of iterations is fixed to $It_{max} = 10,000$.

4.1 Experimental Results

In this section, the experimental results on the four sets of instances described in the previous section are presented. All tests have been executed on a PC equipped with Intel i5 processor and 8 Gb of RAM. Five runs of the ALNS algorithm on each set of instances have been performed. Table 2 reports the average results, over the five runs, on each set of instances for every day of the week (see Table 1). The table contains the results on the best average solution (over the five runs) attained without any use of Spoke Centers, i.e., in which stabilization is not allowed (see Columns 2–5 of the table), the average results for the case in which Spoke Centers can be used (Columns 6–10) and the best solution obtained in all the runs, with or without the use of Spoke Centers (Columns 11–15). In all the cases, the solution with the smallest number of late requests, i.e. batches, is considered as the best solution. In columns 2, 6 and 11 of the table, *dist* is the average total travel distance of the best solutions on the 5 runs of the algorithm. In Columns 3, 7 and 12, B_{Late} reports on the average percentage of tardy batches. A batch is tardy if it contains at least a sample

Table 2 Results on instances of the four Sets.

	Solution without Spokes				Solution with Spokes					Best Solution				
	dist	B_{Late}	max_{Delay}	# visited	dist	B_{Late}	max_{Delay}	# visited	B_{Spoke}	dist	B_{Late}	max_{Delay}	# visited	B_{Spoke}
Set 1														
Monday	1884.20	15%	26.20	1.97	2030.80	6%	24.40	1.72	13.20	2041.00	6%	24.00	1.73	12.00
Tuesday	2167.00	24%	47.00	1.95	2503.40	12%	36.80	1.75	15.80	2418.00	6%	21.00	1.84	10.00
Wednesday	1941.00	17%	31.20	2.01	2127.40	8%	33.60	1.75	11.80	2064.00	7%	9.00	1.79	10.00
Thursday	2179.80	25%	41.80	1.98	2548.60	19 %	66.60	1.75	17.40	2506.00	14%	42.00	1.86	15.00
Friday	1844.00	15%	26.00	1.93	2053.40	5%	14.80	1.93	12.00	2112.00	1%	6.00	1.73	14.00
Saturday	1542.40	7%	8.80	2.05	1708.60	2%	10.80	2.03	3.80	1594.00	0%	0.00	2.13	1.00
Average	1926.40	17%	30.17	1.98	2162.03	9%	31.17	1.82	12.33	2122.50	6%	17.00	1.85	10.33
Set 2														
Monday	1953.00	8%	13.80	2.17	2099.80	5%	36.00	2.00	10.20	1956.00	5%	9.00	2.26	12.00
Tuesday	2230.40	22%	35.00	2.04	2652.60	12%	41.80	1.79	18.20	2596.00	6%	21.00	1.86	18.00
Wednesday	2039.60	6%	17.00	2.30	2156.60	3%	14.20	2.17	6.60	2094.00	1%	3.00	2.24	5.00
Thursday	2239.00	19%	23.20	1.99	2668.60	17%	50.00	1.71	21.80	2727.00	12%	29.00	1.84	21.00
Friday	1955.00	26%	18.40	2.25	1991.60	1%	5.80	2.18	4.40	1933.00	0%	0.00	2.13	2.00
Saturday	1616.80	15%	17.60	2.18	1693.20	0%	0.00	2.11	2.00	1632.00	0%	0.00	2.00	2.00
Average	2005.63	16%	20.83	2.16	2210.40	6%	24.63	2.00	10.53	2156.33	4%	10.33	2.06	10.00
Set 3														
Monday	2011.00	4%	8.60	2.32	2087.40	3%	22.80	2.16	5.80	2038.00	0%	0.00	2.32	3.00
Tuesday	2436.80	14%	20.20	2.19	2853.20	14%	48.40	1.85	19.00	2873.00	10%	46.00	1.81	18.00
Wednesday	2093.20	5%	10.00	2.45	2208.00	3%	17.40	2.24	7.80	2115.00	1%	2.00	2.30	4.00
Thursday	2389.40	17%	20.20	2.18	2938.40	17%	70.40	1.87	22.80	2721.00	9%	91.00	1.97	15.00
Friday	1992.40	3%	8.60	2.31	2073.40	2%	26.00	2.20	5.40	2039.00	0%	0.00	2.27	1.00
Saturday	1625.20	5%	6.60	2.11	1821.00	1%	5.20	2.07	5.20	1757.00	0%	0.00	2.13	2.00
Average	2091.33	8%	12.37	2.26	2330.23	7%	31.70	2.06	11.00	2257.17	3%	23.17	2.13	7.17

(continued)

Set 4

Monday	2121.00	< 0.01%	0.40	2.08	2128.40	< 0.01%	2.20	2.07	1.60	2097.00	0%	0.00	2.03	0.00
Tuesday	2531.80	10%	19.60	2.08	2993.80	25%	71.60	1.84	19.40	2571.00	9%	24.00	2.06	0.00
Wednesday	2160.40	1%	3.00	2.11	2271.60	3%	20.20	2.05	3.80	2147.00	0%	0.00	2.07	0.00
Thursday	2401.60	8%	11.00	2.03	2844.00	25%	70.00	1.73	18.60	2368.00	8%	9.00	2.00	0.00
Friday	2099.00	< 0.01%	0.20	2.04	2136.00	1%	8.00	2.07	1.60	2059.00	0%	0.00	2.00	1.00
Saturday	1656.20	1%	4.00	1.93	1757.00	0%	0.00	1.93	2.40	1726.00	0%	0.00	2.06	1.00
Average	2161.67	3%	6.37	2.04	2355.13	9%	28.67	1.95	7.90	2161.33	3%	5.50	2.04	0.33

delivered late with respect to the deadline. In Columns 4, 8 and 13, the maximum (average) delay of a batch is reported, denoted as max_{Delay}, and, in Columns 5, 9 and 14, # *visited* is the average number of times a draw center is visited by a vehicle. Finally, in Columns 10 and 15, B_{Spoke} is the average number of batches delivered to a Spoke Center (when Spoke Centers can be used).

We note that, apart from the instances of Set 4, the use of Spoke Centers yields solutions with a smaller number of late batches (see Columns 3 and 7 of Table 2). On the other hand, in terms of solution quality (i.e., percentage of tardy batches), the ALNS algorithm attains the best results on the instances of Set 4 without any use of Spoke Centers, finding solutions in which 3% of batches are late on average. On the same set of instances, but with the use of Spoke Centers, the ALNS algorithm yields solutions with 9% of late batches, on average. Such unexpected result is probably due to the large dimension of the instances of Set 4, containing more than 300 requests in the days from Monday to Friday. In fact, the instances of Set 4 related to the days of Tuesday and Thursday have the biggest dimension, with 329 and 356 requests respectively. On these instances, the algorithm finds solutions with the largest percentage of late batches when Spoke Centers can be used (equal to 25%, much bigger than the solutions without spokes on the same set, with at most 10% of late batches). When Spoke Centers are enabled, the algorithm has the best performance on Sets 2 and 3, with 6% and 7% of tardy batches, respectively. Observe that, the instances of these sets are characterized by a smaller number of requests than Set 4 and by requests with larger time-windows than those of Set 1. According to the traveled distance, solutions with Spoke Centers enabled and a smaller number of samples per batch are those with a higher overall traveled distance, *dist*. In solutions without Spoke Centers, each draw center is visited by a vehicle a number of times ranging from 1.98 (on Set 1) to 2.26 on Set 3. Smaller values are attained for instances with Spoke Centers in which the draw centers are visited from 1.82 to 2.06 times, on average. Average computational times of the algorithm on the instance sets ranges from about 145 s in Set 1 to about to 6400 s in Set 4.

In the real-world application, around 40% of the samples arrives late with respect to the lifetime constraints, on average, and each draw center is visited at most once by a vehicle. On the other hand, the ALNS algorithm is able to attain solutions in which about 3% of samples are delivered after their deadline, without any use of Spoke Centers and visiting about twice each draw center. Such a result is interesting at the strategic decision level, since it highlights that most of the samples can be delivered on time even when no stabilization is performed.

5 Conclusions

In this paper, a problem arising from a real-world healthcare application has been presented, in which biological samples must be transported from draw centers to the main hospital within given deadlines. Dedicated centers, i.e., Spokes, can be used to enlarge the deadlines of the samples. An ALNS algorithm has been proposed able to tackle all the characteristics of the problem. Computational results on real-life instances show the effectiveness of the proposed approach.

References

1. Benini, M., Detti, P., Zabalo Manrique de Lara, G.: A MILP model for biological sample transportation in healthcare. In: Paolucci, M.. et al. (eds), Advances in Optimization and Decision Science for Society, AIRO Springer Series, vol. 3, pp. 81–94 (2019)
2. Cordeau, J.F., Laporte, G.: A tabu search heuristic for the static multi-vehicle dial-a-ride problem. Transp. Res. B Methodol. **37**(6), 579–594 (2003)
3. Cortés, C.E., Matamala, M., Contardo, C.: The pickup and delivery problem with transfers: formulation and a branch-and-cut solution method. CEJOR **200**(3), 711–724 (2010)
4. Grasas, A., Ramalhinho, H., Pessoa, L.S., Resende, M.G., Caballé, I., Barba, N.: On the improvement of blood sample collection at clinical laboratories. BMC Health Serv. Res. **14**(1), 12 (2014)
5. Karakoc, M., Gunay, M.: Priority based vehicle routing for agile blood transportation between donor/client sites. In: 2017 International Conference on Computer Science and Engineering (UBMK), pp. 795–799. IEEE, Piscataway (2017)
6. Masson, R., Lehuédé, F., Péton, O.: An adaptive large neighborhood search for the pickup and delivery problem with transfers. Transp. Sci. **47**(3), 344–355 (2013)
7. Potvin, J.Y., Rousseau, J.M.: A parallel route building algorithm for the vehicle routing and scheduling problem with time windows. Eur. J. Oper. Res. **66**(3), 331–340 (1993)
8. Rais, A., Alvelos, F., Carvalho, M.: New mixed integer-programming model for the pickup-and-delivery problem with transshipment. Eur. J. Oper. Res. **235**(3), 530–539 (2014)
9. Ropke, S., Pisinger, D.: An adaptive large neighborhood search heuristic for the pickup and delivery problem with time windows. Transp. Sci. **40**(4), 455–472 (2006)
10. Şahinyazan, F.G., Kara, B.Y., Taner, M.R.: Selective vehicle routing for a mobile blood donation system. Eur. J. Oper. Res. **245**(1), 22–34 (2015)
11. Sapountzis, C.: Allocating blood to hospitals as a multiobjective transportation problem. In: Medical Informatics Europe, vol. 90, pp. 733–739. Springer, Berlin (1990)

Optimal Planning of Waste Sorting Operations Through Mixed Integer Linear Programming

Diego Maria Pinto and Giuseppe Stecca

Abstract Circular economy imposes a new view of operations with the aim of zero waste. To obtain this result it is critical to adopt an holistic approach and to optimize every step of the production and logistics processes. This work investigates the operations of waste recycling centers where materials are collected by a fleet of trucks and then sorted in order to be converted in secondary raw materials. The activity is characterized by low margins, uncertainties in supplies, and difficulties to track flows. In these settings, we propose a mixed integer linear programming model to schedule the sorting operations of each phase of the waste sorting process. The model can be described as a variant of a lot size model with non linear costs (approximated by mean of piece-wise linear functions) with the additional features of scheduling the operations and allocating the appropriate workforce dimension. The model is tested on a real world case study and results demonstrate the validity of the approach.

Keywords Circular economy · Lot sizing · Mixed integer linear programming · Waste recycling

1 Introduction

Waste management is a worthwhile and important challenge concerning both the protection of the environment and the conservation of natural resources. Notably, a considerable attention has been directed over the last decade towards the optimization of planning procedures related to waste management. In particular, performances of municipal solid waste systems have been improving thanks to a noticeable commitment of decision makers and research efforts regarding the

D. M. Pinto (✉) · G. Stecca
CNR-IASI, Rome, Italy
e-mail: diegomaria.pinto@iasi.cnr.it; giuseppe.stecca@iasi.cnr.it

optimization of each system components. In the meantime, some similar kind of optimization models have been drastically reducing transportation costs enhancing the growth of the online shopping of any sort of good. As a result, while logistic companies start serving a new magnitude of customers, also a new dimension of packaging waste started affecting the overall waste system. This leads to the need of a stronger technological and strategic decision support to packaging waste facilities in order to lower all the extra costs involved with the selective collection and sorting of this kind of waste. Not only logistic companies but also every other kind of industry generates a considerable amount of packaging waste. In Europe the Directive 2004/12/EC on packaging and packaging waste laid down the European recycling and recovery targets. In particular, official reporting on packaging waste for all EU Member States was implemented in 2007 and since then Eurostat monitors also the developments of this statistics. For example, in 2016, 170 kg of packaging waste was generated per inhabitant in the EU (varying from 55 kg per inhabitant in Croatia and 221 kg per inhabitant in Germany). Instead, from 2007 to 2016, paper and cardboard was the main packaging waste material in the EU (35.4 million tonnes in 2016) followed by plastic and glass (16.3 million tonnes for each of these waste materials in 2016). Therefore, the need of meeting the recovery and recycling targets imposed by EU law and the rising prices of raw materials used for packaging have resulted in an increasing interest in the recovery of materials from the waste streams. Moreover, the recycling industry is characterized by very low margins and high percentage of operation and logistics costs. For this reason it is critical the optimization of the process in order to turn it in an economically sustainable business. Accordingly, the main research aim of this study is to develop a mixed integer linear programming model for planning and scheduling the packaging waste recycling operations. The model supports also other strategic decisions such as sizing the amount of processed waste and allocating the optimal number of operators for each shift of the waste sorting processes. To the best of our knowledge and literature review, the subject has not gained large academic interest previously, and with this work we intend to expand the operation research academic community understanding on packaging waste management.

In the described setting, waste companies usually serve their industrial customers according to a pull logic for the waste containers collection. Indeed, a company truck picks up the waste container of a customer whenever the company logistic services are contacted by the client for the container pick up. The production demand of the waste facility arises from the need to program and size the sorting operations of a certain quantity of waste in order to balance the availability of the buffer of received material with the production and set-up costs of sorting operations and storage costs of all the inter-operational buffers. Therefore, the simultaneity of the scheduling problem and the lot sizing problem is highlighted. One of the main problems in the field of production planning is indeed the lot sizing. Starting from the study of the main lot-sizing models of the literature, we derived the formulation of a model that could properly reflect the waste business reality. All models are wrong, but some are useful, famous quote by George E. P. Box was taken into consideration when tuning the complexity of the model because, even if we cannot describe exactly the reality, it could be very helpful if a model gets close enough. Hence, the formulation was intended to be linear and it remained so after performing some suitable linearization. The reminder of the paper is organized as follows: the recycling process is described in Sect. 1.1; the literature review is given in Sects. 1.2 and 2 is dedicated to the problem description and the MILP formulation; Sect. 3 presents the experimental results and the instances creation procedure; Sect. 4 gives some conclusions and research perspectives.

1.1   The Waste Sorting Process

A typical recycling plant manages flows of waste materials and process them by separating all the several kinds of mixed materials such as paper, cardboard, iron, wood, plastic and glass. Once those materials are separated from each other, they are considered as secondary raw materials, and each kind of them is then moved to a dedicated stock area. Depending on the type of material some minor transformation may be necessary before stocking them. At the end of a complete sorting the remaining part of the waste constituents that cannot be recycled are then treated as trash and intended to be moved to a garbage dump. The waste sorting is divided into two sorting jobs that are performed in series with an inter-operational buffer between them. Indeed, during the rest of the dissertation we always refer to them as first and second sorting respectively. The first sorting job is performed in order to separate the components of the overall waste mix of a particularly big size. Not only this matter would be too big to fit the conveyor belt on which a finer and more precise second sorting is performed, but it can be actually moved directly to the corresponding stock of secondary raw materials. This operation usually takes place in the trucks dump area where each waste container is unloaded. This area is indeed the first input buffer of the production problem addressed by this paper. The first sorting job draws material from this buffer. As a result of this sorting, the bigger and heavier part of the dumped material is moved to the corresponding stock, and the smaller and lighter part is moved to a second buffer. This buffer feeds a conveyor belt that is used in order to activate and accomplish the second sorting job. This subsequent additional operation achieves a finer separation by making the small mixed waste being carefully inspected by blue collars eventually supported by dedicated devices such as cameras, sensors and any sort of smart mechanism like a magnet used for separating small iron pieces. Depending on the dimension of the second sorting cabin this can hold a maximum number of blue collars. Each kind of matter slides into a dedicated buffer hole, one for each kind of secondary raw material. The waste sorting stage is followed by the ending packaging phase. The fundamental machine for this phase is a press, which is powered by an octopus bucket. Once a minimum workable quantity of a certain material is selected and stocked, an operator maneuvers the bucket, collects the batch of selected waste and places it all inside the press. In this way single-material compact bales are created and stacked pending collection by customers of secondary raw materials.

1.2 Literature Review

The scientific literature related to the waste management is not particularly rich but is expanding. The range of contributions is justified by the variety of technological configurations and decision levels (mainly strategic and operational). At the same time the main type of waste flows considered are municipal solid waste, even though also food waste has been addressed by some research. Indeed the great majority of works are related to the management or the strategic definition of municipal solid waste networks, such as in [3, 8, 10–12]. Instead the conventional operational task addressed by research is about waste shipment and collection trucks route optimization [2, 4, 6, 7]. Besides them, a real case application is presented in [1]. The last review of methods concerning optimal routing of solid waste collection trucks is given in [9]. Moreover, a complete survey of both strategic and tactical issues in solid waste management that have been addressed by operations research methods is presented in [5]. As far as our knowledge is concerned, none of the previous works is close to our operational type of waste management study. As a matter of fact we present in this paper a new kind of operational application of mathematical programming within the waste management paradigm.

2 Problem Definition and Modeling

In this section we describe the main operational features covered by the model and present its formulation. It will clarify how the model is able to cover the principal strategic decisions of the process while properly modeling the typical production dynamics of a reverse logistic setting. First of all it is important to notice that, in the considered industrial case, costs of storage are not measurable directly. In fact, the waste stored in the buffers have neither cost nor value attributable to it that generate a variable cost of storage. In particular, the only cost potentially attributable to the waste concerns its transportation to the plant, but this is a cost paid by those who need to dispose of their waste. Secondly, the quantity stored in the buffers has no observable value since the percentages of the secondary raw materials contained in it are not known before the sorting process is complete. Only once the sorting process is complete, the waste regains its value since it can be sold again as a secondary raw material. Of course, there are storage costs due to energy and staff employed, however, since these are indirect and constant costs, they are not covered by the model. At the same time the level of buffer storage can be such as to constitute a criticality in terms of saturation of the storage capacity. This is particularly evident when a specific level of stock is passed. Therefore, it was considered appropriate to model this dynamic through a storage cost curve which originally included a non-linearity from the exceeding of the critical stock level. The linearity of the model is indeed guaranteed using a piece-wise linear curve that approximates the real cost curve. Critical levels can be estimated considering the size of the buffer areas and the averages of both the specific weights of raw materials and the overall weight of the unloaded containers. In addition to these, the indications about the threshold perceived by the waste company in relation to the customer service level were also considered. Thanks to this information, the filling percentages of the buffers which constitute their critical stock level can be obtained.

In the following, we introduce a mixed integer linear programming (MILP) model which defines the newly introduced problem. The notations that will be used in the MILP, such as parameters and indexes, are the following:

- $j = \{1, \ldots, J\}$: index of the J sorting stages
- $p = \{1, \ldots, P\}$: index of the P time-shifts
- T: time horizon partitioned in time shifts with $t \in \{1, \ldots, T\} = T_1 \cup \ldots \cup T_P$
- C: hourly cost of each operator
- σ_t: working hours for time t determined by the corresponding shift p
- $C_t = C * \sigma_t$: cost of each operator at time t
- f_j: set-up cost of sorting stage j
- a_t: quantity of material in kg unloaded from trucks at time t
- α_j: percentage of waste processed in stage $j - 1$, received in input by buffer j
- S_j: maximum inventory capacity of the sorting stage buffer j
- LC_j: critical stock level threshold of buffer j
- ρ_j: fraction of material allowed to be left at buffer j at the end of time horizon
- K_j: single operator hourly production capacity [kg/h] of sorting stage j
- $SK_{j,t} = K_j * \sigma_t$: operator sorting capacity in sorting stage j, at time t
- M: maximum number of operators available in each time shift
- E_j: minimum number of operators to be employed in each time shift of stage j
- ∂h_j^i: slope of the i-th part of linearization of the buffer j stock cost curve

The model consider the following variables.

- $x_{j,t} \in \mathbb{Z}^+$: operators employed in the sorting stage j at time t
- $u_{j,t} \in \mathbb{R}^+$: processed quantity at stage j at time t
- $y_{j,t} \in \{0, 1\}$: equal to 1 if stage j is activated at time t, 0 otherwise
- $I_{j,t} = I'_{j,t} + I''_{j,t} \geq 0$: stock level of material in buffer j at time t; for each stage j the corresponding $I'_{j,t}$ and $I''_{j,t}$ represent the inventory level before and after reaching the critical threshold respectively.
- $w_{j,t} \in \{0, 1\}$: equal to 1 if $I''_{j,t} > 0$, 0 otherwise. Indeed, this binary variables are used to model the piece-wise linear functions of the buffer stock costs.

Considering a case study where $J = 2$ sorting stages, for the first sorting phase, $u_{1,t} \geq 0$ and $x_{1,t} \in \{0, 1\}$ represent the quantity (in kg) of material to be selected and decision to activate the process respectively at time t. For the second sorting phase, $u_{2,t} \geq 0$ and $x_{2,t} \in \{0, 1\}$ represent the quantity (in kg) of material to be selected and decision to activate the process respectively at time t. $I_{1,t}, I'_{1,t}, I''_{1,t} \geq 0$ are the inventory levels at first phase sorting buffer while $I_{2,t}, I'_{2,t}, I''_{2,t} \geq 0$ are inventory levels at second phase sorting buffer. As previously stated, w_1 and w_2 are used to model the piece-wise linear functions of the buffer stock costs. In detail $w_1 = 0$

if $I'_{1,t} < LC$, 1 if $I'_{1,t} = LC$ and $I''_{1,t} > 0$; similarly $w_2 = 0$ if $I'_{2,t} < LC$, 1 if $I'_{2,t} = LC$ and $I''_{2,t} > 0$.

The model minimizes the sum of sorting and holding costs and is detailed as following:

$$\min Z = \sum_{j \in J} \sum_{t \in T} C_t x_{j,t} + \sum_{j \in J} \sum_{t \in T} f_j y_{j,t} + \sum_{j \in J} \sum_{t \in T} \left(\partial h_j^1 I'_{j,t} + \partial h_j^2 I''_{j,t} \right) \quad (1)$$

s.t.

$$E_j \, y_{j,t} \leq x_{j,t} \leq M \, y_{j,t} \qquad\qquad \forall j \in J, t \in T_p, p \in P \qquad (2)$$

$$\sum_{j \in J} x_{j,t} \leq M \qquad\qquad\qquad\qquad \forall t \in T \qquad\qquad (3)$$

$$u_{j,t} \leq SK_{j,t} \, x_{j,t} \qquad\qquad\qquad \forall j \in J, t \in T \qquad (4)$$

$$I_{1,t} = I_{1,t-1} + a_t - u_{1,t} \qquad\qquad \forall t \in T \setminus 0 \qquad (5)$$

$$I_{j,t} = I_{j,t-1} - u_{j,t} + \alpha_j \, u_{j-1,t} \qquad \forall t \in T \setminus 0, j \in J \setminus 1 \qquad (6)$$

$$I_{j,t} = I'_{j,t} + I''_{j,t} \qquad\qquad\qquad \forall j \in J, t \in T \qquad (7)$$

$$LC_j \, w_{j,t} \leq I'_{j,t} \leq LC_j \qquad\qquad \forall j \in J, t \in T \qquad (8)$$

$$0 \leq I''_{j,t} \leq (S_j - LC_j) \, w_{j,t} \qquad \forall j \in J, t \in T \qquad (9)$$

$$I_{j,T} \leq \rho_j \, LC_j \qquad\qquad\qquad\qquad \forall j \in J \qquad\qquad (10)$$

$$x_{j,t} \in \mathbb{Z}^+ \qquad\qquad\qquad\qquad \forall j \in J, t \in T \qquad (11)$$

$$u_{j,t} \in \mathbb{R}^+ \qquad\qquad\qquad\qquad \forall j \in J, t \in T \qquad (12)$$

$$y_{j,t} \in \{0, 1\} \qquad\qquad\qquad\qquad \forall j \in J, t \in T \qquad (13)$$

The objective function (1) defines the minimization of the sum of the three cost terms, which are sorting, setup, and inventory costs respectively. (2) and (3) bounds the number of workers that can be assigned to each sorting station and to each time shift. Constraints (4) limit the quantity sorted $u_{j,t}$ to the sorting capacity dependent on the number of workers $x_{j,t}$. The remaining constraint sets define and limit the inventories: constraint set (5) defines the inventory for the first buffer, considering the inbound material a_t and the sorted material u_{1t}, while (6) defines the inventory for the other buffers corresponding to $j > 1$. Indeed, constraints (6) describe the waste flow across the sorting stages that follow one another: each subsequent inter-operational buffer j receives by the previous sorting stage $j - 1$ a quantity of waste equal to a α_j percentage of the waste processed in stage $j - 1$. Constraint sets (7), (8), and (9) define the piece-wise linear functions for inventories; in these constraints, level S_j and maximum capacity LC_j are connected with the inventory

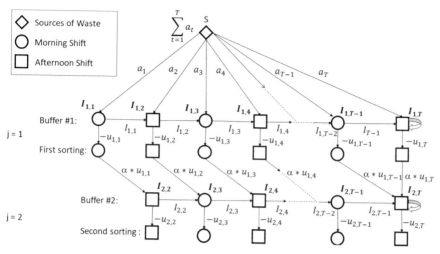

Fig. 1 Illustration of the model architecture when $J = 2$

levels through the variable $w_{j,t}$. The last constraint set (10) imposes the maximum unsorted material allowed to be left at the end of the planning period for each buffer.

Figure 1 shows a graph representation of the model over time, working shifts and sorting stages including buffers stock evolution and sorting processes.

3 Experimental Results

This section holds the main results from the studied scenarios described in the following Sect. 3.1. The model is tested on scenarios that are different in time horizon dimension and in the range of production demand in terms of kg of waste unloaded in the waste plant during each shift. All instances are created by a real-world case study from a waste sorting plant located next to Rome, Italy. We solved each scenario first by applying the Gurobi solver to the MILP model, then applying the typical planning rule regularly used by the management of the waste plant. This has been performed in order to test the model response to each of the feasible instances and to measure the model performances against the company procedure.

3.1 Instances Creation

Thanks to the availability of company data we are able to properly create real-world problem instances. Indeed, used data are rich both in terms of quantity and quality. The considered recycling plant organizes its production in two working shifts of 6

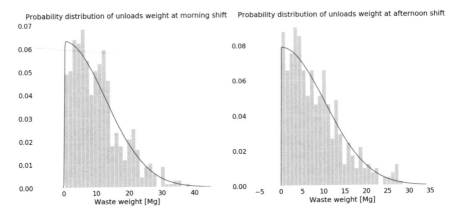

Fig. 2 Distribution of packaging waste weight unloads for each shift

Table 1 Upper and lower bounds of arrivals weight ranges in [kg]

Arrivals	Shift_1 LB	Shift_1 UB	Shift_2 LB	Shift_2 UB
Low	200	3250	200	2600
Moderate	3260	6600	2600	5400
Average	6600	10,800	5400	8650
High	10,800	16,500	8660	13,400
Extreme	16,500	38,000	13,400	28,000

and 7 h respectively. Therefore as a first step we performed a probability distribution analysis of the waste weights unloaded from incoming trucks at each working shift. The Anderson Darling test is used for the assessment of data normality. This statistical test assesses whether a sample comes from a specified distribution and set this hypothesis as the null hypothesis of the test. As a result, the hypothesis of normality is rejected with some significance level if the test statistic exceeds a given critical value. Tests are performed on data with the addiction of their negative counterpart in order to prove a perfect skew-normal distribution as the positive portion of a zero-mean normal distribution. Figure 2 shows a bar plot and the fitting distribution for each work period.

Each test failed to reject the null hypothesis, ergo both arrivals distribution are found. These distributions are then used for constructing reference intervals of instances in terms of arrivals weight. We defined five dimensions of unloads such as *low*, *moderate*, *average*, *high*, and *extreme*. Table 1 displays for each unloads dimension the corresponding minimum and maximum weight of the arrival weight intervals for each daily working shift. Combining these settings with the same number of five different time horizon, we created a 5×5 grid of instances. Time horizon test settings range from 1 to 5 weeks of production planning. Therefore, considering six working days of two shifts each, tests are performed over $T \in \{12, 24, 36, 48, 60\}$.

3.2 Results

We attempt to solve each scenario with both the presented MILP and the company solution strategy using an Intel Core i7-4710HQ CPU @ 2.50 GHz with 8 GB RAM with GUROBI 8 solver, and setting the solver time-limit to 1500 s. The company solution is obtained by tuning the constraints set of the MILP formulation in order to make it represent the actual solution scheme of the company decision makers. As a main distinction, the company does not permit the simultaneity of different sorting stages during the same working shift. Therefore, while the presented model includes the creation of teams of operators dedicated to different and parallelized sorting operations, this is not true for the company solution program. In addition, when the company decision makers activate a working shift, they always allocate the maximum number of available operators. We first give the presented formulation response to each instance for what concerns the optimal solution objective value as illustrated in Fig. 3. It is clear that the objective value linearly increases along both the features of the instances grid. Instead, as shown in Fig. 4, the Gurobi solver runtime is not affected by the arrivals dimension while it does increase as the time horizon magnitude expands.

The model response is then compared with the actual strategy of the waste company. Table 2 presents, for each instance, the differences in terms of optimal objective value and in term of runtime needed to close the gap. We remark that the objective value represents the sum of sorting and holding costs in EURO needed to handle the arriving packaging waste over the considered time horizon. In the table, as *ObjVal_M* regards the MILP model solutions, *ObjVal_C* concerns the actual

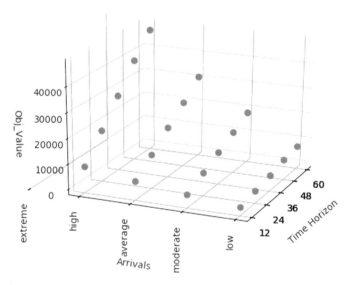

Fig. 3 Optimum obj. value for each instance

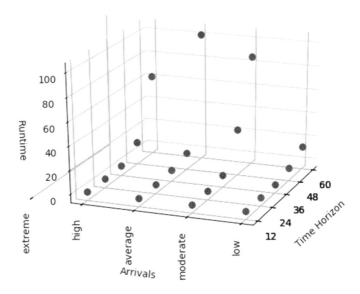

Fig. 4 Runtime for each instance

company solution; the same goes for *Runtime_M* and *Runtime_C*. Both approaches fail to find a solution for each of the *extreme* arrivals instances, thus confirming a comprehensive and stressful test instances grid. At the same time it points out the same feasibility limit of the model with respect to the company solution. For what concerns the feasible instances, the main remarkable result is the optimality gain obtained by the MILP model, which produces a better solution for each scenario. Indeed, the percentage cost reduction ranges from a minimum of 5.5% for the most demanding instance to a maximum of 15.9% for the longer horizon instance. There is also a slight tendency of a higher costs reduction with lower quantity arrivals and longer time plan. Instead, the convergence runtime to get the company solutions is always shorter, except for the average arrivals scenarios. Certainly, on an equal time horizon, these instances take longer to converge for the company solution criteria. In order to extract deeper insights about the solver performance when dealing with other model parameters setting, we performed two types of experiments. Few of the tested instances faced some approximation problems that made the solver stuck to a gap of about 0.25% for an indefinitely long time. Therefore we decided to set the gap termination tolerance to 0.25% in order to prevent this behavior while still providing a reasonable small convergence gap. We first attempted to solve instances with a greater number of J sorting stages, up to 5 subsequent stages. These instances also differ in time horizon but share the same kind of *average* magnitude of arriving waste quantity. Results are presented in the first portion of Table 3, where it is evident how the convergence gap and run-time increase with respect to the depth of the parameters settings, i.e. the number of stages and time horizon. All instances having more than two sorting stages fail to close the gap before reaching the time

Table 2 Optimal objective value and runtime comparison with the company model [C]

Instances TH—arrivals	ObjVal_M	Obj_Val_C	%Obj reduction	Runtime_M	Runtime_C	%Run diff
12—low	1151.036	1263.556	8.90	0.036	0.029	−0.241
12—moderate	3233.546	3421.400	5.50	0.052	0.033	−0.576
12—average	5388.570	6319.804	14.7	0.092	0.081	−0.136
12—high	8262.336	9117.556	9.4	0.271	0.050	−4.420
12—extreme	Infeasible	Infeasible	Infeasible	Infeasible	Infeasible	Infeasible
24—low	2708.048	3094.936	12.5	0.131	0.058	−1.259
24—moderate	6933.418	7659.156	9.5	0.446	0.126	−2.540
24—average	11,502.106	13,507.594	14.8	0.477	0.748	0.362
24—high	18,320.334	20,186.248	9.2	2.217	0.155	−13.303
24—extreme	Infeasible	Infeasible	Infeasible	Infeasible	Infeasible	Infeasible
36—low	4104.140	4564.886	10.1	0.333	0.152	−1.191
36—moderate	10,577.530	11,834.330	10.6	7.372	1.154	−5.388
36—average	17,902.680	20,485.472	12.6	6.530	31.161	0.790
36—high	28,025.228	30,655.698	8.6	1.837	0.361	−4.089
36—extreme	Infeasible	Infeasible	Infeasible	Infeasible	Infeasible	Infeasible
48—low	5952.454	7030.596	15.3	6.667	2.150	−2.101
48—moderate	14,472.414	15,962.360	9.3	64.374	6.642	−8.692
48—average	23,870.748	26,794.730	10.9	12.559	29.889	0.580
48—high	38,239.526	41,147.530	7.1	20.576	2.148	−8.579
48—extreme	Infeasible	Infeasible	Infeasible	Infeasible	Infeasible	Infeasible
60—low	7105.006	8443.570	**15.9**	26.816	2.697	−8.943
60—moderate	18,328.174	20,426.798	10.3	140.476	62.179	−1.259
60—average	30,355.192	34,709.392	12.5	148.196	300.026	0.506
60—high	47,101.504	50,434.394	6.6	98.605	1.611	−60.207
60—extreme	Infeasible	Infeasible	Infeasible	Infeasible	Infeasible	Infeasible

Bold value indicates the maximum value present within the column "Obj reductions"

limit, except for the case of 36 working shifts to schedule with 3 or 4 sorting stages. Still the convergence gap never exceeds 3.5% within the considered time limit of 1500s. The last four columns of Table 3 include the number of variables and constraints of each instance in order to better understand optimization performance related to the instances size. The second portion of Table 3 presents the results of a second test concerning the solver performance over a set of problems where the production and stock costs of the formulation are alternately removed. These are marked in first column with a "v" when the specific cost is present and "x" otherwise. These formulation costs tuning can be suitable whenever the decision maker wants to foster a scattered production schedule with a bigger lot size by removing stock costs, or to encourage a lean production by removing the production costs. Results prove how removing stock costs makes the problem easier to solve with respect to its standard form, while promoting a lean production would make the solver converge by exploring only its first node, a trivial optimal solution is

Table 3 Solver performances over the two additional set of experiments

Instances settings	Status	Gap	Obj_Val	RunTime	NodeCount	NumConstr	NumVars	NumBinVar	NumIntVar
TH = 36, J = 2	Optimal	0.0025	20,608	5.59	30,241	542	504	144	216
TH = 36, J = 3	Optimal	0.0025	22,500	89.68	196,932	795	756	216	324
TH = 36, J = 4	Optimal	0.0025	23,472	353.64	572,552	1048	1008	288	432
TH = 36, J = 5	Time limit	0.0109	23,982	1500.00	755,240	1301	1260	360	540
TH = 48, J = 2	Optimal	0.0025	28,283	36.67	102,454	722	672	192	288
TH = 48, J = 3	Time limit	0.0072	29,771	1500.00	232,8238	1059	1008	288	432
TH = 48, J = 4	Time limit	0.0139	31,155	1500.00	849,918	1396	1344	384	576
TH = 48, J = 5	Time limit	0.0314	33,026	1500.00	382,850	1733	1680	480	720
TH = 60, J = 2	Optimal	0.0025	34,231	301.06	764,941	902	840	240	360
TH = 60, J = 3	Time limit	0.0144	38,793	1500.00	148,9421	1323	1260	360	540
TH = 60, J = 4	Time limit	0.0241	40,767	1500.00	669,998	1744	1680	480	720
TH = 60, J = 5	Time limit	0.0343	39,889	1500.00	337,922	2165	2100	600	900
TH = 36, J = 2, Prod: v, Stock: v	Optimal	0.0025	20,099	4.22	19,205	542	504	144	216
TH = 36, J = 2, Prod: v, Stock: x	Optimal	0.0025	18,605	4.56	11,326,675	326	288	72	144
TH = 36, J = 2, Prod: x, Stock: v	Optimal	0.0025	123	0.98	1	542	504	144	216
TH = 48, J = 2, Prod: v, Stock: v	Optimal	0.0025	27,856	118.43	373,014	722	672	192	288
TH = 48, J = 2, Prod: v, Stock: x	Optimal	0.0025	25,750	6.33	2,188,403	434	384	96	192
TH = 48, J = 2, Prod: x, Stock: v	Optimal	0.0025	228	0.18	1	722	672	192	288
TH = 60, J = 2, Prod: v, Stock: v	Optimal	0.0025	33,845	585.24	1,491,897	902	840	240	360
TH = 60, J = 2, Prod: v, Stock: x	Optimal	0.0025	31,200	9.38	2,196,713	542	480	120	240
TH = 60, J = 2, Prod: x, Stock: v	Optimal	0.0025	217	0.21	1	902	840	240	360

indeed found by setting all variables in order to produce as much as allowed by the formulation constraints.

4 Conclusions

We presented a mixed integer linear programming model for planning and scheduling packaging waste recycling operations. The model supports several strategic decisions that are critical in the business considered. Indeed, within the waste industry, an high percentage of costs arises from sorting operations and logistics. Moreover, the presented formulation is sufficiently flexible for what concerns the sorting plants architectures that is able to replicate. In fact, by easily setting some of its parameters, the formulation showed a good modeling capacity when used for representing a real-world application. Results concerning costs optimization in the considered case study are also encouraging. Indeed, for the company turnover, this economical improvement is highly remarkable, taking into account the low margin of the activity. Future works may consider to introduce more complexity in the formulation, such as adding robustness on parameters values or production capacity dependent on the size of working teams. Another formulation improvement would be obtained by considering the fractions of material moving to the following sorting phase being subject to each arriving materials composition.

Acknowledgments This work has been partially supported by EU POR-FESR program of LAZIO Region on "Circular Economy and Energy" through the project REMIND Reverse Manufacturing Innovation Decision system [Grant No. B86H18000160002].

References

1. Aringhieri, R., Bruglieri, M., Malucelli, F., Nonato, M.: A special vehicle routing problem arising in the optimization of waste disposal: a real case. Transp. Sci. **52**(2), 277–299 (2018)
2. Benjamin, A.M., Beasley, J.: Metaheuristics for the waste collection vehicle routing problem with time windows, driver rest period and multiple disposal facilities. Comput. Oper. Res. **37**(12), 2270–2280 (2010)
3. Berger, C., Savard, G., Wizere, A.: Eugene: an optimisation model for integrated regional solid waste management planning. Int. J. Environ. Pollut. **12**(2–3), 280–307 (1999)
4. Bonomo, F., Durán, G., Larumbe, F., Marenco, J.: A method for optimizing waste collection using mathematical programming: a Buenos Aires case study. Waste Manag. Res. **30**(3), 311–324 (2012)
5. Ghiani, G., Laganà, D., Manni, E., Musmanno, R., Vigo, D.: Operations research in solid waste management: A survey of strategic and tactical issues. Comput. Oper. Res. **44**, 22–32 (2014)
6. Kim, B.I., Kim, S., Sahoo, S.: Waste collection vehicle routing problem with time windows. Comput. Oper. Res. **33**(12), 3624–3642 (2006)
7. Samanlioglu, F.: A multi-objective mathematical model for the industrial hazardous waste location-routing problem. Eur. J. Oper. Res. **226**(2), 332–340 (2013)

8. Stecca, G.: Electrical and electronic equipment recycling analysis and simulation for urban reverse logistics services. In: International Symposium on Flexible Automation Awaji-Island, Hyogo, Japan, 14–16 July, pp. 1–6. ISFA (2014)
9. Sulemana, A., Donkor, E.A., Forkuo, E.K., Oduro-Kwarteng, S.: Optimal routing of solid waste collection trucks: a review of methods. J. Eng. **2018**, 4586376 (2018)
10. Sun, W., Huang, G.: Inexact piecewise quadratic programming for waste flow allocation under uncertainty and nonlinearity. J. Environ. Inf. **16**(2), 80–93 (2015)
11. Wang, S., Huang, G.H., Yang, B.: An interval-valued fuzzy-stochastic programming approach and its application to municipal solid waste management. Environ. Model. Softw. **29**(1), 24–36 (2012)
12. Zhu, H., Huang, G.: SLFP: a stochastic linear fractional programming approach for sustainable waste management. Waste Manag. **31**(12), 2612–2619 (2011)

Selecting and Initializing Representative Days for Generation and Transmission Expansion Planning with High Shares of Renewables

Giovanni Micheli, Maria Teresa Vespucci, Marco Stabile, and Alessia Cortazzi

Abstract Generation and Transmission Expansion Planning (GTEP) to achieve decarbonisation targets in the power sector requires installing relevant shares of production from Renewable Energy Sources (RES). GTEP optimization models determine expansion plans over a long-term horizon so as to minimize the total investment and operation cost. In the presence of high shares of non-dispatchable RES power plants, operation costs must be evaluated taking into account the different operating conditions due to the intermittency of RES power generations. In this work we consider a GTEP model for a power system based on thermal, solar and wind power generation and we introduce a novel approach to select representative days (RDs), discretized in hours, in order to obtain accurate estimates of the operational costs. The RDs are not connected, in order to get a decomposable problem. A procedure is introduced to assign the ON/OFF status of every thermal unit at the beginning of each RD. Numerical results on a test case of suitable dimension show that the proposed method provides expansion plans very close to those obtained by the complete hourly model, while dramatically reducing computational costs.

Keywords Generation and transmission expansion planning · Representative days · Decision trees · Unit commitment

G. Micheli (✉) · M. T. Vespucci
Department of Management, Information and Production Engineering, University of Bergamo, Bergamo, Italy
e-mail: giovanni.micheli@unibg.it; maria-teresa.vespucci@unibg.it

M. Stabile · A. Cortazzi
CESI, Milano, Italy
e-mail: marco.stabile@cesi.it; alessia.cortazzi@esterni.cesi.it

© The Author(s), under exclusive license to Springer Nature Switzerland AG 2021   321
C. Gentile et al. (eds.), *Graphs and Combinatorial Optimization: from Theory to Applications*, AIRO Springer Series 5,
https://doi.org/10.1007/978-3-030-63072-0_25

1 Introduction

Generation and transmission expansion planning (GTEP) is the problem of determining technology, capacity and location of new generation units, as well as new electrical interconnections to be built. The definition of joint expansion plans is one of the most relevant problems in the field of power systems and many approaches have been proposed in recent years to deal with it [1, 2, 8, 10, 16, 20, 21]. Because of the computational restrictions due to the long-term horizon, most of the existing planning models represent power systems with a low level of technical and temporal detail, ignoring operational unit commitment constraints and evaluating operations on a daily or weekly basis. However, this approach is not appropriate for power systems with increasing penetration of generation from renewable energy sources (RES), since in this case a more detailed description of short-term dynamics is to be included in the expansion planning framework in order to accurately address all the challenges related to integrating high shares of intermittent energy sources [13].

To provide a better representation of the short-term operation while maintaining the problem computationally tractable, some energy planning models use a small number of representative days (RDs) instead of modeling every hour of the planning horizon. Different approaches have been proposed in the literature to identify RDs [3–7, 11, 12, 14]. However, the use of RDs raises the crucial issue regarding how these days should be linked in the expansion planning model. Most of the existing methods consider the RDs as temporally consecutive [15], linking these days according to an arbitrary order, from which, however, the model results may be affected. More sophisticated approaches, such as [19], connect RDs by computing the transition matrix, which gives the number of transitions between each pair of RDs. However, the interconnection among days increases computational costs and prevents from exploiting the decomposable structure of the expansion planning problem given by the use of disconnected RDs. Indeed, keeping RDs separate in the long-term planning models allows pursuing scalability and solving large-scale models in reasonable time through decomposition techniques [18].

The simplest approach to deal with temporally disconnected RDs assumes all thermal plants are offline at the beginning of each RD, so that many start-up manoeuvres need to take place in the first hour of each cluster in order to supply load. The consequences of this approach are: (1) an over-estimation of start-up costs and (2) a distortion of the system operation, since units with low start-up costs may result preferable with respect to plants supplying base-load, that usually have lower production costs but higher start-up costs. Thus, in order to provide an accurate solution to the unit commitment problem while maintaining RDs separate, it is necessary to apply a method that could accurately predict the online or offline status of every thermal power plant at the beginning of each RD.

This paper contributes by developing a novel approach to identify RDs and by providing a new method based on decision trees to determine the initial ON/OFF status of thermal power plants in RDs. A simplified model for GTEP is used as a testing framework to assess the performances of the proposed method. Specifically, in this paper expansion plans provided by the *hourly model* (i.e. the model that considers all 8760 h in a year) are compared with those obtained by considering RDs and different heuristic approaches to determine the initial status of thermal plants. Numerical tests show how the proposed method provides better expansion plans with respect to other approaches, while further reducing computational times keeping RDs separate. The structure of the paper is as follows. Section 2 introduces the MILP model for GTEP representing the testing framework of this analysis. The model is referred to power systems in which thermal, wind and solar power technologies are used to supply load (as it is, for instance, in the South of Italy). In Sect. 3 we describe the cluster analysis to select the RDs. Section 4 introduces the method to assign the status at the beginning of each RD for every thermal power plant. In Sect. 5 we present a numerical test to assess the performances of the proposed method with respect to other heuristics. Finally, conclusions are drawn in Sect. 6.

2 Generation and Transmission Expansion Planning Model

In this section, the GTEP problem is formulated as a MILP model that determines the investment schedule so as to minimize the total costs (i.e. investment and oper-ation costs) over the planning period. Specifically, the formulation here introduced is a simplified version of the model described in [9], as no hydro power plants, batteries, decommissioning decisions or uncertainty sources are included in the current formulation. Indeed, in order to assess the performances of the proposed procedure for selecting and initializing RDs, in this work expansion plans obtained with the proposed heuristic are compared with those provided by the hourly model. Due to computational restrictions, expansion plans with the hourly model can be obtained only by considering a simplified model for power systems.

The power system consists of a set Z of zones and the time horizon is discretized in years, with the set of years denoted by Y. The structure of the power system at the beginning of the planning horizon is described by set L_E of power transmission lines connecting zones, set K_E of thermal power plants and parameters sol_{z_0} and $wind_{z_0}$ that represent the solar power capacity and the wind power capacity, respectively, installed in zone $z \in Z$. The decisions to be taken concern investments in new transmission lines, new thermal power plants and new RES generation capacity.

The investments in new RES power generation capacity in zone z in year y are represented by the continuous variables $sol_{z,y}$ (solar power capacity) and $wind_{z,y}$ (wind power capacity). While it is possible to build wind and solar power plants of any capacity, thermal units usually present specified size, which does not allow modeling thermal power capacity expansion by means of continuous variables. Therefore, given the set K_C of candidate thermal power plants, for every $k \in K_C$ and $y \in Y$ we define binary variable $\delta_{k,y}$ to represent the decision to build thermal power plant k in year y and binary variable $\theta_{k,y}$ to express the availability of thermal power plant k in year y. Candidate plant k is available for production in year y (i.e.

$\theta_{k,y} = 1)$ if it has been constructed in any year i, $1 \le i \le y$, which is expressed by constraints

$$\theta_{k,y} = \sum_{i=1}^{y} \delta_{k,i} \quad k \in K_C, \quad y \in Y. \tag{1}$$

The investment decisions regarding new transmission lines are modeled similarly. Given the set L_C of candidate transmission lines, for every $l \in L_C$ and $y \in Y$ we define the binary variable $\delta_{l,y}$ to represent the decision to build transmission line l in year y and binary variable $\theta_{l,y}$ to express the availability of transmission line l in year y. Constraints

$$\theta_{l,y} = \sum_{i=1}^{y} \delta_{l,i} \quad l \in L_C, \quad y \in Y \tag{2}$$

express that candidate transmission line l is available in year y (i.e. $\theta_{l,y} = 1$) if it has been constructed in any year i, $1 \le i \le y$.

We now introduce the constraints that describe in detail the power system short-term operation. In particular, the hourly energy balance and spinning reserve constraints in every zone z are imposed and the unit commitment constraints of the thermal power plants available in year y are taken into account. As mentioned above, the short-term operation is modeled by considering in every year y a small set C^y of RDs. In every hour t of every RD $c \in C^y$ the solar power production in zone z is the fraction $\mu_{z,t}^c$ of the solar power capacity in year y and the wind power production in zone z is the fraction $\rho_{z,t}^c$ of the wind power capacity in year y: therefore the zonal hourly power production in each RD $RES_{z,t}^c$ is given by

$$RES_{z,t}^c = \mu_{z,t}^c \left(sol_{z_0} + \sum_{i=1}^{y} sol_{z,i} \right) + \rho_{z,t}^c \left(wind_{z_0} + \sum_{i=1}^{y} wind_{z,i} \right)$$

$$z \in Z, \quad 1 \le t \le 24, \quad c \in C^y, \quad y \in Y. \tag{3}$$

The power output of thermal unit k in hour t of RD c is expressed as $\underline{P}_k \gamma_{k,t}^c + p_{k,t}^c$, where $\gamma_{k,t}^c$ is the binary decision variable that represent the status of unit k, i.e. ON, if $\gamma_{k,t}^c = 1$ and OFF, if $\gamma_{k,t}^c = 0$, $\underline{P}_k > 0$ is the minimum power output of unit k and $p_{k,t}^c$ is the power output above the minimum, subject to

$$0 \le p_{k,t}^c \le (\overline{P}_k - \underline{P}_k) \gamma_{k,t}^c \quad k \in K_E \cup K_C, \quad 1 \le t \le 24, \quad c \in C^y, \quad y \in Y, \tag{4}$$

where $\overline{P}_k$ is the capacity of unit k. From (4) it follows that $\underline{P}_k \gamma_{k,t}^c + p_{k,t}^c = 0$, if $\gamma_{k,t}^c = 0$, and $\underline{P}_k \le \underline{P}_k \gamma_{k,t}^c + p_{k,t}^c \le \overline{P}_k$, if $\gamma_{k,t}^c = 1$.

If candidate unit $k \in K_C$ is not available in year y, i.e. $\theta_{k,y} = 0$, its status must be OFF in all hours t of all RDs $c \in C^y$ of year y: this is enforced by constraints

$$\gamma_{k,t}^c \leq \theta_{k,y} \quad k \in K_C, \ 1 \leq t \leq 24, \ c \in C^y, \ y \in Y. \tag{5}$$

If candidate unit $k \in K_C$ is available in year y (i.e. $\theta_{k,y} = 1$), its ON/OFF status in every hour t of all RDs $c \in C^y$ is determined by the model so as to guarantee the hourly zonal energy balance constraints, the spinning reserve constraints and the unit commitment constraints. The hourly zonal energy balance is expressed by constraints

$$\sum_{k \in \Omega_z^k} \left(\underline{P}_k \gamma_{k,t}^c + p_{k,t}^c \right) + RES_{z,t}^c + \sum_{l|r(l)=z} x_{l,t}^c + ENP_{z,t}^c = D_{z,t}^c + \sum_{l|s(l)=z} x_{l,t}^c + OG_{z,t}^c$$

$$z \in Z, \ 1 \leq t \leq 24, \ c \in C^y, \ y \in Y \tag{6}$$

where Ω_z^k is the set of thermal power plants located in zone z, while $r(l)$ and $s(l)$ denote the receiving end-zone and the sending-end zone, respectively, of transmission line l. Moreover, for every zone z in hour t of RD c parameter $D_{z,t}^c$ denotes the load, nonnegative slack variable $ENP_{z,t}^c \geq 0$ represents energy not provided, nonnegative slack variable $OG_{z,t}^c \geq 0$ represents over-generation and free variable $x_{l,t}^c$ denotes the power flow on transmission line l. In every hour t of each RD c Eqs. (6) ensure equality between energy sources of zone z (given by thermal, solar and wind generation and incoming energy flows) and energy uses of zone z (given by load and outgoing energy flows). The continuous variables $ENP_{z,t}^c$ and $OG_{z,t}^c$ allow detecting mismatch between supply and demand in the simulated system.

The transmission network operation is represented by a transportation model. Although transportation models do not provide a perfect representation of load flows, in real-scale GTEP problems this choice is justified by the computational burden. In our model, power flows $x_{l,t}^c$ are subject to lower transmission limit $\underline{F}_l$ and upper transmission limit $\overline{F}_l$ by means of the following constraints

$$\underline{F}_l \leq x_{l,t}^c \leq \overline{F}_l \quad l \in L_E, \ 1 \leq t \leq 24, \ c \in C^y, \ y \in Y \tag{7}$$

$$\underline{F}_l \theta_{l,y} \leq x_{l,t}^c \leq \overline{F}_l \theta_{l,y} \quad l \in L_C, \ 1 \leq t \leq 24, \ c \in C^y, \ y \in Y. \tag{8}$$

Inequalities (8) impose consistency between power flows on candidate transmission lines and the binary variables related to investment decisions, not allowing energy flows on candidate lines which have not been built (i.e. $\theta_{l,y} = 0$).

The spinning reserve is the amount of unused capacity in online power plants which can compensate for power shortages or frequency drops within a given period of time. The following constraints guarantee that the thermal power plants available in year y provide the requested spinning reserve level $R_{z,t}^c$ in every zone z and in every hour t of all RDs $c \in C^y$

$$\sum_{k \in K_E} \left(\overline{P}_k - \underline{P}_k \gamma_{k,t}^c - p_{k,t}^c \right) + \sum_{k \in K_C} \left(\overline{P}_k \theta_{k,y} - \underline{P}_k \gamma_{k,t}^c - p_{k,t}^c \right) \geq R_{z,t}^c$$

$$z \in Z, \ 1 \leq t \leq 24, \ c \in C^y, \ y \in Y. \qquad (9)$$

In order to introduce the unit commitment constraints, binary variables $\alpha_{k,t}^c$ and $\beta_{k,t}^c$ are defined for every thermal power plant k and every hour t of all RDs c, where $\alpha_{k,t}^c$ represents the decision about whether or not thermal unit k is to be started up in hour t of RD c and $\beta_{k,t}^c$ represents the decision about whether or not thermal unit k is to be shut down in hour t of RD c. If unit k is started up in hour t of RD c (i.e. $\alpha_{k,t}^c = 1$), it has to stay ON for at least MUT_k hours: this is expressed by constraints

$$\sum_{i=t-MUT_k+1}^{t} \alpha_{k,i}^c \leq \gamma_{k,t}^c \quad k \in K, \ MUT_k \leq t \leq 24, \ c \in C^y, \ y \in Y \qquad (10)$$

which are called *minimum up time* constraints and impose that in an interval of MUT_k consecutive time periods a unit can be started-up at most once. For each RD the minimum up time constraints are enforced for the hours from MUT_k to 24, being the RDs disconnected from each other. Analogously, if unit k is shut down in hour t (i.e. $\beta_{k,t}^c = 1$), it has to stay OFF for at least MDT_k hours: this is expressed by the following *minimum down time* constraints

$$\sum_{i=t-MDT_k+1}^{t} \beta_{k,i}^c \leq 1 - \gamma_{k,t}^c \quad k \in K, \ MDT_k \leq t \leq 24, \ c \in C^y, \ y \in Y. \qquad (11)$$

Consistency must be enforced between binary variables that represent start-up, shut down and status in adjacent hours: this is done by the constraints

$$\gamma_{k,t}^c - \gamma_{k_0}^c = \alpha_{k,t}^c - \beta_{k,t}^c \quad k \in K, \ t = 1, \ c \in C^y, \ y \in Y \qquad (12)$$

$$\gamma_{k,t}^c - \gamma_{k,t-1}^c = \alpha_{k,t}^c - \beta_{k,t}^c \quad k \in K, \ 2 \leq t \leq 24, \ c \in C^y, \ y \in Y. \qquad (13)$$

In constraints (12) the parameter $\gamma_{k_0}^c$ represents the status of unit k at the beginning of RD c. The procedure we propose to determine the values to be assigned to $\gamma_{k_0}^c$ is outlined in Sect. 4. In the expansion plans definition, constraints to control the renewable penetration are imposed: the following inequalities force the total renewable generation in zone z in year y to cover at least ratio $\varphi_{z,y}$ of the total yearly load

$$\sum_{c \in C^y} w_c \sum_{t=1}^{24} RES_{z,t}^c \geq \varphi_{z,y} \left(\sum_{c \in C^y} w_c \sum_{t=1}^{24} D_{z,t}^c \right) \quad z \in Z, \ y \in Y. \qquad (14)$$

In constraints (14) the parameter w_c denotes the weight of RD c. The objective function comprises three terms: (1) the annualized investment costs in new thermal, solar and wind generation; (2) the annualized investment costs in new transmission lines; (3) the operating costs, which consider for each RD the sum of production costs, start-up costs and penalties for energy not provided and over-generation.

$$
\min \ z = \sum_{y \in Y} \left[\sum_{k \in K_C} \frac{IC_k^{th} \delta_{k,y}}{(1+r)^{y-y_0}} + \sum_{z \in Z} \frac{IC_{z,y}^{sol} sol_{z,y} + IC_{z,y}^{wind} wind_{z,y}}{(1+r)^{y-y_0}} \right] +
$$

$$
+ \sum_{y \in Y} \sum_{l \in L_C} \frac{IC_l^{line} \delta_{l,y}}{(1+r)^{y-y_0}} +
$$

$$
+ \sum_{y \in Y} \sum_{c \in C^y} w_c \sum_{t=1}^{24} \left[\sum_{k \in K} CM_{k,y} \left(\underline{P}_k \gamma_{k,t}^c + p_{k,t}^c \right) + \sum_{k \in K} SUC_k \alpha_{k,t}^c + \right.
$$

$$
\left. + c_{ENP} \sum_{z \in Z} ENP_{z,t}^c + c_{OG} \sum_{z \in Z} OG_{z,t}^c \right] \quad (15)
$$

In (15) IC_k^{th} denotes the investment cost of candidate thermal power plant k, while parameters $IC_{z,y}^{sol}$ and $IC_{z,y}^{wind}$ represent the investment cost of new solar and wind power capacity, respectively, in zone z and in year y. Moreover parameter IC_l^{line} denotes the investment cost of candidate transmission line l, $CM_{k,y}$ represents the marginal production cost of thermal power plant k in year y, SUC_k represents the start-up cost of power plant k, while parameters c_{ENP} and c_{OG} denote the penalty costs for energy not provided and over-generation, respectively. Finally, y_0 represents the reference year to which all investment costs are discounted, while r denotes the annual discount rate.

3   The Selection of Representative Days

To reduce computational burden while maintaining a high level of temporal detail, frequently a small number of RDs is considered to evaluate short-term operation. Different approaches have been proposed to identify RDs. For instance, some authors use simple heuristics, such as the selection of days containing the minimum load, the maximum load or the largest daily demand spread [3]. Other works combine heuristic approaches with the random selection of some additional days [6, 7]. More advanced methods are based on clustering algorithms in order to group days with similar load, wind power production or solar power production into clusters [4, 11, 12]: clusters centroid or a specific historical day for each group is then taken as RD. Finally, some works select RDs by minimizing the difference between the historical load duration curve and the one obtained from the load in the RDs [5, 14].

In this section we propose a novel approach to select RDs from a set

$$D_{z,t}^d, \; \mu_{z,t}^d, \; \rho_{z,t}^d, \; z \in Z, \; 1 \le d \le 365, \; 1 \le t \le 24, \tag{16}$$

where the index d refers to the set of historical days and the load data $D_{z,t}^d$ are either historical values (typically related to the last year before the planning horizon) or forecast values for the first year of the planning horizon, while $\mu_{z,t}^d$ and $\rho_{z,t}^d$ are technical production/capacity ratios for solar power production and wind power production, respectively. By performing the cluster analysis on this data set, correlations among production and load, as well as spatial correlations among zones, can be taken into account.

RDs c_1 and c_2 are chosen as the days with minimum and maximum total load in the power system, i.e.

$$D_{z,t}^{c_1}, \; \mu_{z,t}^{c_1}, \; \rho_{z,t}^{c_1}, \;\; c_1 = argmin_d \left(\sum_{z \in Z} \sum_{t=1}^{24} D_{z,t}^d \right), \; z \in Z, \; 1 \le t \le 24 \tag{17}$$

and

$$D_{z,t}^{c_2}, \; \mu_{z,t}^{c_2}, \; \rho_{z,t}^{c_2}, \;\; c_2 = argmax_d \left(\sum_{z \in Z} \sum_{t=1}^{24} D_{z,t}^d \right), \; z \in Z, \; 1 \le t \le 24. \tag{18}$$

Further RDs are selected by the following iterative procedure performed on the modified data set which is obtained from the original one by deleting days c_1 and c_2 and by normalizing the load values $D_{z,t}^d$. For every zone z the load duration curve $LDC_{z,\tau}, \; 1 \le \tau \le 8760$ (i.e., the curve in which the original hourly load data $D_{z,t}^d$, $1 \le d \le 365$ and $1 \le t \le 24$, are in order of decreasing magnitude) is determined so as to be compared in the termination test with the zonal load duration curves corresponding to the RDs and their associated weights. The steps of the iterative procedure are as follows:

1. set $k = 2$;
2. the days of the modified data set are partitioned in k clusters by the k-medoids algorithm;
3. the RD $c_{2+\xi}$, for ξ, $1 \le \xi \le k$, is selected from the original data set as the day corresponding to the centroid of cluster ξ; the weight associated to RD $c_{2+\xi}$ is the number of days in cluster ξ;
4. determine the load duration curve corresponding to the $k + 2$ RDs and their associated weights (a unit weight is associated to c_1 and c_2) and compute the mean absolute percentage error between the original load duration curve and the one corresponding to the current set of RDs;

5. if the system average mean absolute percentage error (i.e., the average between the mean absolute percentage errors of zonal load duration curves) is below the given threshold, stop, otherwise increase k by 1 and go to step 2.

Once the RDs for the first year of the planning horizon are determined, the RDs of the following years are derived by applying annual growth factors to load profiles.

4 Determining the ON/OFF Status in Representative Days

In this section we describe our approach to assign a status to every thermal power plant at the beginning of each RD so as to maintain RDs temporally disconnected. In our work, parameters $\gamma_{k_0}^c$ are computed by estimating on historical data a decision tree [17], which is a classifier expressed as a recursive partition of the instance space. To the best of our knowledge, there are no similar works in the literature. Specifically, we consider for each existing thermal power plant the vector of daily initial statuses $\gamma_{k_0}^d$, which describes the ON/OFF status of thermal power plant k in the last hour of day $d - 1$, $2 \le d \le 365$. In our analysis, parameters $\gamma_{k_0}^d$ are computed by considering the last year before the planning horizon.

Then, for each thermal plant we compute the following features: (1) marginal cost ratio, i.e. the ratio between unit marginal cost and average marginal cost of available thermal plants; (2) start-up cost; (3) minimum up time; and (4) minimum down time. These attributes, as well as parameters $\gamma_{k_0}^d$, are used to train a decision tree in order to estimate a classification rule to determine the initial ON status according to features values. We then apply this classification rule to determine the probability π_k^y of thermal power plant k having an initial ON status in year y. Indeed, since production costs and available thermal units change throughout the years of the planning horizon, each plant k is usually characterized by time-varying marginal cost ratios and thus by different probabilities π_k^y along the planning horizon. Moreover, it is worth mentioning that probabilities π_k^y are obtained for both existing and candidate thermal power plants.

Finally, parameters π_k^y are used to set the probability of extracting 1 in the random selection between 0 (i.e. OFF) and 1 (i.e. ON). For each thermal plant and for every year y, this random selection is repeated for all RDs, in order to assign to each RD $c \in C^y$ a specific initial status $\gamma_{k_0}^c$.

5 Case Studies and Results

As a case study, we chose the South-Italy power system, which is interesting because it does not have big hydro reservoirs, being thermal, wind and solar plants the available technology to supply load. We considered for year 2018 the existing thermal plants in South-Italy, as well as the wind and solar installed capacity, and we studied the least-cost expansion for a single year by applying a load growth factor of 3.5% and imposing a 30% level for renewables penetration.

To identify RDs, we applied the procedure described in Sect. 3 fixing a threshold of 1% and obtaining 7 RDs. In order to determine the initial status of thermal plants in each RD, we considered the thermal plants commitment in 2018. On this data we estimated the decision tree shown in Fig. 1, which assigns to each thermal plant a specific probability to be ON in the initial hour of RDs according to the attributes values. As it can be noticed, the marginal cost ratio is the most relevant attribute in this classification, as it realizes a clear partition: while units with marginal cost ratios greater than 0.933 are usually OFF, the initial status for thermal plants with marginal cost ratios lower than or equal to 0.849 is ON. Instead, in the range of (0.849; 0.933] for marginal cost ratios the initial status is more uncertain and it depends also on the other attributes values. Specifically, while thermal units with high start-up costs are usually OFF, thermal plants with lower start-up costs are more often committed at the beginning of the day, especially if they have enough flexibility.

The seven RDs with an initial ON/OFF status determined as explained in Sect. 4 have been then used to define the least-cost expansion plans in the considered scenario. Specifically, in order to assess the performances of the proposed method, the following four formulations were implemented.

1. P1: This is the hourly model considering 8760 values for load, solar and wind profiles and representing the benchmark for expansion plans definition.
2. P2: This formulation links RDs obtained by our heuristic as commonly done in the literature, i.e. with the RDs considered in an arbitrary order and with the initial status of thermal power plants in RD c being equal to the final status in RD $c - 1$, $2 \leq c \leq 7$.
3. P3: In this model, RDs are not linked, but thermal plants are considered offline at the beginning of each day.
4. P4: This is the complete formulation proposed in this paper.

Specifically, the comparison between P1 and P4 allows evaluating the accuracy of the proposed method. P2 is implemented in order to observe the differences with the common approach in the literature of linking RDs. Finally, by comparing P3 and P4, one can clearly observe the improvement given by the clusters initialization.

Figure 2 shows the expansion plans determined with the four formulations. As it can be noticed, both P2 and P4 provide expansion plans very similar to the optimal ones identified by the hourly model P1, while P3 underestimates the capacity of coal plants and overestimates the capacity of combined cycle power plants (CCGTs). The same results can be observed by analyzing the total energy produced by source. Indeed, while P2 and P4 closely replicate the optimal output provided by formulation P1, model P3 uses thermal units fuelled with natural gas much more than coal plants. This unbalance is due to start-up costs: since all thermal units are OFF at the beginning of each RD, in order to supply load formulation P3 mainly employs CCGTs that present higher production costs but lower start-up costs than coal plants.

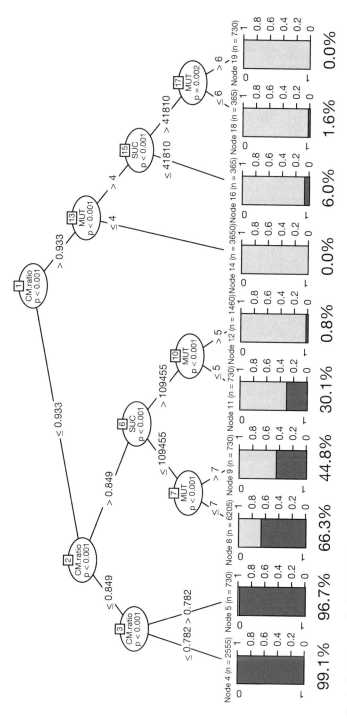

Fig. 1 Decision tree for initial ON/OFF status probability estimation

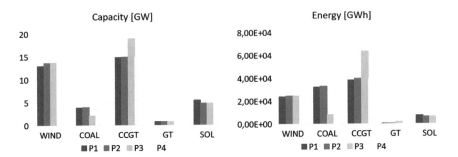

Fig. 2 Installed capacity by source (left) and total generation by source (right) in each formulation

Table 1 Costs [€] and solution time [min] for the four formulations

Formulation	Investment cost	Production cost	Start-up cost	Total cost	Total error	Solution time [min]
P1	$2.12 \cdot 10^9$	$3.04 \cdot 10^9$	$4.45 \cdot 10^7$	$5.21 \cdot 10^9$	–	393.10
P2	$2.19 \cdot 10^9$	$3.11 \cdot 10^9$	$3.46 \cdot 10^7$	$5.33 \cdot 10^9$	2.50%	3.07
P3	$2.21 \cdot 10^9$	$3.39 \cdot 10^9$	$4.42 \cdot 10^8$	$6.04 \cdot 10^9$	16.09%	2.63
P4	$2.16 \cdot 10^9$	$3.09 \cdot 10^9$	$4.00 \cdot 10^7$	$5.29 \cdot 10^9$	1.64%	2.57

Finally, Table 1 compares costs and computational times of expansion plans obtained with the four different formulations. As it can be observed, model P3, considering all thermal plants OFF at the beginning of the RDs and committing more CCGTs than coal plants, overestimates start-up costs, as well as production and total costs. Instead, formulations P2 and P4 provide a very good estimation of optimal system costs (P1), with P4 being the most accurate model. Moreover, it is worth mentioning that while the hourly model takes 393.10 min to be solved, expansion plans with formulation P4 are determined in 2.57 min.

6 Conclusions

This paper introduces a novel approach to select and initialize RDs for GTEP models. Specifically, RDs are identified by iteratively applying the k-medoids algorithm and considering for each data partition the goodness of the load duration curves approximation. RDs are then initialized by assigning to any thermal plant an initial ON/OFF status according to the probability provided by a decision tree estimated on historical data. Thanks to this analysis, it is possible to keep RDs separate in expansion planning models without dramatically overestimating start-up costs and distorting thermal plants commitment decisions. Numerical tests show how the expansion plans provided by the proposed method are very close to the optimal ones identified by the hourly model and more accurate than capacity plans obtained by linking RDs as common in the literature. Moreover, the main advantage of the proposed method over the approach of linking RDs is the scalability: given the separation of RDs, decomposition techniques such as Benders algorithms can be easily implemented in order to decompose the expansion planning model by RD [18]. Thanks to this characteristic, the proposed method is particularly suited to address long-term planning of large-scale power systems while obtaining tractable problems.

References

1. Aghaei, J., Amjady, N., Baharvandi, A., Akbari, M.: Generation and transmission expansion planning: MILP-based probabilistic model. IEEE T. Power Syst. **29**, 1592–1601 (2014)
2. Alizadeh, B., Jadid, S.: A dynamic model for coordination of generation and transmission expansion planning in power systems. Int. J. Electr. Power Energy Syst. **65**, 408–418 (2015)
3. Belderbos, A., Delarue, E.: Accounting for flexibility in power system planning with renewables. Int. J. Electr. Power Energy Syst. **71**, 33–41 (2015)
4. ElNozahy, M., Salama, M., Seethapathy, R.: A probabilistic load modelling approach using clustering algorithms. In: 2013 IEEE Power and Energy Society General Meeting (2013)
5. Fazlollahi, S., Bungener, S., Mandel, P., Becker, G., Marchal, F.: Multi-objectives, multi-period optimization of district energy systems: I. Selection of typical operating periods. Comput. Chem. Eng. **65**, 54–662 (2014)
6. Fripp, M.: Switch: A planning tool for power systems with large shares of intermittent renewable energy. Environ. Sci. Technol. **46**, 6371–6378 (2012)
7. Hart, E., Jacobson, M.: A Monte Carlo approach to generator portfolio planning and carbon emissions. Renew. Energy **36**, 2278–2286 (2011)
8. Krishnan, V., Ho, J., Hobbs, B., Liu, A., McCalley, J., Shahidehpour, M., Zheng, Q.: Co-optimization of electricity transmission and generation resources for planning and policy analysis: review of concepts and modeling approaches. Energy Syst. **7**, 297–332 (2016)
9. Micheli, G., Vespucci, M., Stabile, M., Puglisi, C., Ramos, A.: A two-stage stochastic MILP model for generation and transmission expansion planning with high shares of renewables. Energy Syst. **2020**, 353 (2020)
10. Moghaddam, S.: Generation and transmission expansion planning with high penetration of wind farms considering spatial distribution of wind speed. Int. J. Electr. Power Energy Syst. **106**, 232–241 (2019)
11. Nahmmacher, P., Schmid, E., Hirth, L., Knopf, B.: Carpe diem: a novel approach to select representative days for long-term power system models with high shares of renewable energy sources. Energy **112**, 430–442 (2016)
12. Nick, M., Cherkaoui, R., Paolone, M.: Optimal allocation of dispersed energy storage systems in active distribution networks for energy balance and grid support. IEEE Trans. Power Syst. **29**, 2300–2310 (2014)
13. Poncelet, K., Delarue, E., Six, D., Duerinck, J., D'haeseleer, W.: Impact of the level of temporal and operational detail in energy-system planning models. Appl. Energy **162**, 631–643 (2016)
14. Poncelet, K., Hschle, H., Delaru, E., Virag, A., Dhaeseleer, W.: Selecting representative days for capturing the implications of integrating intermittent renewables in generation expansion planning problems. IEEE Trans. Power Syst. **32**, 1936–1948 (2017)
15. Poncelet, K., van Stiphout, A., Meus, J., Delarue, E., Dhaeseleer, W.: Lusym invest: a generation expansion planning model with a high level of temporal and technical detail. Tech. rep., KU Leuven (2018). TME Working Paper WP EN2018-07
16. Pozo, D., Sauma, E., Contreras, J.: A three-level static MILP model for generation and transmission expansion planning. IEEE Trans. Power Syst. **28**, 201–210 (2013)

17. Rokach, L., Maimon, O.: Data Mining with Decision Trees: Theory and Applications. World Scientific Publishing, Singapore (2008)
18. Schwele, A., Kazempour, J., Pinson, P.: Do unit commitment constraints affect generation expansion? A scalable stochastic model. Energy Syst. 11, 247–282 (2020)
19. Tejada-Arango, D., Domeshek, M., Wogrin, S., Centeno, E.: Enhanced representative days and system states modeling for energy storage investment analysis. IEEE Trans. Power Syst. 33, 6534–6544 (2018)
20. Tohidi, Y., Hesamzadeh, M., Regairaz, F.: Sequential coordination of transmission expansion planning with strategic generation investments. IEEE T. Power Syst. 32, 2521–2534 (2017)
21. You, S., Hadley, S., Shankar, M., Liu, Y.: Co-optimizing generation and transmission expansion with wind power in large-scale power grids implementation in the us eastern interconnection. Electr. Power Syst. Res. 133, 209–218 (2016)

Start-up/Shut-Down MINLP Formulations for the Unit Commitment with Ramp Constraints

Tiziano Bacci, Antonio Frangioni, and Claudio Gentile

Abstract In (Bacci et al. New MINLP formulations for the single-unit commitment problems with ramping constraints, 2019) the first MIP exact formulation was provided that describes the convex hull of the solutions satisfying all the standard operational constraints for the thermal units: minimum up- and down-time, minimum and maximum power output, ramp (including start-up and shut-down) limits, general history-dependent start-up costs, and nonlinear convex power production costs. That formulation contains a polynomial, but large, number of variables and constraints. We present two new formulations with fewer variables defined on the shut-down period and computationally test the trade-off between reduced size and possibly weaker bounds.

Keywords Unit commitment problem · Ramp constraints · MIP formulations · Dynamic programming · Convex costs

1 Introduction

The Unit Commitment (UC) problem is a fundamental problem in power industries. It requires to coordinate the production of a set of power generation units by finding a feasible schedule—satisfying complex operational constraints—of each, over some time period, in order to minimize operational costs while satisfying system-wide constraints. The latter usually comprise the satisfaction of the energy demand, the provision of different types of reserve, and the handling of the transmission network. Operational constraints depend on the type of generation units. Despite

T. Bacci (✉) · C. Gentile
Istituto di Analisi dei Sistemi ed Informatica "A. Ruberti", Rome, Italy
e-mail: tiziano.bacci@iasi.cnr.it; gentile@iasi.cnr.it

A. Frangioni
Dipartimento di Informatica, Università di Pisa, Pisa, Italy
e-mail: frangio@di.unipi.it

the significant increase of contribution of Renewable Energy Sources (RES) units (wind, solar, ...), most power systems are still mainly based on thermal units (comprised nuclear ones) and hydro units. Indeed, these are needed at least to be able to cope with uncertainty in the production output typical of most RES units, which lead to highly complex uncertain (robust and/or stochastic) UC variants [16]. Thus, thermal units remain at the heart of basically every UC model of practical interest. In the last decade, the advances in Mixed-Integer (linear and convex) Programming (MIP) solvers have made MIP approaches an attractive option for solving UC, either as a whole or for specific sub-problems in the context of decomposition approaches (e.g., [2, 13, 14, 16]). This motivated a significant amount of research on the strong combinatorial structure of operational constraints of thermal units. In [10] many of the different types of formulations that appeared in the literature have been surveyed and compared with a large computational experience.

In [1] we gave the first MIP description of the convex hull of the solutions satisfying *all* the standard operational constraints for the thermal units: minimum up- and down-time, minimum and maximum power output, ramp (including start-up and shut-down) limits, general history-dependent start-up costs, and nonlinear convex power production costs. This formulation is inspired by a Dynamic Programming algorithm [4], and contains a polynomial number of variables and constraints. However, the number of variables grows cubically with the number of instants in the time horizon, making the formulation somehow impractical. This is why we also presented two additional MIP formulations which trade a weaker bound for fewer variables. We mention that three independent groups obtained a similar result restricted to linear objective function: the first was [6, 7] and then [8, 9] also appeared with very similar structure but with different proof techniques.

In this paper we continue and extend this line of research by deriving two new MIP formulations for UC that investigate complementary options to reduce the number of variables. In [1] one of the presented formulations was based on variables defining the power produced by a unit when the start-up time has been fixed. Here, we present a nearly-symmetric formulation based on variables defining the power produced by a unit when the shut-down time has been fixed. Despite the near symmetry, the two formulation behave somewhat differently, as our computational results show. Finally, we present and test a further formulation that combines both the "start-up" and the "shut-down" approach.

The structure of the paper is as follows. In Sect. 2 we recall the most popular UC formulation of thermal units. In Sect. 3 we recall the results in the recent paper [1] on the new formulations based on the DP algorithm in [4]. In Sect. 4 we present the new formulation based on shut-down power variables and the combined one. In Sect. 5 we present some preliminary computational experiments to characterize the placement of the new formulations within the state-of-the-art of MIP formulations for UC. Finally, in Sect. 6 we sum up the results, and draw some possible lines for future research on the topic.

2 The Thermal Unit Commitment Problem

Here we briefly recall the MIP formulation of the thermal Unit Commitment problem that became more and more popular in the last years, as it is one of the main innovations which made UC solvable by standard MIP solvers. It has been introduced in [12] and, independently, in [11], and it is usually referred to as the "*3-bin formulation*" from the number of vectors of binary variables that are considered.

Let I be the set of thermal generators, with $m = |I|$, and $T = \{1, \ldots, n\}$ be the set of time periods in the planning horizon. Given two time instants t' and t'', we will denote by $T(t', t'')$ the set of all the periods from t' to t'', extremes included. For each $i \in I$ and $t \in T$, let p_{it} (the *power variables*) be the power level of unit i at period t, and x_{it} (the *commitment variables*) be the binary variable denoting the on/off state of unit i at period t. If $x_{it} = 1$ ("on" state), then the power p_{it} may be nonzero and subject to some technical constraints specified in the following. If $x_{it} = 0$ ("off" state), then $p_{it} = 0$. The 3-bin formulation requires two additional sets of variables: *start-up variables* v_{it} denoting if unit i has been started up at period t (i.e., $x_{it}=1$ and $x_{i,t-1}=0$) and *shut-down variables* w_{it} denoting if i has been shut down at t (i.e., $x_{it}=0$ and $x_{i,t-1}=1$). The basic version of the 3-bin formulation is

$$\min \ \sum_{i \in I} \sum_{t \in T} (x_{it} f_i(p_{it}/x_{it}) + c_i x_{it} + s_i v_{it}) \tag{1}$$

$$\sum_{i \in I} p_{it} = d_t \qquad t \in T \tag{2}$$

$$l_i x_{it} \leq p_{it} \leq u_i x_{it} \qquad i \in I, \ t \in T \tag{3}$$

$$\sum_{s \in T(t-\tau_+^i+1,t)} v_{is} \leq x_{it} \qquad i \in I, \ t \in T(\tau_i^+, n) \tag{4}$$

$$\sum_{s \in T(t-\tau_-^i+1,t)} w_{is} \leq 1 - x_{it} \qquad i \in I, \ t \in T(\tau_i^-, n) \tag{5}$$

$$x_{it} - x_{i,t-1} = v_{it} - w_{it} \qquad i \in I, \ t \in T \tag{6}$$

$$p_{it} - p_{i,t-1} \leq \Delta_i^+ x_{i,t-1} + \bar{l}_i v_{it} \qquad i \in I, \ t \in T \tag{7}$$

$$p_{i,t-1} - p_{it} \leq \Delta_i^- x_{it} + \bar{u}_i w_{it} \qquad i \in I, \ t \in T \tag{8}$$

$$x_{it}, \ v_{it}, \ w_{it} \in \{0, 1\} \qquad i \in I, \ t \in T \tag{9}$$

The objective function (1) is composed of three parts: the variable generation costs evaluated as the Perspective Reformulation [4] of the quadratic function $f_i(p_{it}) = a_i p_{it}^2 + b_i$, the fixed generation costs $c_i x_{it}$, and the start-up costs $s_i v_{it}$. For simplicity, in this formulation we consider only fixed start-up costs; history-dependent start-up costs can be included with some complication [10], and are handled basically "for free" by the DP-based formulations examined here (cf. [1]

for details). The *demand constraints* (2) are the simplest version of system-wide constraints, where d_t is the (forecast) total energy demand at period t; other types may relate reserves and the distribution network (e.g. [10]), but we only consider (2) since our focus is on the description of the individual thermal units, which is logically independent from system-wide constraints. *Minimum and maximum power output constraints* are imposed by (3), where l_i and u_i are the extreme values for the generated power for each unit $i \in I$ (when on). In order to limit the technical stress due to frequent start-up and shut-down operations, *minimum up- and down-time constraints* (4)–(5) establish a minimum number of periods that unit i has to be in on and off state, τ_i^+ and τ_i^-, respectively; for simplicity we have omitted the obvious constraint that may fix the on/off status of the unit depending on its state prior to the beginning of the planning horizon. Constraints (6) establishes the relation among state, start-up, and shut-down variables. *Ramp-up and ramp-down constraints* (7)–(8) limit the maximum increase Δ_i^+ or decrease Δ_i^-, respectively, of the power produced by unit i in two consecutive time instants. These are usually related with *start-up and shut-down limits*, that is the maximum power $\bar{l}_i$ when the unit is started-up and the maximum power $\bar{u}_i$ before the unit is shut-down. For consistency, it must be $l_i \leq \bar{l}_i \leq u_i$ and $l_i \leq \bar{u}_i \leq u_i$.

The above formulation—minus (2), (7), and (8)—is known to be exact only when no ramp-up/down limits are imposed. The question if it is possible to write an exact formulation for UC restricted to a single thermal unit (1UC) in the variable space of the 3-bin formulation is still unsolved. However, 1UC is known to be an easy problem: indeed, in [4] a Dynamic Programming (DP) algorithm was proposed that can solve 1UC with all the above constraints in $O(n^3)$ (and that can be generalized to more complex objectives). Based on that, in [1] we gave the first exact formulation for 1UC that considers all the above mentioned technical features, which is recalled in the next section.

3   DP Formulations

For the description of the DP algorithm we drop the unit index i for notational simplicity. We then define a state-space graph $G = (N, A)$. The nodes in N are of two types: ON_t and OFF_t for each $t \in T$, plus two special nodes, the source s and the sink d. The arcs in A are of two types: *ON-arcs* (OFF_h, ON_k), denoting that the unit is turned on at the beginning of period h and unit remains on until the end of period k, and *OFF-arcs* (ON_k, OFF_r), denoting that the unit is off from period $k+1$ to period $r-1$. Both on- and off-arcs are only constructed, obviously, if they satisfy the minimum (respectively) up- and down-time constraints. Moreover, there are the connections between the source node s and the ON and OFF nodes defined according to the initial state of the unit. That is, if the unit is on since $\tau^0 - 0$ periods, then there is an on-arc from s to each node ON_k such that $k + \tau^0 \geq \tau^+$. If, instead, the unit is off since $-\tau^0$ periods, then there is an off-arc from s to each node

OFF_h such that $h - \tau^0 - 1 \geq \tau^-$. ON-arcs ($OFF_h$, ON_k) are labeled with costs γ_{ON}^{hk} computed as the fixed cost c_i multiplied by $(k - h + 1)$ plus variable costs, i.e. the solution of the restricted Economic Dispatch problems, whose (efficient) computation within the DP algorithm is described in [4]. OFF-arcs are labeled with γ_{OFF}^{kr} corresponding to the start-up cost. All nodes are then connected to the sink node d: OFF-arcs (ON_t, d) and ON-arcs (OFF_t, d). Finally, the single arc (s, d) means that the unit remains with the same status for all the time horizon, and it is an ON- or OFF-arc according to the fact that the unit, is, respectively, on or off at time 0.

The formulation inspired by the DP algorithm for (1UC) consists of two parts:

- the shortest path formulation based on the state-space graph G;
- new power variables and the linking constraints with the previous part.

The shortest path formulation is straightforward: one just introduces the node-arcs incidence matrix of the state-space graph and writes the obvious system of inequalities. Then we can then simply write this part of the formulation as:

$$E^i y_i = \delta^i \ , \quad y_i \geq 0 \ , \tag{10}$$

where E^i is the node-arcs incidence matrix of $G^i = (N^i, A^i)$ (here we reintroduced the unit index $i \in I$), y_i is the vector of arc flow variables, and δ^i is the vector with all zero entries except $\delta_s^i = -1$ and $\delta_d^i = 1$. Within the vector y_i, we denote with y_i^{hk} the variable associated with an ON-arc (OFF_h, ON_k) $\in A^i$. For short, we define as A_{ON}^i as the subset of such ON-arcs, and we denote them simply as "(h, k)" (as the type of the nodes is obvious). For each (h, k) $\in A_{ON}^i$ and $t \in T(h, k)$ we define a variable p_{it}^{hk} to denote the power level for each time instant if the unit i is started-up at time h and shut-down at time k. The following result is proven in [1]:

Theorem 1 *Bacci et al. [1] The following is an exact formulation for (1UC):*

$$min \ \gamma_i^T y_i + \sum_{(h,k) \in A_{ON}^i} \sum_{t \in T(h,k)} y_i^{hk} f_i(p_{it}^{hk}/y_i^{hk}) \tag{11}$$

(10)

$$\left. \begin{aligned} l_i y_i^{hk} &\leq p_{ih}^{hk} \leq \bar{l}_i y_i^{hk} \\ l_i y_i^{hk} &\leq p_{it}^{hk} \leq u_i y_i^{hk} & t \in T(h+1, k-1) \\ l_i y_i^{hk} &\leq p_{it}^{hk} \leq \bar{u}_i y_i^{hk} \\ p_{i,t+1}^{hk} &\leq p_{it}^{hk} + y_i^{hk} \Delta_i^+ & t \in T(h, k-1) \\ p_{it}^{hk} &\leq p_{i,t+1}^{hk} + y_i^{hk} \Delta_i^- & t \in T(h, k-1) \end{aligned} \right\} (h, k) \in A_{ON}^i \tag{12}$$

Constraints (12) express the Economic Dispatch conditions associated with an ON-arc (OFF_h, ON_k) and the objective function (11) is the Perspective Reformulation [3] of the original objective function f_i. This immediately yields the *DP formulation* for the complete UC problem

$$\min \sum_{i \in I} \left(\gamma_i^T y_i + \sum_{(h,k) \in A_{ON}^i} \sum_{t \in T(h,k)} y_i^{hk} f_i(p_{it}^{hk}/y_i^{hk}) \right)$$

$$\sum_{i \in I} \sum_{(h,k) \,:\, t \in T(h,k)} p_{it}^{hk} = d_t \qquad\qquad t \in T \qquad (13)$$

$$(10),\ (12) \qquad\qquad\qquad\qquad i \in I$$

The number of binary and continuous variables in (13) is, respectively, $O(n^2|I|)$ and $O(n^3|I|)$. Although the formulation provides (as expected) a strong bound, its size grows quickly, in particular due to the number of continuous variables. Because of this, in [1] two other formulations were introduced which are also based on the DP approach, but achieve different trade-offs between size and tightness. When restricted to 1UC, both are less tight than the exact formulation (10)–(12). The first one uses the original $O(n|I|)$ power variables p_{it} of the 3-bin formulation, while the second one presents a new type of power variables whose cardinality is intermediate between 3-bin and DP formulations.

Given a unit i, consider the commitment variable x_{it}, the start-up/shut-down variables v_{it}/w_{it} and the set of variables y_i^{hk} for $(h,k) \in A_{ON}^i$. It is easy to see that these variables are related by the following equations:

$$x_{it} = \sum_{(h,k)\,:\,t \in T(h,k)} y_i^{hk}\ ,\quad v_{it} = \sum_{k \geq t} y_i^{tk}\ ,\quad w_{it+1} = \sum_{h \leq t} y_i^{ht}\ . \qquad (14)$$

Consequently, the ramp-up/down constraints assume, respectively, the following form:

$$p_{it} - p_{it-1} \leq \Delta_i^+ \sum_{(h,k)\,:\,t-1 \in T(h,k-1)} y_i^{hk} + \bar{l}_i \sum_{k\,:\,k \geq t} y_i^{tk} - l_i \sum_{h\,:\,h \leq t-1} y_i^{ht-1} \qquad (15)$$

$$p_{it-1} - p_{it} \leq \Delta_i^- \sum_{(h,k)\,:\,t-1 \in T(h,k-1)} y_i^{hk} + \bar{u}_i \sum_{h\,:\,h \leq t-1} y_i^{ht-1} - l_i \sum_{k\,:\,k \geq t} y_i^{tk} \qquad (16)$$

Note that, in case the unit is on at the beginning of time horizon ($\tau_i^0 > 0$), the initial ramp-up/down conditions have to be set by

$$p_{i1} \leq (\Delta_i^+ + p_{i0}) \sum_{k\,:\,1 \leq k} y_i^{0k}\ ,\quad -p_{i1} \leq (\Delta_i^- - p_{i0}) \sum_{k\,:\,1 \leq k} y_i^{0k} \qquad (17)$$

Then minimum and maximum power output constraints can be rewritten as follows:

$$l_i \sum_{(h,k)\,:\,t \in T(h,k)} y_i^{hk} \leq p_{it} \leq u_i \sum_{(h,k)\,:\,t \in T(h,k)} y_i^{hk} \qquad (18)$$

The right-hand side of constraints (18) can be reinforced as follows. Assuming that $\tau_i^+ \geq 2$, if a unit i is switched on at time t then $\sum_{k\,:\,k \geq t} y_i^{tk} = 1$ and the power p_{it} is bounded by $\bar{l}_i$. If the unit is switched off at time t then $\sum_{h\,:\,h \leq t} y_i^{ht} = 1$ and the power p_{it} must not exceed $\bar{u}_i$. In case the unit does not turn on or off but it is committed at time t then $\sum_{(h,k)\,:\,h<t<k} y_i^{hk} = 1$ holds. Consequently,

there exists (h, k) such that $h < t < k$ and $y_i^{hk} = 1$. Because of the maximum power output and the ramp-up/down constraints, the power p_{it} is bounded by $\psi_{it}^{hk} = \min\{u_i, \bar{l}_i + \Delta_i^+(t-h), \bar{u}_i + \Delta_i^-(k-t)\}$. Furthermore, if the unit is initially committed $(\tau_i^0 > 0)$ then $\sum_{k \geq 1} y_i^{0k} = 1$ and we have to set $\psi_{it}^{0k} = \min\{u_i, p_{i0} + \Delta_i^+ \cdot t, \bar{u}_i + \Delta_i^-(k-t)\}$. Hence, if $\tau_i^+ \geq 2$ then (18) can be reinforced as

$$p_{it} \leq \bar{l}_i \sum_{k\,:\,k \geq t} y_i^{tk} + \bar{u}_i \sum_{h\,:\,h \leq t} y_i^{ht} + \sum_{(h,k)\,:\,h < t < k} \psi_{it}^{hk} y_i^{hk} \tag{19}$$

whereas if $\tau_i^+ = 1$ and $y_i^{tt} = 1$, the power p_{it} is bounded by the minimum between $\bar{l}$ and $\bar{u}$, which means that (18) rather becomes

$$p_{it} \leq \bar{l}_i \sum_{k\,:\,k > t} y_i^{tk} + \bar{u}_i \sum_{h\,:\,h < t} y_i^{ht} + \sum_{(h,k)\,:\,h < t < k} \psi_{it}^{hk} y_i^{hk} + \min\{\bar{l}_i, \bar{u}_i\} y_i^{tt} \tag{20}$$

This finally yields the p_t-model

$$\min \; \sum_{i \in I} \left(\gamma_i^T y_i + \sum_{t \in T} \sum_{(h,k)\,:\,t \in T(h,k)} y_i^{hk} \right) f_i \left(p_{it} / \left(\sum_{(h,k)\,:\,t \in T(h,k)} y_i^{hk} \right) \right)$$

$$\begin{aligned}
&(2) \\
&(10), (17) && i \in I \\
&(15), (16) && i \in I, t \in T(2, n) \\
&(18)\text{–}(20) && i \in I, t \in T
\end{aligned} \tag{21}$$

The last formulation introduced in [1] is rather centered on defining variables p_{it}^h denoting the power produced by unit i if committed at time t and if it has been turned on at time instant h; differently from the variable p_{it}^{hk}, in this case the time when the unit will be turned off is not fixed. The relation between p_{it} and p_{it}^h variables is

$$p_{it} = \sum_{h\,:\,h \leq t} p_{it}^h . \tag{22}$$

The ramp-up/down constraints are then reformulated as

$$p_{it}^h - p_{it-1}^h \leq \Delta_i^+ \sum_{k\,:\,k \geq t} y_i^{hk} - l_i y_i^{ht-1} \quad h \in T(1, n-1), \; t \in T(h+1, n) \tag{23}$$

$$p_{it-1}^h - p_{it}^h \leq \Delta_i^- \sum_{k\,:\,k \geq t} y^{hk} + \bar{u}_i y_i^{ht-1} \quad h \in T(1, n-1), \; t \in T(h+1, n) \tag{24}$$

If $\tau_i^0 > 0$, the initial ramp-up/down conditions can be imposed by

$$p_{i1}^0 \leq (\Delta^- + + p_0) \sum_{k\,:\,1 \leq k} y_i^{0k} \quad , \quad -p_{i1}^0 \leq (\Delta^- - - p_0) \sum_{k\,:\,1 \leq k} y_i^{0k} \tag{25}$$

and minimum/maximum power output constraints take the form

$$l_i \sum_{k\,:\,k \geq t} y_i^{hk} \leq p_{it}^h \leq u_i \sum_{k\,:\,k \geq t} y_i^{hk} \quad h \in T(0, n), \; t \in T(h, n) \tag{26}$$

However, for $t = h$ the rightmost inequality in (26) can be substituted by

$$p_{ih}^h \leq \bar{l}_i \sum_{k \,:\, k > h} y_i^{hk} + \min\{\bar{l}_i \,,\, \bar{u}_i\} y_i^{hh} \tag{27}$$

while for $t > h$ one can rather use

$$p_{it}^h \leq \bar{u}_i y_i^{ht} + \sum_{k \,:\, k > t} \psi_{it}^{hk} y_i^{hk} \tag{28}$$

In conclusion, the *Start-Up formulation* (SU) is

$$\min \sum_{i \in I} \left(\gamma_i^T y_i + \sum_{t \in T} \sum_{h \,:\, t \geq h} \sum_{k \,:\, k \geq t} \Big(\sum_{k \,:\, k \geq t} y_i^{hk} \Big) f_i \big(p_{it}^h / \big(\sum_{k \,:\, k \geq t} y_i^{hk} \big) \big) \right) \tag{29}$$

$$(10), \ (23)\text{--}(28) \hspace{5cm} i \in I$$

$$\sum_{i \in I} \sum_{h \,:\, h \leq t} p_{it}^h = d_t \hspace{4cm} t \in T \tag{30}$$

This has $O(n^2 |I|)$ power variables, compared to $O(n|I|)$ of the p_t-model and $O(n^3 |I|)$ of the original DP formulation, with a bound to match [1].

4 Two New Formulations for UC

Mirroring the derivation of (29)–(30), we can construct a nearly symmetric formulation, the *Shut-Down formulation* (SD). This is based on variables $\tilde{p}_{it}^k$ denoting the power produced at time t by a unit i that will be turned *off* at time instant k, i.e.,

$$p_{it} = \sum_{k \,:\, k \geq t} \tilde{p}_{it}^k \tag{31}$$

All the constraints can be derived by using (31); in particular, the ramp-up/down constraints become

$$\tilde{p}_{it}^k - \tilde{p}_{it-1}^k \leq \bar{l}_i y_i^{tk} + \Delta_i^+ \sum_{h \,:\, h \leq t-1} y_i^{hk} \quad k \in T(2, n) \,, \ t \in T(2, k) \tag{32}$$

$$\tilde{p}_{it-1}^k - \tilde{p}_{it}^k \leq -l_i y_i^{tk} + \Delta_i^- \sum_{h \,:\, h \leq t-1} y_i^{hk} \quad k \in T(2, n) \,, \ t \in T(2, k) \tag{33}$$

If $\tau_i^0 > 0$, the initial ramp-up/down conditions can be imposed by

$$\tilde{p}_{i1}^k \leq (\Delta^+ + p_0) y_i^{0k} \,, \quad -\tilde{p}_{i1}^k \leq (\Delta^- - p_0) y_i^{0k} \hspace{1.5cm} k \in T \tag{34}$$

The minimum/maximum power output constraints take the form

$$l_i \sum_{h\,:\,h\leq t} y_i^{hk} \leq \tilde{p}_{it}^k \leq u_i \sum_{h\,:\,h\leq t} y_i^{hk} \qquad k \in T \;,\;\; t \in T(1,k) \tag{35}$$

which for can $t = k$ be strengthened by

$$\tilde{p}_{ik}^k \leq \bar{u}_i \sum_{h\,:\,h<k} y_i^{hk} + \min\{\bar{l}_i\,,\,\bar{u}_i\}y_i^{kk} \tag{36}$$

while for $t < k$ by

$$\tilde{p}_{it}^k \leq \bar{l}_i y_i^{tk} + \sum_{h\,:\,h<t} \psi_{it}^{hk} y_i^{hk} \tag{37}$$

due to the fact that the unit could be turned on at time t ($y_i^{tk} = 1$) or not ($\sum_{h\,:\,h<t} y_i^{hk} = 1$). All in all, the SD formulation is

$$\min \sum_{i\in I}\left(\gamma_i^T y_i + \sum_{t\in T}\sum_{k\,:\,t\leq k}\left(\sum_{h\,:\,h\leq t} y_i^{hk}\right) f_i(\tilde{p}_{it}^k/(\sum_{h\,:\,h\leq t} y_i^{hk}))\right) \tag{38}$$

$$(10),\,(32)-(37) \qquad\qquad\qquad i \in I$$

$$\sum_{i\in I}\sum_{k\,:\,k\geq t} \tilde{p}_{it}^k = d_t \qquad\qquad\qquad t \in T \tag{39}$$

It is now natural to define the *Start-Up/Shut-Down formulation* (SUSD) by basically combining the previous two:

$$\min \sum_{i\in I}\left(\gamma_i^T y_i + \sum_{t\in T}\theta_{it}\right) \tag{40}$$

$$\theta_{it} \geq \sum_{h\,:\,h\leq t}\left(\left(\sum_{k\,:\,k\geq t} y_i^{hk}\right) f_i(p_{it}^h/(\sum_{k\,:\,k\geq t} y_i^{hk}))\right) \qquad i \in I\;,\;\; t \in T \tag{41}$$

$$\theta_{it} \geq \sum_{k\,:\,k\geq t}\left(\left(\sum_{h\,:\,h\leq t} y_i^{hk}\right) f_i(\tilde{p}_{it}^k/(\sum_{h\,:\,h\leq t} y_i^{hk}))\right) \qquad i \in I\;,\;\; t \in T \tag{42}$$

$$(10),\,(30),\,(23)-(28),\,(32)-(37) \qquad\qquad i \in I$$

$$\sum_{h\,:\,h\leq t} p_{it}^h = \sum_{k\,:\,k\geq t} \tilde{p}_{it}^k \qquad\qquad\qquad t \in T \tag{43}$$

Basically, (41) and (42) guarantee that the objective function (40) represents the maximum between these of the SU and the SD formulations. The constraints, and in particular (43), enforce the intersection between the feasible solutions of the two formulations. Thus, the lower bound provided by the continuous relaxation of the SUSD formulation has to be at least as good as the ones of both the SU and the SD models, at the cost of having roughly twice the number of variables of each (but still $O(n^2|I|)$ as opposed to the $O(n^3|I|)$ ones of the original DP formulation).

5 Computational Results

We conducted some preliminary computational experiments to compare the new formulations presented with the DP-based ones introduced in [1] as well as the reference 3-bin model. For our tests, we considered standard benchmark realistic instances with 10, 20 and 50 units and $n = 24$ time periods, available at http://www.di.unipi.it/optimize/Data/UC.html

All the experiments were conducted on a PC with 2.2 GHz Intel Xeon Gold 5120 CPUs and 64 GB of RAM, under a GNU/Linux Ubuntu 18.04.3 LTS operating system. We used CPLEX 12.9 as optimization tool. For a given number of units, we considered ten instances and the average of the results thus obtained are reported.

As in [15], in Table 1 we report the number of variables ("v") and the number of constraints ("c") of each model before any dynamic generation. Column "%v" ("%c") shows the percentage reduction of the number of variables (constraints) after the CPLEX's presolve. Concerning the dimensions of the formulations, the DP model has the largest number of both variables and constraints, while the smallest formulations are the 3-bin and the p_t. The SU and SD models have essentially the same number of variables and constraints, although the latter is slightly larger. On the other hand, the size of the SUSD formulation is smaller than the DP one and larger than the SD model. The results also reveal that the presolve of CPLEX seems to be more efficient on the DP formulation w.r.t the SUSD model, although the size of the former remains significantly larger. Furthermore, the presolve reduce

Table 1 LP size before CPLEX presolve and percentage of reduction after CPLEX presolve for 3-bin, DP, p_t, SU, SD and SUSD formulations

	3-bin				DP				p_t			
Units	v	c	%v	%c	v	c	%v	%c	v	c	%v	%c
10	2e+3	2e+3	23	20	3e+4	8e+4	12	22	3e+3	2e+3	13	14
20	3e+3	4e+3	21	19	7e+4	2e+5	11	19	8e+3	4e+3	12	13
50	8e+3	1e+4	19	17	2e+5	5e+5	11	20	2e+4	1e+4	8	11

	SU				SD				SUSD			
Units	v	c	%v	%c	v	c	%v	%c	v	c	%v	%c
10	7e+3	1e+4	13	19	9e+3	2e+4	9	9	1e+4	3e+4	9	13
20	2e+4	3e+4	12	19	2e+4	3e+4	8	9	3e+4	6e+4	8	13
50	4e+4	7e+4	9	13	5e+4	9e+4	8	9	7e+4	2e+5	6	10

the number of variables and constraints more on the SU model than the SD one. However, the 3-bin has the most consistent reduction of variables.

In Table 2 we compare the formulations by evaluating their continuous relaxations. In particular, for each instance size and for each model, "time" denotes the average time for solving the linear relaxation while "gap" is the average gap (in percentage) between the optimum and the value of the linear relaxation. As it can be expected from the theory, the DP formulation is the one that provides the best gaps; however, being the largest one, it also has a high computing time for the linear relaxation. In general, a clear trend exists between having a larger number of variables, and therefore a larger cost, and having a stronger bound. The interesting comparison is between the SU and SD models that have basically identical size. On the test instances, the SU formulation is somewhat less costly to solve and it provides better bound for large instances. This is somewhat unexpected due to the high degree of symmetry between the two, and worth further investigation. Regarding the SUSD formulation, on these instances it obtains the same gaps as the DP model; it is unclear whether this happens by chance or if it can be proven theoretically. However, it also require more time, despite being smaller (cf. Table 1), which is somewhat surprising. Finally, the 3-bin formulation provides the worst lower bounds, although in much less time than the others.

Table 3 and 4 show the results obtained by setting a relative gap of 0.1% (Table 3) and of 0.01% (Table 4), while solving the integer program, with a time limit of 10,000 s. For each model, column "time" reports the average total time, "opt" the number of instances, over ten, solved within the time limit, "nodes" the number of nodes explored during the B&C, and "gap" the average final gap (in

Table 2 Linear relaxation gaps of the 3-bin, DP, p_t, SU, SD, and SUSD formulations

Units	3-bin		DP		p_t		SU		SD		SUSD	
	Time	Gap	Time	Gap	Time	Gap	Time	Gap	Time	Gap	Time	Gap
10	0.15	2.077	16.81	1.222	0.94	1.335	2.85	1.280	3.94	1.264	46.26	1.222
20	0.34	1.691	93.61	0.811	2.21	0.894	8.40	0.822	12.34	0.880	161.70	0.811
50	1.19	0.906	574.98	0.137	6.88	0.190	34.06	0.146	42.98	0.178	772.12	0.137

Table 3 Computational results with gap 10^{-3} and time limit 10,000 s

Units	3-bin				DP				p_t			
	Time	Opt	Nodes	Gap	Time	Opt	Nodes	Gap	Time	Opt	Nodes	Gap
10	77	10	333	0.08	1150	10	477	0.09	54	10	175	0.07
20	6213	5	1914	0.14	5859	7	824	0.28	1413	9	623	0.14
50	7038	3	909	0.20	8671	3	402	0.48	3409	7	434	0.10

Units	SU				SD				SUSD			
	Time	Opt	Nodes	Gap	Time	Opt	Nodes	Gap	Time	Opt	Nodes	Gap
10	199	10	299	0.09	222	10	297	0.07	1103	10	463	0.08
20	2366	9	1088	0.15	3255	10	1132	0.09	7068	5	692	0.62
50	4978	7	748	0.10	6248	7	735	0.25	9050	1	320	0.36

Table 4 Computational results with gap 10^{-4} and time limit 10,000 s

Units	3-bin				DP				p_t			
	Time	Opt	Nodes	Gap	Time	Opt	Nodes	Gap	Time	Opt	Nodes	Gap
10	83	10	407	0.01	1182	10	568	0.01	121	10	267	0.01
20	7138	4	2728	0.10	6526	6	1147	0.23	1888	9	1088	0.06
50	10,000	0	1469	0.17	10,001	0	473	0.49	9025	1	1538	0.06

Units	SU				SD				SUSD			
	Time	Opt	Nodes	Gap	Time	Opt	Nodes	Gap	Time	Opt	Nodes	Gap
10	206	10	352	0.00	234	10	359	0.01	1173	10	549	0.01
20	3000	8	1642	0.10	4405	9	1780	0.01	8677	4	1080	0.56
50	9053	1	1296	0.08	8528	2	1244	0.23	10,000	0	393	0.36

percentage). The results show that the SD formulation is indeed somewhat more effective (surprisingly) than the SU one. It is also competitive with the p_t-model, which is the best performing one among the formulations presented in [1], in that it solves a few more instances, albeit often (even though not always) at a higher cost. Hence, the trade-off between the higher cost and the tighter bound (cf. Table 2) is positive, at least on these instances. This is not the case for the SUSD model, that has the worst results in general.

6 Conclusions

We have introduced two new formulations for the UC problem with convex cost function. Both models are based on the DP one introduced in [1]. In particular, the Shut-Down (SD) is a nearly symmetric formulation of the Start-Up (SU) model already introduced in [1], while the Start-Up/Shut-Down (SUSD) is a combination of the two. The results of computational experiments show that the SD formulation is surprisingly more effective than its closest sibling, the SU one. The reason is not clear, and surely worth further investigation. On the other hand, the SUSD model so far does not seem effective for solving UC. However, it can be further investigated in at least two aspects. First, the trade-off between size and bound quality is inherently tied with the algorithm that is used to solve the continuous relaxation. A column-and-rows generation approach, such as the Structured Dantzig-Wolfe Decomposition [5], may considerably shift the balance in favour of models that would not be effective using standard linear programming approaches. Second, the experimental results show that the value of the linear relaxation of the SUSD model is equal to the one provided by the DP formulation, that is the strongest model in this sense. It may be interesting to investigate whether this equivalence can be proven theoretically, since the SUSD model has much less variables than the DP one.

Acknowledgments The second author acknowledges the partial financial support by the European Union Horizon 2020 research and innovation programme under grant agreement No 773897 "plan4res". The second and third authors acknowledge the partial financial support by the European Union Horizon 2020 Marie Skłodowska-Curie Actions under the grant agreement No 764759 "MINOA". All the authors acknowledge the partial financial support by the Italian Ministry of Education program MIUR-PRIN under the grant 2015B5F27W "Nonlinear and Combinatorial Aspects of Complex Networks".

References

1. Bacci, T., Frangioni, A., Gentile, C., Tavlaridis-Gyparakis, K.: New MINLP formulations for the single-unit commitment problems with ramping constraints (2019). http://www.optimization-online.org/DB_HTML/2019/10/7426.html
2. Borghetti, A., Frangioni, A., Lacalandra, F., Nucci, C.: Lagrangian heuristics based on disaggregated bundle methods for hydrothermal unit commitment. IEEE T. Power Syst. **18**, 313–323 (2003)
3. Frangioni, A., Gentile, C.: Perspective cuts for a class of convex 0–1 mixed integer programs. Math. Progr. **106**(2), 225–236 (2006)
4. Frangioni, A., Gentile, C.: Solving nonlinear single-unit commitment problems with ramping constraints. Oper. Res. **54**(4), 767–775 (2006)
5. Frangioni, A., Gendron, B.: A stabilized structured Dantzig-Wolfe decomposition method. Math. Progr. **140**, 45–76 (2013)
6. Frangioni, A., Gentile, C.: An extended MIP formulation for the single-unit commitment problem with ramping constraints. In: 17th British-French-German Conference on Optimization, London June 15-17 (2015)
7. Frangioni, A., Gentile, C.: New MIP formulations for the single-unit commitment problems with ramping constraints. Tech. Rep. 15-06, IASI–CNR (2015)
8. Guan, Y., Pan, K., Zhou, K.: Polynomial time algorithms and extended formulations for unit commitment problems. IISE Trans. **50**(8), 735–751 (2018)
9. Knueven, B., Ostrowski, J., Wang, J.: The ramping polytope and cut generation for the unit commitment problem. INFORMS J. Comput. **30**(4), 625–786 (2018)
10. Knueven, B., Ostrowski, J., Watson, J.: On mixed integer programming formulations for the unit commitment problem. INFORMS J. Comput. **32**(4), 855–1186 (2020)
11. Malkin, P., Wolsey, L.: Minimum runtime and stoptime polyhedra. Manuscript (2004)
12. Rajan, D., Takriti, S.: Minimum Up/Down polytopes of the unit commitment problem with start-up costs. Research Report RC23628, IBM (2005)
13. Scuzziato, M., Finardi, E., Frangioni, A.: Comparing spatial and scenario decomposition for stochastic hydrothermal Unit Commitment problems. IEEE Trans. Sustainable Energy **9**(3), 1307–1317 (2018)
14. Taktak, R., D'Ambrosio, C.: An overview on mathematical programming approaches for the deterministic unit commitment problem in hydro valleys. Energy Syst. **8**(1), 57–79 (2017)
15. Tejada-Arango, D.A., Lumbreras, S., Sanchez-Martin, P., Ramos, A.: Which unit-commitment formulation is best? A systematic comparison. IEEE Trans. Power Syst. **35**(4), 2926–2936 (2020)
16. van Ackooij, W., Danti Lopez, I., Frangioni, A., Lacalandra, F., Tahanan, M.: Large-scale Unit Commitment under uncertainty: an updated literature survey. Ann. Oper. Res. **271**(1), 11–85 (2018)

Gaining or Losing Perspective for Piecewise-Linear Under-Estimators of Convex Univariate Functions

Jon Lee, Daphne Skipper, Emily Speakman, and Luze Xu

Abstract We study MINLO (mixed-integer non-linear optimization) formulations of the disjunction $x \in \{0\} \cup [\ell, u]$, where z is a binary indicator of $x \in [\ell, u]$ ($0 \le \ell < u$), and y "captures" $f(x)$, which is assumed to be convex and positive on its domain $[\ell, u]$, but otherwise $y = 0$ when $x = 0$. This model is very useful in non-linear combinatorial optimization, where there is a fixed cost of operating an activity at level x in the operating range $[\ell, u]$, and then there is a further (convex) variable cost $f(x)$. In particular, we study relaxations related to the perspective transformation of a natural piecewise-linear under-estimator of f, obtained by choosing linearization points for f. Using 3-d volume (in (x, y, z)) as a measure of the tightness of a convex relaxation, we investigate relaxation quality as a function of f, ℓ, u, and the linearization points chosen. We make a careful investigation for convex power functions $f(x) := x^p$, $p > 1$.

Keywords Perspective function · Piecewise linear · Indicator variables · Mixed-integer nonlinear optimization

J. Lee (✉) · L. Xu
University of Michigan, Ann Arbor, MI, USA
e-mail: jonxlee@umich.edu; xuluze@umich.edu

D. Skipper
United States Naval Academy, Annapolis, MD, USA
e-mail: skipper@usna.edu

E. Speakman
University of Colorado, Boulder, CO, USA
e-mail: emily.speakman@ucdenver.edu

© The Author(s), under exclusive license to Springer Nature Switzerland AG 2021
C. Gentile et al. (eds.), *Graphs and Combinatorial Optimization:*
from Theory to Applications, AIRO Springer Series 5,
https://doi.org/10.1007/978-3-030-63072-0_27

1 Introduction

1.1 Definitions and Background

Let f be a univariate convex function with domain $[\ell, u]$, where $0 \leq \ell < u$. We assume that f is positive on $[\ell, u]$. We are interested in the mathematical-optimization context of modeling a function, represented by a variable y, that is equal to a given convex function $f(x)$ on an "operating range" $[\ell, u]$ and equal to 0 at 0. We do this using a 0/1 indicator variable z (which conveniently allows for incorporating a fixed cost for x being in the operating range), and we represent the relevant set disjunctively as follows.

We define

$$\hat{D}_f(\ell, u) := \mathrm{conv}\left(\{(0,0,0)\} \bigcup \left\{ (x, y, 1) \in \mathbb{R}^3 : \right. \right.$$

$$\left. \left. f(\ell) + \frac{f(u) - f(\ell)}{u - \ell}(x - \ell) \geq y \geq f(x),\ u \geq x \geq \ell, \right\} \right).$$

Notice that for $x \in \{\ell, u\}$, we have $y = f(x)$. So, the upper bound on y enables us to capture the convex hull of the graph of the convex $f(x)$ on $[\ell, u]$, in the $z = 1$ plane.

Next, we define the *perspective relaxation*

$$\hat{S}_f^*(\ell, u) :=$$

$$\mathrm{cl}\left\{ (x, y, z) \in \mathbb{R}^3 : \left(f(\ell) - \frac{f(u) - f(\ell)}{u - \ell}\ell \right) z + \frac{f(u) - f(\ell)}{u - \ell}x \geq y \geq zf(x/z), \right.$$

$$\left. uz \geq x \geq \ell z,\ 1 \geq z > 0,\ y \geq 0 \right\},$$

where cl denotes the closure operator. Intersecting $\hat{S}_f^*(\ell, u)$ with the hyperplane defined by $z = 0$, leaves the single point $(x, y, z) = (0, 0, 0)$. In this way, the "perspective and closure" construction gives us exactly the value $y = 0$ that we want at $x = 0$. Moreover, $\hat{S}_f^*(\ell, u)$ is precisely the convex closure of $\hat{D}_f(\ell, u)$.

We compare convex bodies relaxing $\hat{S}_f^*(\ell, u)$ via their volumes, with an eye toward weighing the relative tightness of relaxations against the difficulty of solving them. Generally, working with $\hat{S}_f^*(\ell, u)$ implies using a cone solver (e.g., Mosek), while relaxations imply the possibility of using more general NLP or even LP solvers; see [11] for more discussion on this important motivating subject. One key relaxation previously studied requires that the domain of f be all of $[0, u]$, f is convex on $[0, u]$, $f(0) = 0$, and f is increasing on $[0, u]$. For example, convex power functions $f(x) := x^p$ with $p > 1$ have these properties. We define the *naïve relaxation*

$$\hat{S}_f^0(\ell, u) :=$$

$$\left\{ (x, y, z) \in \mathbb{R}^3 \; : \; \left(f(\ell) - \frac{f(u) - f(\ell)}{u - \ell} \ell \right) z + \frac{f(u) - f(\ell)}{u - \ell} x \geq y \geq f(x), \right.$$

$$\left. uz \geq x \geq \ell z, \; 1 \geq z \geq 0, \right\}.$$

1.2   Relation to Previous Literature

The perspective transformation of a convex function is well known in mathematics (see [7], for example). Applying it in the context of our disjunction is also well studied (see [1, 4, 5], with applications to non-linear facility location and also mean-variance portfolio optimization in the style of Markowitz). The idea of using volume to compare relaxations was introduced by Lee and Morris [8] (also see [10] and the references therein). Recently, [11, 12] applied the idea of using volumes to evaluate and compare the perspective relaxation with other relaxations of our disjunction. Piecewise linearization is a very well studied and useful concept for handling non-linearities (see, for example, [3, 9] and also the more recent [13, 14] and the many references therein). It is a natural idea to strengthen a piecewise linearization of a univariate function using the perspective idea, and then to evaluate it using volume computation. This is what we pursue here, concentrating on piecewise-linear under-estimators of univariate convex functions. We also wish to mention and emphasize that our techniques are directly relevant for (additively) separable convex functions (see [2, 6], and of course all of the exact global-optimization solvers which induce a lot of separability).

1.3   Our Contribution and Organization

Our focus is on relaxations related to natural piecewise-linear under-estimators of f. Piecewise linearization is a standard method for efficiently handling non-linearities in optimization. For a convex function, it is easy to get a piecewise-linear under-estimator. But there are a few issues to consider: the number of linearization points, how to choose them, and how to handle the resulting piecewise-linearization.

In particular, we look at the behavior of the perspective relaxation of the piecewise-linear under-estimator, as we vary placement and number of linearization points describing the piecewise-linear under-estimator.

In Sect. 2, we introduce notation for a natural piecewise-linear under-estimator g of f on $[\ell, u]$, using linearizations of f at the $n + 1$ points $\ell =: \xi_0 < \xi_1 < \cdots < \xi_n := u$, $n \geq 1$, we define the convex relaxation $\hat{U}_f^*(\boldsymbol{\xi}) := \hat{S}_g^*(\ell, u)$, and we describe an efficient algorithm for determining its volume (Theorem 1 and Corollary 2). Armed with this efficient algorithm, any global-optimization software could decide between members of this family of formulations (depending on the number and placement of linearization points) and also alternatives (e.g., $\hat{S}_f^*(\ell, u)$ and $\hat{S}_f^0(\ell, u)$, explored in [11]), trading off tightness of the formulations against the relative ease/difficulty of working with them computationally.

In Sect. 3, we give a more detailed analysis for convex power functions $f(x) := x^p$, for $p > 1$. We solve the minimization problem for $\mathrm{vol}(\hat{U}_f^*(\boldsymbol{\xi}))$ when $p = 2$ (Theorem 3), for an arbitrary number of linearization points, thus finding the optimal placement of linearization points for convex quadratics. Further, from this, we recover the associated formula from [11] for $\mathrm{vol}(\hat{S}_f^*(\ell, u))$ (Corollary 4), and we demonstrate that the minimum volume is always less than the volume of the naïve relaxation when $p = 2$ (Corollary 5). When there is only one non-boundary linearization point ξ_1 (i.e., the case of $n = 2$), we demonstrate that $\mathrm{vol}(\hat{U}_f^*(\ell, \xi_1, u))$ has a unique minimizer (Lemma 6, Theorem 7, Corollary 8). From this we could build a reasonable "coordinate-descent style" algorithm for placing linearization points, moving one point at a time to its volume-minimizing location. We demonstrate that for convex power functions x^p and a single non-boundary linearization point, the location of the volume-minimizing perspective relaxation is increasing in p on $(1, \infty)$ (Theorem 9). This starts to give us an idea about where we can efficiently place linearization points for non-quadratics. Using a result from [11], we give an efficient algorithm for computing the volume for the naïve relaxation associated with the piecewise-linear under-estimator (Proposition 10). This can be compared against the volumes for $\hat{S}_f^*(\ell, u)$, $\hat{S}_f^0(\ell, u)$, and $\hat{S}_g^*(\ell, u)$. In the special case of $p = 2$ and equally-spaced linearization points, we obtain a closed-form expression for the volume for the naïve relaxation associated with the piecewise-linear under-estimator (Corollary 11).

2 Piecewise-Linear Under-Estimation and Perspective

Piecewise-linear estimation is widely used in optimization. Lee and Wilson [9] provides some key relaxations using integer variables, even for non-convex functions on multidimensional (polyhedral) domains. We are particularly interested in piecewise-linear *under*-estimation because of its value in global optimization.

Given convex $f : [\ell, u] \to \mathbb{R}_{++}$, we consider linearization points

$$\ell =: \xi_0 < \xi_1 < \cdots < \xi_n := u$$

in the domain of f, and we assume that f is differentiable at these ξ_i.

At each ξ_i, we have the tangent line

$$y = f(\xi_i) + f'(\xi_i)(x - \xi_i), \tag{T_i}$$

for $i = 0, \ldots, n$. Considering tangent lines T_i and T_{i-1} (for adjacent points), we have the intersection point

$$(x, y) = (\tau_i, \ f(\xi_i) + f'(\xi_i)(\tau_i - \xi_i)), \text{ for } i = 1, \ldots, n, \tag{P_i}$$

where

$$\tau_i := \frac{\left[f(\xi_i) - f'(\xi_i)\xi_i\right] - \left[f(\xi_{i-1}) - f'(\xi_{i-1})\xi_{i-1}\right]}{f'(\xi_{i-1}) - f'(\xi_i)}.$$

Finally, we define

$$(x, y) := (\tau_0 := \ell, f(\ell)) \tag{P_0}$$

and

$$(x, y) := (\tau_{n+1} := u, f(u)). \tag{P_{n+1}}$$

It is easy to see that $\ell =: \tau_0 < \tau_1 < \cdots < \tau_{n+1} := u$, and that the piecewise-linear function $g : [\ell, u] \to \mathbb{R}$, defined as the function having the graph that connects the P_i, for $i = 0, 1, \ldots, n + 1$, is a convex under-estimator of f (that agrees with f at the ξ_i, $i = 0, 1, \ldots, n$). In what follows, g is always defined as above (from f and $\boldsymbol{\xi}$).

We wish to compute the volume of the set $\hat{U}_f^*(\boldsymbol{\xi}) := \hat{S}_g^*(\ell, u)$. To proceed, we work with the sequence $\tau_0, \tau_1, \ldots, \tau_{n+1}$ defined above. Below and later, adet denotes the absolute value of the determinant.

Theorem 1

$$\text{vol}(\hat{U}_f^*(\boldsymbol{\xi})) = \frac{1}{6} \sum_{i=1}^n \text{adet} \begin{pmatrix} \tau_0 & \tau_i & \tau_{i+1} \\ g(\tau_0) & g(\tau_i) & g(\tau_{i+1}) \\ 1 & 1 & 1 \end{pmatrix}.$$

Proof We wish to compute the volume of the set $\hat{U}_f^*(\boldsymbol{\xi})$. This set is a pyramid with apex $(x, y, z) = (0, 0, 0)$ and base equal to the intersection of $\hat{U}_f^*(\boldsymbol{\xi})$ with the hyperplane defined by the equation $z = 1$. The height of the apex over the base is unity. So the volume of $\hat{U}_f^*(\boldsymbol{\xi})$ is simply the area of the base divided by 3. We will compute the area of the base by straightforward 2-d triangulation. Our triangles are conv$\{P_0, P_i, P_{i+1}\}$, for $i = 1, \ldots, n$. The area of each triangle is $1/2$ of the absolute determinant of an appropriate 3×3 matrix. The formula follows. $\quad\square$

Corollary 2 *Assuming oracle access to f and f', we can compute $\mathrm{vol}(\hat{U}^*_f(\boldsymbol{\xi}))$ in $\mathcal{O}(n)$ time.*

3 Analysis of Convex Power Functions

Convex power functions constitute a broad and flexible class of increasing convex functions, useful in a wide variety of applications. For convenience, let $\hat{U}^*_p(\boldsymbol{\xi})$ denote $\hat{U}^*_f(\boldsymbol{\xi})$, with $f(x) := x^p$, $p > 1$. Next, we will see that equally-spaced linearization points minimizes the volume of this relaxation when $p = 2$.

Theorem 3 $\xi_i := \ell + \frac{i}{n}(u - \ell)$, *for* $i = 0, 1, \ldots, n$, *minimizes* $\mathrm{vol}(\hat{U}^*_2(\boldsymbol{\xi}))$, *and the minimum volume is* $\frac{1}{18}(u - \ell)^3 + \frac{(u-\ell)^3}{36n^2}$.

Proof The intersection points P_i are $(\frac{\xi_{i-1}+\xi_i}{2}, \xi_{i-1}\xi_i)$. Let $\xi_{n+1} := u$, then $\tau_i = \frac{\xi_{i-1}+\xi_i}{2}$ for $1 \le i \le n + 1$. We have

$$\mathrm{vol}(\hat{U}^*_2(\boldsymbol{\xi})) = \frac{1}{6}\sum_{i=1}^{n} \mathrm{adet}\begin{pmatrix} \tau_0 & \tau_i & \tau_{i+1} \\ g(\tau_0) & g(\tau_i) & g(\tau_{i+1}) \\ 1 & 1 & 1 \end{pmatrix}$$

$$= \frac{1}{12}\sum_{i=1}^{n}(\xi_{i+1} - \xi_{i-1})(\xi_i - \ell)^2$$

$$= \frac{1}{12}\left[\sum_{i=1}^{n}\xi_i\xi_{i-1}(\xi_{i-1} - \xi_i) + u^3 - 2u^2\ell + 2u\ell^2 - \ell^3\right],$$

and

$$\frac{\partial\,\mathrm{vol}(\hat{U}^*_2(\boldsymbol{\xi}))}{\partial\xi_i} = \frac{1}{12}(\xi_{i+1} - \xi_{i-1})(2\xi_i - \xi_{i+1} - \xi_{i-1}), \text{ for } i \in [n-1],$$

$$\frac{\partial^2\,\mathrm{vol}(\hat{U}^*_2(\boldsymbol{\xi}))}{\partial\xi_i^2} = \frac{1}{6}(\xi_{i+1} - \xi_{i-1}), \text{ for } i \in [n-1],$$

$$\frac{\partial^2\,\mathrm{vol}(\hat{U}^*_2(\boldsymbol{\xi}))}{\partial\xi_i\partial\xi_{i+1}} = \frac{1}{6}(\xi_i - \xi_{i+1}), \text{ for } i \in [n-2].$$

Therefore, $\nabla^2\,\mathrm{vol}(\hat{U}^*_2(\boldsymbol{\xi}))$ is a tridiagonal matrix. It is easy to verify that $\nabla^2\,\mathrm{vol}(\hat{U}^*_2(\boldsymbol{\xi}))$ is diagonally dominant because $(\xi_{i+1} - \xi_{i-1}) = (\xi_{i+1} - \xi_i) + (\xi_i - \xi_{i-1})$, thus $\nabla^2\,\mathrm{vol}(\hat{U}^*_2(\boldsymbol{\xi}))$ is positive semidefinite, i.e., $\mathrm{vol}(\hat{U}^*_2(\boldsymbol{\xi}))$ is convex.

The global minimizer satisfies $\nabla\,\mathrm{vol}(\hat{U}^*_2(\boldsymbol{\xi})) = 0$, i.e., $2\xi_i - \xi_{i+1} - \xi_{i-1} = 0$ for $i \in [n-1]$, which proves that the points are equally spaced at the minimizer, and now a simple calculation give the minimum volume as

$$\text{vol}(\hat{U}_2^*(\boldsymbol{\xi})) = \frac{1}{12}\left(\frac{2}{3}(u-\ell)^3 + \frac{1}{3n^2}(u-\ell)^3\right) = \frac{1}{18}(u-\ell)^3 + \frac{(u-\ell)^3}{36n^2}.$$

□

Letting n go to infinity, we recover the volume of the perspective relaxation for the quadratic.

Corollary 4 ([11]) $\text{vol}(\hat{S}_2^*) = \frac{1}{18}(u-\ell)^3.$

We can also now easily see that by using the perspective of our piecewise-linear under-estimator, even with only one (well-placed) non-boundary linearization point, we always outperform the naïve relaxation.

Corollary 5 $\text{vol}(\hat{U}_2^*(\boldsymbol{\xi})) \le \text{vol}(\hat{S}_2^0)$, and with equality only if $n = 1$ and $\ell = 0$.

Proof The naïve-relaxation volume is $\frac{1}{18}(u-\ell)^3 + (u^3 - \ell^3)/36$ (see [11]). Notice that

$$\frac{(u-\ell)^3}{36n^2} \le \frac{(u-\ell)^3}{36} \le \frac{u^3 - \ell^3}{36}.$$

The first inequality is strict when $n > 1$ and the second is strict when $\ell > 0$.   □

Considering $p \ne 2$, even for one non-boundary linearization point, $\text{vol}(\hat{U}_p^*(\boldsymbol{\xi}))$ is not generally convex in ξ_1 for $\boldsymbol{\xi} = (\ell, \xi_1, u)$. Still, it is very useful to make a detailed study of optimal placement of a single non-boundary linearization point, as it relates to optimality conditions for $\boldsymbol{\xi}$. In this direction, we will establish that $\text{vol}(\hat{U}_p^*(\ell, \xi_1, u))$ has a unique minimizer. To pursue this, we need a definition and a technical lemma.

Let

$$h_p(\xi_1) := \left(\frac{u^p - \ell^p}{u^{p-1} - \ell^{p-1}} - \frac{\xi_1^p - \ell^p}{\xi_1^{p-1} - \ell^{p-1}}\right)\left(\frac{\xi_1^p - u^p}{\xi_1^{p-1} - u^{p-1}} - \frac{u^p - \ell^p}{u^{p-1} - \ell^{p-1}}\right).$$

Note that h_p depends on p, ℓ and u, but we only highlight the dependence on p, so as to keep the notation fairly light. We note that $h_p(\xi_1)$ is not generally concave, for $p > 2$, but we do have the following very useful lemma (we defer the long technical proof to the journal version).

Lemma 6

(i) If $1 < p \le 2$, then $h_p(\xi_1)$ is strictly concave in ξ_1;
(ii) If $p > 2$, then $h_p(\xi_1)$ is strictly log-concave in ξ_1.

Theorem 7

(i) If $1 < p \le 2$, then $\text{vol}(\hat{U}_p^*(\ell, \xi_1, u))$ is strictly convex in ξ_1.

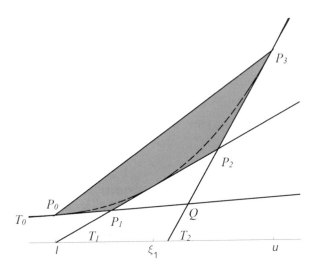

Fig. 1 Proof of Theorem 7

(ii) *If $p > 2$, then* $\mathrm{vol}(\hat{U}_p^*(\ell, \xi_1, u)) = C_1 - C_2 h_p(\xi_1)$, *where the constants* C_1, C_2 *and the strictly log-concave* $h_p(\xi_1)$ *depend only on* p, ℓ, u.

Proof Refer to Fig. 1. For fixed ℓ and u, let T_0, T_1, T_2 be the tangent lines at point ℓ, ξ_1, u; let P_1 be the intersection point of T_0, T_1; let P_2 be the intersection point of T_1, T_2; let Q be the intersection point of T_0, T_2. Then the x coordinates of P_1, P_2, Q are

$$\tau_1 = \frac{p-1}{p} \cdot \frac{\xi_1^p - \ell^p}{\xi_1^{p-1} - \ell^{p-1}}, \quad \tau_2 = \frac{p-1}{p} \cdot \frac{\xi_1^p - u^p}{\xi_1^{p-1} - u^{p-1}}, \quad \tau_Q = \frac{p-1}{p} \cdot \frac{u^p - \ell^p}{u^{p-1} - \ell^{p-1}}.$$

It is easy to see that $\mathrm{vol}(\hat{U}_p^*(\ell, \xi_1, u)) = \frac{1}{3}m(P_0, P_1, P_2, P_3) = \frac{1}{3}m(P_0, Q, P_3) - \frac{1}{3}m(P_1, Q, P_2)$, where $m(P_0, P_1, P_2, P_3)$, $m(P_0, Q, P_3)$ and $m(P_1, Q, P_2)$ denote the area of the quadrangle P_0, P_1, P_2, P_3, the triangle P_0, Q, P_3 and P_1, Q, P_2, respectively. Notice that $m(P_0, Q, P_3)$ is constant for fixed ℓ, u, and $m(P_1, Q, P_2)$ is proportional to $(\tau_Q - \tau_1)(\tau_2 - \tau_Q)$. Therefore, finding the minimum of $\mathrm{vol}(\hat{U}_f^*(\xi))$ is equivalent to finding the maximum of $(\tau_Q - \tau_1)(\tau_2 - \tau_Q) = h_p(\xi_1)$.

Then the result directly follows from the fact that $\mathrm{vol}(\hat{U}_p^*(\ell, \xi_1, u)) = C_1 - C_2 h_p(\xi_1)$, where C_1, C_2 are constant when p, ℓ, u are fixed. □

We immediately have the following very-useful result.

Corollary 8 *For all* $p > 1$, $\mathrm{vol}(\hat{U}_p^*(\ell, \xi_1, u))$ *has a unique minimizer on* (ℓ, u).

Remark 1 If $p > 2$, then $\mathrm{vol}(\hat{U}_p^*(\ell, \xi_1, u))$ is not generally convex in ξ_1, but it is not convex only near ℓ.

Theorem 9 *For fixed ℓ and u, the ξ_1 that minimizes $\mathrm{vol}(\hat{U}_p^*(\ell, \xi_1, u))$ is increasing in p on $(1, \infty)$.*

Proof By Lemma 6, we know that $h_p(\xi_1) = q_1(\xi_1)q_2(\xi_1)$ is concave when $1 < p \leq 2$ and is log-concave when $p > 2$. Therefore, the unique minimizer ξ_1 satisfies $\frac{d}{dx}h_p(\xi_1) = 0$ when $1 < p \leq 2$ and $\frac{d}{dx}\log h_p(\xi_1) = 0$ when $p > 2$. In both cases, the unique minimizer ξ_1 satisfies

$$q_1'(\xi_1)q_2(\xi_1) + q_1(\xi_1)q_2'(\xi_1) = 0.$$

Using $q_2(x)(u^{p-1} - x^{p-1}) = q_1(x)(x^{p-1} - \ell^{p-1})$, we can simplify the optimality condition to

$$G(x, p) := -\frac{x^p + (p-1)\ell^p - p\ell^{p-1}x}{x^{p-1} - \ell^{p-1}} + \frac{x^p + (p-1)u^p - pu^{p-1}x}{u^{p-1} - x^{p-1}} = 0.$$

We are going to use the implicit function theorem to show that $\xi_1 := x(p)$ satisfying $G(\xi_1, p) = 0$ is increasing in p on $(1, \infty)$.

$$
\frac{\partial G(x, p)}{\partial p} = \frac{u^{p-1}(u-x)}{u^{p-1} - x^{p-1}}\left[\frac{(p-1)x^{p-1}\ln\frac{x}{u}}{\left(u^{p-1} - x^{p-1}\right)} + 1\right]
$$

$$
+ \frac{\ell^{p-1}(x-\ell)}{x^{p-1} - \ell^{p-1}}\left[\frac{(p-1)x^{p-1}\ln\frac{x}{\ell}}{\left(\ell^{p-1} - x^{p-1}\right)} + 1\right].
$$

Next, we claim that

$$\frac{(p-1)t^{p-1}\ln t}{1 - t^{p-1}} > -\frac{t^p - 1}{p(t-1)}, \quad \text{for all } t \in (0, 1) \cup (1, \infty).$$

$\square$

Proof of the Claim Let $H(t) := p(p-1)(1-t)t^{p-1}\ln t + (t^{p-1} - 1)(t^p - 1)$. Then

$$
H'(t) = p(p-1)((p-1)t^{p-2} - pt^{p-1})\ln t + p(p-1)(1-t)t^{p-2}
$$

$$
+ (p-1)t^{p-2}(t^p - 1) + pt^{p-1}(t^{p-1} - 1),
$$

$$
\frac{H'(t)}{t^{p-2}} = ((p-1) - pt)p(p-1)\ln t + p(p-1)(1-t)
$$

$$
+ (p-1)(t^p - 1) + p(t^p - t),
$$

$$
\frac{d}{dt}\left(\frac{H'(t)}{t^{p-2}}\right) = -p^2(p-1)\ln t + \frac{p(p-1)^2}{t} - p^3 + p(2p-1)t^{p-1}
$$

$$
= p^2(t^{p-1} - 1 - \ln t^{p-1}) + p(p-1)\left(\frac{p-1}{t} + t^{p-1} - p\right) > 0.
$$

Because $H'(1) = 0$, we have that $H'(t) < 0$ for $t \in (0, 1)$ and $H'(t) > 0$ for $t \in (1, \infty)$. Combined with $H(1) = 0$, we obtain $H(t) > 0$ for $t \in (0, 1) \cup (1, \infty)$, which proves the claim.

Then

$$
\begin{aligned}
\frac{\partial G(x, p)}{\partial p} &> \frac{u^{p-1}(u - x)}{u^{p-1} - x^{p-1}} \left[-\frac{x^p - u^p}{pu^{p-1}(x - u)} + 1 \right] \\
&\quad + \frac{\ell^{p-1}(x - \ell)}{x^{p-1} - \ell^{p-1}} \left[-\frac{x^p - \ell^p}{p\ell^{p-1}(x - \ell)} + 1 \right] \\
&= \frac{1}{p} G(x, p) = 0.
\end{aligned}
$$

Also

$$
\begin{aligned}
\frac{\partial G(x, p)}{\partial x} &= \frac{1}{x} G(x, p) \\
&\quad - \frac{(p - 1)\ell^{p-1}(\ell^p - x^p - px^{p-1}(\ell - x))}{x(x^{p-1} - \ell^{p-1})^2} \\
&\quad - \frac{(p - 1)u^{p-1}(u^p - x^p - px^{p-1}(u - x))}{x(u^{p-1} - x^{p-1})^2} \\
&< 0.
\end{aligned}
$$

The last inequality follows from the strictly convexity of x^p ($p > 1$) and $G(x, p) = 0$.

By the implicit function theorem, we have

$$
\frac{\partial \xi_1}{\partial p} = -\frac{\frac{\partial G(x,p)}{\partial p}}{\frac{\partial G(x,p)}{\partial x}} > 0.
$$

Thus the ξ_1 that minimizes $\mathrm{vol}(\hat{U}_p^*(\ell, \xi_1, u))$ is increasing in p on $(1, \infty)$.   □

Defining g with respect to $f(x) := x^p$ on $[\ell, u]$, with $p > 1$, we can then extend g to the function $\bar{g}$, with domain all of $[0, u]$:

$$
\bar{g}(x) := \begin{cases} \ell^{p-1}x, & x \in [0, \ell); \\ g(x), & x \in [\ell, u]. \end{cases}
$$

In this way, $\bar{g}$ is a piecewise-linear increasing convex function on all of $[0, u]$. In fact, $\bar{g}$ is an under-estimator of the function that is f on $[\ell, u]$ and 0 at 0. In what follows, we calculate the volume of the naïve relaxation of the piecewise-linear under-estimator $\hat{U}_p^0(\xi) := \hat{S}_{\bar{g}}^0(\ell, u)$, by applying Theorem 10 in [11] to $\bar{g}$.

Proposition 10 *For* $\boldsymbol{\xi} = (\ell, \xi_1, \ldots, \xi_{n-1}, u)$, *we can compute* $\mathrm{vol}(\hat{U}_p^0(\boldsymbol{\xi}))$ *in* $\mathcal{O}(n)$ *time.*

Proof Sketch We define the τ_i and g from f, ℓ, u as usual. For $x \in [\ell, u]$, we have

$$\bar{g}(x) = g(x) = g(\tau_i) + \frac{g(\tau_{i+1}) - g(\tau_i)}{\tau_{i+1} - \tau_i}(x - \tau_i), \ \forall\, x \in [\tau_i, \tau_{i+1}], \ i = 0, 1, \ldots, n.$$

Applying Theorem 10 in [11] to $\bar{g}$, we have

$$\mathrm{vol}(\hat{U}_p^0(\boldsymbol{\xi})) = \int_0^{g(\ell)} \left(\int_{\frac{y}{g(u)}}^{\frac{y}{\ell^{p-1}u}} (uz - \ell z)dz + \int_{\frac{y}{\ell^{p-1}u}}^{\frac{y}{\ell^p}} (g^{-1}(y) - \ell z)dz \right) dy$$

$$+ \int_{g(\ell)}^{g(u)} \left(\int_{\frac{y}{g(u)}}^{\frac{g^{-1}(y)}{u}} (uz - \ell z)dz + \int_{\frac{g^{-1}(y)}{u}}^{1} (g^{-1}(y) - \ell z)dz \right) dy$$

$$- \frac{1}{6}(f(u) - f(\ell))(u - \ell)$$

$$= \int_0^{g(\ell)} \left(\int_{\frac{y}{g(u)}}^{\frac{y}{\ell^{p-1}u}} (uz - \ell z)dz + \int_{\frac{y}{\ell^{p-1}u}}^{\frac{y}{\ell^p}} (g^{-1}(y) - \ell z)dz \right) dy$$

$$+ \sum_{i=0}^{n} \int_{g(\tau_i)}^{g(\tau_{i+1})} \left(\int_{\frac{y}{g(u)}}^{\frac{g^{-1}(y)}{u}} (uz - \ell z)dz + \int_{\frac{g^{-1}(y)}{u}}^{1} (g^{-1}(y) - \ell z)dz \right) dy$$

$$- \frac{1}{6}(f(u) - f(\ell))(u - \ell)$$

$$= \frac{(u - \ell)\ell^{1+p}}{6u} - \frac{(u - \ell)u^p}{6} - \frac{1}{2}(u^p - \ell^p)\ell - \frac{1}{6}(u^p - \ell^p)(u - \ell)$$

$$+ \sum_{i=0}^{n} \int_{\tau_i}^{\tau_{i+1}} \left(-\frac{1}{2u}z^2 + z \right) \frac{g(\tau_{i+1}) - g(\tau_i)}{\tau_{i+1} - \tau_i} dz$$

$$= \frac{(u - \ell)\ell^{1+p}}{6u} - \frac{(u - \ell)u^p}{6} - \frac{1}{2}(u^p - \ell^p)\ell - \frac{1}{6}(u^p - \ell^p)(u - \ell)$$

$$+ \sum_{i=0}^{n} \left(-\frac{1}{6u}(\tau_{i+1}^3 - \tau_i^3) + \frac{1}{2}(\tau_{i+1}^2 - \tau_i^2) \right) p\xi_i^{p-1}.$$

$\square$

Corollary 11 *For* $p = 2$, *and the equally spaced points* $\xi_i = \ell + \frac{i}{n}(u - \ell)$, *for* $i = 1, \ldots, n - 1$,

$$\mathrm{vol}(\hat{U}_2^0(\boldsymbol{\xi})) = \frac{3u^3 + \ell^3 - 4u^2\ell}{12} - \frac{(u - \ell)(u^2 - \ell^2)}{6} + \frac{(u - \ell)^4}{24n^2u} - \frac{(u - \ell)\ell^3}{12u}.$$

Acknowledgments J. Lee was supported in part by ONR grant N00014-17-1-2296. D. Skipper was supported in part by ONR grant N00014-18-W-X00709. E. Speakman was supported by the Deutsche Forschungsgemeinschaft (DFG, German Research Foundation)—314838170, GRK 2297 MathCoRe. Lee and Skipper gratefully acknowledge additional support from the Institute of Mathematical Optimization, Otto-von-Guericke-Universität, Magdeburg, Germany.

References

1. Aktürk, M.S., Atamtürk, A., Gürel, S.: A strong conic quadratic reformulation for machine-job assignment with controllable processing times. Oper. Res Lett. **37**(3), 187–191 (2009).
2. Berenguel, J.L., Casado, L.G., García, I., Hendrix, E.M., Messine, F.: On interval branch-and-bound for additively separable functions with common variables. J. Global Optim. **56**(3), 1101–1121 (2013)
3. Charnes, A., Lemke, C.E.: Minimization of nonlinear separable convex functionals. Naval Res. Logist. Q. **1**, 301–312 (1955), 1954
4. Frangioni, A., Gentile, C.: Perspective cuts for a class of convex 0–1 mixed integer programs. Math. Program. **106**(2), 225–236 (2006)
5. Günlük, G., Linderoth, J.: Perspective reformulations of mixed integer nonlinear programs with indicator variables. Math. Program. B **124**, 183–205 (2010)
6. Hijazi, H., Bonami, P., Ouorou, A.: An outer-inner approximation for separable mixed-integer nonlinear programs. INFORMS J. Comput. **26**(1), 31–44 (2014)
7. Hiriart-Urruty, J.B., Lemaréchal, C.: Convex Analysis and Minimization Algorithms. I: Fundamentals. Grundlehren der Mathematischen Wissenschaften, vol. 305. Springer, Berlin (1993).
8. Lee, J., Morris, Jr. W.D.: Geometric comparison of combinatorial polytopes. Discret. Appl. Math. **55**(2), 163–182 (1994)
9. Lee, J., Wilson, D.: Polyhedral methods for piecewise-linear functions. I. The Lambda method. Discret. Appl. Math. **108**(3), 269–285 (2001)
10. Lee, J., Skipper, D., Speakman, E.: Algorithmic and modeling insights via volumetric comparison of polyhedral relaxations. Math. Program. B **170**, 121–140 (2018)
11. Lee, J., Skipper, D., Speakman, E.: Gaining or losing perspective (2019). https://arxiv.org/abs/2001.01435 [Journal version of Lee, J., Skipper, D., Speakman, E.: Gaining or losing perspective. In: Le Thi, H.A., Le, H.M., Dinh, T.P. (eds), Optimization of Complex Systems: Theory, Models, Algorithms and Applications, pp. 387–397. Springer, Berlin (2020)]
12. Lee, J., Skipper, D., Speakman, E.: Gaining or losing perspective. In: Le Thi, H.A., Le, H.M., Dinh, T.P. (eds), Optimization of Complex Systems: Theory, Models, Algorithms and Applications, pp. 387–397. Springer, Berlin (2020)
13. Toriello, A., Vielma, J.P.: Fitting piecewise linear continuous functions. Eur. J. Oper. Res. **219**(1), 86–95 (2012)
14. Vielma, J.P., Ahmed, S., Nemhauser, G.: Mixed-integer models for nonseparable piecewise-linear optimization: unifying framework and extensions. Oper. Res. **58**(2), 303–315 (2010)

Recognizing Cartesian Products of Matrices and Polytopes

Manuel Aprile, Michele Conforti, Yuri Faenza, Samuel Fiorini, Tony Huynh, and Marco Macchia

Abstract The 1-*product* of matrices $S_1 \in \mathbb{R}^{m_1 \times n_1}$ and $S_2 \in \mathbb{R}^{m_2 \times n_2}$ is the matrix in $\mathbb{R}^{(m_1+m_2) \times (n_1 n_2)}$ whose columns are the concatenation of each column of S_1 with each column of S_2. Our main result is a polynomial time algorithm for the following problem: given a matrix S, is S a 1-product, up to permutation of rows and columns? Our main motivation is a close link between the 1-product of matrices and the Cartesian product of polytopes, which relies on the concept of slack matrix. Determining whether a given matrix is a slack matrix is an intriguing problem whose complexity is unknown, and our algorithm reduces the problem to irreducible instances. Our algorithm is based on minimizing a symmetric submodular function that expresses mutual information in information theory. We also give a polynomial time algorithm to recognize a more complicated matrix product, called the 2-*product*. Finally, as a corollary of our 1-product and 2-product recognition algorithms, we obtain a polynomial time algorithm to recognize slack matrices of 2-level matroid base polytopes.

Keywords Cartesian product · Slack matrix · Mutual information · Submodular optimization · 2-level polytopes

M. Aprile (✉) · S. Fiorini
Université libre de Bruxelles, Brussels, Belgium
e-mail: sfiorini@ulb.ac.be

M. Conforti
Università degli Studi di Padova, Padova, Italy
e-mail: conforti@math.unipd.it

Y. Faenza
Columbia University, New York, NY, USA
e-mail: yf2414@columbia.edu

T. Huynh
Monash University, Melbourne, VIC, Australia

M. Macchia
Klarna AB, Berlin, Germany

© The Author(s), under exclusive license to Springer Nature Switzerland AG 2021
C. Gentile et al. (eds.), *Graphs and Combinatorial Optimization:*
from Theory to Applications, AIRO Springer Series 5,
https://doi.org/10.1007/978-3-030-63072-0_28

361

1 Introduction

Determining if an object can be decomposed as the 'product' of two simpler objects is a ubiquitous theme in mathematics and computer science. For example, every integer $n \geq 2$ has a unique factorization into primes, and every finite abelian group is the direct sum of cyclic groups. Moreover, algorithms to efficiently *find* such 'factorizations' are widely studied, since many algorithmic problems are easy on indecomposable instances. In this paper, our objects of interest are matrices and polytopes.

For a matrix S, we let S^ℓ be the ℓth column of S. The 1-*product* of $S_1 \in \mathbb{R}^{m_1 \times n_1}$ and $S_2 \in \mathbb{R}^{m_2 \times n_2}$ is the matrix $S_1 \otimes S_2 \in \mathbb{R}^{(m_1+m_2) \times (n_1 n_2)}$ such that for each $j \in [n_1 \cdot n_2]$,

$$(S_1 \otimes S_2)^j := \begin{pmatrix} S_1^k \\ S_2^\ell \end{pmatrix},$$

where $k \in [n_1]$ and $\ell \in [n_2]$ satisfy $j = (k-1)n_2 + \ell$. For example,

$$(1\ 0) \otimes (0) = \begin{pmatrix} 1 & 0 \\ 0 & 0 \end{pmatrix}, \qquad \begin{pmatrix} 1 & 0 \\ 2 & 3 \end{pmatrix} \otimes \begin{pmatrix} 1 & 0 & 0 \\ 0 & 1 & 1 \end{pmatrix} = \begin{pmatrix} 1 & 1 & 1 & 0 & 0 & 0 \\ 2 & 2 & 2 & 3 & 3 & 3 \\ 1 & 0 & 0 & 1 & 0 & 0 \\ 0 & 1 & 1 & 0 & 1 & 1 \end{pmatrix}.$$

Two matrices are *isomorphic* if one can be obtained from the other by permuting rows and columns. A matrix S is a *1-product* if there exist two non-empty matrices S_1 and S_2 such that S is isomorphic to $S_1 \otimes S_2$. The following is our first main result.

Theorem 1 *Given $S \in \mathbb{R}^{m \times n}$, there is an algorithm that is polynomial[1] in n, m which correctly determines if S is a 1-product and, in case it is, outputs two matrices S_1, S_2 such that $S_1 \otimes S_2$ is isomorphic to S.*

The proof of Theorem 1 is by reduction to symmetric submodular function minimization using the concept of *mutual information* from information theory. Somewhat surprisingly, we do not know of a simpler proof of Theorem 1.

Our main motivation for Theorem 1 is geometric. If $P_1 \subseteq \mathbb{R}^{d_1}$ and $P_2 \subseteq \mathbb{R}^{d_2}$ are polytopes, then their *Cartesian product* is the polytope $P_1 \times P_2 := \{(x_1, x_2) \in \mathbb{R}^{d_1} \times \mathbb{R}^{d_2} \mid x_1 \in P_1, x_2 \in P_2\}$.

Notice that if P is given by an irredundant inequality description, determining if $P = P_1 \times P_2$ for some polytopes P_1, P_2 amounts to determining whether the constraint matrix can be put in block diagonal structure. If P is given as a list of vertices, then the algorithm of Theorem 1 determines if P is a Cartesian product.

---

[1] A straightforward implementation of our algorithm would run in $O(m^3(m+n))$ time. However, throughout the paper we do not explicitly state the running times of our algorithms nor try to optimize them.

Furthermore it turns out that the 1-product of matrices corresponds to the Cartesian product of polytopes if we represent a polytope via its slack matrix, as we now describe.

Let $P = \text{conv}(\{v_1, \ldots, v_n\}) = \{x \in \mathbb{R}^d \mid Ax \leq b\}$, where $\{v_1, \ldots, v_n\} \subseteq \mathbb{R}^d$, $A \in \mathbb{R}^{m \times d}$ and $b \in \mathbb{R}^m$. The *slack matrix* associated to these descriptions of P is the matrix $S \in \mathbb{R}_+^{m \times n}$ with $S_{i,j} := b_i - A_i v_j$, for $i \in [m]$ and $j \in [n]$. That is, $S_{i,j}$ is the slack of point v_j with respect to the inequality $A_i x \leq b_i$.

Slack matrices were introduced in a seminal paper of Yannakakis [15], as a tool for reasoning about the extension complexity of polytopes (see [5]).

Our second main result is the following corollary to Theorem 1.

Theorem 2 *Given a polytope P represented by its slack matrix $S \in \mathbb{R}^{m \times n}$, there is an algorithm that is polynomial in m, n which correctly determines if P is affinely equivalent to a Cartesian product $P_1 \times P_2$ and, in case it is, outputs two matrices S_1, S_2 such that S_i is the slack matrix of P_i, for $i \in [2]$.*

Some comments are in order here. First, our algorithm determines whether a polytope P is affinely equivalent to a Cartesian product of two polytopes. Since affine transformations do not preserve the property of being a Cartesian product, this is a different problem than that of determining whether P equals $P_1 \times P_2$ for some polytopes P_1, P_2, which turns out to be much easier than the problem solved by Theorem 2. Second, the definition of 1-product can be extended to a more complex operation which we call 2-product. Theorems 1 and 2 can be extended to handle 2-products, see Theorem 12.

Slack matrices are fascinating objects, that are not fully understood. For instance, given a matrix $S \in \mathbb{R}_+^{m \times n}$, the complexity of determining whether S is the slack matrix of some polytope is open. In [7], the problem has been shown to be equivalent to the Polyhedral Verification Problem (see [9]): given a vertex description of a polytope P, and an inequality description of a polytope Q, determine whether $P = Q$.

Polytopes that have a 0/1-valued slack matrix are called 2-*level polytopes*. These form a rich class of polytopes including stable set polytopes of perfect graphs, Birkhoff, and Hanner polytopes (see [1, 2, 11] for more examples and details). We conjecture that slack matrix recognition is polynomial-time solvable for 2-level polytopes.

Conjecture 3 *Given $S \in \{0, 1\}^{m \times n}$, there is an algorithm that is polynomial in m, n which correctly determines if S is the slack matrix of a polytope.*

Conjecture 3 seems hard to settle: however it has been proven for certain restricted classes of 2-level polytopes, most notably for stable set polytopes of perfect graphs [1]. As a final result, we apply Theorem 1 and its extension to 2-products to show that Conjecture 3 holds for 2-level matroid base polytopes (precise definitions will be given later).

Theorem 4 *Given $S \in \{0, 1\}^{m \times n}$, there is an algorithm that is polynomial in m, n which correctly determines if S is the slack matrix of a 2-level matroid base polytope.*

Paper Outline In Sect. 2 we study the properties of 1-products and 2-products in terms of slack matrices. In Sect. 3 we give algorithms to efficiently recognize 1-products and 2-products, as well as showing a unique decomposition result for 1-products. Finally, in Sect. 4 we apply the previous results to slack matrices of matroid base polytopes, obtaining Theorem 4.

The results presented in this paper are contained in the PhD thesis of the first author [1]. We refer to [1] and to the arXiv version of the paper [3] for details and omitted proofs.

2   Properties of 1-Products and 2-Products

Here we study the 1-product of matrices defined in the introduction, as well as the 2-product. We remark that the notion of 2-product can be generalized to k-products for every $k \geq 3$ (see [1] for more details). The k-product operation is similar to the glued product of polytopes in [12], except that the latter is defined for 0/1 polytopes, while we deal with general matrices.

We show that, under certain assumptions, the operations of 1- and 2-product preserve the property of being a slack matrix. For a matrix S, $col(S)$ denotes the set of column vectors of S. We recall the following characterization of slack matrices, due to [7].

Theorem 5 (Gouveia et al. [7]) *Let $S \in \mathbb{R}^{m \times n}$ be a nonnegative matrix of rank at least 2. Then S is the slack matrix of a polytope if and only if $conv(col(S)) = aff(col(S)) \cap \mathbb{R}^m_+$. Moreover, if S is the slack matrix of polytope P then P is affinely equivalent to $conv(col(S))$.*

Throughout the paper, we will assume that the matrices we deal with are of rank at least 2, so we may apply Theorem 5 directly.

2.1   1-Products

We show that the 1-product operation preserves the property of being a slack matrix.

Lemma 6 *Let $S \in \mathbb{R}^{m \times n}_+$ and $S_i \in \mathbb{R}^{m_i \times n_i}_+$ for $i \in [2]$ satisfy $S = S_1 \otimes S_2$. Then S is the slack matrix of a polytope P if and only if there exist polytopes P_i, $i \in [2]$ such that S_i is the slack matrix of P_i and P is affinely equivalent to $P_1 \times P_2$.*

Sketch For $i \in [2]$, let $C_i := col(S_i)$. From Theorem 5, and since $col(S_1 \otimes S_2) = col(S_1) \times col(S_2) = C_1 \times C_2$, Lemma 6 follows from (i) $aff(C_1 \times C_2) = aff(C_1) \times aff(C_2)$ and (ii) $conv(C_1 \times C_2) = conv(C_1) \times conv(C_2)$. Both statements are straightforward to prove.   □

2.2 2-Products

We now define the operation of 2-product, and show that, under certain natural assumptions, it also preserves the property of being a slack matrix.

Consider two real matrices S_1, S_2, and assume that S_1 (resp. S_2) has a 0/1 row x_1 (resp. y_1), that is, a row whose entries are 0 or 1 only. We call x_1, y_1 *special* rows. For any matrix M and row r of M, we denote by $M - r$ the matrix obtained from M by removing row r. The row x_1 determines a partition of $S_1 - x_1$ into two submatrices according to its 0 and 1 entries: we define S_1^0 to be the matrix obtained from S_1 by deleting the row x_1 and all the columns whose x_1-entry is 1, and S_1^1 is defined analogously. Thus,

$$S_1 = \left(\begin{array}{c|c} S_1^0 & S_1^1 \\ 0 \cdots 0 & 1 \cdots 1 \end{array} \right) \leftarrow x_1 .$$

Similarly, y_1 induces a partition of $S_2 - y_1$ into S_2^0, S_2^1. Here we assume that none of S_1^0, S_1^1, S_2^0, S_2^1 is empty, that is, we assume that the special rows contain both 0's and 1's.

The 2-*product* of $S_1 \in \mathbb{R}^{m_1 \times n_1}$ with special row x_1 and $S_2 \in \mathbb{R}^{m_2 \times n_2}$ with special row y_1 is defined as:

$$S = (S_1, x_1) \otimes_2 (S_2, y_1) := \left(\begin{array}{c|c} S_1^0 \otimes S_2^0 & S_1^1 \otimes S_2^1 \\ 0 \cdots 0 & 1 \cdots 1 \end{array} \right) .$$

Similarly as before, we say that S is a 2-product if there exist matrices S_1, S_2 and 0/1 rows x_1 of S_1, y_1 of S_2, such that S is isomorphic to $(S_1, x_1) \otimes_2 (S_2, y_1)$. Again, we will abuse notation and write $S = (S_1, x_1) \otimes_2 (S_2, y_1)$.

For a polytope P with slack matrix S, consider a row r of S corresponding to an inequality $a^\mathsf{T} x \leq b$ that is valid for P. We say that r is 2-*level* with respect to S, and that $a^\mathsf{T} x \leq b$ is 2-*level* with respect to P, if there exists a real $b' < b$ such that all the vertices of P either lie on the hyperplane $\{x \mid a^\mathsf{T} x = b\}$ or the hyperplane $\{x \mid a^\mathsf{T} x = b'\}$.

We notice that, if r is 2-level, then r can be assumed to be 0/1 after scaling. Moreover, adding to S the row $\mathbf{1} - r$ (that is, the complement of 0/1 row r) gives another slack matrix of P. Indeed, such a row corresponds to the valid inequality $a^\mathsf{T} x \geq b'$.

The latter observation is crucial for our next lemma: we show that, if the special rows are chosen to be 2-level, the operation of 2-product essentially preserves the property of being a slack matrix. We remark that having a 2-level row is a quite natural condition. For instance, for 0/1 polytopes, any non-negativity yields a 2-level row in the corresponding slack matrix. By definition, all facet-defining inequalities of a 2-level polytope are 2-level. Finally, we would like to mention that the following result could be derived from results from [12] (see also [4]), but we give here a new, direct proof.

Lemma 7 Let $S \in \mathbb{R}_+^{m \times n}$ and let $S_i \in \mathbb{R}_+^{m_i \times n_i}$ for $i \in [2]$ such that $S = (S_1, x_1) \otimes_2 (S_2, y_1)$ for some 2-level rows x_1 of S_1, y_1 of S_2. The following hold:

(i) If both S_1 and S_2 are slack matrices, then S is a slack matrix.
(ii) If S is a slack matrix, let $S_1' := S_1 + (\mathbf{1} - x_1)$ (that is, S_1 with the additional row $\mathbf{1} - x_1$), and similarly let $S_2' := S_2 + (\mathbf{1} - y_2)$. Then both S_1' and S_2' are slack matrices.

Proof

(i) Let $P_i := \mathrm{conv}(\mathrm{col}(S_i)) \subseteq \mathbb{R}^{m_i}$ for $i \in [2]$. Recall that S_i is the slack matrix of P_i, by Theorem 5. Without loss of generality, x_1 and y_1 can be assumed to be the first rows of S_1, S_2 respectively. We overload notation and denote by x_1 the first coordinate of x as a point in $\mathbb{R}^{m_1}$, and similarly for $y \in \mathbb{R}^{m_2}$. Let H denote the hyperplane of $\mathbb{R}^{m_1+m_2}$ defined by the equation $x_1 = y_1$.

   We claim that S is a slack matrix of the polytope $(P_1 \times P_2) \cap H$. By Lemma 6, S is a submatrix of the slack matrix of $(P_1 \times P_2) \cap H$. But the latter might have some extra columns: hence we only need to show that intersecting $P_1 \times P_2$ with H does not create any new vertex.

   To this end we notice that no new vertex is created if and only if there is no edge e of $P_1 \times P_2$ such that $H = \{(x, y) \mid x_1 = y_1\}$ intersects e in its interior. Let e be an edge of $P_1 \times P_2$, and let (v_1, v_2) and (w_1, w_2) denote its endpoints, where $v_1, w_1 \in \mathrm{col}(S_1)$ and $v_2, w_2 \in \mathrm{col}(S_2)$. By a well-known property of the Cartesian product, $v_1 = w_1$ or $v_2 = w_2$. Suppose that (v_1, v_2) does not lie on H. By symmetry, we may assume that $v_{11} < v_{21}$. This implies $v_{11} = 0$ and $v_{21} = 1$, which in turn implies $w_{11} \leq w_{21}$ (since $v_1 = w_1$ or $v_2 = w_2$). Thus (w_1, w_2) lies on the same side of H as (v_1, v_2), and H cannot intersect e in its interior. Therefore, the claim holds and S is a slack matrix.

(ii) Assume that $S = (S_1, x_1) \otimes_2 (S_2, y_1)$ is a slack matrix. We show that $S_1' = S_1 + (\mathbf{1} - x_1)$ is a slack matrix, using Theorem 5. The argument for S_2' is symmetric. It suffices to show that $\mathrm{aff}(\mathrm{col}(S)) \cap \mathbb{R}_+^{m_1} \subseteq \mathrm{conv}(\mathrm{col}(S))$ since the reverse inclusion is obvious.

   Let $x^* \in \mathrm{aff}(\mathrm{col}(S_1')) \cap \mathbb{R}_+^{m_1}$. One has $x^* = \sum_{i \in I} \lambda_i v_i$ for some coefficients $\lambda_i \in \mathbb{R}$ with $\sum_{i \in I} \lambda_i = 1$, where $v_i \in \mathrm{col}(S_1')$ for $i \in I$. We partition the index set I into I_0 and I_1, so that $i \in I_0$ (resp. $i \in I_1$) if v_i has its x_1 entry equal to 0 (resp. 1). For simplicity, we may assume that x_1 is the first row of S_1', and $\mathbf{1} - x_1$ the second. Then, the first coordinate of x^* is $x_1^* = \sum_{i \in I_1} \lambda_i \geq 0$, and the second is $x_2^* = \sum_{i \in I_0} \lambda_i \geq 0$. Notice that $x_1^* + x_2^* = \sum_{i \in I} \lambda_i = 1$.

   Now, we extend x^* to a point $\tilde{x} \in \mathrm{aff}(\mathrm{col}(S))$ by mapping each $v_i, i \in I$ to a column of S, as follows. For each $a \in \{0, 1\}$, fix an arbitrary column c_a of S_2^a, then map each v_i with $i \in I_a$ to the column of S consisting of v_i, without its second component, followed by c_a. We denote such a column by u_i, for $i \in I$, and let $\tilde{x} := \sum_{i \in I} \lambda_i u_i$.

We claim that $\tilde{x} \in \mathbb{R}_+^m$. This is trivial for any component corresponding to a row of S_1, since those are components of x^* as well. Consider a component $\tilde{x}_j$ corresponding to a row of S_2, and denote by $c_{a,j}$ the corresponding component of c_a, for $a = 0, 1$. We have:

$$\tilde{x}_j = \sum_{i \in I_0} \lambda_i c_{0,j} + \sum_{i \in I_1} \lambda_i c_{1,j} = x_2^* c_{0,j} + x_1^* c_{1,j} \geq 0.$$

Now, Theorem 5 applied to S implies that $\tilde{x} \in \mathrm{conv}(\mathrm{col}(S))$. That is, we can write $\tilde{x} = \sum_{i \in I'} \mu_i u_i'$ where $u_i' \in \mathrm{col}(S)$ and $\mu_i \in \mathbb{R}_+$ for $i \in I'$ and $\sum_{i \in I'} \mu_i = 1$. For each $i \in I'$, let $v_i' \in \mathrm{col}(S_1')$ denote the column vector obtained from u_i' by restricting to the rows of S_1 and inserting as a second component $1 - u_{i,1}' = 1 - v_{i,1}'$. We claim that $x^* = \sum_{i \in I'} \mu_i v_i'$, which implies that $x^* \in \mathrm{conv}(\mathrm{col}(S_1'))$ and concludes the proof. The claim is trivially true for all components of x^* except for the second, for which one has $x_2^* = 1 - x_1^*$ since $v_{i,2}' = 1 - v_{i,1}'$ for all $i \in I'$ by definition of v_i'.

$\square$

3 Algorithms

In this section we study the problem of recognizing 1-products. Given a matrix S, we want to determine whether S is a 1-product, and find matrices S_1, S_2 such that $S = S_1 \otimes S_2$. Since we allow the rows and columns of S to be permuted in an arbitrary way, the problem is non-trivial.

At the end of the section, we extend our methods to the problem of recognizing 2-products. We remark that the results in this section naturally extend to a more general operation, the k-product, for every constant k (see [1] for more details).

We begin with a preliminary observation, which is the starting point of our approach. Suppose that a matrix S is a 1-product $S_1 \otimes S_2$. Then the rows of S can be partitioned into two sets R_1, R_2, corresponding to the rows of S_1, S_2 respectively. We write that S is a 1-product *with respect to* the partition R_1, R_2. A column of the form (a_1, a_2), where a_i is a column vector with components indexed by R_i ($i \in [2]$), is a column of S if and only if a_i is a column of S_i for each $i \in [2]$. Moreover, the number of occurrences of (a_1, a_2) in S is just the product of the number of occurrences of a_i in S_i for $i \in [2]$. Under uniform probability distributions on the columns of S, S_1 and S_2, the probability of picking (a_1, a_2) in S is the product of the probability of picking a_1 in S_1 and that of picking a_2 in S_2. We will exploit this intuition below.

3.1 Recognizing 1-Products via Submodular Minimization

First, we recall some notions from information theory, see [6] for a more complete exposition. Let A and B be discrete random variables with ranges $\mathscr{A}$ and $\mathscr{B}$, respectively. The *mutual information* of A and B is:

$$I(A; B) = \sum_{a \in \mathscr{A}, b \in \mathscr{B}} \Pr(A = a, B = b) \cdot \log_2 \frac{\Pr(A = a, B = b)}{\Pr(A = a) \cdot \Pr(B = b)}.$$

The mutual information of two random variables measures how close is their joint distribution to the product of the two corresponding marginal distributions.

Let $C_1, \ldots, C_m$ be discrete random variables. For $X \subseteq [m]$ we consider the random vectors $C_X := (C_i)_{i \in X}$ and $C_{\overline{X}} := (C_i)_{i \in \overline{X}}$, where $\overline{X} := [m] \setminus X$. The function $f : 2^{[m]} \to \mathbb{R}$ such that

$$f(X) := I(C_X; C_{\overline{X}}) \tag{1}$$

will play a crucial role. We will use the following facts, which are proved in [6, 10].

Proposition 8

(i) *For all discrete random variables A and B, we have $I(A; B) \geq 0$, with equality if and only if A and B are independent.*

(ii) *If $C_1, \ldots, C_m$ are discrete random variables, then the function f as in (1) is submodular.*

Let S be an $m \times n$ matrix. Let $C := (C_1, \ldots, C_m)$ be a uniformly chosen random column of S. That is, $\Pr(C = c) = \mu(c)/n$, where $\mu(c)$ denotes the number of occurrences in S of the column $c \in \mathrm{col}(S)$.

Let $f : 2^{[m]} \to \mathbb{R}$ be defined as in (1). We remark that the definition of f depends on S, which we consider fixed throughout the section. The set function f is non-negative (by Proposition 8.(i)), symmetric (that is, $f(X) = f(\overline{X})$) and submodular (by Proposition 8.(ii)).

The next lemma shows that we can determine whether S is a 1-product by minimizing f.

Lemma 9 *Let $S \in \mathbb{R}^{m \times n}$, and $\emptyset \neq X \subsetneq [m]$. Then S is a 1-product with respect to $X, \overline{X}$ if and only if C_X and $C_{\overline{X}}$ are independent random variables, or equivalently (by Proposition 8.i), $f(X) = 0$.*

Proof First, we prove "$\Longrightarrow$". Suppose that S is a 1-product with respect to $X, \overline{X}$ for some non-empty and proper set X of row indices of S. Let $S = S_1 \otimes S_2$ be the corresponding 1-product, where $S_i \in \mathbb{R}^{m_i \times n_i}$ for $i \in [2]$.

For every column $c = (c_X, c_{\overline{X}}) \in \text{col}(S)$, we have $\mu(c) = \mu_1(c_X)\mu_2(c_{\overline{X}})$, where μ_i denotes the multiplicity of a column in S_i, for $i \in [2]$. Hence

$$\Pr(C_X = c_X, \ C_{\overline{X}} = c_{\overline{X}}) = \Pr(C = c) = \frac{\mu(c)}{n}$$

$$= \frac{\mu_1(c_X) \cdot n_2}{n} \cdot \frac{n_1 \cdot \mu_2(c_{\overline{X}})}{n} = \Pr(C_X = c_X) \cdot \Pr(C_{\overline{X}} = c_{\overline{X}}),$$

where we used $n = n_1 n_2$. This proves that C_X and $C_{\overline{X}}$ are independent.

We now prove "$\Longleftarrow$". Let $a_1, \ldots, a_k$ denote the different columns of the restriction $S|_X$ of matrix S to the rows in X, and $b_1, \ldots, b_\ell$ denote the different columns of $S|_{\overline{X}}$ (the restriction of matrix S to the rows in $\overline{X}$). Since C_X and $C_{\overline{X}}$ are independent, we have that, for every column $c = (a_i, b_j)$ of S,

$$\mu(a_i, b_j) = n \cdot \Pr(C_X = a_i, C_{\overline{X}} = b_j) = n \cdot \Pr(C_X = a_i)\Pr(C_{\overline{X}} = b_j) = \frac{\mu_X(a_i)\mu_{\overline{X}}(b_j)}{n},$$

where $\mu_X(\cdot)$ and $\mu_{\overline{X}}(\cdot)$ denote multiplicities in $S|_X$ and $S|_{\overline{X}}$ respectively.

Now, let M denote the $k \times \ell$ matrix such that $M_{i,j} := \mu(a_i, b_j)$ for $i \in [k]$ and $j \in [\ell]$. We have shown that M is a non-negative integer matrix with a rank-1 non-negative factorization of the form uv^{T}, where $u_i := \mu_X(a_i)/n$ and $v_j := \mu_{\overline{X}}(b_j)$, for $i \in [k]$ and $j \in [\ell]$.

Next, one can easily turn this non-negative factorization into an integer one. Suppose that u_i is fractional for some $i \in [k]$. Writing u_i as $u_i = p_i/q_i$, where $p_i \in \mathbb{Z}_{\geq 0}$ and $q_i \in \mathbb{Z}_{>0}$ are coprime, we see that q_i divides v_j since $u_i v_j$ is integer, for every $j \in [\ell]$. Then the factorization $q_i u \cdot \frac{1}{q_i} v^{\mathsf{T}} = u'(v')^{\mathsf{T}}$ is such that v' is integer and u' has at least one more integer component than u. Iterating this argument, we obtain that $M = \overline{u}\,\overline{v}^{\mathsf{T}}$ where $\overline{u}, \overline{v}$ have non-negative integer entries.

Finally, let S_1 be the matrix consisting of the column a_i repeated $\overline{u}_i$ times for $i \in [k]$, and construct S_2 from $\overline{v}$ in an analogous way. Clearly $S = S_1 \otimes S_2$, and in particular S is a 1-product with respect to the row partition $X, \overline{X}$, which concludes the proof. □

Notice that the previous proof also gives a way to efficiently reconstruct S_1, S_2 once we identified X such that $f(X) = 0$. In particular, if the columns of S are all distinct, then S_1 consists of all the distinct columns of $S|_X$, each taken once, and S_2 is obtained analogously from $S|_{\overline{X}}$. The last ingredient we need is that every (symmetric) submodular function can be minimized in polynomial time:

Theorem 10 (Queyranne [14]) *There is a polynomial time algorithm that outputs a set X such that $X \notin \{\emptyset, [m]\}$ and $f(X)$ is minimum, where $f : 2^{[m]} \to \mathbb{R}$ is any given symmetric submodular function.*

As a direct consequence, we obtain Theorem 1.

Proof of Theorem 1 It is clear that $f(X)$ can be computed in polynomial time for any X. It suffices then to run Queyranne's algorithm to find X minimizing f. If $f(X) > 0$, then S is not a 1-product. Otherwise, $f(X) = 0$ and S_1, S_2 can be reconstructed as described in the proof of Lemma 9.   □

We conclude the section with a decomposition result which will be useful in the next section. We call a matrix *irreducible* if it is not a 1-product. The result below generalizes the fact that a polytope can be uniquely decomposed as a Cartesian product of "irreducible" polytopes.

Lemma 11 *Let $S \in \mathbb{R}^{m \times n}$ be a matrix whose columns are all distinct. Then there are matrices $S_1, \ldots, S_t$ such that $S = S_1 \otimes \cdots \otimes S_t$, each S_i is irreducible, and the choice of the S_i's is unique up to isomorphism.*

3.2 Extension to 2-Products

We now extend the previous results to obtain a polynomial algorithm to recognize 2-products. Recall that, if a matrix S is a 2-product, then it has a special row that divides S into submatrices S^0, S^1, which are 1-products *with respect to the same partition*. Hence, our algorithm starts by guessing the special row, and obtaining the corresponding submatrices S^0, S^1. Let f_0 (resp. f_1) denote the function f as defined in (1) with respect to the matrix S^0 (S^1), and let $\tilde{f} = f_0 + f_1$. Notice that $\tilde{f}$ is submodular, and is zero if and only if each f_i is. Let X be a proper subset of the non-special rows of S (which are the rows of S^0 and S^1). It is an easy consequence of Lemma 9 that S^0, S^1 are 1-products with respect to X if and only if $\tilde{f}(X) = 0$. Then S is a 2-product with respect to the chosen special rows if and only if the minimum of $\tilde{f}$ is zero.

Once a feasible partition is found, S_1, S_2 can be reconstructed by first reconstructing all of $S_1^0, S_1^1, S_2^0, S_2^1$ and then concatenating them and adding the special rows. We obtained the following:

Theorem 12 *Let $S \in \mathbb{R}^{m \times n}$. There is an algorithm that is polynomial in m, n and determines whether S is a 2-product and, in case it is, outputs two matrices S_1, S_2 and special rows x_1 of S_1, y_1 of S_2, such that $S = (S_1, x_1) \otimes_2 (S_2, y_1)$.*

4 Application to 2-Level Matroid Base Polytopes

In this section, we use the results in Sect. 3 to derive a polynomial time algorithm to recognize the slack matrix of a 2-level matroid base polytope.

We start with some basic definitions and facts about matroids, and we refer the reader to [13] for missing definitions and details. We regard a matroid M as a couple $(E, \mathcal{B})$, where E is the ground set of M, and $\mathcal{B}$ is its set of bases. A matroid

$M = (E, \mathcal{B})$ is *uniform* if $\mathcal{B} = \binom{E}{k}$, where k is the rank of M. We denote the uniform matroid with n elements and rank k by $U_{n,k}$.

Consider matroids $M_1 = (E_1, \mathcal{B}_1)$ and $M_2 = (E_2, \mathcal{B}_2)$, with non-empty ground sets. If $E_1 \cap E_2 = \emptyset$, the *1-sum* $M_1 \oplus M_2$ is defined as the matroid with ground set $E_1 \cup E_2$ and base set $\mathcal{B}_1 \times \mathcal{B}_2$. If, instead, $E_1 \cap E_2 = \{p\}$, where p is neither a loop nor a coloop in M_1 or M_2, we let the *2-sum* $M_1 \oplus_2 M_2$ be the matroid with ground set $(E_1 \cup E_2) \setminus \{p\}$, and base set $\{(B_1 \cup B_2) \setminus \{p\} \mid B_i \in \mathcal{B}_i$ for $i \in [2]$ and $p \in B_1 \triangle B_2\}$. A matroid is *connected* if it cannot be written as the 1-sum of two matroids, each with fewer elements.

The *base polytope* $B(M)$ of a matroid M is the convex hull of the characteristic vectors of its bases. It is easy to see that, if $M = M_1 \oplus M_2$, then $B(M)$ is the Cartesian product $B(M_1) \times B(M_2)$, hence its slack matrix is a 1-product thanks to Lemma 6. If $M = M_1 \oplus_2 M_2$, then a slightly less trivial polyhedral relation holds, providing a connection with the 2-product of slack matrices. We will explain this connection below.

Our algorithm is based on the following decomposition result, that characterizes those matroids M such that $B(M)$ is 2-level (equivalently, such that $B(M)$ admits a 0/1 slack matrix).

Theorem 13 ([8]) *The base polytope of a matroid M is 2-level if and only if M can be obtained from uniform matroids through a sequence of 1-sums and 2-sums.*

The general idea is to use the algorithms from Theorems 1 and 12 to decompose our candidate slack matrix as 1-products and 2-products, until each factor corresponds to the slack matrix of a uniform matroid. The latter can be easily recognized. Indeed, the base polytope of the uniform matroid $U_{n,k}$ is the (n, k)-hypersimplex $B(U_{n,k}) = \{x \in [0, 1]^E \mid \sum_e x_e = k\}$. If $2 \leq k \leq n - 2$, the (irredundant, 0/1) slack matrix S of $B(U_{n,k})$ has $2n = 2|E|$ rows and $\binom{n}{k}$ columns of the form $(v, \mathbf{1} - v)$ where $v \in \{0, 1\}^n$ is a vector with exactly k ones, hence can be recognized in polynomial time (in its size). We denote such matrix by $S_{n,k}$. If $k = 1$, or equivalently $k = n - 1$, $S = S_{n,1} = S_{n,n-1}$ is just the identity matrix I_n. The case $k = 0$ or $k = n$ corresponds to a non-connected matroid whose base polytope is just a single vertex, and can be ignored for our purposes.

We now focus on the relationship between 1-sums and 1-products. As already remarked, if S_1, S_2 are the slack matrices of $B(M_1)$, $B(M_2)$ respectively, then $S_1 \otimes S_2$ is the slack matrix of $B(M_1) \times B(M_2) = B(M_1 \oplus M_2)$. The next lemma shows that the converse holds.

Lemma 14 *Let M be a matroid and let S be the slack matrix of $B(M)$. If $S = S_1 \otimes S_2$ for some matrices S_1, S_2, then there are matroids M_1, M_2 such that $M = M_1 \oplus M_2$ and S_i is the slack matrix of $B(M_i)$ for $i \in [2]$.*

Now, we deal with slack matrices of connected matroids and with the operation of 2-product. We will need the following result, which provides a description of the base polytope of a 2-product $M_1 \oplus_2 M_2$ in terms of the base polytopes of M_1, M_2. Its proof can be derived from [8], or found in [2].

Lemma 15 *Let M_1, M_2 be matroids on ground sets E_1, E_2 respectively, with $E_1 \cap E_2 = \{p\}$, and let $M = M_1 \oplus_2 M_2$. Then $B(M)$ is affinely equivalent to*

$$(B(M_1) \times B(M_2)) \cap \{(x, y) \in \mathbb{R}^{E_1} \times \mathbb{R}^{E_2} \mid x_p + y_p = 1\}.$$

Lemma 15 implies that if $M = M_1 \oplus_2 M_2$ and S_i is a slack matrix of $B(M_i)$ for $i \in [2]$, then the slack matrix of $B(M)$ is actually $(S_1, x_p) \otimes_2 (S_2, \overline{y}_p)$, where x_p is the row corresponding to $x_p \geq 0$, and $\overline{y}_p$ the row corresponding to $y_p \leq 1$. If the special rows $x_p, \overline{y}_p$ have this form, we say that they are *coherent*.

The only missing ingredient is now a converse to the above statement.

Lemma 16 *Let $M = (E, \mathscr{B})$ be a connected matroid and let S be the slack matrix of $B(M)$. Assume there are S_1, S_2 such that $S = (S_1, x_1) \otimes_2 (S_2, \overline{y}_1)$, for some 2-level rows $x_1, \overline{y}_1$, and let $S_1' = S_1 + (I - x_1)$ and similarly for S_2'. Assume that S_1 or S_1' is equal to $S_{d,k}$ for some $d > k \geq 1$. Then there is a matroid M_2 such that $M = U_{d,k} \oplus_2 M_2$ and S_2' is a slack matrix of $B(M_2)$.*

We are now ready to prove the main result of this section, namely, Theorem 4.

Proof of Theorem 4 We first check whether $S = S_{d,k}$ for some d and k, in which case we are done.

Then, we run the algorithm to recognize 1-products, and if S is a 1-product, we decompose it in irreducible factors $S_1, \ldots, S_t$ and test each S_i separately. This can be done efficiently thanks to Theorem 1, and using Lemma 14 we have that S is the slack matrix of $B(M)$ if and only if S_i is the slack matrix of $B(M_i)$ for each i, and $M = M_1 \oplus \cdots \oplus M_t$.

We can now assume that S is irreducible, and apply the algorithm from Theorem 12 until we decompose S as a repeated 2-product of matrices $S_1, \ldots, S_t$ where $S_i = S_{d_i, k_i}$ for $i \in [t]$ (of course, if this is not possible, we conclude that S is not a slack matrix of a base polytope). There is one last technicality we have to deal with, before we can conclude that S is the slack matrix of a matroid polytope. Indeed, as noticed above, we need to ensure that each pair of special rows involved in a 2-product is coherent. This might not be possible since the form of the special row of S_i is fixed whenever S_i is an identity matrix. However, one can model this problem as a simple coloring problem on a tree: the tree naturally arises from the decomposition on S that we performed, there are two colors, one for each possible form of the special row, and some nodes have a fixed color (see [1] for the details). One can efficiently determine whether the tree can be properly colored given these constraints, thus concluding the algorithm. $\qquad\square$

Acknowledgments This project was supported by ERC Consolidator Grant 615640-ForEFront and by a gift by the SNSF.

References

1. Aprile, M.: On some problems related to 2-level polytopes. Ph.D. thesis, École Polytechnique Fédérale de Lausanne (2018). https://manuel-aprile.github.io/my_website/file/Aprile_thesis. pdf
2. Aprile, M., Cevallos, A., Faenza, Y.: On 2-level polytopes arising in combinatorial settings. SIAM J. Discret. Math. **32**(3), 1857–1886 (2018)
3. Aprile, M., Conforti, M., Faenza, Y., Fiorini, S., Huynh, T., Macchia, M.: Recognizing Cartesian products of matrices and polytopes (2020). arXiv preprint arXiv:2002.02264
4. Conforti, M., Pashkovich, K.: The projected faces property and polyhedral relations. Math. Program. **156**(1–2), 331–342 (2016)
5. Conforti, M., Cornuéjols, G., Zambelli, G.: Extended formulations in combinatorial optimization. Ann. Oper. Res. **204**(1), 97–143 (2013)
6. Cover, T.M., Thomas, J.A.: Elements of Information Theory. Wiley, London (2012)
7. Gouveia, J., Grappe, R., Kaibel, V., Pashkovich, K., Robinson, R.Z., Thomas, R.R.: Which nonnegative matrices are slack matrices? Linear Algebra Appl. **439**(10), 2921–2933 (2013)
8. Grande, F., Sanyal, R.: Theta rank, levelness, and matroid minors. J. Comb. Theory B **123**, 1–31 (2017)
9. Kaibel, V., Pfetsch, M.E.: Some algorithmic problems in polytope theory. In: Algebra, Geometry and Software Systems, pp. 23–47. Springer, Berlin (2003)
10. Krause, A., Guestrin, C.: Near-optimal nonmyopic value of information in graphical models. In: Proceedings of the Twenty-First Conference on Uncertainty in Artificial Intelligence, pp. 324–331. AUAI Press (2005)
11. Macchia, M.: Two level polytopes: geometry and optimization. Ph.D. thesis, Université libre de Bruxelles (2018)
12. Margot, F.: Composition de polytopes combinatoires: une approche par projection, vol. 4. PPUR Presses Polytechniques (1995)
13. Oxley, J.G.: Matroid Theory, vol. 3. Oxford University Press, Oxford (2006)
14. Queyranne, M.: Minimizing symmetric submodular functions. Math. Program. **82**(1–2), 3–12 (1998)
15. Yannakakis, M.: Expressing combinatorial optimization problems by linear programs. J. Comput. Syst. Sci. **43**, 441–466 (1991)

Special Subclass of Generalized Semi-Markov Decision Processes with Discrete Time

Alexander Frank

Abstract The class of Generalized Semi-Markov Decision Processes (GSMDPs) covers a large area of stochastic modelling. For continuous time steps modelled problems are discussed in some articles, but not for the discrete case. Several events can be triggered in the same time step and the evaluation of them is more complex than for continuous time with an agreement, that two events can not be triggered at the same time point.

In this paper a specification for discrete GSMDPs is defined and analysed. The exponential cost, solving these problems exactly, are reduced to a polynomial number by two randomized approaches. Runtimes and relative results, compared to almost exact solutions, are shown and some extensions for the common class of discrete GSMDPs are mentioned.

Keywords GSMDP · Discrete time steps · Events · Acyclic phase type distribution · Randomized algorithms · Stochastic games · Unknown time event planning

1 Introduction

Many planning problems with stochastic uncertainty can be modelled as Markovian Decision Processes. The resulting agent assigns an optimal action in a given state and released time by paying attention to the gaining rewards and the future states, because of their condition to be memoryless. Those processes are used in stochastic games, network planning, robotics and further more. Discrete- and continuous-time Markov Decision Processes (MDPs and CTMDPs) can be solved efficiently with policy iteration or linear programming, [7].

A. Frank (✉)
TU Dortmund, Dortmund, Germany
e-mail: alexander.frank@tu-dortmund.de

© The Author(s), under exclusive license to Springer Nature Switzerland AG 2021
C. Gentile et al. (eds.), *Graphs and Combinatorial Optimization: from Theory to Applications*, AIRO Springer Series 5,
https://doi.org/10.1007/978-3-030-63072-0_29

One more universal class of decision problems is given by Generalized Semi-Markov Decision Processes (GSMDPs). The formalism in this article is similar to the definition by Younes and Simmons [8], based on previous definitions of GSMPs by Glynn [4]. In this class of problems we have several events, which can be triggered. Those events cause transitions from state to state and achieve some rewards. It is possible that for a period of time no event is triggered and the agent only knows the progressing clocks, so there are different sojourn times. By adding a choice of actions affecting the active set of clocks the agent has to make a decision for the underlying problem.

Another approach with events called alarms is discussed in [1] for continuous-time Markov chains (CTMC) with alarms.

There are only a few articles about continuous time GSMDPs. [3] examine a generalized model of Stochastic Automate (SA) with clocks, which are triggered asynchronously, activating transitions in the SA. By the Kronecker product clock states are combined to handle their interaction. Similarly [8], defined asynchronous events by continuous phase-type distributions (PHDs). Events are triggered and affect the underlying Markovian problem. They bring all active events in relationship and calculate their coherent probabilities to trigger one of them without losing the current progress of the others. Based on that article and a previous of [6], an approximative planner for solving deliberation scheduling problems was build using results for GSMDPs in [5].

To the best of my knowledge, there are no research articles written up to now about discrete time GSMDPs. The fact that there are discrete time steps, in which several events can be released, leads to a high dimensional problem. Even if the events have a certain order to be worked off, the agent has to consider over an exponential number of possible event combinations. This paper is focused on a special subclass of discrete time GSMDPs. The first limitation is that once an action is chosen in a state it is fixed until at least one event is triggered. The second limitation is that all progress for all events is lost if at least one single event is triggered.

The complexity is still PSPACE hard, but in this paper two randomized algorithms in polynomial runtime are introduced and analysed. Some backgrounds and a completed formulation for discrete time GSMDPs are given. After that, the two randomized algorithms are explained and in the last section the results are discussed.

2 Definitions

In this section, basic definitions and problem formulations are introduced. In general $\mathbb{P}(X)$ is the probability of X and $\mathbb{E}(X)$ is the expected reward. Bold letters are for linear functions (like $\mathbf{P}$) and calligraphic letters (like $\mathscr{S}$) are used for sets. $\mathbf{1}$ and $\mathbf{0}$ are vectors only consisting of 0 or rather 1 (where I have to say that the dimension is always logically conceivable).

2.1   Markov Decision Process

A tuple of $(\mathscr{S}, \mathscr{A}, \mathbf{P}, \mathbf{R}, \mathbf{p}_0)$ defines a discrete Markov Decision Process (MDP) where $\mathscr{S}$ is a finite set of states, $\mathscr{A}$ is a finite set of actions, $\mathbf{P} : \mathscr{S} \times \mathscr{A} \times \mathscr{S} \to [0, 1]$ is the transition function for moving from s to s' choosing action a and is used as a set of stochastic matrices $\mathbf{P}(s, a, s') \equiv \mathbf{P}^a(s, s')$, so that the condition $\sum_{s' \in \mathscr{S}} \mathbf{P}^a(s, s') = 1$ is fulfilled for all $s \in \mathscr{S}$. Furthermore $\mathbf{R} : \mathscr{S} \times \mathscr{A} \times \mathscr{S} \to \mathbb{R}$ is the reward function and is also used as a set of matrices $\mathbf{R}(s, a, s') \equiv \mathbf{R}^a(s, s')$. At least $\mathbf{p}_0 \in \mathbb{R}^{1 \times |\mathscr{S}|}$ is the initial distribution over all states.

By mapping actions to states the agent produces a policy $\pi(s, t) = a$, depending also on the past time $t \in [0, T]$. It is called a pure policy if time has no relevance.

An optimal policy maximizes the gained (discounted) rewards in the time horizon $[0, T]$. There are some options to solve MDPs like policy iteration and linear programming. These methods are exact and solve MDPs in a polynomial time. Much more information about MDPs can be found in [7].

2.2   Generally Semi-Markov Decision Process

GSMDPs are defined as a tuple of $(\mathscr{S}, \mathscr{A}, \mathscr{E}, \mathbf{C}, \mathbf{P}, \mathbf{R}, F)$. As in Sect. 2.1 , $\mathscr{S}$ and $\mathscr{A}$ are sets of states and actions. $\mathscr{E}$ is an extension and a set of independent events which are triggered with a probability given by $F(t, e)$ for a discrete passed time $t \in \mathbb{N}$ since activation of the event. $\mathscr{E}_0$ includes the trivial event e_0, that nothing happens. The function $\mathbf{C} : \mathscr{S} \times \mathscr{A} \times \mathscr{E} \to \{0, 1\}$ specifies if an event $e \in \mathscr{E}$ is active $\mathbf{C}(s, a, e) = 1$ or inactive $\mathbf{C}(s, a, e) = 0$ in a given state and a chosen action. The transition function $\mathbf{P} : \mathscr{S} \times \mathscr{E} \to \mathscr{S}$ declares the full known following state, if an event $e \in \mathscr{E}$ is triggered in $s \in \mathscr{S}$. Also the rewards depend on the occurred events $\mathbf{R} : \mathscr{S} \times \mathscr{E}_0 \times \mathscr{S} \to \mathbb{R}$, however it is sufficient to know the current state and event.

The agent has to make decisions identifying the active events. Then discrete time steps are made until the first event is triggered. Furthermore, all other active events can be triggered in the same time step. For a given order (or rather with decreasing priority) the system is affected by the events so that the status of events can be changed, transitions switch the state and rewards are gained. This happens in a so called zero-step-phase, where a path $\gamma = < s_0, x_1, s_1, x_2, \ldots, x_{|\mathscr{E}|}, s_{|\mathscr{E}|} >$, consisting of triggering events $x_i = e_i$ and running or disabled events $x_i = \overline{e_i}$, makes uninterrupted state transitions. All possible paths γ during the zero-step-phase are parts of the set of regular zero-steps Γ^*. Every path $\gamma \in \Gamma$ is also well defined for a given initial state s_0 by the formula $\gamma = < x_1, x_2, \ldots, x_{|\mathscr{E}|} >$. The reward for a zero-step path γ is computed additively

$$\mathbf{R}(\gamma) = \sum_{i: x_i = e_i} \mathbf{R}(s_i, e_i, s_{i+1}) \quad \text{or} \quad \mathbf{R}(\gamma) = \mathbf{R}(s_0, e_0, s_0). \tag{1}$$

Fig. 1 Example: transition
graph for a fixed action a, 4
states and 3 events

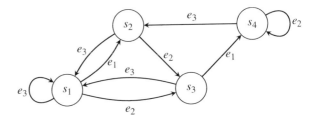

An optimal decision earns optimal rewards heeding to the next status of the system. So the policy $\pi : \mathscr{S} \times \mathbb{N}^{|\mathscr{E}|}$ maps an action to the given state and the passed event times since an event gets active and has not been triggered before.

Therefore $\Gamma^*(s, a, s')$ is defined as the set of all regular zero-step paths starting in s and ending in s' by choosing action a. Also $\mathbf{t}$ means the actual progress in the current time step i. The set of all system paths is then defined by

$$\Sigma := \{< s_0, a_1, \gamma_1, s_2, \ldots, s_{T-1}, a_T, \gamma_T, s_T > \ |s_i \in \mathscr{S}, a_i \in \mathscr{A}, \gamma_i \in \Gamma^*(s_{i-1}, a_i, s_i)\}.$$

Now the mathematical formula for the optimization criteria can be written as

$$\max_{\pi \in \Pi} \sum_{\sigma \in \Sigma} \mathbb{P}(\sigma | \pi) \sum_{i=0}^{T} \beta^i \sum_{\gamma \in \Gamma^*(s_i, \pi(s_i, \mathbf{t}), s_{i+1})} \mathbb{P}(\gamma | \pi(s_i, \mathbf{t}), \mathbf{t}) \cdot \mathbf{R}(\gamma) \qquad (2)$$

Figure 1 shows a small example for the transition graph with four states and three events without decision making by given a single action a. $\mathbf{C}(\cdot, a, \cdot)$ is visualized by the set of edges, so all events not belonging to an edge are blocked (like $\mathbf{C}(s_2, a, e_1) = 0$). As an illustration let s_4 be a semi-self-regulating state, s_2 a critical, working state and the rest (s_1, s_3) failure states. The event e_3 stands for a hardware crush, e_2 for a autonomous software update (with possible system errors) and e_1 is a finished repair of a mechanic. If a maintenance is made every day for a machine with this behaviour, several events can be triggered per day.

Starting in failure state s_1 with a successful mechanic the system switches in the critical, working state s_2. After that a software update is also made autonomously and crushes the system in a failure state s_3. So in the next decision period (next day) the machine is also in a failure state s_1 or s_3 corresponding to an additional hardware error.

All reachable states starting in s_1 are $\{s_1, s_2, s_3\}$ with a different number of paths leading to the states:

$$|\Gamma^*(s_1, a, s_1)| = 5, \quad |\Gamma^*(s_1, a, s_2)| = 1, \quad |\Gamma^*(s_1, a, s_3)| = 2$$

2.3 Resetting Discrete GSMDPs

Discrete GSMDPs are very difficult to solve, due to an exponential huge definition amount the agent has to handle. Even if the passed time has an upper bound for each event forcing it to be triggered, the system is too huge to be solved in polynomial time.

A resetting discrete GSMDP ($GSMDP_0$) has the same definition as a GSMDP with two more restrictions:

(a) If one or more events are triggered, before they are inactivated (per action or in transitions of zero-steps), all progress of each event is set to 0 after the zero-step-phase.

(b) When entering a state after one or more events are released the agent has to make a decision for an action. This action is not able to be changed until the next regular event is triggered.

Due to these two restrictions the agent only has to find a policy $\pi : \mathscr{S} \to \mathscr{A}$. The vector for the progress time in GSMDPs can be seen as $\mathbf{t} := t \cdot \mathbf{C}(s, a, \cdot)$. Nevertheless the problem is further hard to solve, due to the evaluation of zero-steps.

At least every time the agent has to make a decision the progress vector $\mathbf{t} \in \mathbb{N}^{1 \times |\mathscr{E}|}$ is $\mathbf{0}$. This criteria makes it possible to create an approximating model in polynomial time. In the conclusion some approaches for future algorithms are presented, solving discrete GSMDPs without specifications.

3 Randomized Approaches

Now the basic definitions are explained and a closer look at the analysis of $GSMDP_0$s is possible. The first question is: What happens in a discrete time step of our model? The behaviour of the model in a discrete time step is defined as zero-step-phase. Nevertheless there is also the opportunity that no event is released in this time step. The set of all regular zero-steps is Γ^* with $|\Gamma^*| \leq 2^{|\mathscr{E}|}$, on the other hand Γ is the set of all paths, regular or not.

At least two randomized approaches to solve discrete $GSMDP_0$s approximate in polynomial time are given. Both solve every instance exact if their input value for the bounding capacity is unlimited.

3.1 Zero-Step-Phase

This phase is the main focus of this paper, because the zero-step-phases make GSMPDs so difficult. For the analysis of the zero-step-phases the current state

$s \in \mathscr{S}$, only a single available action $a \in \mathscr{A}$ is considered and the past time since no event has been triggered $t \in \mathbb{N}$ is known. Also for every event $e \in \mathscr{E}$ a distribution function is known, given by an acyclic discrete Phase-Type distribution (ADPH). These functions can be defined with a tuple $(\mathbf{q}_e, \tilde{\mathbf{Q}}_e)$ of an initial vector and a part of a probability matrix (without absorbing transitions). The advantage of ADPHs is that the probability of a triggering event can be easily computed with

$$\mathbb{P}(e|t) = 1 - ||\mathbf{q}_e \tilde{\mathbf{Q}}_e^t||_1 = 1 - \sum_{i=1}^n \mathbf{q}_e \cdot \left(\tilde{\mathbf{Q}}_e^t(\cdot, i) \right). \tag{3}$$

More information and different formalisms for ADPHs are introduced in [2]. Inasmuch as the focus of this paper is on the randomized algorithms transforming $GSMDP_0$s to manageable MDPs no other distributions are analysed. Nevertheless the results can be derived if computable distributions for the events are given.

Not all events are active for a fixed combination of (s, a). The set of active events in s under a is defined as

$$\mathscr{E}_{act}(s, a) := \{e \in \mathscr{E} \mid \mathbf{C}(s, a, e) = 1\}. \tag{4}$$

The probability that no event is triggered for the progress time t is

$$\mathbb{P}(e_0|\mathbf{t}) = \prod_{e \in \mathscr{E}_{act}} ||\mathbf{q}_e \tilde{\mathbf{Q}}_e^{\mathbf{t}(e)}||_1 \tag{5}$$

In the other case one or more events are released. All in all there are $2^{|\mathscr{E}_{act}|}$ possible combinations of events, which are triggered or stay in progress. For a given zero-step path $\gamma \in \Gamma$ the correctness, if γ is also in Γ^*, has to be evaluated. Also the probability $\mathbb{P}(\gamma|a, \mathbf{t})$ (7) and rewards $\mathbf{R}(\gamma)$ (1) of a path can be computed. The special path $\gamma_0 := < s_0, \overline{e_1}, s_0, \overline{s_0}, \ldots, \overline{e_{|\mathscr{E}|}}, s_0 >$ is defined for no triggering event.

Definition 1 A path $\gamma \in \Gamma$ is regular for an action $a \in \mathscr{A}$ and a progress vector $\mathbf{t}$ (for a decreasing priority), if and only if

$$\forall i \in \{1, \ldots, |\mathscr{E}|\} : (x_i = e_i) \Rightarrow \left(\forall j < i : \mathbf{C}(s_j, a, e_i) = 1 \right). \tag{6}$$

So it has to be verified that the event is not set inactive before the priority of this event is high. As an example you want to buy several things online in one session, but when you come to the fourth article, it is already sold out.

Definition 2 The probability of a regular zero-step path $\gamma \in \Gamma^*$ depends on the probability that e is triggered in γ and Eq. (3), so it is $\mathbb{P}(e_i|\gamma, a, \mathbf{t}) = \mathbb{P}(e_i|\mathbf{t}(e_i)) \cdot \prod_{j=0}^{i-1} \mathbf{C}(s_j, a, e_i)$. The probability for the path now is given by

$$\mathbb{P}(\gamma \mid a, \mathbf{t}) = \prod_{i:x_i=e_i} \mathbb{P}(e_i|\gamma, a, \mathbf{t}) \cdot \prod_{i:x_i=\overline{e_i}} (1 - \mathbb{P}(e_i|\gamma, a, \mathbf{t})) \tag{7}$$

The additionally gained rewards are already defined in (1). For the planing of the agent it is important to calculate correctness, probabilities and rewards for all $\gamma \in \Gamma^*$. If multiple paths end in a state s' the probabilities can be summarized as

$$\mathbb{P}(s'|s, a, \mathbf{t}) = \sum_{\gamma \in \Gamma^*(s,a,s')} \mathbb{P}(\gamma|a, \mathbf{t}). \tag{8}$$

On the other hand the rewards are summarized with weights in relation to their probabilities

$$\mathbf{R}(s, a, \mathbf{t}, s') = \mathbb{P}(s'|s, a, \mathbf{t})^{-1} \cdot \sum_{\gamma \in \Gamma^*(s,a,s')} \mathbb{P}(\gamma|a, \mathbf{t}) \cdot \mathbf{R}(\gamma). \tag{9}$$

Since all possibilities and rewards are computed, the agent has total knowledge about the future status of the system. With these information an optimal decision can be made to collect discounted rewards.

3.2 Randomized Γ-Method

The first approach to avoid an exponential number of zero-steps is to limit the set of active unset events like in Algorithm 1. Generally there are $|\mathcal{E}_{act}|$ events which can be triggered or not, leading to a set Γ_{act} with $|\Gamma_{act}| = 2^{|\mathcal{E}_{act}|}$ different paths (also with irregular paths).

The main idea of the Γ-method (1) is to fix so many events randomly in step *randomize* of the algorithm depending on their probability, that the set of the other events $\mathcal{E}_{rest}$ fulfils $2^{|\mathcal{E}_{rest}|} \leq \Omega$. The more the probability of an event is near to 0 or 1, the more it is fixed randomly by the method. That means

$$\mathbb{P}(e \text{ is fixed} \mid \mathcal{E}_{act}, \tilde{s}, a, \mathbf{t}, \Omega) = \frac{|\mathbb{P}(e \mid \mathbf{t}(e)) - 0.5|}{\sum_{\tilde{e} \in \mathcal{E}_{act}} |\mathbb{P}(\tilde{e} \mid \mathbf{t}(\tilde{e})) - 0.5|}. \tag{10}$$

In *randomize* a total number of $\lceil |\mathcal{E}_{act}| - \log_2 \Omega \rceil$ events is selected to be fixed to 0 or 1 without double selection. Also the fixed value is randomly chosen equal to the triggering probability (3). So with the *randomize* function $\mathcal{E}_{act}$ is split into $\mathcal{E}_{rest}$, $\mathcal{E}_0$ and $\mathcal{E}_1$, where $\mathcal{E}_0$, $\mathcal{E}_1$ define sets of events specified to be triggered ($e = 1$) or not ($e = 0$). Now the main loop of the following algorithm has an upper bound of Ω.

The update steps in Algorithm 1 are similar to Eqs. (8) and (9). For that the sum in the equations is only a combination between two elements: the saved entries $L(s')$, representing a summary of all paths before, and the new incoming path γ. The sequence of the update steps is crucial, cause the results of (8) are required in (9).

This method has a running time in $\Theta(\Omega \cdot |\mathcal{E}|^2 + |\mathcal{S}|)$. Moreover $\Omega = 2^{|\mathcal{E}_{act}|}$ leads to the exact solution and calculates all possible paths in the zero steps.

Algorithm 1 Zero-steps over randomized paths

algorithm: zero_steps_Γ

input : $(\mathscr{S}, \mathscr{E}_{act}, \mathbf{P}, \mathbf{C}, \mathbf{R}, F)$, $(\tilde{s}, a, \mathbf{t}) \in \mathscr{S} \times \mathscr{A} \times \mathbb{N}_0^{|\mathscr{E}|}$, $\Omega \in \mathbb{N}$
output: L list of states, probabilities and rewards

$L(s, \cdot, \cdot) \leftarrow [s, 0, 0]$; $// \ \forall s \in \mathscr{S} \ (\mathscr{E}_{rest}, \mathscr{E}_0, \mathscr{E}_1) \leftarrow randomize(\mathscr{E}_{act}, \tilde{s}, a, \mathbf{t}, \Omega)$
  $// \ \texttt{explained in Sect.3.2} \ \Gamma' \leftarrow \mathscr{P}(\mathscr{E}_{rest}) \backslash \{\gamma_0\}$

for $\gamma \in \Gamma'$ **do**
  if γ *is regular (6)* **then**
     $L(s')(2)$ is updated with (8) $// \ \gamma \ \texttt{has destination state} \ s' \ L(s')(3) \ \texttt{is}$
     updated with (9) .
  end
end

3.3 Randomized $\mathscr{E}$-Method

The other Algorithm 2 based on a totally different structure. This time all steps for a single event are evaluated and saved in a sorted list. The higher prior sorting key is the actual state and the lower one is for blockings from $\mathbf{C}$.

So at the time point when event $e \in \mathscr{E}_{act}$ is evaluated all list entries become an update on the one hand for triggering and on the other for staying in progress. The list size will grow by factor up to 2 in every iteration, so again we limit the size to Ω.

In general two entries which are at the same state after an iteration step are not able to be combined, because they walked different paths and passed different $\mathbf{C}(\cdot, a, \cdot)$. With an additional function *combine_entries*, which searches for and combines same acting entries for all future iterations, the list is kept small. Sufficient is the same state in $L\{\cdot\}(1)$ and the same relevant blockings for future iterations in $L\{\cdot\}(2)$ This guarantees that the list size will increase to a maximum of $2^{|\mathscr{E}_{act}|/2}$ and after that it shrinks in every iteration until there are not more than $|\mathscr{S}|$ entries. The combination of the last entries in the list item is equal to the proceeding in Sect. 3.2 with Eqs. (8) and (9).

The method *insert*(L, new, Ω) is relevant for the sorted list L, because it searches for the correct place in L for an item, while it verifies that the capacity of Ω is not exceeded. Otherwise *insert* calls another method to delete a random item in L depending on its current probability $L\{\cdot\}(3)$.

The last line in the algorithm is to correct the influence of γ_0. On the one hand it is possible to stay in the initial state $\tilde{s}$ on the other hand it is possible to join the state per a chain of transitions (zero-steps). But in the first case the progress is increased by 1 and otherwise $\mathbf{t}$ is reset to 0. Hence both cases has to be separated.

As well as the Γ-method the Algorithm 2 solves the zero-steps exactly if Ω is great enough. The running time of the $\mathscr{E}$-method Algorithm 2 is defined in $\Theta(|\mathscr{E}| \cdot \Omega^3 + |\mathscr{S}|)$.

Algorithm 2 Zero-steps over events

---

algorithm: zero_steps_$\mathcal{E}$

input : $(\mathcal{S}, \mathcal{E}_{act}, \mathbf{P}, \mathbf{C}, \mathbf{R}, F)$, $(\tilde{s}, a, \mathbf{t}) \in \mathcal{S} \times \mathcal{A} \times \mathbb{N}_0^{|\mathcal{E}|}$, $\Omega \in \mathbb{N}$
output: L list of reachable s, blockings, probabilities and rewards

$L\{1\} \leftarrow [\tilde{s}, \mathcal{E}_{act}, 0, 0]$; **for** $e \in \mathcal{E}_{act}$ **do**
$\quad$ **for** $l \in L$ and e unevaluated for l **do**
$\quad\quad$ **if** e in $l(2)$ inactive **then**
$\quad\quad\quad | \quad$ update l with e blocked;
$\quad\quad$ **else**
$\quad\quad\quad | \quad s_{new} \leftarrow \mathbf{P}(l(1), e)$;
$\quad\quad\quad | \quad new \leftarrow [s_{new}, l(2) \cdot \mathbf{C}(s_{new}, a, \cdot), l(3) \cdot \mathbb{P}(e), l(4) + \mathbf{R}(l(1), e, s_{new})]$;
$\quad\quad\quad | \quad l(3) \leftarrow l(3) \cdot (1 - \mathbb{P}(e))$;
$\quad\quad\quad | \quad L \leftarrow insert(L, new, \Omega)$ $\qquad\qquad$ // explained in Sect. 3.3;
$\quad\quad$ **end**
$\quad$ **end**
$\quad L \leftarrow combine\_entries(L)$ $\qquad\qquad\qquad$ // explained in Sect. 3.3;
end
Correct the entry of $l\{\cdot\}(1) == \tilde{s}$ $\qquad\qquad$ // explained in Sect. 3.3;

3.4 Transformation to a MDP

The chance that no event triggers in a state s by choosing action a in $t \in \mathbb{N}$ time steps converges to 0, because of the ADPHs. For a negligible error of $\varepsilon > \mathbb{P}(e_0|t)$ the progress time t reaches an upper bound of $\theta(s, a) \in \mathbb{N}$ for every tuple of state and action. The new set of states $\tilde{\mathcal{S}}$ consists of

$$\tilde{\mathcal{S}} = \bigcup_{s \in \mathcal{S}} (s, 0) \quad \cup \bigcup_{s \in \mathcal{S}, a \in \mathcal{A}, t \in \mathbb{N}_{\leq \theta(s,a)}} (s, a, t).$$

Therefore the zero-step-phase has to be evaluated for every tuple (s, a, t) . With the results of the zero-step-phases a MDP $(\tilde{\mathcal{S}}, \mathcal{A}, \tilde{\mathbf{P}}, \tilde{\mathbf{R}})$ can be built with the expected transition probabilities and rewards. The entries in the list describe the probabilities $\mathbb{P}(s'|s, a, t) = \tilde{\mathbf{P}}((s, a, t), a, (s', 0))$ and the collected rewards $\tilde{\mathbf{R}}((s, a, t), a, (s', 0))$. By adding expensive penalties for switching a chosen action a' in a given tuple (s, a, t) while no transition takes place the second restriction is guaranteed. At least transitions from the states $(s, 0) \rightarrow (s, a, 1)$ and $(s, a, t-1) \rightarrow (s, a, t)$ have to be computed with $(t - 1)$ and Eq. (5).

By using one of the presented methods (Sects. 3.2 or 3.3) a MDP is build in a time based on the method and $\tilde{\mathcal{S}}$. ADPHs can be built so that $\tilde{\mathcal{S}}$ is enormous, but in general the reached upper bound θ is super exponentially.

4 Experiments

In this section the results for both randomized methods are shown. Their solutions will be compared to each other and to the exact ones using the $\mathcal{E}$-method with $\Omega = 2^{|\mathcal{E}|}$. The lack of literature causes no comparison to state of the art algorithms.

The test instances are randomized in transitions and transition probabilities, the order of ADPHs is normally distributed with expectancy value equal to 5 and variance equal to 1. The entries of ADPHs are randomized exponentially. The rewards are equally distributed just about $[-|\mathcal{E}|, |\mathcal{E}|]$ and also the entries $C(s, a, e) \in \{0, 1\}$ have the same probability.

The instances are build for $|\mathcal{S}| = \{50, 100\}$, different number of actions $|\mathcal{A}| = \{2, 4, 6, 8\}$ and various events $|\mathcal{E}| \in \{15, 20, 25\}$ (for greater $\mathcal{E}$s the exact solution can not be computed with the used computers).

Both randomized approaches run 10 times for a single instance and for 10 different instances. Over all results for a fixed number of states, actions and events the average values are calculated and presented. Here the $\mathcal{E}$-method is shorten with E and the Γ-method is named G.

In Fig. 2 the results for different actions and different Ωs are presented, with a grid size ($|\mathcal{S}|$) of 50 and 20 events. There is no obvious pattern for a specific influence by the number of actions. Whereas the $\mathcal{E}$-method has mostly a smaller relative error to the exact solution, which is also cut with an error below ε, than the Γ-method for an equal Ω. There are more unnoticed zero-step-paths with a higher number of events. In general the quality of the solutions gets better for a greater Ω. Thus it should be mentioned that for $\Omega = |\mathcal{S}| \cdot |\mathcal{E}|$ always all relative errors are lower than 10^{-5}.

The next Fig. 3 shows the average results for fixed sets of $\mathcal{S}$ and $\mathcal{A}$. It underlines the statements that the goodness of the algorithms for a fixed Ω decreases and

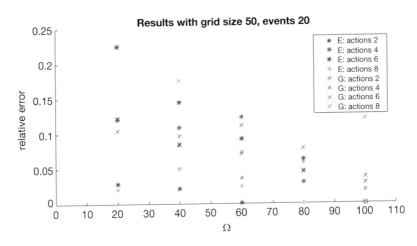

Fig. 2 Average results for tests with $|\mathcal{S}| = 50$

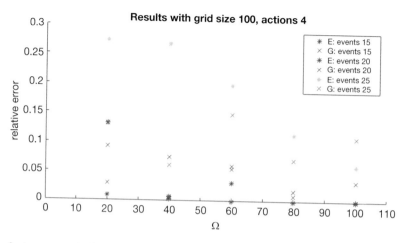

Fig. 3 Average results for tests with $|\mathscr{S}| = 100$ and $|\mathscr{A}| = 4$

Table 1 Relative run times for $|\mathscr{S}| = 100$ and $|\mathscr{A}| = 4$

Method and $|\mathscr{E}|$	$\Omega = 20$	$\Omega = 40$	$\Omega = 60$	$\Omega = 80$	$\Omega = 100$
$\mathscr{E}$-method, $|\mathscr{E}| = 15$	0.985	0.973	0.981	0.985	1.004
Γ-method, $|\mathscr{E}| = 15$	0.791	1.379	1.385	2.543	2.531
$\mathscr{E}$-method, $|\mathscr{E}| = 20$	0.840	0.953	0.971	0.988	0.992
Γ-method, $|\mathscr{E}| = 20$	0.368	0.689	0.689	1.364	1.365
$\mathscr{E}$-method, $|\mathscr{E}| = 25$	0.607	0.863	0.929	0.946	0.957
Γ-method, $|\mathscr{E}| = 25$	0.204	0.303	0.305	0.613	0.610

the size of $\mathscr{E}$ influences the quality. Furthermore, several tests proof similar to the runtime of the algorithms neither the size of $\mathscr{S}$ nor $\mathscr{A}$ is relevant for the quality of both approaches.

The run times in Table 1 are relative to the time used by an exact model to be created and solved. The missing force causes the polynomial approach to posses possibly longer time as presented in the last column. It is evaluated that the Γ-method is faster than the $\mathscr{E}$-method, if the upper bound Ω is considerably lower than $2^{|\mathscr{E}|}$, otherwise if both algorithms compute almost the exact zero-steps, Γ-method takes much longer. Consequently, that the list for the $\mathscr{E}$-method is naturally limited by $\sqrt{2^{|\mathscr{E}|}}$ and the Γ-method has no smaller natural bound than $2^{|\mathscr{E}|}$.

5 Conclusion

The experiments indicate, that both approaches have distinct advantages. Generally small run times are possible with the Γ-method causing a loss of quality compared to the exact solution. On the other hand the results of the $\mathscr{E}$-method for the same Ω

are closer to exact ones, but it takes more time. Both algorithms have a polynomial time to evaluate the zero-steps and can be used to transform a $GSMDP_0$ to an approximating MDP. By expanding the state space with copies in several time layers, a computable MDP is created in polynomial time. Edges are exclusive in the copies of the same basic state and to other basic states (not to their copies, so that "one way trees" are created).

Future work has to focus on the class of normal GSMDPs, which are more complex. Hereby an exponential number of combinations of the states and different progress times exists. Hence new approaches will be needed to decrease the decision space by aggregating progress times or selecting representative states. If it will be successful, a modified version of the algorithms can be used to evaluate the zero-step-phase and transform the problem into a MDP.

Acknowledgments I would like to thank my father, Harald Frank, my colleagues, Clara Scherbaum and Alexander Puzicha, and my close friend, Stephan Blömker, for their assistance, proofreading and inspiration.

References

1. Baier, C., Dubslaff, C., Korenčiak, L., kučera, A., Řehák, V.: Mean-payoff optimization in continuous-time markov chains with parametric alarms. ACM Trans. Model. Comput. Simul. **29**(4) (2019). https://doi.org/10.1145/3310225
2. Bobbio, A., Horvth, A., Scarpa, M., Telek, M.: Acyclic discrete phase type distributions: properties and a parameter estimation algorithm. Perform. Eval. **54**(1), 1–32 (2003)
3. Buchholz, P., Kriege, J., Scheftelowitsch, D.: Model checking stochastic automata for dependability and performance measures. In: 2014 44th Annual IEEE/IFIP International Conference on Dependable Systems and Networks, pp. 503–514. IEEE, Piscataway (2014)
4. Glynn, P.W.: A GSMP formalism for discrete event systems. Proc. IEEE **77**(1), 14–23 (1989)
5. Krebsbach, K.D.: Deliberation scheduling using GSMDPs in stochastic asynchronous domains. Int. J. Approx. Reason. **50**(9), 1347–1359 (2009). Special Track on Uncertain Reasoning of the 19th International Florida Artificial Intelligence Research Symposium (FLAIRS 2006)
6. Musliner, D.J., Goldman, R.P., Krebsbach, K.D.: Deliberation scheduling strategies for adaptive mission planning in real-time environments. In: AAAI Spring Symposium: Metacognition in Computation (2005)
7. Puterman, M.L.: Markov Decision Processes: Discrete Stochastic Dynamic Programming. Wiley, London (1994)
8. Younes, H.L., Simmons, R.G.: Solving generalized semi-Markov decision processes using continuous phase-type distributions. In: AAAI, vol. 4, p. 742 (2004)

Coupling Machine Learning and Integer Programming for Optimal TV Promo Scheduling

Ruggiero Seccia, Gianmaria Leo, Mehrnoosh Vahdat, Qiannan Gao, and Hanadi Wali

Abstract Optimal TV promo Scheduling is the process of scheduling promos over breaks in order to maximize promos and programs' viewership. It is a complex task to tackle since viewership is an uncertain quantity to estimate affected by uncontrollable events, many business requirements need to be satisfied and unexpected events may require the definition of a new schedule in a very short time. In this work, a new efficient framework for solving the Optimal TV promo Scheduling problem is introduced by formulating the problem as an integer optimization problem where the viewership is estimated through Machine Learning models. Different objective functions are defined and benchmarked. Numerical results on real word instances show the effectiveness of the resulting framework in solving the Optimal TV promo Scheduling problem in a very short amount of time leading to good or optimal solutions and improving schedules KPI provided by business experts.

Keywords Tv promo scheduling · Integer programming · Machine learning · Reach estimation

R. Seccia (✉)
Department of Computer, Control and Management Engineering Antonio Ruberti, Sapienza University of Rome, Roma, Italy
e-mail: ruggiero.seccia@uniroma1.it

G. Leo · M. Vahdat · Q. Gao
IBM Data Science Elite Team, Munich, Germany
e-mail: gianmaria.leo@ibm.com; mehrnoosh.vahdat@ibm.com; qiannangao1@ibm.com

H. Wali
IBM Data Science Center of Excellence, Riyadh, Saudi Arabia
e-mail: hanadiw@sa.ibm.com

© The Author(s), under exclusive license to Springer Nature Switzerland AG 2021
C. Gentile et al. (eds.), *Graphs and Combinatorial Optimization: from Theory to Applications*, AIRO Springer Series 5,
https://doi.org/10.1007/978-3-030-63072-0_30

387

1 Introduction

TV Media companies divide the available airtime into several TV programs, separated by breaks. Within breaks, two main kinds of spots can be placed: *commercials*, which promote products or services offered by third-party companies, and *promos*, which advertise TV programs or other products offered by the media group itself. Usually, selling airtime to commercial advertisers represents one of the main sources of revenue for media companies, while airing promos does not provide any direct increase in the company revenue. Airing promos, instead, is responsible for increasing and carry the viewership along channels and programs. Since the price for airing commercials is determined by the viewership of the specific airtime to be placed in, since the higher the viewership of the program/channel where the commercial is placed, the higher the price for airing the commercial [6], promos indirectly affect media group's revenue by influencing the viewership. In this setting, Optimal TV Promo Scheduling is meant to increase the viewership of programs and channels of the media group so to increase the income from selling airtime to commercials.

Overall the promo scheduling process is a challenging task for several reasons. Firstly, estimating the impact of each promo-break combination to the overall viewership requires the definition of accurate forecasting models. Secondly, many business rules must be taken into consideration while scheduling promos, some of them are strict requirements while others can be relaxed and violated if justified. Finally, the frequent occurrence of unexpected events that result in sudden changes in the overall airing schedule combined with the limited time available to accommodate these unexpected changes, makes the scheduling procedure even more complicated.

In this work, to measure the viewership of TV contents, we focus on *Reach*, which is defined as the percentage of unique households reached by the aired content. Concerning promos, we make a distinction between two types: promos that promote TV programs, called *Promos for programs*, and promos that advertise non-channel products of the media group (e.g. digital platforms streaming a TV program), called *Non-channel promos*. The program within which the promo is aired will be called the *host program*, while the program promoted by the promo will be called the *client program*.

Currently, the promo scheduling process is very inefficient, considering that the number of variables and the set of constraints to take into account is very large and the process is mostly done manually. In this study, we tackle the problem of Optimal TV Promo Scheduling by coupling techniques belonging to the field of Machine Learning and Integer Programming. Firstly, two forecasting models for estimating the viewership of *non-channel promos*, and the viewership of programs promoted by *promos for programs* are defined. On top of that, an integer optimization (IP) problem is designed in order to determine the promo schedule such that the overall viewership is maximized while satisfying all the business constraints. Numerical results illustrate the effectiveness of this framework in solving real-word instances in a very short amount of time by finding optimal or near to optimal solutions always outperforming business experts' schedules. This work is focused on scheduling promos day-by-day over a single channel, however, the framework can be extended to consider longer time horizons and the influence between several channels.

2 Literature Review

Scheduling for broadcast television industry has already been widely studied [16]. Most of the literature on this topic has been focused on solving the scheduling problem for airing commercials instead of promos. The first attempt in this field is represented by Brown [3] who proposed a programming logic algorithm for finding a feasible solution to the problem of scheduling commercials over breaks without optimizing any specific function. In [1] a formulation for scheduling commercials in a balanced way is proposed and then solved through ad-hoc heuristics. Similarly, in [2] more complex real-world conditions are considered in scheduling commercials and a heuristic method to solve it is proposed. More recently, [15] applies statistical methods to estimate the viewership of commercials and then integrate these estimations in an IP problem to maximize the income from scheduling commercials. However, to solve the problem in a reasonable amount of time, warm-start techniques were needed. Other applications of mathematical programming for scheduling commercials can be found in [7, 14, 17] and the reference therein.

Although scheduling commercials is closely related to the problem of scheduling promos, they mainly differ in the purpose and the requirements they need to satisfy. While scheduling commercials aims at allocating breaks such that revenue is maximized, scheduling promos focus on maximizing viewership of promoted programs. In addition, the set of constraints required in the two cases differ. Finally, when it comes to viewership prediction, while in scheduling commercials we are interested in estimating the viewership of the advertisement itself, when scheduling promos we want to estimate the viewership obtained by the client program, which represents a much more challenging task. While most of the literature has been focused on solving scheduling problems for commercials, very few studies have been focused on scheduling problems for promos. In particular, [5, 13] developed genetic algorithms for maximizing the total gross rating point (GRP) derived by the promo scheduling without specifying how the GRP is estimated.

This work represents the first attempt to solve the promo scheduling problem in a very short time through exact algorithms and not heuristic methods. Good solutions and small time to achieve them are both guaranteed without recurring to sub-optimal methods. Moreover, compared to previous works, the application of Machine Learning techniques for forecasting the client program viewership leads to very accurate models providing more reliable predictions. Finally, the distinction between soft and hard constraints allows the scheduler operator to have some control of the solution obtained by tuning the penalty parameters γ while guaranteeing those stricter business requirements.

3 Problem Formulation

To solve the Optimal Promo Scheduling problem, two main frameworks are considered: a Machine Learning framework which estimates the consequences of placing a promo in a specific break; an optimization framework that, given the ML predictions, solves an integer optimization problem to find the best pairs break-promo.

3.1 Machine Learning Framework

As a first step, the estimated viewership that will result from placing a promo in each specific break needs to be estimated with high accuracy. As already discussed, we distinguish promos between *Non-channel promos*, namely promos advertising media group's products, and *promos for programs*, namely promos advertising a specific program. For *Non-channel promos* we focus on estimating the Reach of the promo itself, called *Promo Reach* (PR), since the reach of the advertised product is not available and not easy to compute, while for *Promos for programs* we focus on estimating the Reach of the promoted program, called *Client Reach* (CR).

The estimation of PR is achieved by considering promo and host program's information such as time of airing and duration of the content. Prediction of CR, instead, is a more challenging task and is achieved by considering also client program information, historical data, future information already available (e.g. the number of days before the promoted program is aired) and the PR obtained by placing the promo in a specific break (derived from the first prediction model).

The output of the ML framework is a matrix Φ, where ϕ_{pt} represent the final viewership obtained by placing the promo p in the break t, regardless whether it regards the CR obtained by *promos for programs* or the PR obtained by *non-channel promos*.

3.2 Mathematical Formulation

In each break many promos can be aired. Once the prediction models are defined, the main objective of the optimization framework is to choose which promos air in each break. To formulate the optimization problem, the binary variables $\delta_{pt} \in \{0, 1\}$ are defined, where $\delta_{pt} = 1$ if promo p is assigned to break t and 0 otherwise. In

Table 1 Parameters definition

			Set of *non-channel promos*
P	Set of promos	$P_2 \subset P$	$(P_1 \cap P_2 = \emptyset)$
T	Set of breaks	$A \subset P$	Set of generic promos
Q	Set of client programs	$B \subset P$	Set of specific promos $(G \cap Z = \emptyset)$
K	Set of genres	d_p	Duration of promo p
$P(c) \subseteq P$	Subset of promos for campaign $c \in C$	D_t	Duration of break t
$P(q) \subseteq P$	Subset of promos for client program $c \in C$	r	Required ratio of self promos
$P(k) \subseteq P$	Subset of promos for genre $k \in K$	k_p	Minimum number of times promo p is aired
$S \subset P$	Subset of self promos	l_p, u_p	*Nominal* minimum and maximum number
$P_1 \subset P$	Set of *promos for programs*		of times promo p can be aired

Table 1 the mathematical notation is presented. Some of the terms here introduced will be better discussed in the following.

Before considering the objective function, we focus on the constraints definition. In particular, there are two different kinds of constraints: *hard constraints*, namely constraints that cannot be violated; and *soft constraints*, namely constraints that should be preferably satisfied but that are not strict requirements. While the former set of constraints is defined through standard mathematical constraints, the latter is defined as penalty/rewards terms in the objective function.

Hard Constraints

The polytope $\mathscr{P} \subseteq \mathbb{R}^{|P| \times |T|}$ defining the feasible region is expressed by the following set of constraints:

$$\sum_{p \in P} \delta_{pt} d_p \leq D_t \qquad \forall t \in T \tag{1}$$

$$\sum_{t \in T} \delta_{pt} \geq k_p \qquad \forall p \in P \tag{2}$$

$$\sum_{p \in P(j)} \delta_{pt} \leq 1 \qquad \forall t \in T, \forall j \in J = \{C, Q, K\} \tag{3}$$

where: constraint (1) implies that in each break the sum of promos' duration cannot be higher than the break duration itself; constraint (2) implies that each promo needs to be aired a minimum number of times per day; constraint (3) implies that at most one promo of each specific campaign, or client or genre can be placed in each break (*non-channel promos* do not have a genre) to avoid audience over-exposition to similar contents within the same break.

Soft Constraints

Ratio of Self Promos (SR) Promos for programs can be divided into two categories: promos for programs aired on the same channel (called self promos) and promos for programs aired on other channels (called cross promos). Each channel wants to broadcast more self promos than cross promos according to some business strategy. This is modeled by introducing a specific target r on the percentage of self promos aired in each day (in our study, $r = 0.4$) and asking to obtain values as close as possible to this percentage:

$$g_{SR}(\delta) := - \left| \sum_{p \in S} \sum_{t \in T} \delta_{pt} - r \sum_{p \in P} \sum_{t \in T} \delta_{pt} \right| \tag{4}$$

Placement of Self Promos (SP) As a good business practice, each break should start and end with a self promo so to give more emphasis to the channel where the promos are aired. This constraint is considered in the post-processing of the solution when the order of promos in each specific break is defined. However, this implies that we should guarantee at least two self promos per break, which cannot be always satisfied (e.g. if the available time is not enough to air two promos in the same slot). As a consequence, given $h_t(\delta) = 1$ if $\sum_{p \in S} \delta_{pt} \leq 1$ and 0, we model it as follows:

$$g_{SP}(\delta) := - \sum_{t \in T} h_t(\delta) \tag{5}$$

Specifics Versus Generics (SG) Promos can be distinguished in two kinds: *specific promos* are those promos promoting a specific content (e.g. a specific episode of a Tv program), while *generic promos* are promos advertising some content without any reference to a specific event (e.g. promote a Tv series without any reference to a specific episode). As a good practice, media companies want to give higher priority to specific promos over generic promos since their content is more relevant and attractive for the audience. This constraint is modeled in the objective function by adding the term:

$$g_{SG}(\delta) := \sum_{p \in Z} \sum_{t \in T} \delta_{pt} - \sum_{p \in G} \sum_{t \in T} \delta_{pt} \tag{6}$$

The following two constraints are not directly dictated by the business problem we are modeling but derive from the formulation of the problem we defined.

Promos Variation (PV) To avoid unbalanced schedules, we want each promo to be aired possibly no more than its desired number of times, called θ_p, which depends on the relevance of the product advertised by the promo (i.e. the average of the potential viewership values). To determine the maximum frequency of each promo that the planning should target, we define the constant θ_p for each promo, as follows:

$$\theta_p := \left\lceil l_p + (u_p - l_p) \frac{\mu_p - m}{M - m} \right\rceil \qquad \forall p \in P \qquad (7)$$

where: $\mu_{p \in P} := \frac{\sum_{t \in T} \phi_{pt}}{T}$, $m := \min_{p \in P}\{\mu_p\}$ and $M := \max_{p \in P}\{\mu_p\}$. Then, the rule is modeled by introducing the following function:

$$g_{PV}(\delta) := -\sum_{p \in P} \max \left\{ 0, \sum_{t \in T} \delta_{pt} - \theta_p \right\} \qquad (8)$$

Fill Gap (FG) To fill as much as possible all the available spaces within each break, we introduce the term:

$$g_{FG}(\delta) := -\left| \sum_{t \in T} \left(D_t - \sum_{p \in P} \delta_{pt} d_{pt} \right) \right| \qquad (9)$$

Objectives

The objective function is composed by a first term, $f(\delta)$, representing the final objective we want to achieve (i.e. maximize the final viewership of each program) and the set of penalties and rewards introduced when considering the soft rules. Let $G(\delta) := \sum_{j \in \{SR,SC,SG,PV,FG\}} \gamma_j g_j(\delta)$, with $\gamma_j \geq 0$ representing the cost of violating the corresponding rules, then the objective function can be written as

$$F(\delta) := f(\delta) + G(\delta) \qquad (10)$$

Concerning the first term $f(\delta)$, different options are considered.

Sum of the Viewership The promo placement impact is measured as the sum of the predicted reaches for all the breaks considered by the planning. The reaches refer to both the programs advertised and the promos advertising broadcasting products:

$$f(\delta) := \sum_{p \in P} \sum_{t \in T} \delta_{pt} \phi_{pt} \qquad (11)$$

Average of the Viewership The promo placement impact is measured as the sum of the predicted mean reaches. Analogously to the former objective, the reaches refer to the programs advertised and the promos advertising non-channel products:

$$f(\delta) := \sum_{q \in Q} \frac{\sum_{p \in P(q)} \sum_{t \in T} \delta_{pt} \phi_{pt}}{\sum_{p \in P(q)} \sum_{t \in T} \delta_{pt}} \qquad (12)$$

The term (11) is a linear function, easy to optimize, while the term (12) leads to an increase in the complexity of the optimization problem in terms of formulation and solution but is more in line with the real business objective, by aiming to increase the average reach of each client program considering all aired promos that are promoting it. The solutions returned when considering these two different terms are analyzed in Sect. 4.2.

3.3   Formulation as Mixed Integer Linear Program (MILP)

The problem can be formulated as the following Integer Program (IP):

$$\max \left\{ F(\delta) : \delta \in \mathscr{P} \cap \{0, 1\}^{|P| \times |T|} \right\} \qquad (13)$$

The two objective functions discussed in the previous subsection give rise to two distinct IP problems denoted respectively as **IP-SUM** and **IP-AVG**. Hence, the following proposition holds:

Proposition 1 **IP-SUM** *and* **IP-AVG** *can be reformulated as MILP problems.*

Proof **IP-SUM**: by definition, f and $\mathscr{P}$ are given by linear expressions; since G is the sum of piecewise linear functions, it can be reformulated as a MILP. **IP-AVG**: it is easy to check that f can be written as the sum of products between binary and continuous variables; hence, it can be linearized by introducing continuous variables and indicator constraints.

4   Computational Experience

In this section, we report and discuss the numerical results of the Machine Learning frameworks described in Sect. 3.1, and the solution of the optimization problems defined in Sect. 3.2. Airtime data concerning a TV channel during 2018 were provided by a large media company whose name is not disclosed for privacy policy. The Machine Learning assets have been implemented in Python with Jupyter notebooks by leveraging well-known open-source libraries, like `pandas`, `sklearn` and `xgboost`. The optimization models have been developed in Python with the IBM Decision Optimization `docplex` APIs, then solved with IBM Decision Optimization on Cloud APIs[9], hosting IBM ILOG CPLEX 12.9 in an environment with 10 cores and 60 GB of RAM. The entire solution has been deployed in IBM Cloud Pak for Data[8].

4.1 Machine Learning Results

For both PR and CR estimation, features selection strategies (e.g. correlation analysis and feature importance from decision trees) coupled with outliers elimination were performed to filter out all those noisy and not influential features. For each estimation task, different forecasting models were implemented and benchmarked. Most of the 2018 data were used for training the models while few weeks were kept out as test set. To avoid overfitting, the models were trained through a hyperparameter tuning and a cross-validation was performed with an expanding window since we are dealing with time series data.

Concerning the results from predicting PR, XGBoost [4] over-performed the other models with an R^2 slightly higher than 90% and a mean absolute percentage error (MAPE) around 10%. The duration of the host program and features related to the airing time resulted among the most important features. Concerning the results from predicting CR, by stacking XGBoost with a shallow Neural Network [11, 12], we retained the high accuracy of predictions, while capturing the changes of CR when placing the corresponding promo in different breaks. The results indicate the R^2 of around 90% and the MAPE of 6%. Among the most important features, historical data of the client program followed by promos' statistics like the frequency of promos per day stood out.

4.2 Optimization Results

The two models, **IP-SUM** and **IP-AVG**, are tested over 7 real-world daily instances, denoted day-i for $i = 1, \ldots 7$, which have been provided by the broadcasting company. Each instance is composed on average by 65 breaks, 42 promos and 40 programs. Table 2 compares the quality of the optimized solutions (denoted as $\delta^\star$) with the ones provided by the business experts of the company (denoted as $\delta\prime$). We report the values of the relevant KPIs defined in Sect. 3.2 and the relative benefit generated by the optimization computed as:

$$\text{Gain} := \frac{KPI(\delta^\star) - KPI(\delta')}{1 \times 10^{-6} + |KPI(\delta')|} \tag{14}$$

We refer to f_{SUM} and f_{AVG} as the formulas (11) and (12) respectively and divide values by $F(\delta^\star)$ as defined in (10) in order to make results easier to be interpreted. Note that instances day-$\{1, 3\}$ have been omitted from the comparison since the solution returned by the business export turned out to be infeasible and difficult to

Table 2 Comparing planning solutions. "_" is used to denote that the KPI is not computed for the specific instance

Day	KPI	IP-SUM			IP-AVG		
		$\delta = \delta'$	$\delta = \delta^\star$	Gain (%)	$\delta = \delta'$	$\delta = \delta^\star$	Gain (%)
Day-2	$F(\delta)$	$1.23 \cdot 10^6$	$4.26 \cdot 10^6$	247.055	$1.80 \cdot 10^7$	$1.93 \cdot 10^7$	7.141365956
	$f_{SUM}(\delta(P_1))/F(\delta^\star)$ (%)	$2.07 \cdot 10^1$	$6.61 \cdot 10^1$	218.996	–	–	–
	$f_{AVG}(\delta(P_1))/F(\delta^\star)$ (%)	$2.89 \cdot 10^0$	$2.88 \cdot 10^0$	-0.466	$9.88 \cdot 10^{-6}$	$1.01 \cdot 10^{-5}$	1.800845758
	$f_{SUM}(\delta(P_2))/F(\delta^\star)$ (%)	$2.89 \cdot 10^1$	$4.39 \cdot 10^1$	51.711	–	–	–
	$f_{AVG}(\delta(P_2))/F(\delta^\star)$ (%)	$1.78 \cdot 10^0$	$2.03 \cdot 10^0$	13.81	$1.97 \cdot 10^{-6}$	$2.36 \cdot 10^{-6}$	13.34280389
	$\gamma_{PV} g_{PV}(\delta)$	$-6.00 \cdot 10^4$	$0.00 \cdot 10^0$	100	$-6.00 \cdot 10^4$	$0.00 \cdot 10^0$	100
	$\gamma_{SR} g_{SR}(\delta)$	$-1.01 \cdot 10^6$	$-2.55 \cdot 10^6$	-152.778	$-1.01 \cdot 10^6$	$-1.68 \cdot 10^5$	83.33333333
	$\gamma_{SG} g_{SG}(\delta)$	$2.63 \cdot 10^2$	$1.67 \cdot 10^3$	534.286	$2.63 \cdot 10^2$	$3.75 \cdot 10^2$	42.85714269
	$\gamma_{FG} g_{FG}(\delta)$	–	–	–	$5.67 \cdot 10^3$	$6.07 \cdot 10^3$	6.992729675
Day-4	$F(\delta)$	$1.20 \cdot 10^6$	$3.78 \cdot 10^6$	214.787	$2.02 \cdot 10^7$	$2.19 \cdot 10^7$	8.566773525
	$f_{SUM}(\delta(P_1))/F(\delta^\star)$ (%)	$2.84 \cdot 10^1$	$6.40 \cdot 10^1$	125.296	–	–	–
	$f_{AVG}(\delta(P_1))/F(\delta^\star)$ (%)	$3.82 \cdot 10^0$	$3.82 \cdot 10^0$	0.053	$9.89 \cdot 10^{-6}$	$1.01 \cdot 10^{-5}$	1.941738043
	$f_{SUM}(\delta(P_2))/F(\delta^\star)$ (%)	$2.89 \cdot 10^1$	$8.39 \cdot 10^1$	190.086	–	–	–
	$f_{AVG}(\delta(P_2))/F(\delta^\star)$ (%)	$1.96 \cdot 10^0$	$2.45 \cdot 10^0$	25.38	$1.69 \cdot 10^{-6}$	$2.11 \cdot 10^{-6}$	15.44647433
	$\gamma_{PV} g_{PV}(\delta)$	$-2.10 \cdot 10^5$	$0.00 \cdot 10^0$	100	$-2.10 \cdot 10^5$	$0.00 \cdot 10^0$	100
	$\gamma_{SR} g_{SR}(\delta)$	$-1.28 \cdot 10^6$	$-2.46 \cdot 10^6$	-92.187	$-1.28 \cdot 10^6$	$-2.30 \cdot 10^5$	82.03125
	$\gamma_{SG} g_{SG}(\delta)$	$3.45 \cdot 10^2$	$1.28 \cdot 10^3$	271.739	$3.45 \cdot 10^2$	$-3.23 \cdot 10^2$	-193.4782603
	$\gamma_{FG} g_{FG}(\delta)$	–	–	–	$6.15 \cdot 10^3$	$6.40 \cdot 10^3$	4.085864129

Day-5

$F(\delta)$	$1.52 \cdot 10^6$	$4.56 \cdot 10^6$	200.475	$1.97 \cdot 10^7$	$2.19 \cdot 10^7$	10.89539736
$f_{\text{SUM}}(\delta(P_1))/F(\delta^\star)$ (%)	$3.16 \cdot 10^1$	$7.15 \cdot 10^1$	125.965	–	–	–
$f_{\text{AVG}}(\delta(P_1))/F(\delta^\star)$ (%)	$3.17 \cdot 10^0$	$3.18 \cdot 10^0$	0.449	$9.90 \cdot 10^{-6}$	$1.01 \cdot 10^{-5}$	1.669128173
$f_{\text{SUM}}(\delta(P_2))/F(\delta^\star)$ (%)	$1.91 \cdot 10^1$	$4.24 \cdot 10^1$	121.508	–	–	–
$f_{\text{AVG}}(\delta(P_2))/F(\delta^\star)$ (%)	$1.08 \cdot 10^0$	$1.13 \cdot 10^0$	4.689	$1.13 \cdot 10^{-6}$	$1.03 \cdot 10^{-6}$	-4.81641963
$\gamma_{\text{PV}} g_{\text{PV}}(\delta)$	$-3.00 \cdot 10^5$	$0.00 \cdot 10^0$	100	$-3.00 \cdot 10^5$	$0.00 \cdot 10^0$	100
$\gamma_{\text{SR}} g_{\text{SR}}(\delta)$	$-1.64 \cdot 10^6$	$-3.25 \cdot 10^6$	-97.932	$-1.64 \cdot 10^6$	$-2.24 \cdot 10^5$	86.37469586
$\gamma_{\text{SG}} g_{\text{SG}}(\delta)$	$5.33 \cdot 10^2$	$1.68 \cdot 10^3$	215.493	$5.33 \cdot 10^2$	$-2.63 \cdot 10^2$	-149.2957744
$\gamma_{\text{FG}} g_{\text{FG}}(\delta)$	–	–	–	$7.14 \cdot 10^3$	$7.79 \cdot 10^3$	9.183566249

Day-6

$F(\delta)$	$7.27 \cdot 10^5$	$3.89 \cdot 10^6$	435.317	$1.95 \cdot 10^7$	$2.06 \cdot 10^7$	5.175828193
$f_{\text{SUM}}(\delta(P_1))/F(\delta^\star)$ (%)	$1.47 \cdot 10^1$	$6.85 \cdot 10^1$	365.605	–	–	–
$f_{\text{AVG}}(\delta(P_1))/F(\delta^\star)$ (%)	$3.46 \cdot 10^0$	$3.48 \cdot 10^0$	0.652	$9.83 \cdot 10^{-6}$	$1.01 \cdot 10^{-5}$	2.22507555
$f_{\text{SUM}}(\delta(P_2))/F(\delta^\star)$ (%)	$1.05 \cdot 10^1$	$5.28 \cdot 10^1$	403.324	–	–	–
$f_{\text{AVG}}(\delta(P_2))/F(\delta^\star)$ (%)	$1.09 \cdot 10^0$	$1.08 \cdot 10^0$	-0.849	$1.03 \cdot 10^{-6}$	$9.88 \cdot 10^{-7}$	-2.020219587
$\gamma_{\text{PV}} g_{\text{PV}}(\delta)$	$0.00 \cdot 10^0$	$0.00 \cdot 10^0$	0	$0.00 \cdot 10^0$	$0.00 \cdot 10^0$	0
$\gamma_{\text{SR}} g_{\text{SR}}(\delta)$	$-6.64 \cdot 10^5$	$-2.62 \cdot 10^6$	-295.181	$-6.64 \cdot 10^5$	$-1.84 \cdot 10^5$	72.28915663
$\gamma_{\text{SG}} g_{\text{SG}}(\delta)$	$1.88 \cdot 10^2$	$1.34 \cdot 10^3$	612	$1.88 \cdot 10^2$	$-2.03 \cdot 10^2$	-207.9999989
$\gamma_{\text{FG}} g_{\text{FG}}(\delta)$	–	–	–	$3.18 \cdot 10^3$	$5.76 \cdot 10^3$	0.814359943

(continued)

Table 2 (continued)

Day	KPI	IP-SUM			IP-AVG		
		$\delta = \delta'$	$\delta = \delta^\star$	Gain (%)	$\delta = \delta'$	$\delta = \delta^\star$	Gain (%)
Day-7	$F(\delta)$	$1.01 \cdot 10^6$	$2.91 \cdot 10^6$	188.737	$1.90 \cdot 10^7$	$2.02 \cdot 10^7$	6.600385907
	$f_{\text{SUM}}(\delta(P_1))/F(\delta^\star)$ (%)	$2.88 \cdot 10^1$	$6.77 \cdot 10^1$	135.2	–	–	–
	$f_{\text{AVG}}(\delta(P_1))/F(\delta^\star)$ (%)	$4.58 \cdot 10^0$	$4.61 \cdot 10^0$	0.742	$9.87 \cdot 10^{-6}$	$1.01 \cdot 10^{-5}$	1.775034119
	$f_{\text{SUM}}(\delta(P_2))/F(\delta^\star)$ (%)	$1.60 \cdot 10^1$	$3.97 \cdot 10^1$	147.919	–	–	–
	$f_{\text{AVG}}(\delta(P_2))/F(\delta^\star)$ (%)	$1.37 \cdot 10^0$	$1.70 \cdot 10^0$	23.934	$9.86 \cdot 10^{-7}$	$1.01 \cdot 10^{-6}$	0.994860825
	$\gamma_{\text{PV}} g_{\text{PV}}(\delta)$	$-6.00 \cdot 10^4$	$0.00 \cdot 10^0$	100	$-6.00 \cdot 10^4$	$0.00 \cdot 10^0$	100
	$\gamma_{\text{SR}} g_{\text{SR}}(\delta)$	$-9.34 \cdot 10^5$	$-1.81 \cdot 10^6$	-94.218	$-9.34 \cdot 10^5$	$-1.84 \cdot 10^5$	80.29978587
	$\gamma_{\text{SG}} g_{\text{SG}}(\delta)$	$3.00 \cdot 10^2$	$1.19 \cdot 10^3$	295	$3.00 \cdot 10^2$	$-7.50 \cdot 10^1$	-124.9999996
	$\gamma_{\text{FG}} g_{\text{FG}}(\delta)$	–	–	–	$4.25 \cdot 10^3$	$4.20 \cdot 10^3$	-1.112091791

recover. Moreover, the KPI associated to the soft rule SP (placement of self promos) is not reported in the table as it is always equal to zero in all the compared solutions.

We observe that the optimization of **IP-SUM** raises the impact of the promo placement, measured by $F(\delta)$, on average of 257%: the main drivers of these results are $f_{SUM}(\delta(P_1))$ and $f_{SUM}(\delta(P_2))$ that increases the impact on average of resp. 194% and 183%. This means that **IP-SUM** is able to increase the quality of the planning by providing a remarkably effective allocation of promos. Moreover, the optimized solutions tend to increase the violation of soft-rule SR (ratio of self promos), whereas the rules SG (specific versus generic promos) and PV (promos variation) are better preserved. This suggests that a better diversification of programs and products advertised by specific promos, at the cost of a higher violation of the targeted self promo ratio, can remarkably boost the impact of the planning. However, we highlight that the optimization of **IP-SUM** increases the average viewership of programs and products, given by $f_{AVG}(\delta(P_1))$ and $f_{AVG}(\delta(P_2))$, on average resp. of 0.29% and 13.39%. This result may give an indication of the conversion rate of *promos for programs* (i.e. the percentage of customers that after having watched the promo, watch the linked program as well) from the effectiveness of the planning towards the viewership of programs and products, which is influenced by heterogeneous exogenous factors out of the promo placements, like creativity, authorship, quality of contents, etc.

We further confirm our results by analyzing the outcomes obtained for **IP-AVG**, that aims to maximize the average viewership. In fact, we observe that the total impact of the optimization of **IP-AVG**, measured by $F(\delta)$, increases on average of 7.68%, where the average viewership of programs and products, given by $f_{AVG}(\delta(P_1))$ and $f_{AVG}(\delta(P_2))$, contribute with an average increase of resp. 1.88% and 4.75%. Furthermore, we observe that the optimization of **IP-AVG** offers more balanced adherence and lower violation of the soft rules, except for SG: we may conclude that generic promos would better consolidate and increase the average viewership. Finally, we remark that the tuning of the parameters controlling the soft rules can change the impact of the optimized solutions. Indeed, the introduced models offer also the opportunity to analyze and validate in advance the effect of corporate strategies towards the business practice of the promo planning.

Table 3 reports the size of the introduced programs, together with some computational aspects of their resolution. In particular, *# bin var*, *# cont var*, *# lin cons* and *# ind cons* denote resp. the number of binary and continuous variables, and the number of linear and indicator constraints. The KPIs *time (sec)* and *gap (%)* are resp. the real time and the relative integrality gap obtained with CPLEX: **IP-SUM** is solved at the optimum on average in less than 5 s, while sub-optimal solutions to **IP-AVG** with *gap (%)* less than 1% are obtained within the same time limit. To analyze how results can be improved by adopting an extended computational time window, still acceptable for the business users, we set a limit of 300 s for **IP-AVG**. In fact, we observe that *gap (%)* can be reduced on average from 1 to 0.09% by extending the computation time from 5 to 300 s. While we keep default parameters of CPLEX 12.9 for **IP-SUM**, to boost solutions of **IP-AVG**, we tuned the CPLEX parameters [10]: we first set the *parallel mode switch* to *deterministic* in order to

Table 3 Computational results

		Day-1	Day-2	Day-3	Day-4	Day-5	Day-6	Day-7
IP-SUM	Time (s)	3.918	3.272	5.448	8.524	8.789	0.663	3.633
	gap (%)	0	0	0	0	0	0	0
	# bin var	2880	3293	3210	3108	4113	2923	2340
	# cont var	24	25	25	24	24	22	23
	# lin cons	4054	4327	4513	4312	5485	4271	3321
	# ind cons	72	77	75	74	96	79	60
IP-AVG	Time (s)	291.278	lim	lim	lim	lim	lim	lim
	Gap (%)	0	0.028	0.022	0.019	0.136	0.166	0.261
	# bin var	2900	3313	3232	3129	4133	2942	2359
	# cont var	2852	3261	3182	3079	4061	2885	2322
	# lin cons	6882	7563	7670	7367	9522	7134	5620
	# ind cons	5708	6529	6367	6163	8150	5786	4639

stabilize performance over multi-threading, then we set the MIP emphasis switch to *finding hidden feasible solutions* and the *RINS heuristic frequency* to 10, in order to generate multiple high-quality solutions, given the weak bounds induced by the linearization and the business need of having multiple good scheduling options. Finally, we can conclude that both the programs can be effectively solved, both in terms of solution quality and computational time with respect to the business needs leading to improvements in the schedule defined by business experts.

5 Conclusions

In this work, by coupling techniques belonging to the Machine Learning field with Integer Programming, we have introduced a new framework for solving the Optimal Promo Scheduling problem. Two different objective functions are introduced and their properties analysed on real-word instances. The resulting framework turns out to be effective in defining optimal or near to optimal schedules in a very short amount of time if compared with previous works on this topic and by always obtaining better solutions than those found by business experts. Furthermore, by specifying the penalty terms γ_j, the solution can be tuned and adjusted according to the strategic requirements specific to each TV media company.

Regarding future directions of investigation, it might be interesting to study specific cuts that enable closing the gap when considering the **IP-AVG** objective function so to guarantee the solution of the problem at the optimal value. Moreover, a combination of the two different objective functions, **IP-SUM** and **IP-AVG**, could be furtherly investigated so as to reach better schedules able to balance the two objectives. Finally, some constraints could be reformulated so as to reduce the computational complexity while solving the MILP problems.

References

1. Bollapragada, S., Bussieck, M.R., Mallik, S.: Scheduling commercial videotapes in broadcast television. Oper. Res. **52**(5), 679–689 (2004)
2. Bollapragada, S., Garbiras, M.: Scheduling commercials on broadcast television. Oper. Res. **52**(3), 337–345 (2004)
3. Brown, A.: Selling television time: an optimisation problem. Comput. J. **12**(3), 201–207 (1969)
4. Chen, T., Guestrin, C.: Xgboost: A scalable tree boosting system. In: Proceedings of the 22nd ACM SIGKDD International Conference on Knowledge Discovery and DATA MINING, pp. 785–794. ACM, New York (2016)
5. Dalila B.M.M. Fontes Paulo A., P., Fontes, F.A.: A decision support system for TV self-promotion scheduling. Int. J. Adv. Trends Comput. Sci. Eng. **8**(2), 134–139 (2019). https://doi.org/10.30534/ijatcse/2019/06822019
6. Fontes, D., Pereira, P., Fontes, F.: A decision support system for TV self-promotion scheduling. Int. J. Adv. Trends Comput. Sci. Eng. **8**, 134–139 (2019). https://doi.org/10.30534/ijatcse/2019/06822019
7. Ghassemi Tari, F., Alaei, R.: Scheduling TV commercials using genetic algorithms. Int. J. Prod. Res. **51**(16), 4921–4929 (2013)
8. IBM: IBM Cloud Pak for Data (2019). https://www.ibm.com/support/producthub/icpdata/docs/
9. IBM: IBM Decision Optimization on Cloud (2019). https://developer.ibm.com/docloud/documentation/decision-optimization-on-cloud/
10. IBM: IBM ILOG CPLEX Optimization Studio V12.9.0 Documentation (2019). https://www.ibm.com/support/knowledgecenter/SSSA5P_12.9.0
11. LeCun, Y., Bengio, Y., Hinton, G.: Deep learning. Nature **521**(7553), 436–444 (2015)
12. Palagi, L., Seccia, R.: Block layer decomposition schemes for training deep neural networks. J. Global Optim. 77, 97–124 (2020)
13. Pereira, P.A., Fontes, F.A., Fontes, D.B.: A decision support system for planning promotion time slots. In: Operations Research Proceedings 2007, pp. 147–152. Springer, Berlin (2008)
14. Reddy, S.K., Aronson, J.E., Stam, A.: Spot: scheduling programs optimally for television. Manag. Sci. **44**(1), 83–102 (1998)
15. Seshadri, S., Subramanian, S., Souyris, S.: Scheduling spots on television (2015)
16. Singh, M., Pant, M., Kaul, A., Jha, P.: Advertisement scheduling models in television media: a review. In: Soft Computing: Theories and Applications, pp. 505–514. Springer, Berlin (2018)
17. Zhang, X.: Mathematical models for the television advertising allocation problem. Int. J. Oper. Res. **1**(3), 302–322 (2006)

A Distributed Algorithm for Spectral Sparsification of Graphs with Applications to Data Clustering

Fabricio Mendoza-Granada and Marcos Villagra

Abstract Spectral sparsification is a technique that is used to reduce the number of non-zero entries in a positive semidefinite matrix with little changes to its spectrum. In particular, the main application of spectral sparsification is to construct sparse graphs whose spectra are close to a given dense graph. We study spectral sparsification under the assumption that the edges of a graph are allocated among sites which can communicate among each other. In this work we show that if a graph is allocated among several sites, the union of the spectral sparsifiers of each induced subgraph give us an spectral sparsifier of the original graph. In contrast to other works in the literature, we present precise computations of the approximation factor of the union of spectral sparsifiers and give an explicit calculation of the edge weights. Then we present an application of this result to data clustering in the Number-On-Forehead model of multiparty communication complexity when input data is allocated as a sunflower among sites in the party.

Keywords Spectral sparsification · Dense graphs · Distributed algorithms · Communication complexity · Data clustering

1 Introduction

Spectral sparsification is a technique introduced by Spielman and Teng [11] that is used to approximate a graph G by a sparse graph H. The notion of approximation used by spectral sparsification is that the spectra of both H and G must be close up to a constant factor. Batson, Spielman and Srivastava [1] proved that every graph G has an spectral sparsifier with a number of edges linear in the number of vertices of G and provided an algorithm achieving such bound. There are several algorithms in the literature that construct spectral sparsifiers of graphs with a trade-off between

F. Mendoza-Granada (✉) · M. Villagra
Núcleo de Investigación y Desarrollo Tecnológico (NIDTEC), Facultad Politécnica - Universidad Nacional de Asunción, San Lorenzo, Paraguay

© The Author(s), under exclusive license to Springer Nature Switzerland AG 2021　　　403
C. Gentile et al. (eds.), *Graphs and Combinatorial Optimization:*
from Theory to Applications, AIRO Springer Series 5,
https://doi.org/10.1007/978-3-030-63072-0_31

running time and number of edges of H. To the best of our knowledge, Lee and Sun [8] has the best probabilistic algorithm for spectral sparsification with a running time that is almost linear and constructs spectral sparsifiers with $O(qn/\epsilon^2)$ edges, where n is the number of vertices of G, ϵ is an approximation factor and $q \geq 10$ is a constant.

There are situations where algorithms need to work with data that is not centralized and allocated in different sites. One way to deal with decentralized data is to design communication protocols so that the sites can communicate among them. The efficiency of a communication protocol can be measured by the number of bits shared among the sites and such a measure is known as the *communication complexity* of the protocol [10]. When data comes in the form of a graph, the edges greatly affects communication complexity, and hence, computing spectral sparsifiers of graphs in distributed systems is of great importance.

In this work we present a distributed algorithm for spectral sparsification of graphs in the communication complexity model. In this model, we are only interested in the communication costs among sites and we assume that each site has arbitrary computational power. The idea behind this protocol is that, given an input graph G, spectral sparsifiers of induced subgraphs of G can be computed in each site first, and then any given site computes the union of such graphs which results in a spectral sparsifier of G. Even though other works have used the idea of taking the union of spectral sparsifiers like Chen et al. [2], they have not shown a precise calculation of the approximation factor. The main contribution of this work, presented in Theorem 1, is an estimation of the approximation factor and an explicit calculation of the edge weights in the union of spectral sparsifiers. In order to compute the approximation factor we introduce an idea that we call "overlapping cardinality partition," which is a way to partition the edge set of a graph with respect to the number of times each edge is allocated among sites. Overlapping cardinality partition is a technical tool that allows us to express the Laplacian matrix of the union of induced subgraphs of G as a linear combination of the Laplacian matrices of graphs induced from the partition.

In a second part of this paper, we present in Sect. 4 an application of Theorem 1 in distributed data clustering in the Number-On-Forehead model of communication complexity. In particular, if we assume the existence of a sunflower structure [3–5] on the input data, we show how a communication protocol can detect the presence of the sunflower and take advantage of its kernel to reduce the communication costs.

The rest of this paper is organized as follows. In Sect. 2 we present the main definitions and notation used throughout this work. In Sect. 3 we present the main result of this work, and in Sect. 4 we present our application to data clustering.

2   Preliminaries and Notation

In this section we will introduce some definitions and notations that will be used throughout this paper.

2.1 Spectral Graph Theory

Let $G = (V, E, w)$ be an undirected and weighted graph with n vertices and m edges. Let $\{E_i\}_{i \geq 1}$ be family of subsets of E. We denote by $G_i = (V, E_i, w_i)$ the subgraph induced by E_i, where $w_i : E_i \to \mathbb{R}^+$ is defined as $w_i(e) = w(e)$ for all $e \in E_i$ and 0 otherwise. Every graph G has an associated matrix called its *Laplacian* matrix, or simply Laplacian, which is define as

$$L_G = D_G - W_G,$$

where W_G is the weighted adjacency matrix and D_G is the weighted degree matrix. We will omit the subindex G from L_G, W_G and D_G when it is clear from the context.

The normalized Laplacian is defined as $\mathscr{L} = D^{-1/2} L D^{-1/2}$. The Laplacian matrix (and normalized Laplacian) is positive semidefinite (PSD) with its first eigenvalue λ_1 always equals zero with multiplicity equal to the number of connected components of G [9]. Indeed, if there exists a multicut of size k in G then the k-th smallest eigenvalue λ_k of L gives useful information to find a multicut.

One of the fastest methods to approximate an optimal multicut in a graph is the so-called *spectral clustering* algorithm. This technique uses k eigenvectors of L or $\mathscr{L}$ associated to the first k smallest eigenvalues in order to construct a matrix X with the eigenvectors as columns, and then, it applies a simpler clustering algorithm (like *k-means*) to the rows of X [9]. Lee et al. [7] proved that λ_k approximates the optimal value of a multicut of size k in G and the eingevectors give the corresponding partition over V.

2.2 Spectral Sparsification

Spectral sparsification is a technique used to reduce the density of a given PSD matrix changing its spectra only by a constant factor of approximation. Given a matrix M, spectral sparsification constructs another matrix which is "similar" to M in some well-defined way. We will use a notion of similarity defined in [11]. A subgraph H of G is called an ϵ-spectral sparsifier of G if for any $x \in \mathbb{R}^n$ we have that

$$(1 - \epsilon)x^T L_G x \leq x^T L_H x \leq (1 + \epsilon)x^T L_G x.$$

The importance of a spectral sparsifier lies on the sparseness of L_H, for example, some computations are easier over an sparse matrix. There are deterministic and probabilistic algorithms to find spectral sparsifiers of a given graph. The algorithm of Batson, Spielman and Srivastava [1] is currently the best deterministic algorithm. The algorithm of [1] constructs a graph with $O(\frac{qn}{\epsilon^2})$ edges in $O(\frac{qmn^{5/q}}{\epsilon^{4+4/q}})$ time, where ϵ is the approximation factor and $q \geq 10$ is a constant.

3   A Distributed Algorithm for Spectral Sparsification

In this section we present our main result. In particular, given a graph G and a family of induced subgraphs of G, we show that the union of spectral sparsifiers of the induced subgraphs is a spectral sparsifier of G. In contrast to other work, however, we give explicit bounds on the approximation factor and a construction of the new weight function.

First we introduce some definitions which will help us understand the overlapping of data among the sites. We denote by $[n]$ the set $\{1, 2, \ldots, n\}$.

Definition 1 (Occurrence Number) Let $\mathscr{E} = \{E_1, \ldots, E_t\}$ be a family of subsets of $[n]$. For any $a \in [n]$, the occurrence number of a in $\mathscr{E}$, denoted $\#(a)$, is the maximum number of sets from $\mathscr{E}$ in which a appears.

Example 1 Let $n = 7$ and $\mathscr{E} = \{\{1, 2, 3\}, \{2, 3, 4\}, \{4, 5, 1\}, \{3, 2, 6\}, \{4, 7, 1\}, \{2, 3\}, \{5, 6, 7\}, \{1, 3, 5\}, \{2, 4\}\}$. Here we have that $\#(1) = 4$, $\#(2) = 5$, $\#(3) = 5$, and so on.   □

Definition 2 (Overlapping Cardinality) Let $\mathscr{E} = \{E_1, \ldots, E_t\}$ be a family of subsets of $[n]$ for some fixed n and $E = \bigcup_{i=1}^{t} E_i$. The *overlapping cardinality* of a subset $E' \subseteq E$ in $\mathscr{E}$ is a positive integer k such that for each $a \in E'$ its ocurrence number $\#(a) = k$; otherwise the overlapping cardinality of E' in $\mathscr{E}$ is 0.

The overlapping cardinality identifies the maximum number of times the elements of a subset appears in a family of subsets.

Example 2 Let $n = 7$ and $\mathscr{E}$ be as in Example 1. Here we have that $E = \bigcup_{i=1}^{t} E_i = [n]$. Now consider the sets $\{1, 4\}$ and $\{1, 2, 3\}$.

- The overlapping cardinality of $\{1, 4\}$ in $\mathscr{E}$ is 4, because $\#(1) = \#(4) = 4$.
- The overlapping cardinality of $\{1, 2, 3\}$ in $\mathscr{E}$ is 0 because the occurrence number of one of the elements of the set is different from the others, namely, $\#(1) = 4$, $\#(2) = 5$, and $\#(3) = 5$.   □

Now we use the idea of overlapping cardinality to construct a partition on the set $\mathscr{E}$ of subsets of $[n]$.

Definition 3 (Overlapping Cardinality Partition) Given a family $\mathscr{E}$ as in Definition 2, an *overlapping cardinality partition* over E on $\mathscr{E}$ is a partition $\{E'_1, \ldots, E'_k\}$ of E where each E'_i has overlapping cardinality c_i on $\mathscr{E}$. We call the sequence $(c_1, c_2, \ldots, c_k)$, with $1 \leq c_1 < c_2 < \cdots < c_k$, the *overlapping cardinalities* over the family $\mathscr{E}$.

Example 3 Take $\mathscr{E}$ from examples 1 and 2. An overlapping cardinality partition is

$$\{\{6, 7\}, \{5\}, \{1, 4\}, \{2, 3\}\}.$$

Here, $\{6, 7\}$ has overlapping cardinality equal to 2 because $\#(6) = \#(7) = 2$. The subset $\{5\}$ has overlapping cardinality equal to 3 because $\#(5) = 3$. In Example 2 we saw that the subset $\{1, 4\}$ has overlapping cardinality 4. Finally, the subset $\{2, 3\}$ has overlapping cardinality equal to 5 because $\#(2) = \#(3) = 5$.   □

Our main technical lemma shows that the Laplacian of an input graph can be rewritten as a linear combination of Laplacians corresponding to induced subgraphs constructed from an overlapping cardinality partition of the set of edges.

For the rest of this section we make the following assumptions. Let $G = (V, E, w)$ be an undirected and weighted graph with a function $w : E \to \mathbb{R}^+$, let $\mathscr{E} = \{E_1, \ldots, E_t\}$ be a collection of subsets of E such that $\bigcup_{i=1}^{t} E_i = E$ where $E_i \neq \emptyset$ and $G_i = (V, E_i, w_i)$ is an induced subgraph of G where $w_i : E_i \to \mathbb{R}^+$ and $w_i(e) = w(e)$ for all $e \in E_i$ and 0 otherwise.

Lemma 1 *If* $1 \leq c_1 < c_2 < \cdots < c_k$ *are the overlapping cardinalities over the family* $\mathscr{E}$ *with an overlapping cardinality partition* $\{E'_{c_j}\}_{j \leq k}$, *then* $\sum_{i=1}^{t} L_{G_i} = \sum_{j=1}^{k} c_j L_{G'_{c_j}}$ *where* $L_{G'_{c_j}}$ *is the Laplacian of* $G'_{c_j} = (V, E'_{c_j}, w'_{c_j})$.

Proof First notice that, for all $e = xy \in E'_{c_j}$ there exists a subfamily of $\mathscr{E}$ with cardinality equal to c_j such that e belongs to every member of it and its associated subgraph. Take any $xy \in E'_{c_j}$ for some $j \in \{1, \ldots, k\}$. There exists c_j induced subgraphs $G_{i_1}, \ldots, G_{i_{c_j}}$ of G that has xy as an edge, and all other induced subgraphs $G_{k_1}, \ldots, G_{k_\ell}$ do not have xy as an edge, where $c_j + \ell = t$. This means that

$$\sum_{i=1}^{t} L_{G_i}(x, y) = c_j \cdot L_{G'_{c_j}}(x, y) = -c_j \cdot w(x, y). \tag{1}$$

Now, let $d_G(x)$ denote the degree of x in G. We know that $d_G(x) = \sum_y w(x, y)$ where $xy \in E$. Since $\{E'_{c_j}\}_{j \leq k}$ is a partition of E, we can rewrite the degree of x as

$$d_G(x) = \sum_{xy_{c_1} \in E'_{c_1}} w(x, y_{c_1}) + \cdots + \sum_{xy_{c_k} \in E'_{c_k}} w(x, y_{c_k}).$$

Then, the degree of x in the graph G'_{c_j} is

$$L_{G'_{c_j}}(x, x) = \sum_{xy_{c_j} \in E'_{c_j}} w(x, y_{c_j}) = d_{G'_{c_j}}(x).$$

If we take an edge $xy_{c_j} \in E'_{c_j}$, where x is fixed, we know that xy_{c_j} appears only in the induced subgraphs $G_{i_1}, \ldots, G_{i_{c_j}}$, and hence, we obtain

$$\sum_{i=1}^{t} \left(\sum_{xy_{c_j} \in E'_{c_j}} w_i(x, y_{c_j}) \right) = c_j \cdot d_{G'_{c_j}}(x). \tag{2}$$

If we take another edge $uv \in E'_{c_m}$, with $m \neq j$, note that uv does not belong to any of the graphs $G_{i_1}, \ldots, G_{i_{c_j}}$ and each Laplacian matrix $L_{G_{i_1}}, \ldots, L_{G_{i_{c_j}}}$ has 0 in its (u, v)-entry. Therefore, adding uv to Eq.(1) we have that

$$\sum_{i=1}^{t} \left(L_{G_i}(x, y) + L_{G_i}(u, v) \right) = c_j \cdot L_{G'_{c_j}}(x, y) + c_m \cdot L_{G'_{c_m}}(u, v).$$

Extending this argument to all equivalent classes in $\{E'_{c_j}\}_{j \leq k}$, for each non-diagonal entry (x, y), with $xy \in E$, it holds

$$\sum_{i=1}^{t} L_{G_i}(x, y) = \sum_{j=1}^{k} c_j \cdot L_{G'_{c_j}}(x, y). \tag{3}$$

A similar argument can be made for the diagonal entries with Eq.(2), thus obtaining

$$\sum_{i=1}^{t} \left(\sum_{xy_{c_1} \in E'_{c_1}} w_i(x, y_{c_1}) + \cdots + \sum_{xy_{c_k} \in E'_{c_k}} w_i(x, y_{c_k}) \right) = \sum_{i=1}^{t} L_{G_i}(x, x) = \sum_{j=1}^{k} c_j \cdot L_{G'_{c_j}}(x, x). \tag{4}$$

Equations (3) and (4) imply the lemma. $\square$

Now we will use Lemma 1 to show that the spectral sparsifier of $\sum_{j=1}^{k} c_j L_{G'_{c_j}}$ is an spectral sparsifier of the Laplacian L_G of an input graph G.

Theorem 1 *Let* $(1 = c_1 < c_2 < \cdots < c_k)$ *be the overlapping cardinalities over the family* $\mathcal{E}$ *with* $\{E'_{c_j}\}_{j \leq k}$ *its associated overlapping cardinality partition and* $L_{G_1}, \ldots, L_{G_t}$ *the Laplacians of* $G_1, \ldots, G_t$. *If* $H_i = (V, D_i, h_i)$ *is an* ϵ-*spectral sparsifier of* G_i, *then* $H = (V, \bigcup_{i=1}^{t} D_i, h)$ *is an* ϵ'-*spectral sparsifier of* G *where* $h(e) = \frac{\sum_{i=1}^{t} h_i(e)}{c_1 c_k}$ *and* $\epsilon' \geq 1 - \frac{1-\epsilon}{c_k}$.

Proof Let L_{H_i} be the Laplacian of H_i. By hypothesis we have that for every $i \in [t]$ and $x \in \mathbb{R}^V$

$$(1 - \epsilon) x^T L_{G_i} x \leq x^T L_{H_i} x \leq (1 + \epsilon) x^T L_{G_i} x.$$

Then we may take the summation over all $i \in [t]$ to get

$$(1 - \epsilon) \sum_{i=1}^{t} x^T L_{G_i} x \leq \sum_{i=1}^{t} x^T L_{H_i} x \leq (1 + \epsilon) \sum_{i=1}^{t} x^T L_{G_i} x. \tag{5}$$

Now, lets consider the left hand side of Eq. (5). Using Lemma 1 we get

$$(1 - \epsilon) \sum_{i=1}^{t} x^T L_{G_i} x = (1 - \epsilon) \sum_{i=1}^{k} c_i \cdot x^T L_{G'_{c_i}} x$$

$$\geq (1 - \epsilon) c_1 \sum_{i=1}^{k} x^T L_{G'_{c_i}} x$$

$$= (1 - \epsilon) c_1 x^T L_G x, \tag{6}$$

where the last equality follows from the fact that $\{E'_{c_j}\}_{j \leq k}$ is a partition of E. Similarly for the right hand side of Eq. (5) we have that

$$(1 + \epsilon) \sum_{i=1}^{t} x^T L_{G_i} x \leq (1 + \epsilon) c_k x^T L_G x. \tag{7}$$

Therefore, by multiplying Eqs. (6) and (7) by $\frac{1}{c_1 c_k}$ we obtain

$$(1 - \epsilon) \frac{x^T L_G x}{c_k} \leq x^T L_H x \leq (1 + \epsilon) \frac{x^T L_G x}{c_1},$$

where $x^T L_H x = (\sum_{i=1}^{t} x^T L_{H_i} x)/(c_1 c_k)$.

To finish the proof, note that we want $1 - \epsilon' \leq (1 - \epsilon)/c_k$ and $(1 + \epsilon)/c_1 \leq 1 + \epsilon'$ with $\epsilon \leq \epsilon' < 1$. In order to solve this, we choose an $\epsilon' \geq 1 - \frac{1-\epsilon}{c_k}$. First notice that $1 - \epsilon' \leq 1 - 1 + \frac{1-\epsilon}{c_k} = \frac{1-\epsilon}{c_k}$. Then we have that $\frac{1+\epsilon}{c_1} \leq \frac{1+\epsilon'}{c_1} = 1 + \epsilon'$. $\qquad\square$

From Theorem 1, a distributed algorithm for computing spectral sparsifiers is natural. Just let every site compute a spectral sparsifier of its own input and then each site sends its result to a coordinator that will construct the union of all spectral sparsifiers.

4 Data Clustering in the Number-on-Forehead Model

In this section we will show an application of Theorem 1 to distributed data clustering in the Number-On-Forehead model of communication complexity for the case when the input data is allocated as a sunflower among sites.

Clustering is an unsupervised machine learning task that involves finding a partition over a given set of points $x_1, \ldots, x_n \in \mathbb{R}^d$. Such a partition must fulfill two conditions, (1) every two points in the same set must be "similar" in some way and (2) every two points on different sets must be far from being similar. Each equivalence class from the partition is also called a *cluster*. Clustering can be accomplished by different kinds of techniques, where *spectral clustering* [9] is one of the fastest methods.

It is easy to see clustering as a graph problem, where each point corresponds to a vertex in a complete graph and the cost of each edge is interpreted as a similarity between points. Thus, finding a set of optimal clusters in data is equivalent to finding an optimal multicut in a graph. Since the optimal multicut depends on the spectrum of the graph's Laplacian [7] and we want to keep the communication costs low, each site must be capable of constructing sparse induced subgraphs of its own data while preserving the spectrum of its graph Laplacian.

In our communication protocol, each site is assigned an induced subgraph of G, and we want each site to be aware of all clusters in the data. Consequently, each site must be capable of running a clustering algorithm on its own data, communicate its results to the other sites, and then use the exchanged messages to construct an approximation to the original graph G. This is where the distributed spectral sparsification algorithm is relevant.

First, we will construct a protocol to verify if the input data in every site is a sunflower. If the input is indeed allocated in a sunflower structure, then a party can take advantage of the sunflower to find an approximation of clusters in the data.

4.1 Models of Communication and Their Complexity Measure

We will introduce some standard notations from communication complexity—we refer the interested reader to the textbook by Kushilevitz and Nisan [6] for more details. Let $P_1, P_2, \ldots, P_s$ be a set of sites where a site P_j has an input $x_j \in \{0, 1\}^r$, with r a positive integer. In a multiparty communication protocol, with $s \geq 3$, the sites want to jointly compute a function $f : \{0, 1\}^r \times \cdots \times \{0, 1\}^r \to Z$ for some finite codomain Z. In the *Number-On-Forehead* model of communication, or NOF model, each site only has access to the other sites's input but not its own, that is, a site P_j has access to $(x_1, \ldots, x_{j-1}, x_{j+1}, \ldots, x_s)$. In order to compute f the sites must communicate, and they do so by writing bits on a blackboard which can be accessed by all sites in the party. This is the so-called *blackboard model* of communication.

The maximum number of bits exchanged in the protocol over the worst-case input is the *cost* of the protocol. The *deterministic communication complexity* of the function f is the minimum cost over all protocols which compute f.

Let $G = (V, E)$ be an input graph and $\{E_j\}_{j \leq s}$ be a family of subsets of E. In order to study communication protocols for graph problems we assume that E_j is the input data to site P_j. In the NOF model, we let $F_j = \{E_1, E_2, \ldots, E_{j-1}, E_{j+1}, \ldots, E_s\}$ be the set of edges which P_j can access. Given a site P_j, the *symmetric difference on* P_j, denoted Δ_j, is defined as the symmetric difference among all sets P_j has access to, that is, Δ_j is the symmetric difference between each set in F_j.

For the rest of this paper, we use as a shorthand $\mathscr{E}$ for the set $\{E_1, \ldots, E_s\}$ of subsets of the set of edges E of an input graph $G = (V, E)$ with $\bigcup_{i=1}^{s} E_i = E$, and $\mathscr{F}$ for the set $\{F_1, \ldots, F_s\}$ where $F_j = \{E_1, \ldots, E_{j-1}, E_{j+1}, \ldots, E_s\}$. Here $\mathscr{F}$ captures the idea of the NOF model where every site have access to the other's sites input but not its own.

4.2　Sunflowers and NOF Communication

A *sunflower* or Δ-*System* is a family of sets $\mathscr{A} = \{A_1, \ldots, A_t\}$ where $(A_i \cap A_j) = \bigcap_{k=1}^t A_k = K$ for all $i \neq j$. We call K the kernel of $\mathscr{A}$. The family $\mathscr{A}$ is a *weak Δ-System* if $|A_i \cap A_j| = \lambda$ for all $i \neq j$ for some constant λ [5]. It is known that if $\mathscr{A}$ is a weak Δ-System and $|\mathscr{A}| \geq \ell^2 - \ell + 2$, where $\ell = \max_{i=1}^t \{A_i\}$, then $\mathscr{A}$ is a Δ-System [3].

We start with a simple fact that ensures the existence of Δ-Systems with the same kernel in the NOF model if input data in a communication protocol is allocated as a sunflower among sites.

Lemma 2 *If $s = |\mathscr{E}| \geq 3$ and $\mathscr{E}$ is a Δ-System with kernel K, then any F_i is a Δ-System with kernel K.*

The following lemma states a sufficient condition for the existence of a Δ-System in the input data in the NOF model with the requirement, however, that we need at least four or more sites

Lemma 3 *Let $s = |\mathscr{E}| \geq 4$. If, for all $i \in [s]$, we have that F_i is a Δ-System, then $\mathscr{E}$ is a Δ-System.*

Proof Suppose that $\mathscr{E}$ is no a Δ-System, and we want to prove that for some $1 \leq i \leq s$, F_i is not a Δ-System.

With no loss of generality, suppose that there exists exactly two sets E_i and E_j that certify that $\mathscr{E}$ is not a Δ-System; that is, there exists E_i and E_j such that $E_i \cap E_j = K'$, and, for any $a \neq i$ and $b \neq j$, it holds that $E_a \cap E_j = E_b \cap E_i = K$, with $K \neq K'$. Now take any F_c, with c different from i and j. Then F_c cannot be a Δ-System because E_i and E_j belong to F_c and there is at least another set in F_c because $|\mathscr{E}| \geq 4$. □

Lemma 3 implies that we only need to know if all sites in a communication protocol have access to a Δ-System to ensure that an entire family of input sets is a Δ-System, provided there are at least 4 sites.

Proposition 1 *There exists a protocol that verifies if $\mathscr{E}$, with $|\mathscr{E}| \geq 4$, is a Δ-System or not with $s - 1$ bits of communication exchanged.*

With Proposition 1, a multiparty communication protocol with a number of sites $s \geq 4$ can check for the existence of a sunflower structure in its input data. Furthermore, if input data is allocated among sites as a sunflower, then, by Lemma 2, any site immediately knows the kernel of the sunflower.

4.3   Data Clustering with Sunflowers

In this section, we present a NOF communication protocol that exploits the sunflower structure in input data. First, we start by defining an overlapping coefficient of the edges of G which can be seen as a measure of how well spread out are the edges among sites.

Definition 4 The overlapping coefficient on site P_j is defined as $\delta(j) = \frac{|\bigcap_{i \neq j} E_i|}{|\bigcup_{i \neq j} E_i|}$ and the greatest overlapping coefficient is defined as $\delta = \max_{j \in [s]} \delta(j)$.

The following proposition presents a simple protocol that makes every site aware of the entire input graph.

Proposition 2 *Let P_j be a site and let $\mathcal{E}$ be a weak Δ-System with each $|E_k| = \ell$ for $k = 1, 2, \ldots, s$, with a kernel of size λ. Suppose that $s \geq \ell^2 - \ell + 3$. If site P_j sends all the edges in Δ_j, then every other site will know the entire graph G. The number of edges this communication protocol sends is at most $|\bigcup_{i \neq j} E_i|(1-\delta)+\ell$.*

Proof We will prove this proposition by showing how each site constructs the graph G. First, a given site P_j computes Δ_j and writes it on the blackboard. Since $s \geq \ell^2 - \ell + 3$, by the result of Deza [3], we known that $\mathcal{E}$ is a sunflower with kernel K and by Lemma 2 this kernel is the same in all sites. At this point all sites $i \neq j$ know Δ_j, therefore, they can construct G by its own using the kernel K of $\mathcal{E}$. In one more round, one of the sites $i \neq j$ writes E_j so that site P_j can also construct G.

In order to compute the communication cost of the protocol, first notice that $\delta = \lambda/(|\bigcup_{i \neq j} E_i|) = \lambda/(|\Delta_j| + \lambda)$, where we used the fact that the union of all edges in every site equals the union of the symmetric difference and the kernel K. Then we have that $\delta|\Delta_j| = \lambda - \delta\lambda$, which implies $|\Delta_j| = \frac{\lambda - \delta\lambda}{\delta} = |\bigcup_{i \neq j} E_i|(1 - \delta)$, where the last equality follows from the fact that $|\bigcup_{i \neq j} E_i| = \lambda/\delta$. Finally, after E_j was sent to the blackboard the communication cost is $|\bigcup_{i \neq j} E_i|(1-\delta)+\ell$.   □

Theorem 2 *Let $\mathcal{E}$ be a weak Δ-system with each $|E_k| = \ell$ for $k = 1, 2, \ldots, s$, and suppose that $s \geq \ell^2 - \ell + 3$. There exists a communication protocol such that after two rounds of communication every site knows an ϵ-spectral sparsifier of the entire graph G with communication cost $O\left(\log\left(\frac{n}{\epsilon^2}\sqrt{1-\delta}\right)\right)$.*

Proof From [3] we know that $\mathcal{E}$ is a sunflower with a kernel K of size λ and, by Lemma 2, K is equal in all sites. First, a site P_j computes a spectral sparsifier $H_j = (V, \hat{\Delta}_j)$ of the induced subgraph $G_j = (V, \Delta_j)$ using the spectral sparsification algorithm of [8]. This way we have that $|\hat{\Delta}_j| = O(n/\epsilon^2)$ where $0 < \epsilon \leq 1/120$. Then site P_j writes $\hat{\Delta}_j$ on the blackboard. Any other site $i \neq j$ constructs an ϵ-spectral sparsifier $H'_i = (V, \hat{E}_j)$ of $G'_i = (V, E_j)$. By Theorem 1, the graph $H = (V, \hat{\Delta}_j \cup \hat{E}_j)$ is a ϵ'-spectral sparsifier of G. In a second round, a given site P_i

writes $\hat{E}_j$ on the blackboard. Finally, site P_j receives $\hat{E}_j$ and by Theorem 1 it can also construct an ϵ'-spectral sparsifier for G. Finally, the communication complexity is upper-bounded by $O\left(\log\left(\frac{n}{\epsilon^2}(1-\delta)\right) + \log\left(\frac{n}{\epsilon^2}\right)\right) = O\left(\log\left(\frac{n}{\epsilon^2}\sqrt{1-\delta}\right)\right)$.

$\square$

Acknowledgments We give our thanks to the reviewers of CTW 2020 for their comments that helped improve this paper. This work is supported by Conacyt research grants POSG17-62 and PINV15-208.

References

1. Batson, J.D., Spielman, D.A., Srivastava, N.: Twice-Ramanujan sparsifiers. In: Proceedings of the 41st Annual ACM Symposium on Theory of Computing (STOC), pp. 255–262 (2009)
2. Chen, J., Sun, H., Woodruff, D., Zhang, Q.: Communication-optimal distributed clustering. In: Proceedings of the 30th International Conference on Neural Information Processing Systems (NIPS), pp. 3727–3735 (2016)
3. Deza, M.: Solution d'un problème de Erdös-Lovász. J. Comb. Theory B **16**(2), 166–167 (1974)
4. Erdös, P., Chao, Rado, R.: Intersection theorems for systems op finite sets. Q. J. Math. **12**(1), 313–320 (1961)
5. Kostochka, A.: Extremal problems on Δ-systems. Numbers, Information and Complexity pp. 143–150 (2000)
6. Kushilevitz, E., Nisan, N.: Communication Complexity. Cambridge University Press, Cambridge (2006)
7. Lee, J.R., Gharan, S.O., Trevisan, L.: Multiway spectral partitioning and higher-order Cheeger inequalities. J. ACM **61**(6), 37 (2014)
8. Lee, Y.T., Sun, H.: Constructing linear-sized spectral sparsification in almost-linear time. SIAM J. Comput. **47**(6), 2315–2336 (2018)
9. von Luxburg, U.: A tutorial on spectral clustering. Stat. Comput. **17**(4), 395–416 (2007)
10. Nisan, N.: The communication complexity of threshold gates. Combinatorics Paul Erdös Eighty **1**, 301–315 (1993)
11. Spielman, D.A., Teng, S.H.: Spectral sparsification of graphs. SIAM J. Comput. **40**(4), 981–1025 (2011)

Printed in the United States
by Baker & Taylor Publisher Services